Advances in Mechanical Systems Dynamics 2020

Advances in Mechanical Systems Dynamics 2020

Editors

Alberto Doria
Giovanni Boschetti
Matteo Massaro

MDPI • Basel • Beijing • Wuhan • Barcelona • Belgrade • Manchester • Tokyo • Cluj • Tianjin

Editors

Alberto Doria
University of Padova
Italy

Giovanni Boschetti
University of Padova
Italy

Matteo Massaro
University of Padova
Italy

Editorial Office
MDPI
St. Alban-Anlage 66
4052 Basel, Switzerland

This is a reprint of articles from the Special Issue published online in the open access journal *Applied Sciences* (ISSN 2076-3417) (available at: https://www.mdpi.com/journal/applsci/special_issues/Mechanical_Systems_Dynamics_2020).

For citation purposes, cite each article independently as indicated on the article page online and as indicated below:

LastName, A.A.; LastName, B.B.; LastName, C.C. Article Title. *Journal Name* **Year**, *Volume Number*, Page Range.

ISBN 978-3-0365-2870-0 (Hbk)
ISBN 978-3-0365-2871-7 (PDF)

Contents

About the Editors

Alberto Doria is a full professor at the Department of Industrial Engineering, University of Padova, Padova, Italy. His interests include vibrations, energy harvesting, multi-physics systems, and vehicle dynamics.

Giovanni Boschetti is an associate professor at the Department of Management and Engineering, University of Padova, Vicenza, Italy. His interests include industrial and collaborative robotics, robot and mechanism design, performance evaluation, and cable-driven robots.

Matteo Massaro is an associate professor at the Department of Industrial Engineering, University of Padova, Padova, Italy. His interests include dynamics and optimal control of road vehicles, driving simulators, and multibody modelling.

MDPI

Editorial

Advances in Mechanical Systems Dynamics 2020

Alberto Doria [1,*], Giovanni Boschetti [2] and Matteo Massaro [1]

[1] Department of Industrial Engineering, University of Padova, 35131 Padova, Italy; matteo.massaro@unipd.it
[2] Department of Management and Engineering, University of Padova, 36100 Vicenza, Italy; giovanni.boschetti@unipd.it
* Correspondence: alberto.doria@unipd.it; Tel.: +39-049-827-6803

1. Introduction

The fundamentals of mechanical system dynamics were established before the beginning of the industrial era. The 18th century was very important for science and was characterized by the development of classical mechanics. This development went on in the 19th century and new important applications related to industrialization were found and studied. The development of computers in the 20th century revolutionized mechanical system dynamics owing to the development of numerical simulation. Nowadays we are in the presence of the fourth industrial revolution. Mechanical systems are increasingly integrated with electrical, fluidic, and electronic systems and the industrial environment is going to be characterized by the cyber-physical systems of Industry 4.0. In this framework, the state-of-the–art will be represented soon by integrated mechanical systems, supported by accurate dynamic models able to predict their dynamic behavior. Therefore, mechanical systems dynamics will play a central role in the coming years. This Special Issue aims at disseminating the latest research findings, and ideas in the field of mechanical systems dynamics, with particular emphasis on novel trends and applications.

2. Present Trends in Mechanical Systems Dynamics

The 20 papers collected in this Special Issue refer to different areas of engineering (automotive, manufacturing, and civil engineering) and to very different applications. Nevertheless, they clearly highlight the presence of some important trends in the methods, developments, and challenges of the present research on mechanical systems dynamics.

2.1. Multi-Physics Modeling and Simulation

Until a few years ago, multi-physics modeling and simulation were restricted to some specific fields of research, such as electromechanical problems, vibroacoustic problems, and flow-induced vibrations. The development of powerful codes for multi-physics simulation and the need for detailed analyses of the performance of new devices that require a multi-physics approach (e.g., vibration energy harvesters) have strongly increased the interest in multi-physics analysis. Many papers collected in this Special Issue deal with new developments and/or applications of modeling and simulation.

Papers [1,2] deal with multi-physics modeling and simulation of tires. In [1], the well-known Tire Magic Formula is coupled with a thermal model. In [2], the Fourier equations for thermal analysis are coupled with a tire structural model.

Paper [3] deals with a new and specific application: a liquid handling robot. Computational fluid dynamics was employed to capture the motion of the fluid inside a tank carried by a walking robot.

In [4], the noise vibration and harshness (NVH) of a full engine is carried out coupling a multi-body model of the moving components, with FE models of the structural elements and an acoustic model based on the wave-based technique (WBT). The advantages of coupled simulations are highlighted.

Citation: Doria, A.; Boschetti, G.; Massaro, M. Advances in Mechanical Systems Dynamics 2020. *Appl. Sci.* **2021**, *11*, 2352. https://doi.org/10.3390/app11052352

Received: 3 March 2021
Accepted: 4 March 2021
Published: 6 March 2021

Finally, in [5] a novel vibration energy absorbing mechanism for cables is presented. Since it includes non-linear magnetic springs and hydraulic dampers, a coupled model is developed. The advantages of this multi-physics damper are shown.

2.2. Non-Linear Analysis

Non-linear phenomena are inherent in many natural and engineering systems. Very typical examples of non-linear behavior can be found in tire mechanics and in robot dynamics. Until recently, the presence of non-linearities hindered the analysis and the design of mechanical systems, but nowadays the development of numerical methods allows the exploitation of non-linear phenomena in many fields of industry. In this Special Issue, there are many papers that address non-linear problems.

Papers [4,6] deal with non-linear contact problems in piston-liner interaction and in rolling bearings, respectively.

Paper [7] addresses the problem of vibratory conveying of cylindrical parts, a non-linear numerical model is developed that takes into account the transition between pure rolling and rolling with sliding and the impacts of the cylindrical parts with the edge of the conveyor.

In [5], non-linear stiffness properties are exploited to dampen cable vibrations. Finally, in [8], non-linear effects are estimated to find chain transmission efficiency.

2.3. Identification of Dynamic Systems

The development of the dynamic models of cyber-physical systems requires detailed knowledge of the parameters of the systems. Therefore, identification techniques are very important and have to cope with many problems, such as the presence of non-linearities, noisy data, and new material properties. In most cases, the identification of a mechanical system is based on modal testing, but new technologies based on moving sensors and swarms of sensors are under development. In this Special Issue, four papers focus on identification problems.

Paper [9] deals with the identification of the visco-elastic properties of materials used to dampen vibrations in industrial and civil engineering.

In [10], experimental modal analysis is used to identify the stiffness and damping properties of the joints of industrial robots, which can influence the performance of automatic robotic assembly and material removal processes.

The friction parameters (Stribeck friction model) of a parallel kinematics mechanism with prismatic joints are identified in [11].

Finally, in [4], the numerical modes of vibration of an engine are validated and updated, with an experimental modal analysis.

2.4. Control Strategies and Motion Planning

Presently, more and more mechanical systems are being controlled, with the aim of adjusting their dynamics. Trajectory and motion planning are increasingly relevant in robots and in other mechanical systems. Indeed, the goal of achieving ever-higher speeds is extending into all fields of mechanics. In order to preserve accuracy and repeatability, proper strategies should be adopted in order to generate trajectories that could be executed at high speed, while avoiding excessive motor accelerations and mechanical structure vibrations.

In [12], an active rear axle independent steering system of a ground vehicle is developed adopting a hierarchical synchronization control strategy. In [13], a driver-aid system to improve directional stability of articulated vehicles is developed. Finally, in [14], the driver–steering wheel interaction in emergency situations is dealt with in the framework of the development of advanced driver assistance systems.

2.5. Man–Machine Interaction

Man–machine interaction is becoming increasingly important in the automotive sector. Indeed, the design of advanced driver assistance systems (ADAS) in four-wheeled vehicles is resulting in complex interactions between the driver and vehicle while in two-wheeled vehicles the importance of the rider–vehicle interaction on the vehicle stability has been recognized. In this Special Issue, the investigation of the driver–steering interaction is experimentally carried out in [14], with an instrumented steering wheel, while in [15] the effect of the rider's passive vibration on the stability of a two-wheeled vehicle is considered, with a focus on the weave and wobble modes. Finally, in [16], the most used biomechanical models are compared and employed to build a multibody rider model suitable for multibody applications.

3. Fields of Application

3.1. Machine Elements

Mechanisms, gears, and transmissions are still key elements of advanced industrial systems, automatic machines, and robots. To improve the performance of the system, detailed models of machine elements, taking into account non-linearities or time-variant properties, are needed. This Special Issue includes three papers that cover the modeling and simulation of machine elements.

The first of these deals with the dynamics of cylindrical parts in a vibratory conveyor [7]. A non-linear numerical model is developed, taking into account pure rolling, rolling with sliding, and the impacts on the edges. A comparison between numerical and experimental results guarantees the effectiveness of the method.

A method for the dynamic parameter identification in parallel mechanisms is proposed in [11]. Non-negligible friction has been incorporated in the dynamic model, and a bound-constrained optimization technique was employed to minimize the residual errors while maintaining the physical feasibility of the solutions.

A novel approach for estimating the efficiency of roller chain power transmission systems is proposed in [8]. It is based on sliding friction losses and damping force. The effects of rotational speed, load, derailleur system, and damping coefficient on transmission efficiency were analyzed. The test was set up to verify the estimated efficiency, and the results highlight a good correlation, demonstrating the validity of the estimation.

3.2. Noise and Vibration Control

Vibrations and noise are a potential problem for any application that includes moving components. For many years, the increase in working speed has been a typical trend of machines, but often the increased speed leads to high levels of vibrations and noise, which have to be controlled. The Internet of things and the fourth industrial revolution are characterized by networks of sensors that monitor industrial plants, machines, and vehicles. On the one hand, the sensors' performance and durability can be affected by high amplitude vibrations; on the other hand, environment vibrations can be exploited to feed the distributed sensors adopting vibration energy harvesting technologies based on piezoelectric, electromagnetic, and capacitive phenomena. In the Special Issue, there are four papers that focus on noise and vibration control. Paper [6] deals with machinery vibrations and presents a simplified approach for the analysis of varying compliance vibrations of rolling bearings. New viscoelastic materials are frequently adopted to control and dampen vibrations, in [9], two generalized formulations of viscoelastic materials are deeply analyzed and a comparison with experimental results is made. In [4], the problem of noise vibrations and harshness (NVH) in engines is considered. A coupled model, including an FE model of the structural parts and a multi-body model of valve train dynamics, is developed. An interesting comparison between results obtained with un-coupled and coupled models is made. Finally, in [5], a novel vibration absorbing mechanism for cables is presented. It exploits multi-physical components (magnetic springs and hydraulic cylinders). The mechanical design and the mathematical model of the mechanism are developed, and numerical results are presented; future applications will include stayed bridges.

3.3. Robotics

Robotics is a science in continuous evolution and the efforts of the researchers consist of improving more and more its ability to help operators in their work and to solve their problems. Many works have addressed the performance evaluation and the safety improvement of industrial manipulators. This Special Issue includes three papers dealing with robotics. One of these deals with the issue of liquid handling [3]. In these applications, the problem of sloshing has to be considered as a parameter that can affect the stability of the system. In the paper, a proper motion planning algorithm is proposed to solve the liquid transport problem.

In [10], a compliant joint dynamic model of an industrial robot, based on a modal approach, has been proposed. A novel testing method has been developed, in order to excite the mode of vibration of one-by-one joints, in order to identify each joint stiffness. By means of the developed dynamic model, it is possible to predict the variation of the natural frequencies in the workspace.

Finally, in [17], a recovery strategy for cable robots in case of cable failure is presented and discussed. The paper, making use of a proper performance index, improves on previous works by considering the actuator dynamics. Simulation results prove the feasibility and the effectiveness of the presented approach.

3.4. Cars and Heavy Vehicles

This Special Issue includes six papers dealing with cars and heavy vehicles.

The modeling and simulations of tires are still classic, especially when it comes to the modeling of the effect of temperature on the tire performance. In [1], an extension of the well-known Magic Formula is derived in order to include the effect of temperature on car tires: new coefficients are added to account for changes in longitudinal and cornering stiffness as well as peak factors. In addition, the thermal model is extended.

The handling of trucks is considered in [18], with a focus on the effects of taper-leaf vs. multi-leaf suspensions on handling. The drift test, the ramp steer test, the step steer, and brake-in-turn tests are simulated in a multi-body environment, after validation of the numerical model in steady-turning conditions.

Independent active rear steering is the most flexible approach to rear steering of road vehicles. A two-level hierarchical synchronization control strategy is proposed in [12]. The upper controller adopts a virtual synchronization controller based on a dynamic model of the virtual rear axle steering mechanism to reduce the synchronization error between the rear wheel steering angles, while the lower controller is designed to realize an accurate tracking control of the steering angle for each wheel and reject disturbances. Experimental results on the prototype vehicle are shown.

Heavy-wheeled vehicles with articulated hydraulic steering systems are widely used (e.g., in construction and road building). However, they often suffer from poor directional stability. A driver-aid system aimed at improving the directional stability is presented in [13], based on the steering wheel rotation, a frame articulation sensor, and an additional hydraulic circuit for straightening the frame.

Farming tracked vehicles on deformable soils are considered in [19]. A multibody approach is employed. The two flexible tracks are modeled as groups of rigid links constrained by bushing elements able to replicate the elastic behavior of the track. A soil model is used to describe the terrain behavior subjected to the grouser's action.

The interaction of the driver with the vehicle's steering has been investigated in the past both in the case of cars and motorcycles. In [14], the focus on an instrumented steering wheel (ISW) is employed to investigate the driver's response to a lateral disturbance (the so-called "kick-plate" maneuver) on the vehicle dynamics. A notable feature of the work is that the ISW employed is capable of measuring the force and moment applied by each hand separately. An involuntary torque peak is observed after the disturbance before the voluntary (recovery) torque is applied, both in one- and two-hands maneuvers.

3.5. Light Vehicles

In the near future, light vehicles equipped with an electric or hybrid propulsion system will give a large contribution to sustainable mobility in urban areas. This Special Issue includes four papers dealing with light vehicles.

The thermodynamic modeling of motorcycle tires is considered in [2], where a physical model accounting for sources of heating such as friction power at the road interface and the cyclic generation of heat because of rolling and to asphalt indentation, and for the cooling effects because of the air forced convection, to road conduction and to turbulences in the inflation chamber is proposed. A special focus is the modeling of the contact patch and the need for real-time capabilities.

The characterization of mountain bike tires is addressed in [20], which is a topic rarely covered in the literature, which is mainly focused on cars and motorcycle tires. The work presents laboratory measurements of inflated tire profiles, tire contact patch footprints, force and moment data, as well as static lateral and radial stiffness.

It is well known that the inertial properties of riders affect the stability of two-wheeled vehicles and that there are different reference biomechanical databases for the estimation of such parameters. A review and a comparison focused on multibody applications is presented in [16].

Finally, an analysis of the weave and wobble vibrations of motorcycles is reported in [15], with a focus on the effect of a flexible fork on wobble stability and the effect of the change in the shape of the weave mode on its stability.

4. Conclusions

This Special Issue contains twenty interesting research papers focused on advances in mechanical systems dynamics, covering a wide area of applications. In most of the papers of the Special Issue, numerical results are corroborated by experimental results.

This collection shows the actuality of this topic and gives hints about future developments and applications.

Author Contributions: Conceptualization, A.D., G.B., and M.M.; writing—original draft preparation, A.D., G.B., and M.M.; writing—review and editing, A.D., G.B., and M.M.; supervision, A.D. All authors have read and agreed to the published version of the manuscript.

Funding: This research received no external funding.

Institutional Review Board Statement: Not applicable

Informed Consent Statement: Not applicable

Conflicts of Interest: The authors declare no conflict of interest.

References

1. Harsh, D.; Shyrokau, B. Tire Model with Temperature Effects for Formula SAE Vehicle. *Appl. Sci.* **2019**, *9*, 5328. [CrossRef]
2. Farroni, F.; Mancinelli, N.; Timpone, F. A Real-Time Thermal Model for the Analysis of Tire/Road Interaction in Motorcycle Applications. *Appl. Sci.* **2020**, *10*, 1604. [CrossRef]
3. Usman, M.; Sajid, M.; Uddin, E.; Ayaz, Y. Investigation of Zero Moment Point in a Partially Filled Liquid Vessel Subjected to Roll Motion. *Appl. Sci.* **2020**, *10*, 3992. [CrossRef]
4. Zheng, X.; Luo, X.; Qiu, Y.; Hao, Z. Modeling and NVH Analysis of a Full Engine Dynamic Model with Valve Train System. *Appl. Sci.* **2020**, *10*, 5145. [CrossRef]
5. Qin, Z.; Wu, Y.; Huang, A.; Lyu, S.; Sutherland, J. Theoretical Design of a Novel Vibration Energy Absorbing Mechanism for Cables. *Appl. Sci.* **2020**, *10*, 5309. [CrossRef]
6. Tomovic, R. A Simplified Mathematical Model for the Analysis of Varying Compliance Vibrations of a Rolling Bearing. *Appl. Sci.* **2020**, *10*, 670. [CrossRef]
7. Comand, N.; Doria, A. Dynamics of Cylindrical Parts for Vibratory Conveying. *Appl. Sci.* **2020**, *10*, 1926. [CrossRef]
8. Zhang, S.; Tak, T. Efficiency Estimation of Roller Chain Power Transmission System. *Appl. Sci.* **2020**, *10*, 7729. [CrossRef]

9. Genovese, A.; Carputo, F.; Maiorano, A.; Timpone, F.; Farroni, F.; Sakhnevych, A. Study on the Generalized Formulations with the Aim to Reproduce the Viscoelastic Dynamic Behavior of Polymers. *Appl. Sci.* **2020**, *10*, 2321. [CrossRef]
10. Bottin, M.; Cocuzza, S.; Comand, N.; Doria, A. Modeling and Identification of an Industrial Robot with a Selective Modal Approach. *Appl. Sci.* **2020**, *10*, 4619. [CrossRef]
11. Rosyid, A.; El-Khasawneh, B. Identification of the Dynamic Parameters of a Parallel Kinematics Mechanism with Prismatic Joints by Considering Varying Friction. *Appl. Sci.* **2020**, *10*, 4820. [CrossRef]
12. Deng, B.; Zhao, H.; Shao, K.; Li, W.; Yin, A. Hierarchical Synchronization Control Strategy of Active Rear Axle Independent Steering System. *Appl. Sci.* **2020**, *10*, 3537. [CrossRef]
13. Lopatka, M.; Rubiec, A. Concept and Preliminary Simulations of a Driver-Aid System for Transport Tasks of Articulated Vehicles with a Hydrostatic Steering System. *Appl. Sci.* **2020**, *10*, 5747. [CrossRef]
14. Comolli, F.; Gobbi, M.; Mastinu, G. Study on the Driver/Steering Wheel Interaction in Emergency Situations. *Appl. Sci.* **2020**, *10*, 7055. [CrossRef]
15. Passigato, F.; Eisele, A.; Wisselmann, D.; Gordner, A.; Diermeyer, F. Analysis of the Phenomena Causing Weave and Wobble in Two-Wheelers. *Appl. Sci.* **2020**, *10*, 6826. [CrossRef]
16. Bova, M.; Massaro, M.; Petrone, N. A Three-Dimensional Parametric Biomechanical Rider Model for Multibody Applications. *Appl. Sci.* **2020**, *10*, 4509. [CrossRef]
17. Boschetti, G.; Minto, R.; Trevisani, A. Improving a Cable Robot Recovery Strategy by Actuator Dynamics. *Appl. Sci.* **2020**, *10*, 7362. [CrossRef]
18. Zhao, L.; Zhang, Y.; Yu, Y.; Zhou, C.; Li, X.; Li, H. Truck Handling Stability Simulation and Comparison of Taper-Leaf and Multi-Leaf Spring Suspensions with the Same Vertical Stiffness. *Appl. Sci.* **2020**, *10*, 1293. [CrossRef]
19. Mocera, F.; Somà, A.; Nicolini, A. Grousers Effect in Tracked Vehicle Multibody Dynamics with Deformable Terrain Contact Model. *Appl. Sci.* **2020**, *10*, 6581. [CrossRef]
20. Dressel, A.; Sadauckas, J. Characterization and Modelling of Various Sized Mountain Bike Tires and the Effects of Tire Tread Knobs and Inflation Pressure. *Appl. Sci.* **2020**, *10*, 3156. [CrossRef]

Article

Efficiency Estimation of Roller Chain Power Transmission System

Sheng-Peng Zhang and Tae-Oh Tak *

Department of Mechanical and Biomedical Engineering, Kangwon National University, Chun Cheon 24341, Korea; zsp363527125@gmail.com
* Correspondence: totak@kangwon.ac.kr

Received: 30 September 2020; Accepted: 29 October 2020; Published: 31 October 2020

Abstract: In the present study, a novel approach to estimating the efficiency of roller chain power transmission systems is proposed based on sliding friction losses and damping force. The dynamics model is taken into account between chain links with lateral offset owing to the derailleur system. Frictional losses were calculated according to Coulomb's law of friction, and the damping force was dependent on the damping coefficient. The effects of rotational speed, load, derailleur system, and damping coefficient on transmission efficiency were analyzed. The test stand of the roller chain power transmission system was set up to verify the estimated efficiency, and the results showed a good correlation, demonstrating the validity of the chain power transmission efficiency estimation.

Keywords: roller chain; chain efficiency; sliding friction; lateral offset; damping force

---

1. Introduction

Chains are widely used in bicycles, motorcycles, personal mobility devices, and various industrial equipment since they provide reliable, efficient, and durable power transmission with relatively low economic cost. Figure 1 presents a schematic of a roller chain, the most common and popular type of chain, in which the inner link is connected by a pin on the outer link, and the pin is covered by a bush and roller that contacts with the mating sprocket. The transmission efficiency of the chain drive system, which is the ratio of the input power of the driving sprocket to the output power of the driven sprocket, depends on the configuration of the chain and the operation conditions such as torque and speed. In the process of power transmission, energy loss occurs due to sliding friction between the pin, the roller, and the bush during the rotation of articulation, vibration in the chain spans, impact when the roller engages and disengages with the sprocket, and contacting force between the inner and the outer side plates of the derailleur system for gear shifting.

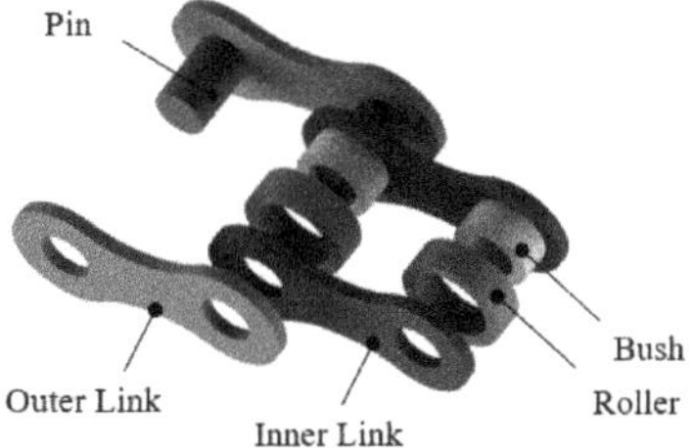

Figure 1. Schematic of roller chain and its components.

Even though the roller chain drive system has been utilized for many years, little is known concerning how to estimate efficiency, which is affected by various factors, such as transmission speed, load, lateral offset between driving and driven sprockets, deflection of the tight side of the chain, etc. In 1999, Kidd et al. [1], performed an experimental examination in which tension was statically loaded for a bicycle chain with a sprocket under conditions of in-plane loading and out-plane loading. However, since only steady-state analysis for tension was carried out, the mechanical interaction of the roller chain and its components was not addressed. James C. Conwell and Johnson G.E. [2] in 1995 investigated the tension of link and impact force on roller sprockets in roller chains under various operation conditions and summarized the results by regression analysis. The impact force increased as the rotational speed and tension of link increased. However, the results did not provide thorough theoretical basis. In 2001, [3] James B. Spicer found that chain transmission is primarily affected by frictional energy loss between 97 and 99%, and compared and validated this result under various rotational speed, power, and lubrication conditions. However, regarding the aspect of chain drive to transmission efficiency with lateral offset, that study only theoretically modeled frictional loss, and experimental verification was not performed. Furthermore, [1,3,4] designed and analyzed the effects of lateral offset for a roller chain system, but inner contact of the chain link with lateral offset owing to the derailleur type system was not reported. Hollingworth, N.E. [5] and Hills, D.A. [6] in 1986 modeled and analyzed the link tension and theoretical efficiency of a chain and formulated the contact force of the pin and bush articulation. However, it is only used in a cranked chain, and does not apply to a roller chain. Although the roller chain system is lubricated, friction occurs at the contacts between the sprockets and the rollers. One advantage of presence of friction force is that it damps out high peak forces during the contact [7,8]. Additionally, Coulomb friction modeling is simple because it only requires the kinetic coefficient of friction. Burgess, S.C. [9] in 1998 calculated chain power transmission efficiency based on the energy loss between the chain links and sprockets and investigated the effects of size of sprockets. It was found that average chain transmission efficiency was 98.8%. However, this model is limited to a race cycling bike chain links that only include pin and sleeve without roller. In 2001, Lodge [10] and Burgess [11] assessed the efficiency of roller chains by modeling frictional losses due to pin sliding against the bush and bush sliding against the roller, in which sliding friction was based on Coulomb's law of friction. The results of efficiency were in the range of 9–99% in high torque transmission. Troedsson, I. [12] and Vedmar, L. [13] in 2001 modeled a complete chain transmission with two sprockets and a chain with tight and slack spans. Damping force is considered as an important factor in the tight span that transmits the most power, and it is assumed to be proportional to the relative velocity between the two end points at the tight side. In 2004, Stuart Burgess and Chris Lodge [14] demonstrated that the efficiency of the chain is significantly influenced by the center distance between two sprocket axes, and showed that chain center distance is a key design parameter for motorcycles, although experimental verification was not achieved. Wragge-Morley, R. et al. [15], in 2018, estimated the drive transmission efficiency of the roller chain according to the sliding friction force among the pin, bush, and roller of a chain; however, the experimental test was only validated in the condition of low rotational speed. In 2018, Aleksey Egorov et al. developed a method that determined the transmission energy consumed to accelerate a rotating chain system [16–19]. In addition, chain efficiency was estimated by measuring the time of angular acceleration in the rotation axis of the drive shaft with and without load during the speeding-up of a chain drive. This method, however, could not discern the characteristics of a steady-state for a chain system, such as various affected parameters.

In computing the roller chain system efficiency, energy loss should be always considered. While some energy stored as elastic deformation of components is recovered, other energy losses due to friction and damping cannot be recovered. James B. Spicer et al. [3] estimated the bicycle chain drive efficiency based on the energy losses due to sliding friction. Theoretical efficiency of the chain system was found to be between 97 and 99%, however, the results from the experiment and theoretical analysis showed some discrepancy. In order to estimate the energy losses based on damping force, Troedsson, I. and Vedmar, L. [12] proposed a method to determine the load distribution during power transmission considering the damping factor. The damping force was modeled according to the damping coefficient and rotational speed of the chain span.

The aim of the present study is to estimate the efficiency of the roller chain transmission system considering energy losses due to sliding friction and damping force in the tight span. The sliding frictional force between the pin, the bush, and the roller was established. The dynamics between links with lateral offset owing to the derailleur system are also modeled. In addition, the damping force between the driving sprocket and the driven sprocket in the tight span is considered theoretically according to the choice of damping coefficient. The effects of rotational speed, torque, offset angle, and damping coefficient are analyzed. The test stand of the roller chain power transmission system is set up to verify the estimated efficiency. The results demonstrate the experimental measurements of the efficiency of the roller chain drive under various operation conditions to validate the theoretical estimation.

2. Chain Drive Efficiency

Chain drive efficiency, which is the ratio of the input power of the driving sprocket to the output power of the driven sprocket, depends on torque and speed. In other words, power is transferred from the driving shift to the driven shift. Therefore, part of the total energy extracted from the system cannot be recovered due to such factors as frictional loss and damping. The sliding friction force includes the coplanar drive transmission, which is calculated as the surface contact pressure, and the lateral offset drive, which is the two-points contact owing to the derailleur system.

2.1. Sliding Friction Force

Figure 2 shows the interaction between the chain and the sprocket, where the roller is engaged with the sprocket tooth, resulting in normal contact force N_1 and friction force F_1 between the pin and the bush. Articulation from the chain span to the seated sprocket turns a maximum angle θ. Normal contact force N_1 between the pin and the bush is generated, where T_c represents the link tension with the chain span, θ_N is the angle of normal contact force; R_p and R_{bi} are the radius of the pin and the bush inner surface, respectively; θ is the articulation angle of the pin between axes of the chain span and the seated link in the sprocket; and t is the width of the bush.

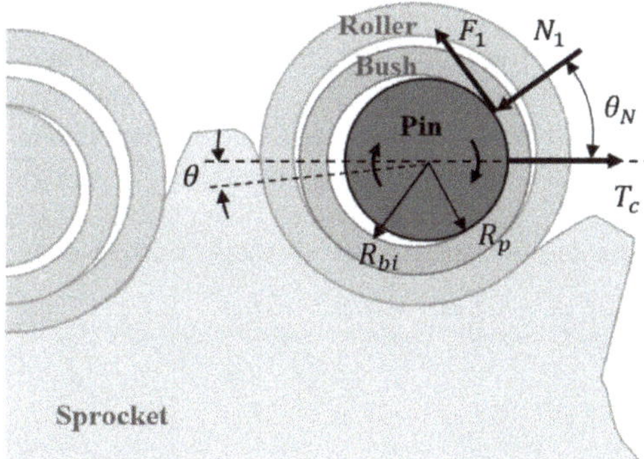

Figure 2. Schematic of contact for pin articulation.

Hollingworth, N.E. [5] and Hills, D.A. [6] determined that sliding frictional losses during pin articulation are dependent on the pin sliding against the bush. Static or slip motion between the pin and the bush depends on the magnitude of the friction angle of θ_N compared with the articulation angle θ. The frictional force is sufficiently large enough to ensure pin stiction under the condition of $\tan\theta < \tan\theta_N$. In the case of the pin slipping, it should be satisfied with the condition of $\tan(\theta) \geq$

μ_1 and $\mu_1 = F_1/N_1 = \tan(\theta_N)$, where μ_1 is the friction coefficient between the pin and the bush. The relationship of contact force between the pin and the bush can be written as follows:

$$N_1 = T_c \cdot \cos(\theta_N) = T_c \cdot \frac{1}{\sqrt{1+\tan^2(\theta_N)}} = T_c \cdot \frac{1}{\sqrt{1+\mu_1{}^2}} \tag{1}$$

Normally, a normal force can be found by integrating the contact pressure over the area. The link tension is then related to the surface contact pressure and area. James B. Spicer [3] calculated pressure related with contact surface between the pin and the bush. It was assumed that the contact pressure is a constant, and the contact angle between the pin and the bush is also a constant. Here, the normal contact force N_1 between the pin and the bush is considered to be a normal force. The relationship of contact pressure and normal force is shown as Equation (2), where it is assumed that contact pressure $P \cdot \cos(\theta)$ acts radially and is a constant:

$$N_1 = R_p \cdot t \cdot \int_{-\frac{\pi}{2}}^{\frac{\pi}{2}} P \cdot \cos(\theta) d\theta \tag{2}$$

Then, friction force between the pin and the bush can be given by Equation (3):

$$F_1 = \mu_1 \cdot (P \cdot dA) \tag{3}$$

The tension in the chain links owing to centripetal acceleration is considered to have little effect on friction force. The energy losses W_{pin} of pin articulation over the contact area between the pin and the bush are given by:

$$W_{pin} = \frac{\pi}{2} \cdot \left(\frac{1}{\sqrt{1+\mu_1{}^2}} \cdot T_c \right) \cdot \mu_1 \cdot R_{bi} \cdot \theta \tag{4}$$

Figure 3 shows that the chain is driven under lateral offset owing to the derailleur system. The contact force between the pin and the bush with lateral offset angle γ is generated. This analysis differs from that in Kidd [1] and James B. Spicer [3], in that their analysis did not consider the contact of chain links. In fact, contact between the pin and the bush is not generated along the vertical direction of the inner bush surface, but rather occurs at the boundary of the bush plate, shown as normal force in one side N_0^1 and other side N_0^2 due to lateral offset. When the link engages the sprocket, the sliding friction force of pin articulation F_0^1 and F_0^2 are generated. The pressure angles of both normal forces are assumed to be equal, and the normal forces can be expressed as follows:

$$N_0^1 = N_0^2 = T_c \cdot \cos(\theta'_N) = T_c \cdot \frac{1}{\sqrt{1+\tan^2(\theta'_N)}} = T_c \cdot \frac{1}{\sqrt{1+\mu_1{}^2}} \tag{5}$$

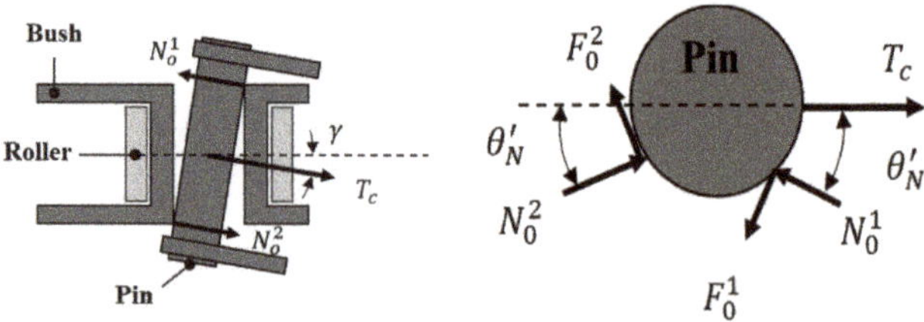

Figure 3. Schematic of contact for chain links with lateral offset (above view) on the left side and contact of pin articulation (side view) on the right side.

The energy losses W_{pin} of pin articulation with a lateral offset between the pin and the bush are given by:

$$W'_{pin} = 2 \cdot \left(\frac{1}{\sqrt{1+\mu_1{}^2}} \cdot T_c \right) \cdot \mu_1 \cdot r_{bi} \cdot \theta \tag{6}$$

Figure 4 presents the schematic of forces and motion between the bush and the roller. The tooth contact force transfers to the bush when the roller is engaged with the sprocket tooth. Contact for bush articulation, which is located at the pin-bush contact and the bush-roller contact, is generated. Different from the relationship between the pin and the bush, the surface contact force between the bush and the roller occurred with lateral offset or coplanar offset when the link engages the teeth, due to the independent roller. Therefore, the friction F'_1 between the pin and the bush, and F_2 between the bush and the roller, are generated. W_{bush} is the energy loss of bush articulation owing to friction; R_{ri} is the inner radius of the roller; T_1 is the link tension seated in the sprocket; N_s represents the sprocket tooth contact force; β is the nominal pressure angle; and φ is the friction angle for normal contact force between the bush and the roller.

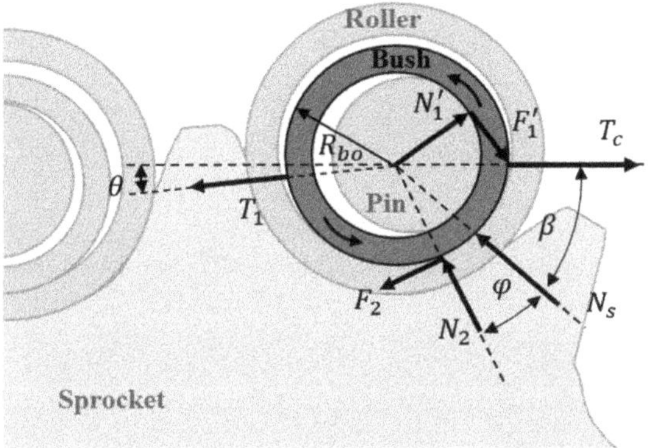

Figure 4. Schematic of contact for bush articulation.

According to the friction forces, the frictional losses between the pin-bush and the bush-roller can be calculated. The sliding friction force will occur at the bush and roller interface first because it is located at a small radius. Therefore, it is assumed that there is no sliding force between the tooth and rollers, and the friction loss occurs between the roller and the bush. For bush articulation, this is because the tension of the chain in the tight span is much greater than that in the slack span.

For the slip condition of pin articulation, it should be satisfied with the condition of $\tan(\theta) \geq \mu_1, \mu_2$ and $\mu_2 = F_2/N_2 = \tan(\varphi)$, where μ_2 represents the friction coefficient between the bush and the roller. We assume the following:

$$N_s = \frac{N_2}{\cos(\varphi)} \tag{7}$$

and

$$\frac{N_s}{T_c} = \frac{\sin(\theta)}{\sin(\beta-\theta)} \tag{8}$$

Substituting Equation (7) into Equation (8), the normal contact force N_2 between the bush and the roller can be written as follows:

$$N_2 = T_c \cdot \frac{\cos(\varphi) \cdot \sin(\theta)}{\sin(\beta-\theta)} \tag{9}$$

Due to the surface contact of bush articulation, the energy loss between the bush and the roller in the tight span using the relationship between pressure and normal force can be given by the following:

$$W_{bush} = \frac{\pi}{2} \cdot \frac{\sin\theta}{\sqrt{1+\mu_2{}^2} \cdot \sin(\beta-\theta)} \cdot T_c \cdot \cos(\gamma) \cdot \mu_2 \cdot R_{bo} \cdot \theta \quad (10)$$

2.2. Damping Force

Viscous damping is an important characteristic in the friction force. The damping coefficient is dependent on lubrication, contacting surface, damping force, and contacting speed. In this study, the magnitude of the damping coefficient refers to the work of Troedsson, I. and Vedmar, L [12]. Power due to damping force in the chain drive system is mainly transmitted through the tight span. In the work of Yu, S. [20] and Lodge, C, J. and Burgess, S, C. [10], magnitude of tension at the slack and tight span were investigated, which showed that tension in the slack span is much smaller than tension in the tight span. Therefore, the damping force is assumed to occur at the tight span, and the damping force in the slack span is neglected, since tension at the tight span is larger than that at the slack span. The damping force is assumed to be proportional to the relative velocity between the end points from the driving sprocket to the driven sprocket at the tight span in Figure 5. Therefore, the damping force is expressed as:

$$F_d = D_{chain}\left[R_{s,1}\dot{\sigma_1} \cdot \cos(\sigma_1 - \theta_{in,1}) - R_{s,2}\dot{\sigma_2} \cdot \cos(\sigma_2 - \theta_{in,2})\right] \quad (11)$$

where D_{chain} is the damping coefficient in the tight span of the chain; subscripts 1 and 2 describe the driving and the driven sprocket, respectively; $R_{s,1}$ and $R_{s,2}$ are the radius of the driving and the driven sprocket, respectively; σ_1 is the angle between the circle line of the first seated tooth into the links and coordinate line of the driving sprocket; σ_2 is the angle between the circle line of the last seated tooth into the links and coordinate line of the driven sprocket; $\dot{\sigma_1}$ is the rotational speed with respect to σ_1; $\dot{\sigma_2}$ is the rotational speed with respect to σ_2; and $\theta_{in,1}$ and $\theta_{in,2}$ are the chain angles related to the x-axle of the first links in contact with the sprocket in the driving and the driven sprocket, respectively.

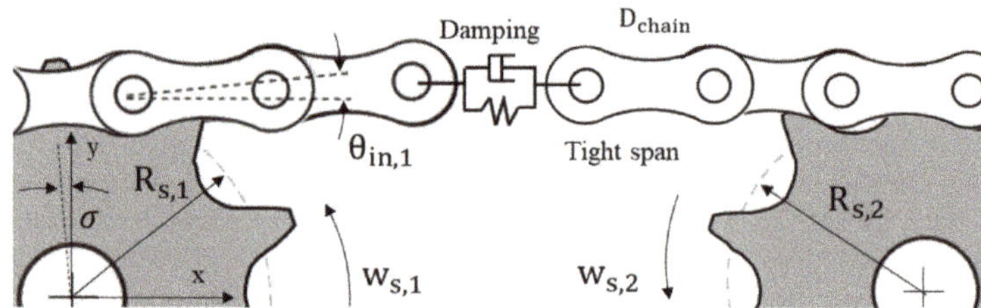

Figure 5. Schematic of damping force between two sprockets.

2.3. Efficiency of Chain Drive System

The magnitude of the efficiency of the chain drive system is given by the output energy divided by the overall energy, including output energy and transmission loss energy. The total transmission energy losses in the chain drive can be calculated by summing the sliding frictional loss and damping force in the tight span. The sliding friction force includes the coplanar drive transmission, which is calculated as the surface contact pressure and the lateral offset drive, which is the two-points contact.

Using the power of the driven sprocket, the transmission efficiency η of the chain in the coplanar drive can be given by Equation (12):

$$\eta = \frac{T_o \cdot w_{s,2}}{T_o \cdot w_{s,2} + N_{s,1} \cdot w_{s,1} \cdot (W_{pin} + W_{bush}) + F_d \cdot w_{s,1} \cdot R_{s,1}} \quad (12)$$

Transmission efficiency η', considering lateral offset, can be written as Equation (13):

$$\eta' = \frac{T_o \cdot w_{s,2}}{T_o \cdot w_{s,2} + N_{s,1} \cdot w_{s,1} \cdot (W'_{pin} + W_{bush}) + F_d \cdot w_{s,1} \cdot R_{s,1}} \quad (13)$$

where η is the chain drive efficiency with the coplanar drive; T_o is the torque of the driven sprocket; $w_{s,1}$ and $w_{s,2}$ are the rotational speed of the driving and the driven sprocket, respectively; and $N_{s,1}$ is the number of teeth on the driving sprocket.

3. Results

The chain system consists of the driving sprocket, the driven sprocket, and the roller chain. The driving sprocket and the driven sprocket are selected with teeth number 32T-32T, which is the maximum ratio in a sprocket, and avoids the chain from falling off of the sprocket due to the lateral offset angle. The type of roller chain is specified in ISO 606 [21]. The friction coefficient for lubricated steel–steel interfaces is assumed to be 0.11 [22]. Driving torque is applied to the driving sprocket, and braking torque is exerted on the driven sprocket. Efficiency was calculated with different parameter values of rotational speed, torque, lateral offset angle, and damping coefficient. The range of parameters is as follows: A rotational speed 40–80 RPM with an interval of 10 RPM, torque 5–9 nm with an interval of 1 nm, a lateral offset angle of 0/0.0174/0.0349/0.0523 rad (0/1/2/3 deg), and a damping coefficient of 40/60/80 ns/m. The list of drive train components and their parameters is presented in Table 1.

Table 1. Drive train components and parameters for roller chain system.

Parts	Parameters	Symbol [Unit]	Value
Roller chain	Friction coefficient between the pin and the bush/among the pin, the bush, and the roller	μ_1/μ_2 [-]	0.11/0.11
	Radius of inner bush	r_{bi} [m]	1.74×10^{-3}
	Maximum rotation angle of chain link	α_m [rad]	0.19
	Average mass of one chain link	m [kg]	0.0024
	Chain pitch	p [m]	12.7×10^{-3}
	Radius of inner roller	r_{bo} [m]	2.83×10^{-3}
	Damping coefficient	D_{chain} [Ns/m]	40/60/80
	Chain angle related to x-axle of link in driving/driven sprocket	$\theta_{in,1/2}$ [rad]	0.00012/0.00012
	Tension in the chain span	T_c [N]	-
Sprocket	Radius of driving/driven sprockets	$R_{s,1/2}$ [m]	$62.43/62.43 \times 10^{-3}$
	Teeth number of driving-driven sprocket	$N_{s,1-2}$ [-]	32–32
	Rotational speed of driving/driven sprocket	$w_{s,1/2}$ [rad/s]	4.18/5.23/6.28/ 7.32/9.37
	Coordinate for position of driving/driven sprocket	$\sigma_{1/2}$ [rad]	3.09/1.52
	Lateral offset angle between sprockets	γ [rad]	0/0.0174/0.0349/0.0523
	Power of driven sprocket	$P_{s,2}$ [W]	-

Figure 6 shows the following: Transmission efficiency as a function of rotational speed and offset angle; the damping coefficient, in which the range of efficiency is 86.3–93.1%; and demonstrates that efficiency decreases with increasing rotational speed and damping coefficient for a given lateral offset angle. In addition, for a given rotational speed, chain efficiency decreases with increasing damping coefficient and lateral offset angle.

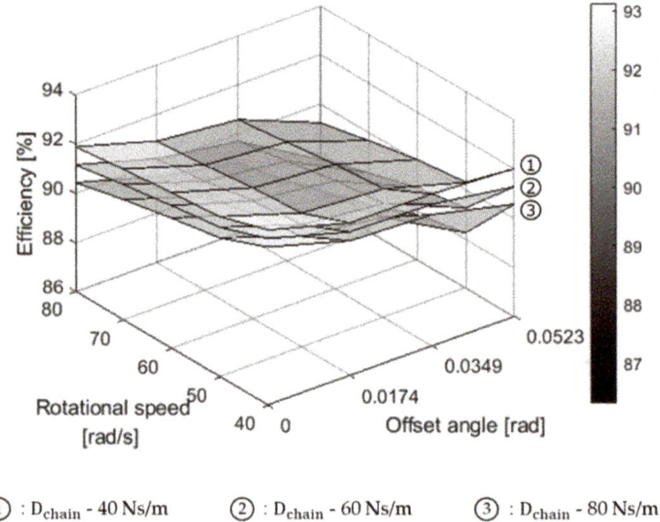

① : D_{chain} - 40 Ns/m ② : D_{chain} - 60 Ns/m ③ : D_{chain} - 80 Ns/m

Figure 6. Transmission efficiency as a function of rotational speed, offset angle, and damping coefficient.

Figure 7 presents the effect of torque, lateral offset angle, and damping coefficient to transmission efficiency, demonstrating that efficiency increases with torque, and the range of efficiency is 90.8–93.0%. Similar to the results from Figure 6, chain efficiency decreases with increasing damping coefficient and offset angle for a given torque.

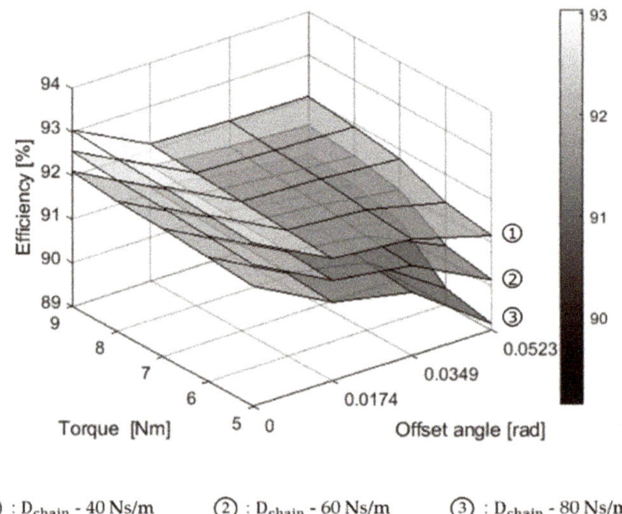

① : D_{chain} - 40 Ns/m ② : D_{chain} - 60 Ns/m ③ : D_{chain} - 80 Ns/m

Figure 7. Transmission efficiency as function of torque, offset angle, and damping coefficient.

4. Experimental Measurement

An experiment is performed in order to measure and verify the estimated efficiency. The layout of the experiment is shown schematically in Figure 8. Driving sprocket axes and driven sprocket axes are connected via the test chain. In the driving sprocket axes, an electric motor with a maximum rotational speed of 1700 RPM and a maximum torque capability of 13 Nm provides propulsion. Driving torque from the electric motor is measured through a torque transducer (accuracy ± 0.3%), which has a rated capacity of 0–100 kgf-m. In the driven sprocket axes, a friction disk brake that can regulate speed and torque is attached. The torque of the driven sprocket axes is measured with the torque transducer of the same type as that with the driving axes. Support bearings are used in the rotational axes between the electric motor, torque transducers, brake disk, and driving/driven sprocket axes. In order to take into consideration friction loss due to support bearings, friction loss of individual support bearings is measured with varying rotational speeds in the range of 50 RPM–280 RPM. The lateral offset angle between sprockets can be adjusted by moving the driven sprocket axes laterally.

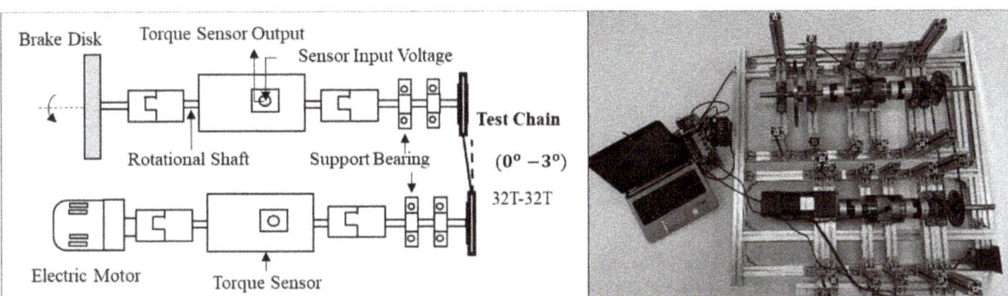

Figure 8. Schematic of setup and test stand equipment.

Efficiency is measured with different operating conditions. The lateral offset angle is 0/0.0174/0.0349/0.0523 rad with an increment of 1 degree, the torque range is 5–9 Nm with an increment of 1 nm, and the rotational speed is in the range of 40–80 RPM with an increment of 10 RPM. Each test takes approximately 20 s and is repeated seven times.

5. Experiment Results

Figures 9 and 10 show the chain drive efficiency under different conditions (rotational speeds and lateral offset angle) and compares them with theoretical results. The experimental results demonstrate that chain drive efficiency decreased with increasing rotational speed for a given offset angle; whereas, for a given rotational speed, chain drive efficiency decreased with increasing offset angles. It is found that both results exhibit the same tendency in terms of the effect of rotational speed and lateral offset angle. These results agree with the theoretical results. The efficiency is dependent on rotation speed and the offset angle in the Equations (11) and (12). Figure 10 shows the difference between the theoretical and experimental results with an error range of 0.35–1.81%. The results reveal that, when the rotational speed is high, the error value is small. However, error increases with decreasing rotational speed. This is attributable to that, when the test was carried out with low rotational speed and offset angle, vibration and impact occur between the chain and the sprocket.

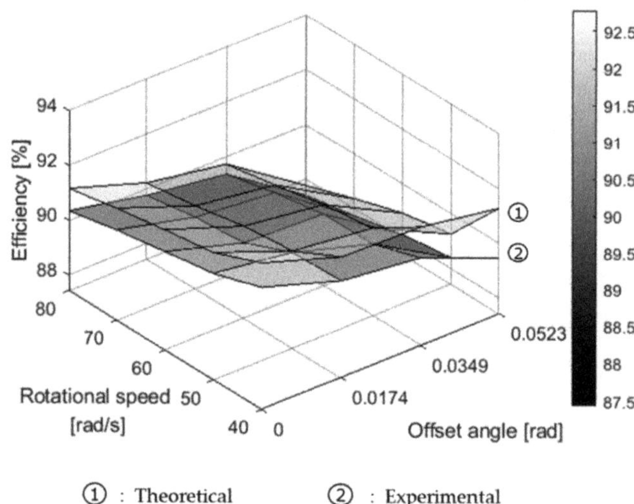

Figure 9. Transmission efficiency of theoretical and experimental results.

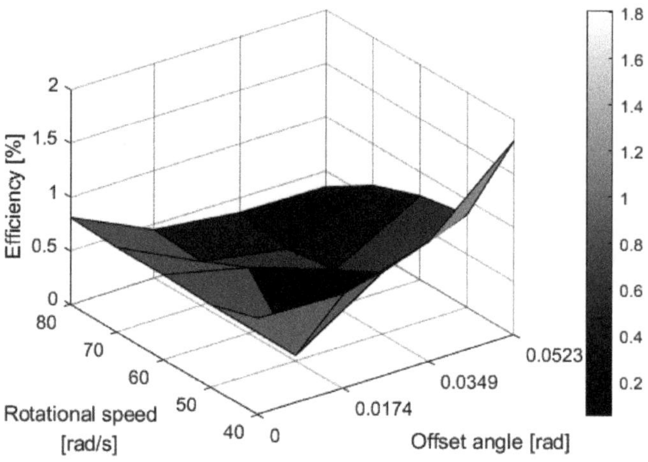

Figure 10. The difference between theoretical and experimental results.

Figures 11 and 12 show the chain drive efficiency under different output torques and offset angles, and compares them with the theoretical results. The results demonstrate that the chain drive efficiency decreased with increasing output torque for a given offset angle; whereas, for a given output torque, the chain drive efficiency decreased with the increasing offset angle. It is found that both results possess the same trend in terms of the effect of torque and lateral offset angle. Figure 12 presents the difference between theoretical and experimental results, with an error range between 0.23–3.77%. The reason for this is that since the test is carried out with small torque and large offset angle, vibration and impact become more apparent. Thus, the contact surface becomes large between the chain links and sprockets.

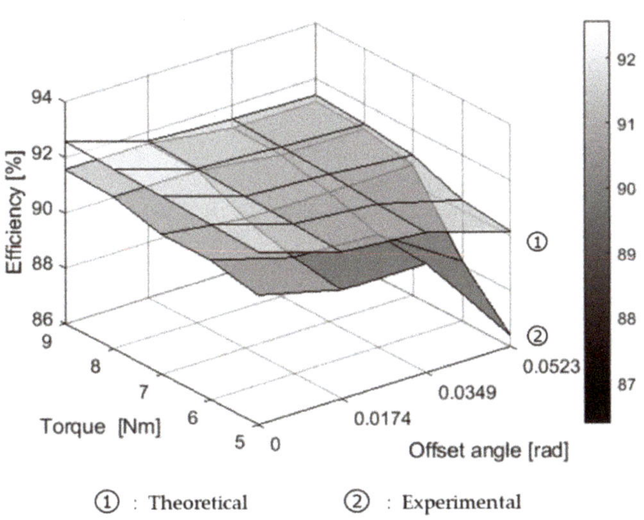

Figure 11. The transmission efficiency of theoretical and experimental results.

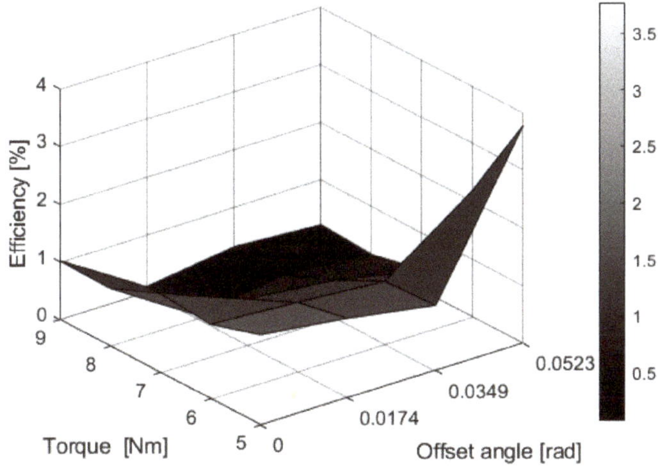

Figure 12. The difference between theoretical and experimental results.

Figure 13 shows the efficiency under the different center distance between two sprockets axes and a different offset angle and compares them with the theoretical results. The experiment was carried out with an interval of 0.05 m in the test stand. The results demonstrate that the efficiency of the chain drive system increased with decreasing center distance between sprocket axes for a given offset angle. The range of measurement results is 89.34–93.24%.

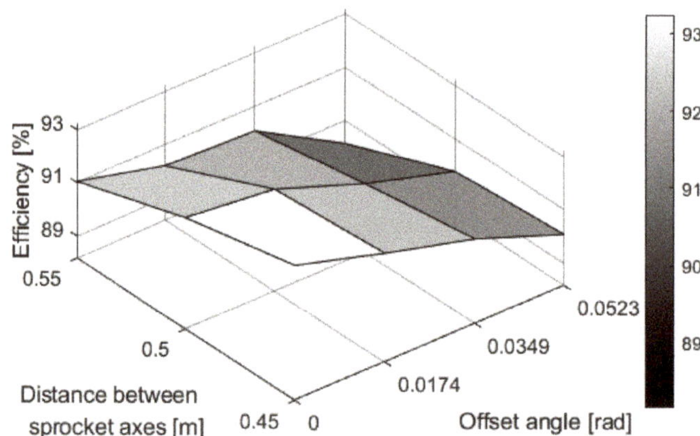

Figure 13. Chain drive efficiencies for different center distances between sprockets with an offset angle.

6. Conclusions

A novel approach for estimating chain transmission efficiency, which takes into account combined sliding friction losses due to pin articulation, bush articulation, and damping force in the tight span, was proposed. The frictional losses due to pin articulation motion and bush articulation motion were calculated according to Coulomb's law of friction. The dynamics model was applied to investigate chain links with lateral offset owing to the derailleur system. The damping force, which exists between the driving sprocket and driven sprocket in the tight span, was also considered according to the choice of damping coefficient.

The theoretical value of chain transmission efficiency was calculated with a range of 86.3–93.1%, depending on the driving conditions. Transmission loss in each parameter was analyzed by the various operation conditions of the chain drive system, including rotational speed, torque, lateral offset angle, and damping coefficient. Tests were performed to measure efficiency and compared with the theoretical results, with a difference range of 0.23–3.77%, which demonstrates good agreement with theoretical values.

Author Contributions: T.-O.T. conceived and designed the experiments; S.-P.Z. performed the experiments, analyzed the data and wrote the paper. All authors have read and agreed to the published version of the manuscript.

Funding: This work was supported by Ministry of Trade, Industry and Energy (20001447).

Conflicts of Interest: The authors declare no conflict of interest.

References

1. Kidd, M.D.; Loch, N.E.; Reuben, R.L. Experimental Examination of Bicycle Chain Forces. *Exp. Mech.* **1999**, *39*, 278–283. [CrossRef]
2. Conwell, J.C.; Johnson, G.E. Experimental investigation of link tension and roller-sprocket impact in roller chain drives. *Mech. Mach. Theory* **1995**, *31*, 533–544. [CrossRef]
3. James, B.S.; Christopher, J.K.R.; Michael, J.E.; Johanna, R.B.; Masahiko, F.; Masao, T. Effects of Frictional Loss on Bicycle Chain Drive Efficiency. *J. Mech. Des.* **2001**, *123*, 598–605.
4. Wang, C.C.; Tseng, C.H.; Fong, Z.H. A method for improving bicycle shifting performance. *Int. J. Vehicle Des.* **1997**, *18*, 100–117.
5. Hollingworth, N.E.; Hills, D.A. Force in a heavy-duty drive chain during articulation. *J. Mech. Eng. Sci.* **1986**, *200*, 367–374. [CrossRef]
6. Hollingworth, N.E.; Hills, D.A. Theoretical efficiency of a cranked link chain drive. *J. Mech. Eng. Sci.* **1986**, *200*, 375–377. [CrossRef]

7. Jalon, J.G.D.; Bayo, E. *Kinematic and Dynamic Simulation of Multibody Systems*; Springer: Berlin/Heidelberg, Germany, 1994.
8. Marques, F.; Flores, P.; Pimenta Claro, J.C.; Hamid, M.L. Modeling and analysis of friction including rolling effects in multibody dynamics: A review. *Multibody Syst. Dyn.* **2019**, *45*, 223–244. [CrossRef]
9. Burgess, S.C. Improving cycling performance with large sprockets. *Sports Eng.* **1998**, *1*, 107–113. [CrossRef]
10. Lodge, C.J.; Burgess, S.C. A model of the tension and transmission efficiency of a bush roller chain. *Proc. Inst. Mech. Eng.* **2001**, *216*, 385–394. [CrossRef]
11. Lodge, C.J.; Burgess, S.C. An investigation into the selection of optimum chain and sprocket size. *J. Eng. Des.* **2004**, *15*, 536–580. [CrossRef]
12. Troedsson, I.; Vedmar, L. A method to determine the dynamic load distribution in a chain drive. *Proc. Inst. Mech. Eng.* **2001**, *215*, 569–579. [CrossRef]
13. Troedsson, I.; Vedmar, L. A method to determine the static load distribution in a chain drive. *J. Mech. Des.* **1999**, *121*, 402–408. [CrossRef]
14. Burgess, S.; Lodge, C. Optimisation of the chain drive system on sports motorcycles. *Sports Eng.* **2004**, *7*, 65–73. [CrossRef]
15. Wragge-Morley, R.; Yon, J.; Lock, R.; Alexander, B.; Burgess, S. A novel pendulum test for measuring roller chain efficiency. *Meas. Sci. Technol.* **2018**, *29*. [CrossRef]
16. Egorov, A.; Kozlov, K.; Belogusev, V. A method for evaluation of the chain drive efficiency. *J. Appl. Eng. Sci.* **2015**, *34*, 13–14. [CrossRef]
17. Kozlov, K.; Belogusev, V.; Egorov, A.; Syutov, N. Development of method of evaluate friction losses of chain drives. In Proceedings of the 17th International Scientific Conference Engineering for Rural Development, Jelgava, Latvia, 23–25 May 2018; pp. 930–936.
18. Egorov, A.; Kozlov, K.; Belogusev, V. Experimental identification of the electric motor moment of inertia and its efficiency using the additional inertia. *ARPN J. Eng. Appl. Sci.* **2016**, *11*, 10582–10588.
19. Kozlov, K.E.; Egorov, A.V.; Belogusev, V.N. Experimental evaluation of chain transmissions lubricants quality using a new method based on additional inertia moment use. *Procedia Eng.* **2017**, *206*, 617–623. [CrossRef]
20. Huo, J.; Yu, S.; Yang, J.; Li, T. Static and dynamic characteristic of the chain drive system of a heavy duty apron feeder. *Open Mech. Eng. J.* **2013**, *7*, 121–128. [CrossRef]
21. *Short-Pitch Transmission Precision Roller and Bush Chains: Attachments and Associated Chain Sprockets*; ISO606; ISO: Vernier, Switzerland, 2015.
22. Lee, C.H.; Polycarpou, A.A. Static Friction Experiments and Verification of an Improved Elastic-Plastic Model Including Roughness Effects. *J. Tribol.* **2007**, *129*, 754–760. [CrossRef]

Article

Improving a Cable Robot Recovery Strategy by Actuator Dynamics

Giovanni Boschetti *, Riccardo Minto and Alberto Trevisani

Department of Management and Engineering, University of Padova, 36100 Vicenza, Italy; riccardo.minto@unipd.it (R.M.); alberto.trevisani@unipd.it (A.T.)

* Correspondence: giovanni.boschetti@unipd.it

Received: 29 August 2020; Accepted: 15 October 2020; Published: 21 October 2020

Abstract: Cable-driven parallel robots offer several benefits in terms of workspace size and design cost with respect to rigid-link manipulators. However, implementing an emergency procedure for these manipulators is not trivial, since stopping the actuators abruptly does not imply that the end-effector rests at a stable position. This paper improves a previous recovery strategy by introducing the physics of the actuators, i.e., torque limits, inertia, and friction. Such features deeply affect the reachable acceleration during the recovery trajectory. The strategy has been applied to a simulated point-mass suspended cable robot with three translational degrees of freedom to prove its effectiveness and feasibility. The acceleration limits during the recovery phase were compared with the ones obtained with the previous method, thus confirming the necessity of contemplating the properties of the actuators. The proposed strategy can be implemented in a real-time environment, which makes it suitable for immediate application to an industrial environment.

Keywords: cable robots; motion planning; cable failure; recovery strategy

1. Introduction

Cable-driven robots, commonly called cable robots, are an important field in robotics. The parallel structure of cable robots leads to high payload capacity and high speeds of the end-effector throughout their workspace [1] due to low-moving masses. Moreover, cable robots stand out from other parallel robots since their links are constituted by cables wounded around actuated pulleys, allowing not only for a simple and cheap design but also for large workspace manipulation tasks. These advantages lead to a wide spreading of cable robots, which not only have been suggested for industrial applications [2] but also for home assistance and rehabilitation purposes [3] and for entertainment [4].

With respect to rigid-link parallel robots, the use of cables introduces an additional constraint due to their ability to resist tension but not compression, thus each wire can pull but not push the end-effector. This leads to safety concerns when cable robots are adopted in crowded areas or interact with humans, especially since an emergency stop procedure for these devices is not as simple to implement as in rigid-link robots. Although the damage occurring due to failures of cable-driven manipulators can be very severe, research on failure analysis and fault tolerance has not been sufficiently explored [5] in comparison to serial manipulators. Indeed, implementing an emergency procedure for these manipulators is not trivial, since stopping the actuators abruptly does not imply that the end-effector rests at a stable position, thus making failure analysis particularly challenging for cable robots.

An analysis of failure modes was provided in [6] as an application of the proposed wrench-based analysis, which can determine if a failure is due to a negative tension, i.e., compression, or if the tension limits of a wire have been exceeded. In [7], the faults due to wire jamming, i.e., wire at a constant length, are also presented and the author proposed a certain degree of redundancy to provide fail-safe operation and fault tolerance.

A very basic solution to improve safety in case of cable failure at the design stage can be admittedly achieved by increasing the number of cables far beyond the minimum needed: indeed, cable robots with many degrees of redundancy present a higher fault tolerance, particularly with a broken cable. However, this approach is far from being the optimal one, since, in addition to being more expensive, cables tend to obstruct the workspace and it can even be difficult to avoid interference between the cables. Therefore, it is important to safely control the robot in case of failure without increasing cable redundancy excessively. In [8], an emergency strategy for a two-dimensional cable robot is presented. This strategy aims to guide and stop the end-effector in a safe position on the basis of two different approaches: minimization of the kinetic energy and potential fields. In [9], a recovery strategy for a spatial 3-DOF four-wires cable robot is presented: the proposed approach aims at safely recovering the end-effector after a cable failure by controlling the robot with an elliptic trajectory even outside its residual Static Equilibrium Workspace (SEW) (i.e., a set of the end-effector poses, for a given motor mount configuration, where static equilibrium can be obtained while applying tension to all cables, assuming infinite maximum cable lengths and tensions [10]).

Indeed, cable failure induces a variation in the SEW which should be taken into account by the recovery strategy algorithm. This is shown in [11], where a feasible motion leads the end-effector toward a Safe Pose (SP), where static equilibrium can be achieved. The end-effector is then stopped at the SP. The approach splits the recovery trajectory into two parts, each with different motion requirements, called the Connecting Path and Straight-Line Path, respectively. The first one is needed to change the direction of motion of the end-effector since it is not guaranteed that it is moving towards the SP; the latter is a linear trajectory path used to reach and stop the end-effector at the SP. Furthermore, the authors exploited the Wrench Exertion Capability (WEC) [12] index which belongs to the family of performance indexes that take into account the direction of motion [13,14]. In this work, we considered the computation method presented in [15], which allows for computing cable tensions along the straight-line trajectory on-line. The algorithm offers great flexibility since it can be applied to any cable robot topology (i.e., under-actuated [16], fully actuated [17] or redundant cable robots [18]), in the presence of any number of broken cables, provided that a not-null static equilibrium workspace is available after cable failure. However, this approach does not take into account the physical features and the dynamic limits of the actuators, i.e., limited exertable torque, moment of inertia and presence of friction, which could affect the feasibility of the recovery strategy.

Therefore, the aim of this work is to take into account the actuator limits to improve the recovery strategy introduced in [11]. The formulation of the optimization problems is improved by adding the constraints related to the actuator dynamics. Moreover, this work provides an alternative real-time approach for each part of the recovery trajectory. These algorithms can be more suitable than time-consuming linear programming approaches, enabling the use of this strategy in real-time applications. The recovery strategy has been tested in a simulation environment to show its feasibility. Lastly, the trajectories and acceleration constraints obtained by using the new formulation and without considering the actuator dynamics are compared.

The paper is organized as follows: Section 2 introduces the cable robot model adopted and Section 3 presents the recovery strategy proposed considering the physics and dynamics of the actuators. The simulated tests and results are discussed in Section 4. Lastly, Section 5 concludes the work.

2. Cable Robot Model

The proposed strategy was applied to a 3-DOF suspended cable robot with four cables, as represented in Figure 1. This configuration is particularly challenging since suspended cable robots rely on gravity to maintain equilibrium. The end-effector was considered as a point mass with only translational DOFs, since its size can be considered negligible with respect to the magnitude of the workspace. This assumption allows for simplifying the dynamics and kinematics of the end-effector without a lack of generality; indeed, the most important goal of the after failure strategy consists in avoiding collision of the end-effector. Furthermore, the strategy can be applied only if the robot is at least fully constraint after failure, therefore we must assure a minimum number of cables to control all the DOFs. In this specific case, four cables are needed before failure. Hence, for the considered robot, we suppose that only one cable breaks at a time.

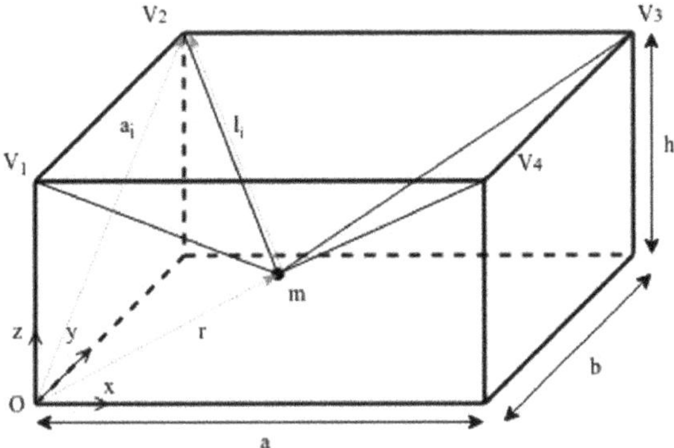

Figure 1. Structure of the considered suspended cable robot.

2.1. Robot Dynamics

The overall wrench w applied on the end-effector of a cable robot, meant as the forces $\mathbf{f}$ and torques $\mathbf{t}$, is the sum of the cable forces w_c and the external forces w_{ext}, e.g., the gravitational load. By defining a nxm matrix, called the structure matrix $\mathbf{S}$, as

$$\mathbf{S}^{nxm} = \begin{bmatrix} u_1 & u_2 & \cdots & u_m \end{bmatrix} \tag{1}$$

where each column u_i is the unit vector oriented from the end-effector towards the i-th pulley. A relation between the cable tensions $\boldsymbol{\tau}$ and the cable forces w_c can be obtained

$$w_c = \mathbf{S}\boldsymbol{\tau} \tag{2}$$

A more complete formulation [12] can be obtained by considering the external wrench w_e in the structure matrix $\mathbf{S}$ by defining a new matrix $\mathbf{W}$

$$w = w_c + w_e = \mathbf{S}\boldsymbol{\tau} + w_e = \begin{bmatrix} \mathbf{S} & w_e \end{bmatrix} \left\{ \boldsymbol{\tau}^T \; 1 \right\}^T = \mathbf{W}\boldsymbol{\tau}_w \tag{3}$$

Therefore, it is possible to evaluate the cable tensions $\boldsymbol{\tau}_w$, which are required to move the end-effector.

2.2. Actuator Dynamics

After the evaluation of $\boldsymbol{\tau}_w$, it is necessary to study the actuator dynamics to evaluate the actuator torques. Each considered actuator comprises an electric motor, a drum and a rotating pulley. The dynamics of each actuator i can be described as follows

$$j_m \ddot{\theta}_i = -b\dot{\theta}_i + c_i + \rho \tau_i \tag{4}$$

where j_m is its moment of inertia, θ_i is the position of the motor, and therefore $\dot{\theta}_i$ and $\ddot{\theta}_i$ are the motor velocity and acceleration, respectively. b is the friction coefficient, c_i is the motor torque and ρ the drum radius. j_m, b and ρ have been assumed as constant, since the mass of the cables are negligible with respect to the magnitude of the system. It is possible to gather (4) in matrix form for all the actuators as follows

$$j_m\ddot{\boldsymbol{\theta}} = -b\dot{\boldsymbol{\theta}} + \boldsymbol{c} + \rho\boldsymbol{\tau} \tag{5}$$

A relation between the wire length l_i and θ_i, and therefore $\dot{\theta}_i$ and $\ddot{\theta}_i$, can be defined as

$$\begin{aligned} |l_i| &= l_{i0} + \rho\theta_i \\ |\dot{l}_i| &= \rho\dot{\theta}_i \\ |\ddot{l}_i| &= \rho\ddot{\theta}_i \end{aligned} \tag{6}$$

By obtaining $\dot{\theta}_i$ and $\ddot{\theta}_i$ from (6) and substituting in (5), it is possible to evaluate the motor torque as

$$\boldsymbol{c} = \frac{j_m}{\rho}\ddot{\mathbf{l}} + \frac{b}{\rho}\dot{\mathbf{l}} - \rho\boldsymbol{\tau} \tag{7}$$

Equation (7) defines the relation between actuator torques and the cable tensions. Moreover, the addition of other transmissions to the considered system does not change the equation form.

3. After Failure Strategy Approach

The proposed after failure strategy approach aims to drive the end-effector inside the residual SEW in a minimum time. Indeed, when a cable breaks, the robot workspace changes and the end-effector might lie outside the residual SEW. Therefore, as a cable failure occurs, the SP must be defined as the final landing position inside the residual SEW. Several factors influence the coordinate choice, such as the robot pose after failure or the presence of obstacles or dangerous zones in the robot workspace. Without lack of generality, the SP should be placed at a suitable height with respect to the workspace. Regarding the x, y coordinates, they are defined as the barycenter of the area obtained as the intersection of the horizontal plane passing through the points at the chosen height and the residual SEW.

Once the SP has been identified, the goal of the study is to plan a linear trajectory that allows one to reach it from the point where the failure occurred. However, since the end-effector velocity could not be null or directed towards the desired direction, it could be not feasible to plan a straight line path immediately after failure. Thus, it is necessary to firstly define a Connecting Path to change the direction of motion towards the SP. Once this requirement is satisfied, the linear trajectory can be planned.

The Connecting Path approach will be presented in Section 3.1, and the linear trajectory planning in Section 3.2.

3.1. Connecting Path

As stated before, the direction of motion d of the end-effector after failure should be oriented towards the SP. Thus, any velocity components orthogonal to d should be null or a suitable small value. To reach this goal, during the first part of the recovery approach a suitable force should be applied to the end-effector.

3.1.1. Linear Programming Problem

To easily identify any velocity components orthogonal to the direction d, it is suggested to introduce a mobile reference frame placed on the end-effector with the x-axis aligned with d. This frame is obtained after a sequence of three rotations

- The first two rotations $\boldsymbol{R}_z(\alpha)$ and $\boldsymbol{R}_y(\beta)$ are applied to align the x-axis with the direction d of the desired velocity component;

- The third rotation $R_x(\gamma)$ is such that the undesired velocity component (i.e., the orthogonal one) is aligned with the y-axis.

The overall rotation matrix $\mathbf{R}$ is therefore defined as

$$\mathbf{R} = R_z(\alpha)R_y(\beta)R_x(\gamma) \tag{8}$$

The end-effector velocity vector $\mathbf{v}'$ can be expressed in the mobile reference frame as:

$$\mathbf{v}' = \mathbf{R}^T\mathbf{v} = \begin{bmatrix} v'_x \\ v'_y \\ v'_z \end{bmatrix}^T = \begin{bmatrix} v'_x \\ v'_y \\ 0 \end{bmatrix}^T \tag{9}$$

As stated in (9) and represented in Figure 2a, v'_z is always null, v'_y is always positive and v'_x will be positive or negative depending on the initial velocity value. The desired velocity should therefore be obtained by reducing v'_y to a suitable small value, while v'_z is kept to a null one. This can be achieved if the exerted force $\mathbf{f}'$ is defined in the mobile reference system such as

$$\mathbf{f}' = \begin{bmatrix} f'_x \\ f'_y \\ f'_z \end{bmatrix} \text{ with } \begin{cases} f'_x > 0 \\ f'_y < 0 \\ f'_z = 0 \end{cases} \tag{10}$$

as represented in Figure 2b.

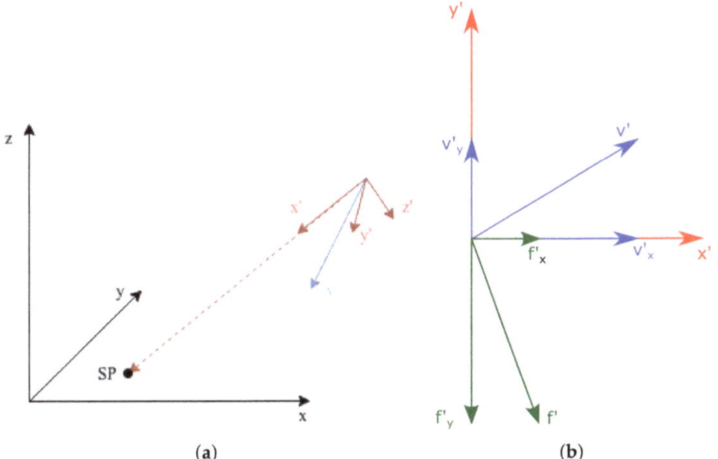

Figure 2. 3d view of the velocity components (**a**) in the mobile reference frame and forces needed to change the velocity direction in the $x' - y'$ plane (**b**).

The exerted force $\mathbf{f}'$ can be defined in terms of $\boldsymbol{\tau}_w$ to obtain the required value to change the velocity direction

$$\mathbf{f}' = \mathbf{R}^T\mathbf{f} = \mathbf{R}^T\mathbf{W}\boldsymbol{\tau}_w = \mathbf{W}_\mathbf{R}\boldsymbol{\tau}_w = \begin{bmatrix} \boldsymbol{w}^T_{Rx} \\ \boldsymbol{w}^T_{Ry} \\ \boldsymbol{w}^T_{Rz} \end{bmatrix} \boldsymbol{\tau}_w \tag{11}$$

The novelty of the proposed approach consists of introducing constraints on the exertable actuator torques. It is indeed possible to evaluate the actuator speed $\dot{\theta}_i$ in terms of the end-effector speed v from (6) as follows

$$\dot{\theta}_i = \frac{\dot{l}_i}{\rho} = \frac{-\mathbf{u}_i^T\mathbf{v}}{\rho} \tag{12}$$

Analogously, the actuator acceleration $\ddot{\theta}_i$ can be expressed as:

$$\ddot{\theta}_i = \frac{\ddot{l}_i}{\rho} = \frac{-\mathbf{u}_i^T\ddot{\mathbf{r}} - \dot{\mathbf{u}}_i^T\mathbf{v}}{\rho} \tag{13}$$

Substituting (12) and (13) in (5) the actuator dynamics can be expressed as:

$$c_i + \tau_i\rho + b\frac{\mathbf{u_i}^T\mathbf{v}}{\rho} = -j_m\frac{\mathbf{u_i}^T\ddot{\mathbf{r}}}{\rho} - j_m\frac{\dot{\mathbf{u}}_\mathbf{i}^T\mathbf{v}}{\rho} \tag{14}$$

and the torque c_i of each actuator can be related to $\mathbf{S}$ by obtaining the acceleration of the end-effector $\ddot{r}$ from its dynamics:

$$\mathbf{M}\ddot{\mathbf{r}} = S\tau + \mathbf{G} \tag{15}$$

where $\mathbf{M}$ is the mass matrix of the end-effector and $\mathbf{G}$ the weight of the platform, thus obtaining:

$$c_i = -\frac{b}{\rho}\mathbf{u_i}^T\mathbf{v} - \frac{j_m}{\rho}\dot{\mathbf{u}}_\mathbf{i}\mathbf{v} - \tau_i\rho - \frac{j_m}{m\rho}\mathbf{u_i}^T S\tau - \frac{j_m}{m\rho}\mathbf{u_i}^T\mathbf{G} \tag{16}$$

Lastly, the equations of the torques c_i of the actuators can be grouped in matrix form

$$\mathbf{c} = -\frac{b}{\rho}\mathbf{S}^T\mathbf{v} - \frac{j_m}{\rho}\dot{\mathbf{S}}^T\mathbf{v} + \left(-\mathbf{I}\rho - \frac{j_m}{m\rho}\mathbf{S}^T\mathbf{S}\right)\tau - \frac{j_m}{m\rho}\mathbf{S}^T\mathbf{G} \tag{17}$$

The maximum breaking effect can be found by solving the linear programming problem defined as follows

$$maximize : (-sign(v'_y)f'y) = (-sign(v'y)\boldsymbol{w}_{\boldsymbol{R}_y}^T\boldsymbol{\tau}_w)$$

$$s.t. \begin{cases} w_{Rx}^T\tau_w > 0 \\ w_{Rz}^T\tau_w = 0 \\ 0 \le \boldsymbol{\tau}_{min} \le \boldsymbol{\tau} \le \boldsymbol{\tau}_{max} \\ \boldsymbol{\tau}_f = 0 \\ \mathbf{c}_{\mathbf{max}} + \frac{b}{\rho}\mathbf{S}^T\mathbf{v} + \frac{j_m}{\rho}\dot{\mathbf{S}}^T\mathbf{v} + \frac{j_m}{m\rho}\mathbf{S}^T\mathbf{G} > \left(-\mathbf{I}\rho - \frac{j_m}{m\rho}\mathbf{S}^T\mathbf{S}\right)\boldsymbol{\tau} \\ \mathbf{c}_{\mathbf{min}} + \frac{b}{\rho}\mathbf{S}^T\mathbf{v} + \frac{j_m}{\rho}\dot{\mathbf{S}}^T\mathbf{v} + \frac{j_m}{m\rho}\mathbf{S}^T\mathbf{G} < \left(-\mathbf{I}\rho - \frac{j_m}{m\rho}\mathbf{S}^T\mathbf{S}\right)\boldsymbol{\tau} \end{cases} \tag{18}$$

where $\boldsymbol{\tau}_f$ the tension referred to the broken cable, which is therefore null, $\boldsymbol{\tau}_{min}$ and $\boldsymbol{\tau}_{max}$ vectors containing the lower and upper bound of each cable tension and $\mathbf{c}_{\mathbf{min}}$ and $\mathbf{c}_{\mathbf{max}}$ the lower and upper bound of each actuator torque. The algorithm calculates a new tension configuration $\boldsymbol{\tau}$ at each step to achieve a suitable deceleration until v_y takes a small enough value. After this, the direction of motion d is oriented at the safe point and a straight line path can be carried out.

3.1.2. Real-Time Approach

The presented solution, based on a linear programming algorithm, is not suited for a real-time application, due to the excessive time needed for computation. However, a real-time approach provides a more flexible and less constrained solution, despite requiring more computation power than an off-line one and providing an approximate solution. To evaluate the maximum force that can be applied to the end-effector to change its velocity direction, this approach applies the WEC index geometric formulation proposed in [15]. Indeed, the WEC index, as stated in [12], is a useful tool to evaluate the maximum wrench that can be exerted along a given direction d while keeping null all the other wrench components. The adopted geometric approach is based on the definition of two n-dimensional polytopes: $\mathbf{T}$, which represents the acceptable cable tension configurations, and $\boldsymbol{\Omega}$, which represents the corresponding forces exerted on the moving platform [15]. The latter can be obtained from $\mathbf{S}^T$ and $\mathbf{T}$ and provides a graphical evaluation of the maximum and minimum force applicable along a given direction.

However, this approach requires the definition of the direction to analyse, which cannot be predetermined when defining the Connecting Path. Hence, this work applies the following iterative procedure to define the appropriate direction:

- In the first iteration, the algorithm rotates the mobile reference frame of 90° along the z-axis, in order to align the x-axis with the direction opposing v'_y;
- The WEC index geometric formulation is used to evaluate the maximum and minimum force along the new x-axis;
- The evaluated force limits are verified in terms of cable tensions and actuator torques, i.e., the constraints in (18) must be satisfied;
- If the constraints are not satisfied, a new iteration is carried out, rotating the initial mobile frame along the z-axis of an angle lower than the previous one, i.e., <90°, until the constraints are fulfilled.

Despite its iterative nature, the proposed approach has showed to be an order of magnitude faster than the linear programming approach in several tests.

3.2. Motion Planning along a Straight Path

It is not possible to predetermine the motion planning along d, since the end-effector position and velocity when the cable breaks are not known a priori. Nonetheless, it is possible to use the WEC index to evaluate the acceleration limits [11] along the desired direction d, thus obtaining the maximum braking acceleration to stop the end-effector in the SP.

3.2.1. Wec Analysis

As stated in (3), the total wrench exerted on the mobile platform is related to the cable tensions by $\mathbf{W}$. By referring to the mobile reference frame applied to the end-effector, i.e., with x-axis along d, we can define

$$\mathbf{W_R} = \mathbf{R}^T\mathbf{W} = \begin{bmatrix} \boldsymbol{w}_{Rx}^T \\ \boldsymbol{w}_{Ry}^T \\ \boldsymbol{w}_{Rz}^T \end{bmatrix} \tag{19}$$

where $\boldsymbol{w}_{Rx}^T$ represents the total force applied along d, which should be maximized to reduce the braking time and, analogously, $\boldsymbol{w}_{Ry}^T$ and $\boldsymbol{w}_{Rz}^T$ represent the total force applied along the y and z axis, respectively.

As presented in [12], the maximum exertable force $w_{f,d}$ can be computed by solving the following linear programming problem

$$\begin{gathered} maximize: (w_{f,d} = \boldsymbol{w}_x^T \boldsymbol{\tau}_w) \\ s.t. \begin{cases} \boldsymbol{w}_y^T \boldsymbol{\tau}_w = 0 \\ \boldsymbol{w}_z^T \boldsymbol{\tau}_w = 0 \\ 0 \le \boldsymbol{\tau}_{min} \le \boldsymbol{\tau} \le \boldsymbol{\tau}_{max} \end{cases} \end{gathered} \tag{20}$$

Similarly to the linear programming problem described in Section 3.1, it is possible to adopt (20) for the considered problem by adding three constraints: one that represents the cable failure, i.e., imposing a null $\boldsymbol{\tau}_f$, and two for the torque limits. The linear programming optimization problem is therefore

$$maximize : (w_{f,d} = \boldsymbol{w}_x^T \boldsymbol{\tau}_w)$$
$$s.t. \begin{cases} \boldsymbol{w}_y^T \boldsymbol{\tau}_w = 0 \\ \boldsymbol{w}_z^T \boldsymbol{\tau}_w = 0 \\ 0 \leq \boldsymbol{\tau}_{min} \leq \boldsymbol{\tau} \leq \boldsymbol{\tau}_{max} \\ \boldsymbol{\tau}_f = 0 \\ \mathbf{c}_{\mathbf{max}} + \frac{b}{\rho}\mathbf{S}^T\mathbf{v} + \frac{j_m}{\rho}\dot{\mathbf{S}}^T\mathbf{v} + \frac{j_m}{m\rho}\mathbf{S}^T\mathbf{G} > \left(-\mathbf{I}\rho - \frac{j_m}{m\rho}\mathbf{S}^T\mathbf{S}\right)\boldsymbol{\tau} \\ \mathbf{c}_{\mathbf{min}} + \frac{b}{\rho}\mathbf{S}^T\mathbf{v} + \frac{j_m}{\rho}\dot{\mathbf{S}}^T\mathbf{v} + \frac{j_m}{m\rho}\mathbf{S}^T\mathbf{G} < \left(-\mathbf{I}\rho - \frac{j_m}{m\rho}\mathbf{S}^T\mathbf{S}\right)\boldsymbol{\tau} \end{cases} \tag{21}$$

with the same nomenclature used in Section 3.1. By solving the linear programming approach, the maximum and, in a similar way, the minimum force limits along d are evaluated.

The acceleration limits of the end-effector in each point, which is a fundamental requirement for the motion planning, can be estimated as follows

$$a_{lim} = \frac{w_{f,d}}{m} \tag{22}$$

where m is the mass of the end-effector.

3.2.2. Real Time Approach

Since the presented WEC formulation can only evaluate the force limits, and thus the acceleration ones, in only one point in time, it is necessary to analyse all the points composing the straight line path, which are not known beforehand.

This work adopted a real-time approach to evaluate the acceleration limits since it is more flexible to changes in the SP position and robot configuration; moreover, the resulting limits are less strict, thus reducing the time needed for the recovery strategy. However, the real-time approach requires faster algorithms, thus a new strategy, based on the evaluation of the WEC presented in Section 3.1.2, has been implemented. Since the fast evaluation of the WEC does not consider the constraints on the actuator torque, the adopted procedure is the following:

- Evaluation of the fast WEC and the acceleration limits along the direction d;
- Given the acceleration limits, using (7) it is possible to evaluate the maximum and minimum value of the actuator torque corresponding to the acceleration limits;
- If the torque limits are not respected, the acceleration limits are reduced to fulfill the constraints. This step is iterative and is not completed until the torque constraints are respected.

This procedure is applied for a variable number of points along d: this number is a compromise between the desired accuracy and the required computation speed.

Figure 3 shows an example of the maximum and minimum acceleration values along d, in red and blue, respectively, with the x-axis representing the distance from the SP.

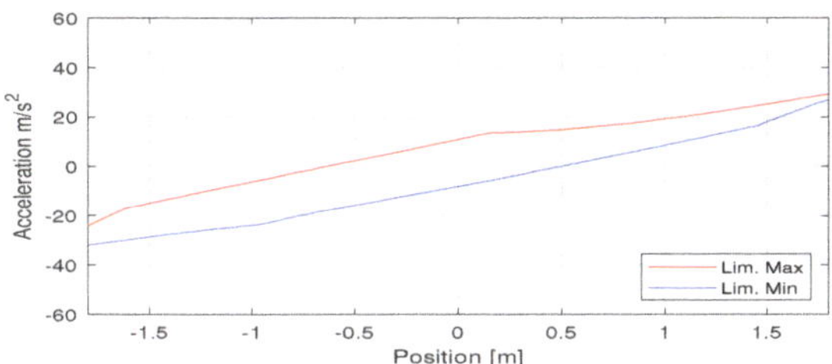

Figure 3. Maximum and minimum acceleration values along d.

Negative values of the distance, i.e., points which are placed beyond the SP from the starting position, have been considered, since it is possible that the end-effector speed is not null at the SP to respect the acceleration constraints, thus it is necessary to go beyond the point and reverse the motion direction. This is shown by observing the values of the minimum acceleration limit for distances greater than 40 m in Figure 3: indeed, the acceleration values are positive, i.e., the end-effector speed must increase to guarantee a positive tension on the wires. Therefore, it is possible that, as the distance between the end-effector and the SP increases, so does the possibility that the end-effector does not stop at the SP.

This effect is considered in the motion-planning algorithm, to return the end-effector to the SP in the case that it goes beyond it.

3.3. Motion Planning

The last step of the recovery approach is the motion planning along d. A polynomial motion law is adopted for its simple and fast application; however, it cannot constrain the acceleration along all the path, but only on the end-points.

The adopted procedure starts from the definition of a fifth-degree polynomial motion law, constraining the end-values of the end-effector position, speed and acceleration

$$\begin{cases} x(0) = x_0 \\ v(0) = v_0 \\ a(0) = a_0 \\ x(T) = 0 \\ v(T) = 0 \\ a(T) = 0 \end{cases} \tag{23}$$

where T is the total motion time, x_0 and v_0 are the position and the velocity, respectively, after the Connecting Path motion, and a_0 the chosen starting acceleration. The value of a_0, and therefore of T, is firstly set as equal to maximum possible acceleration at half of the distance, considering a motion law composed by a positive constant acceleration section, followed by a negative constant one.

The obtained motion law is tested to verify that it respects the acceleration constraints. Three scenarios can be identified:

- The acceleration limits are respected and the trajectory is feasible and does not require any changes;
- The trajectory must be modified to respect the acceleration limits, but the end-effector stops at the SP;
- The trajectory must be modified to respect the acceleration limits and the end-effector goes beyond the SP and has to return.

Figure 4 represents these scenarios: in red, a feasible trajectory, in blue, a trajectory that stops at the SP, and in green, a trajectory beyond the SP. As anticipated before, these three trajectories have different starting points.

If the resulting trajectory is feasible, i.e., the acceleration limits are respected, the motion law can be applied without any changes. However, since the other two cases do not respect the acceleration limits, the motion law should be adapted depending on the scenario.

- *Trajectory stopping at the SP*:
  As the acceleration exceeds the limits, the value of acceleration adopted is set equal to the minimum limit in that position, instead of the one obtained from the polynomial motion law. While keeping the acceleration equal to the minimum value, it is possible to reach a point P along d that allows the end-effector to reach and stop at the SP if the acceleration is set as constant and equal to the minimum acceleration limit in that point. P can be identified since the position, speed and acceleration values are known. However, keeping a constant acceleration from P to SP means an abrupt change in the acceleration in P, which is not feasible. Therefore, a new polynomial law is defined starting from P to the SP, constraining the initial and final values of the position, speed and acceleration; the required time for this section can be estimated as the time needed to stop with constant acceleration. Moreover, this last section is also feasible, since it follows a constant acceleration motion law, which, in turn, is feasible.
- *Trajectory beyond the SP*:
  For greater distances from the SP, after setting the acceleration value equal to the limit, it is not possible to identify a point P that allows the end-effector to stop at the SP with a constant acceleration. It is preferable to decrease the acceleration value, i.e., setting it equal to the minimum limit for each position, until the end-effector stops. Then, similarly to the initial situation, it is possible to design a fifth-degree polynomial motion law that stops the end-effector in the SP. Due to its definition, this second motion law does not assure to keep the acceleration value lower than the maximum acceleration limit, thus it is possible to identify again the three cases presented above. However, since the end-effector is nearer to the SP after this first trajectory, it is less likely that the required acceleration does not respect the acceleration limits.

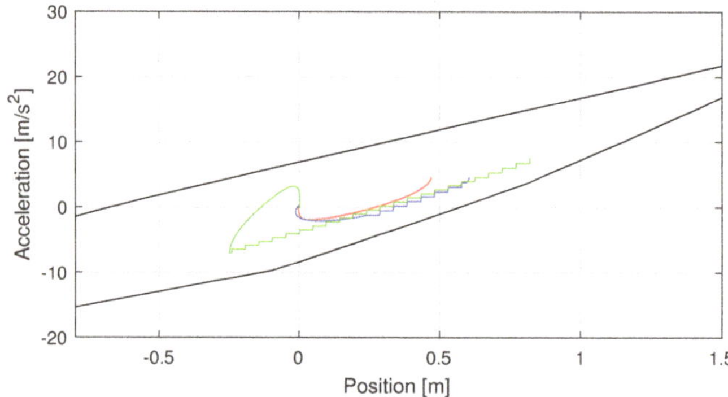

Figure 4. Three different trajectories: depending on the starting point, to respect the acceleration limits, the trajectory is feasible (**red**), must be modified but the end-effector stops at the SP (**blue**), or goes beyond and must turn back (**green**).

The computed acceleration needed for the motion law is evaluated along d, therefore it should be expressed in the spatial reference system to be applied to the motors as

$$\ddot{\mathbf{r}} = \mathbf{R}^T \ddot{\mathbf{r}}' \tag{24}$$

which is the acceleration that should be used to control the end-effector in the absolute reference system.

4. Strategy Simulation

The recovery approach is here tested in a simulation model. A first goal is to make a comparison between the acceleration limits obtained with and without considering the actuators dynamics. Furthermore, the results shows the effectiveness of the proposed methodology.

4.1. Simulation Model

The simulation model was developed in Mathworks Simulink, representing both the cable robot and its controller. Two main parts were designed, focusing on the cable robot model and on the recovery strategy, respectively.

The cable robot model is defined as presented in Section 2, with the actuator torques as input data, evaluating their position and velocity, which allow to evaluate the wire length and velocity, as stated in (6). Moreover, the simulation environment presents additional outputs, i.e., the end-effector position in the absolute reference frame and velocity, which cannot be measured in an experimental phase, but must be evaluated using a direct kinematics algorithm.

The end-effector position and velocity along each reference axis is described by a linear system

$$\begin{cases} \dot{v} = \frac{F}{m} - \frac{b}{m} v \\ \dot{x} = v \end{cases} \tag{25}$$

where v and x are the speed and the position of the end-effector, respectively, and F the exerted force. These equations can be expressed in matrix form as

$$\begin{aligned} \begin{Bmatrix} \dot{v} \\ \dot{x} \end{Bmatrix} &= \begin{bmatrix} -\frac{b}{m} & 0 \\ 1 & 0 \end{bmatrix} \begin{Bmatrix} v \\ x \end{Bmatrix} + \begin{bmatrix} \frac{1}{m} \\ 0 \end{bmatrix} F \\ &= \mathbf{A} \begin{Bmatrix} v \\ x \end{Bmatrix} + \mathbf{b} F \end{aligned} \tag{26}$$

thus, from the three components of F, it is possible to evaluate the spatial position of the end-effector. F can be expressed as the vector sum of the cable forces, evaluated as

$$\mathbf{F_i} = \mathbf{u_i} \tau_i \tag{27}$$

where the unit vectors $\mathbf{u_i}$ are evaluated from the structure matrix alongside the cable length $\mathbf{l_i}$, through

$$\mathbf{l_i} = \mathbf{a_i} - \mathbf{r_i} \tag{28}$$

where $\mathbf{a_i}$ and $\mathbf{r_i}$ are the pulley position and end-effector position with respect to the absolute reference system respectively, with reference to Figure 1.

To simulate the recovery strategy, the algorithm first evaluates the time-instance at failure and which cable has broken. Subsequently, the recovery approach algorithm acquires all the information regarding the robot trajectory, i.e., the end-effector position, the cable length and velocity. The output of the recovery approach is represented by the acceleration of the end-effector $\ddot{\mathbf{r}}$, both for the connecting path and straight line path in case of failure; the model then converts this output in the cable tensions and motor torques required for keeping the recovery trajectory.

Lastly, a feedback loop is required to know the end-effector position in every time-instance needed for the recovery approach to know the position corresponding to a certain value of the acceleration limits. The end-effector position is evaluated with a direct kinematics algorithm using the motor position, since it is impossible to measure it in the absolute reference system without introducing additional measuring systems.

4.2. Simulation Results

Table 1 represents the values adopted for the considered cable robot; these values have been chosen as an example and they do not change the main results presented in the model.

Table 1. Values for the input parameters.

Parameter	Value	Unit
$L1$	2.4	m
$L2$	1.6	m
h	1	m
m	3	kg
j_m	5×10^{-4}	kg m^2
b	1.6×10^{-3}	-
c_{max}	2	Nm
c_{min}	-2	Nm
τ_{max}	50	N
τ_{min}	5	N

4.2.1. Motion Planning

Figure 5 represents the resulting maximum acceleration limits allowed when moving towards the SP from any point of the workspace. The chosen SP is chosen inside the residual SEW resulting from the failure of cable 3, with the spatial coordinates in the absolute reference system set as

$$\begin{cases} x = 0.55\,\text{m} \\ y = 0.60\,\text{m} \\ z = 0.30\,\text{m} \end{cases}$$

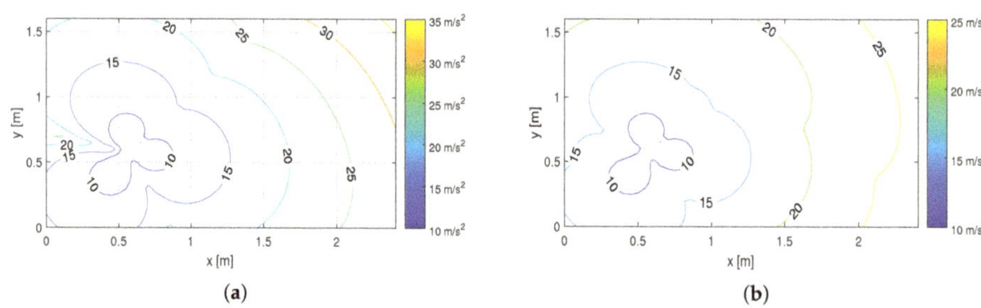

Figure 5. Comparison between the maximum acceleration limits obtained without considering the actuator dynamics (**a**) and considering the actuator dynamics (**b**).

Figure 5a presents the maximum acceleration limits evaluated without considering the actuators' dynamics, whereas Figure 5b represents the acceleration limits when considering the actuators constraints. The proposed method leads to lower acceleration limits, with a maximum acceleration equal to 28 m/s^2, in comparison to the 35 m/s^2 obtained with the previous one. The two figures present a similar behaviour, i.e., smaller acceleration limits near the SP and maximum near the origin of the broken cable. This is due to the different cable configuration in the two zones: near the SP, the z component of the cable length vector $\mathbf{l}_i$ prevails, therefore the cable tension does not contribute as much on the acceleration of the end-effector in the $xy-$plane. On the contrary, near the origin of cable 3, the layout is more favourable for the planar acceleration. Since this behaviour depends only on the cable pose and not on their tension value, the proposed constraints do not affect it. Lastly, the limits are evaluated at the same quota of the SP for simplicity, but different heights lead to similar behaviour.

Similarly, Figure 6 represents the minimum acceleration limits allowed when moving towards the SP, with Figure 6a presenting the values obtained without considering the actuator dynamics, and Figure 6b the values obtained considering the proposed constraints.

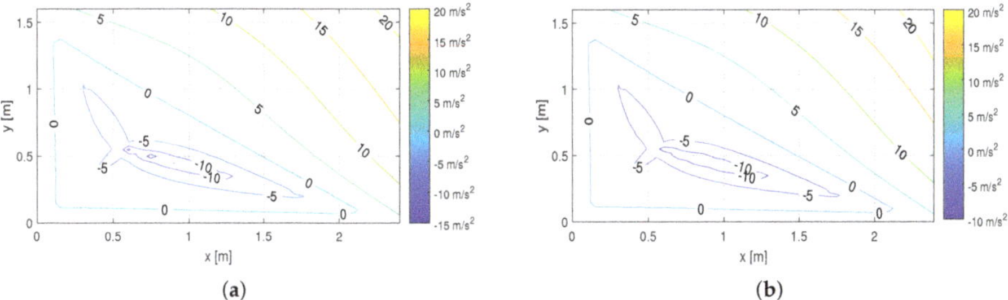

Figure 6. Comparison between the minimum acceleration limits obtained without considering the actuator dynamics (**a**) and considering the actuator dynamics (**b**).

The two figures can distinguish the boundaries of the static stability zone. Indeed, negative acceleration limits correspond to a static stability zone, since a perturbation of the end-effector will result in a static end-effector; whereas outside this area, i.e., with positive acceleration limit, the end-effector must be pulled towards the SP to stop it.

Lastly, Figure 7 represents the acceleration limits along generic lines directed towards the SP.

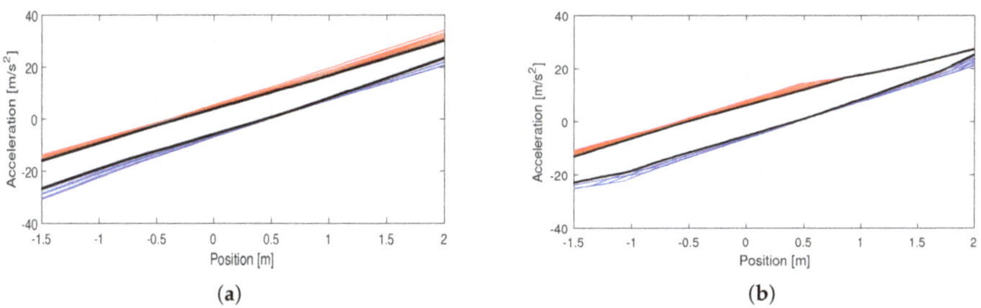

Figure 7. Comparison between the maximum acceleration limits (red) and minimum ones (blue) along a line d towards the SP obtained without considering the actuator dynamics (**a**) and considering the actuator dynamics (**b**).

Each line is evaluated inside a spherical sector with a radius of 2.5 m, centered in SP and limited in the xy-plane between 0° and 45° and in the xz-plane between 0° and 10°. In the example, nine different directions d were considered, with the acceleration limits in each line evaluated at intervals of 0.05 m. Each figure represents, in the x-axis, the distance from the SP and, in the y-axis, the the acceleration values; the maximum acceleration limits are represented by the red lines, the minimum by the blue ones.

The off-line evaluation of the acceleration limits defines them as the minimum maximum acceleration limit and the maximum minimum one for several lines inside a sector of the workspace, and this is represented by the black lines in Figure 7. A comparison between the off-line and on-line evaluation of the acceleration limits is now perceivable, since the former imposes stricter limits, with a maximum difference of about 8 $\mathrm{m/s^2}$ when constraining the actuator torque.

Regarding the motion planning, Figure 8 presents four different motion law, with an initial velocity of 0.64 m/s and incremental distances from the SP, respectively, 1.3, 1.7 and 2.2 m, chosen to represent the different cases presented in Section 3.3.

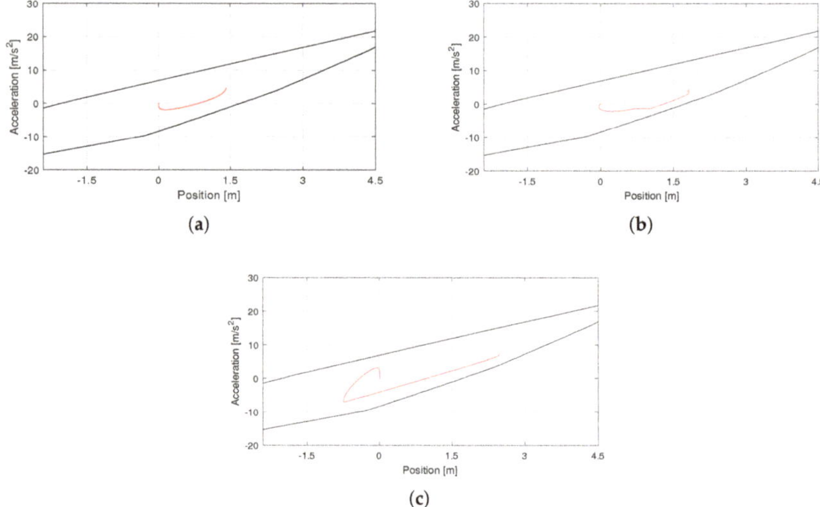

Figure 8. Motion law obtained with the proposed method when changing the distance from SP.(**a**) Distance: 1.3 m; (**b**) Distance: 1.7 m; (**c**) Distance: 2.2 m.

Figure 8a confirms that a polynomial law is sufficient to stop at the SP when the acceleration limits are observed, but as the distance increases, the limits are stricter. As observed in Figure 8b, indeed, when the acceleration value is equal to the inferior limit, its value is kept constant until it can follow a new polynomial law, stopping the end-effector before passing over the SP. However, in Figure 8c the end-effector must pass over the SP, thus initially stopping the end-effector at 0.7 m.

It should be noted that the discontinuous behaviour of the acceleration is due to the chosen resolution, which can be modified depending on the used system.

4.2.2. Robot Behaviour

As stated previously, the following results are provided when considering a failure in cable 3, when the end-effector position r and velocity $\dot{r}$ are, respectively,

$$\begin{cases} \mathbf{r} = (1.85, 0.55, 0.4)\ \mathrm{m} \\ \dot{\mathbf{r}} = (-0.5, 0.4, 0)\ \mathrm{m/s} \end{cases}$$

resulting in the aforementioned SP position placed at coordinates (0.55, 0.6, 0.3) m, thus allowing one to stop the end-effector at the SP without infringing the acceleration limits. The resulting trajectory can be observed in Figure 9 representing, in blue, the nominal trajectory before failure, in red, the Connecting Path, and in black, the linear path towards the SP; Figure 10 presents the trajectory in the $x-y-z$ three-dimensional space. The dashed line in Figure 9 represents the boundary of the static stability zone.

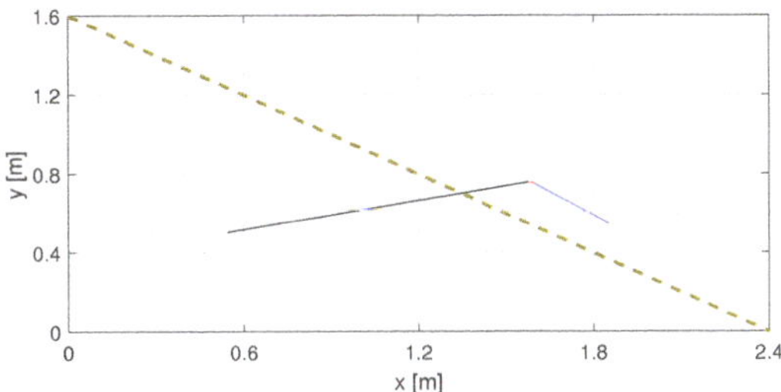

Figure 9. Trajectory of the end-effector in the $x - y$ plane: in blue, the nominal trajectory before failure, in red, the connecting path and in black, the straight line path.

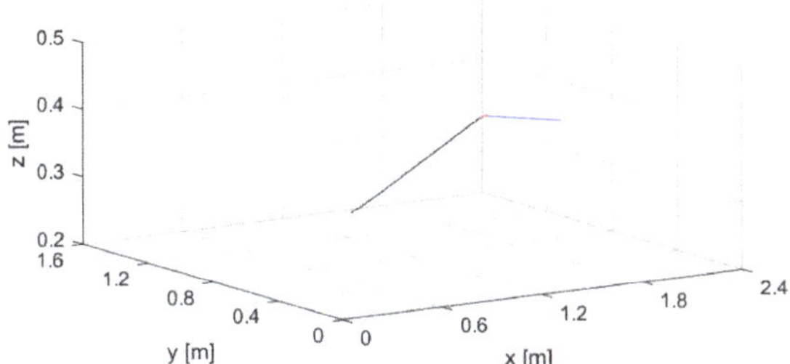

Figure 10. Three-dimensional representation of the end-effector trajectory: in blu,e the nominal trajectory before failure, in red, the connecting path and in black, the straight line path.

The end-effector coordinates during the recovery approach are represented in Figure 11, adopting the RGB-color norm to represent the x (red), y (green) and z (blue) coordinates.

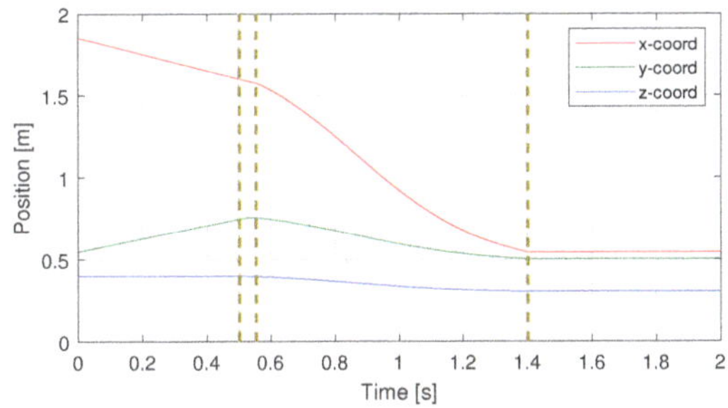

Figure 11. End-effector coordinates during the recovery approach: x (**red**), y (**green**) and z (**blue**).

The three dashed lines identify the different phases of the recovery approach: the first one represents the linear motion before failure, followed by the Connecting Path and the straight line path in the third phase; the stationary end-effector at the SP concludes the motion. The recovery strategy affects all three coordinates and they all reach the respective SP coordinate in the same time.

The different phases of the recovery approach can be observed also from the behaviour of the cable tensions observable in Figure 12.

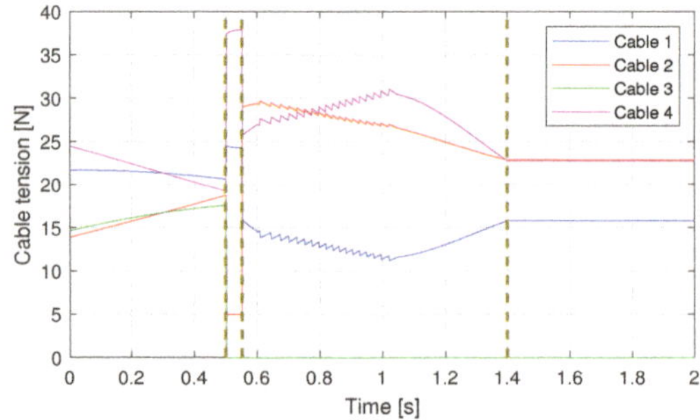

Figure 12. Cable tension during the recovery approach for cable 1 (**blue**), cable 2 (**red**), cable 3 (**green**) and cable 4 (violet).

The blue line represents the tension of cable 1, the red one, the tension of the cable 2, the green one, the tension of cable 3 and the violet one, the tension of cable 4. Focusing on the green line, it is possible to identify the temporal instance of the cable failure as the instant when the cable tension drops to 0. Thus, it is possible to distinguish the moment when the recovery strategy starts (0.5 s). At 0.5 s, we can observe a discontinuity in the cable tensions, due to the instant change in the robot stability; a second discontinuity takes place at the end of the connecting path, caused by the different values of acceleration at the beginning of the straight-line path, which is required to respect the acceleration limits along d.

The actuator torques present a similar behaviour, as observed in Figure 13, with a similar color-code: blue for motor 1, red for motor 2, green for motor 3 and violet for motor 4. As the cable 3 fails, the motor torque of actuator 3 is equal to 0. A negative torque can be adopted with an appropriate strategy to recover the broken cable, further avoiding damage, however, such strategies are out of the scope of this work.

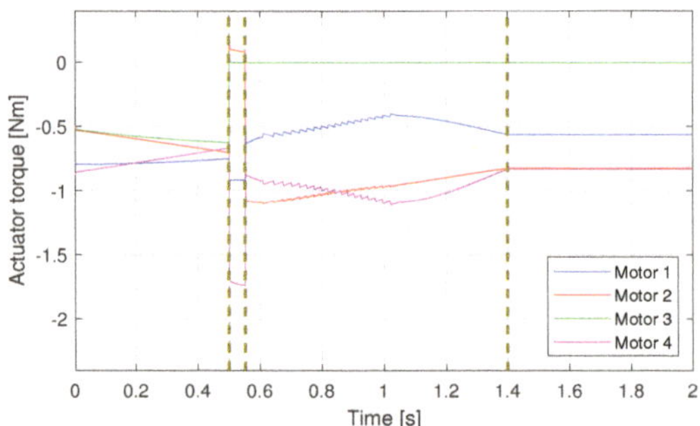

Figure 13. Motor torques during the recovery approach for motor 1 (**blue**), motor 2 (**red**), motor 3 (**green**) and motor 4 (violet).

Motor 2 presents a positive, although small, torque and the adopted convention states that the cables are pulled only when the torques are negative, therefore a positive torque means that the end-effector should be pushed by the cable, which is unfeasible. However, the positive small torque is physically meaningful, since it is applied to compensate the friction and the motor inertia, allowing the actuator to be pulled by the other cables.

These behaviours were also observed for different tension and torque limits, showing how the proposed approach is affected by the imposed limits. The first tests reduced the maximum cable tension to 32 N (Figure 14a) and to 37 N (Figure 14b). Such lower limits, in comparison to the original one, affect the tensions behaviour, since they can no longer reach values over 37 N during the Connecting Path. Therefore, as seen in Figures 14, at the end of the connecting section, the end-effector position and velocity will be different between the two scenarios, thus influencing the robot behaviour during the linear path.

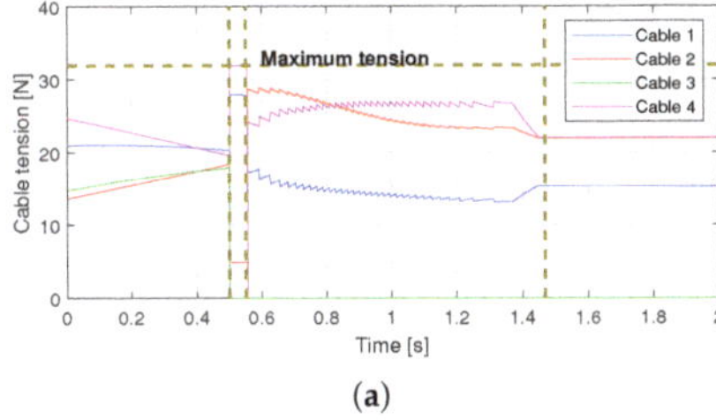

(a)

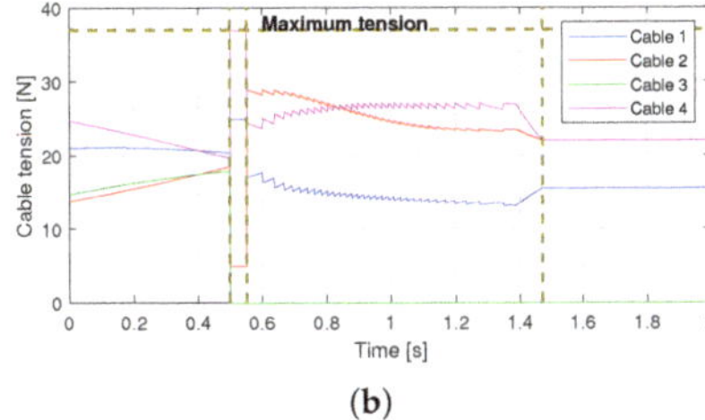

(b)

Figure 14. Cable tensions behaviour for different τ_{max} limits. Cable 1 is represented in blue, cable 2 in red, cable 3 in green and cable 4 in violet. (**a**) Cable tensions for τ_{max} = 32 N; (**b**) Cable tensions for τ_{max} = 37 N.

Similarly, different braking torque limits, i.e., -1.1 and -1.4 Nm, affect the end-effector behaviour during the straight-line path, as shown in Figure 15, respectively. In the first case, represented in Figure 15a, the torque of actuator 4 is set constant and equal to -1.1 Nm during the Connecting Path, whereas with a less strict torque limit, the torque initially decreases and then is again set constant and equal to the limit of -1.4 Nm. As seen previously, different behaviours during the first part of the recovery approach lead to different end-effector positions and velocity when starting the linear trajectory, thus leading to different torque values during the second part of the strategy.

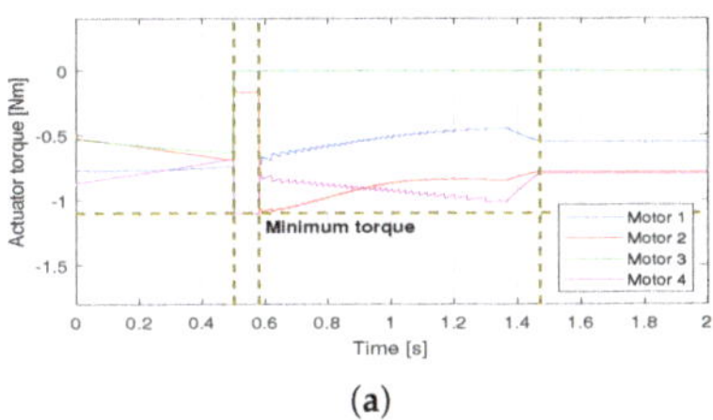

(a)

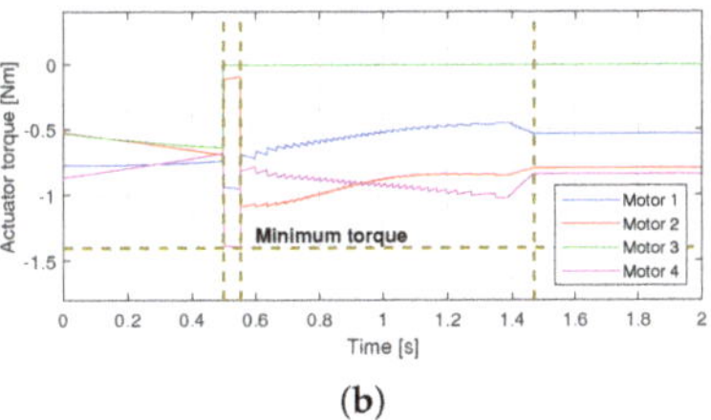

(b)

Figure 15. Actuator torques behaviour for different c_{min} limits. Actuator 1 is represented in blue, actuator 2 in red, actuator 3 in green and actuator 4 in violet. (**a**) Actuator torque for $c_{min}= -1.1$ Nm; (**b**) Actuator torque for $c_{min}= -1.4$ Nm.

By reducing the minimum torque, the duration of the Connecting Path, which is the most affected by the constraints on the actuator torques, increases by 13%, with a duration in the first scenario equal to 0.058 s and a duration of 0.051s with a less strict constraint. Lastly, similar behaviours are observed for different broken cables and initial end-effector position and velocity.

By looking at these simulations, it is clear that the proposed strategy allows the end-effector to reach the SP, while keeping, positive cable tensions for each set of parameters (minimum torque and maximum tension) during the whole recovery process.

5. Conclusions

This work presents an improved recovery strategy for cable robots. This method considers how the physics and dynamics of the actuator (i.e., the inertia, friction and torque limits) affect the robot behaviour. Moreover, the paper describes an iterative approach which is straightforward exploitable to industrial systems. A simulation model was tested to validate the feasibility of the proposed approach to move the end-effector towards a safe static position. Moreover, the simulation results show that the cable tensions are kept positive during the recovery process, and this proves the feasibility of the proposed method.

Future works will validate the proposed approach to prototypes, evaluating the efficiency and accuracy of the real-time approaches proposed and the effectiveness of the recovery strategy.

Author Contributions: Conceptualization, G.B., R.M. and A.T.; Investigation G.B., R.M. and A.T.; Methodology, G.B., R.M. and A.T.; Writing, G.B., R.M. and A.T. All authors have read and agreed to the published version of the manuscript.

Funding: This research received no external funding.

Conflicts of Interest: The authors declare no conflict of interest.

References

1. Merlet, J.P. *Parallel Robots*; Springer Science & Business Media: New York, NY, USA, 2006; Volume 128.
2. Bostelman, R.; Albus, J.; Dagalakis, N.; Jacoff, A.; Gross, J. Applications of the NIST RoboCrane. In Proceedings of the 5th International Symposium on Robotics and Manufacturing, Maui, HI, USA, 25–27 April 1994.
3. Rosati, G.; Gallina, P.; Rossi, A.; Masiero, S. Wire-based robots for upper-limb rehabilitation. *Int. J. Assist. Robot. Mechatronics* **2006**, *7*, 3–10.
4. Cone, L.L. Skycam-an aerial robotic camera system. *Byte* **1985**, *10*, 122.
5. Ghaffar, A.; Hassan, M. Failure analysis of cable based parallel manipulators. In *Applied Mechanics and Materials*; Trans Tech Publications Ltd.: Aedermannsdorf, Switzerland, 2015; Volume 736, pp. 203–210.
6. Bosscher, P.; Ebert-Uphoff, I. Wrench-based analysis of cable-driven robots. In Proceedings of the IEEE International Conference on Robotics and Automation, New Orleans, LA, USA, 26 April–1 May 2004; pp. 4950–4955.
7. Notash, L. Failure recovery for wrench capability of wire-actuated parallel manipulators. *Robotica* **2012**, *30*, 941–950. [CrossRef]
8. Boumann, R.; Bruckmann, T. Development of Emergency Strategies for Cable-Driven Parallel Robots after a Cable Break. In Proceedings of the International Conference on Cable-Driven Parallel Robots, Krakow, Poland, 30 June–4 July 2019; pp. 269–280.
9. Berti, A.; Gouttefarde, M.; Carricato, M. Dynamic recovery of cable-suspended parallel robots after a cable failure. In *Advances in Robot Kinematics 2016*; Springer: Champaign, IL, USA, 2018; pp. 331–339.
10. Riechel, A.T.; Ebert-Uphoff, I. Force-feasible workspace analysis for underconstrained, point-mass cable robots. In Proceedings of the IEEE International Conference on Robotics and Automation, New Orleans, LA, USA, 26 April–1 May 2004; pp. 4956–4962.
11. Boschetti, G.; Passarini, C.; Trevisani, A. A recovery strategy for cable driven robots in case of cable failure. *Int. J. Mech. Control* **2017**, *18*, 41–48.
12. Boschetti, G.; Trevisani, A. Cable robot performance evaluation by wrench exertion capability. *Robotics* **2018**, *7*, 15. [CrossRef]
13. Lorenz, M.; Brinker, J.; Prause, I.; Corves, B. Power manipulability analysis of redundantly actuated parallel kinematic manipulators with different types of actuators. In Proceedings of the 2016 IEEE International Conference on Robotics and Automation (ICRA), Stockholm, Sweden, 16–21 May 2016; pp. 2129–2136.
14. Boschetti, G. A Novel Kinematic Directional Index for Industrial Serial Manipulators. *Appl. Sci.* **2020**, *10*, 5953. [CrossRef]
15. Boschetti, G.; Passarini, C.; Trevisani, A.; Zanotto, D. A fast algorithm for wrench exertion capability computation. In *Cable-Driven Parallel Robots*; Springer: Champaign, IL, USA, 2018; pp. 292–303.
16. Carricato, M.; Abbasnejad, G. Direct geometrico-static analysis of under-constrained cable-driven parallel robots with 4 cables. In *Cable-Driven Parallel Robots*; Springer: Berlin/Heidelberg, Germany, 2013; pp. 269–285.
17. Zhang, N.; Shang, W. Dynamic trajectory planning of a 3-DOF under-constrained cable-driven parallel robot. *Mech. Mach. Theory* **2016**, *98*, 21–35. [CrossRef]

18. Pott, A.; Mütherich, H.; Kraus, W.; Schmidt, V.; Miermeister, P.; Verl, A. IPAnema: A family of cable-driven parallel robots for industrial applications. In *Cable-Driven Parallel Robots*; Springer: Berlin/Heidelberg, Germany, 2013; pp. 119–134.

Publisher's Note: MDPI stays neutral with regard to jurisdictional claims in published maps and institutional affiliations.

Article

Study on the Driver/Steering Wheel Interaction in Emergency Situations

Francesco Comolli [1], Massimiliano Gobbi [2,*] and Gianpiero Mastinu [2]

[1] SMARTMechanical Company s.r.l., Via Tonale 9, 24061 Milan, Italy; francesco.comolli@smartmechanical-company.it

[2] Department of Mechanical Engineering, Politecnico di Milano, Via La Masa 1, 20156 Milano, Italy; gianpiero.mastinu@polimi.it

* Correspondence: massimiliano.gobbi@polimi.it

Received: 30 August 2020; Accepted: 7 October 2020; Published: 11 October 2020

Abstract: Advanced driver assistance systems (ADAS) are becoming increasingly prevalent. The tuning of these systems would benefit from a deep knowledge of human behaviour, especially during emergency manoeuvres; however, this does not appear to commonly be the case. We introduced an instrumented steering wheel (ISW) to measure three components of force and three components of the moment applied by each hand, separately. Using the ISW, we studied the kick plate manoeuvre. The kick plate manoeuvre is an emergency manoeuvre to recover a lateral disturbance inducing a spin. The drivers performed the manoeuvre either keeping two hands on the steering wheel or one hand only. In both cases, a few instants after the lateral disturbance induced by the kick plate occurred, a torque peak was applied at the ISW. Such a torque appeared to be unintentional. The voluntary torque on the ISW occurred after the unintentional torque. The emergency manoeuvre performed with only one hand was quicker, since, if two hands were used, an initial fighting of the two hands against each other was present. Therefore, we propose to model the neuro-muscular activity in driver models to consider the involuntary muscular phenomena, which has a relevant effect on the vehicle dynamic response.

Keywords: ADAS; driver model; load cell; steering wheel

---

1. Introduction

The driver acts on the steering wheel by applying forces and moments with the hands, activating the muscles with voluntary and involuntary actions. Pick and Cole [1] presented a neuro-muscular system (NMS) model that replicates the driver's actions while driving. They highlighted the involuntary actions as a consequence of an inner control loop that allows the driver to maintain a certain position in the space of the hands or arms, despite external disturbances. Similarly, in [2], Wei et al. developed a NMS model of the driver for the human–vehicle shared control in autonomous vehicles. The knowledge of the driver's actions on the steering wheel allows the study of the interaction between the driver's hands and the advanced driver assistance systems (ADAS) [3] to lead improvements in their activation procedures and make them less intrusive during their action. Saito et al. [4] developed an ADAS control able to intervene only in necessary situations.

In [5], the authors highlighted the importance of the steering feeling while driving. The authors studied the influence of the steering geometry on the applied steering torque. Cai et al. [6] presented a new method to classify the driver behaviours by analysing a set of data collected during normal driving. They obtained a driving fingerprint map that characterized the driver's behaviour. By using such a classification, it is possible to train ADAS to customize their intervention. An instrumented steering wheel could also improve the performance of haptic guidance steering systems, which are actually ADAS for driver assistance [7].

In order to intervene correctly, ADAS require a deep knowledge of human–vehicle interactions, and the research on driver models is very active. Jiang et al. [8] proposed an electric power steering (EPS) logic developed adopting human simulated intelligent control (HSIC). The authors noticed that the information on the steering effort allows an increase in the performance of the proposed EPS control logic. Okamoto and Panagiotis [9] studied the performance of some data-driven human driver models that are able to predict the torque applied by the driver on the steering wheel. In [10,11], driver models were developed that imposed the steering wheel angle using a neural network.

Similarly, in [12–14], the driving styles were classified to optimize the ADAS intervention. Kolekar et al. [15] developed a human-like driver model that was able to capture the steering behaviour for both routine and emergency situations. Zhu et al. [16] identified the driver behaviour characteristics of novel, normal, and skilled drivers and applied the model parameters to a driving simulation, obtaining personalized control in order to successfully follow a path. You et al. [17] identified driver parameters using Kalman filters, but concluded that the driver parameters were not time constant, but varied during the driving task. Gote et al. [18] presented a method to identify the driver characteristics using inverse dynamic optimal control. In [19], the authors studied a clustering method for driver's steering intention classification, based on certain steering action parameters, such as steering angle and applied torque. In [20,21], the authors proposed two different methods for driver behaviour analysis, considering the available vehicle operation information.

Some authors, such as [22–24], analysed vehicle dynamic signals to identify the driver behaviours, while other authors preferred to use data measured at the human-machine interface, such as electrocardiography (ECG) sensors [25] or wearable sensors [26]. In [27], the authors proposed the use of physiological data of the driver to analyse his state while driving.

None of the presented works studied the forces and the moments applied by the driver's hands, which constitute the direct interaction between the driver and the vehicle. The knowledge of such interactions drives a better comprehension of the phenomena that occur while driving.

The driver applies forces and moments on the steering wheel in all directions, and not only tangentially. The driver's action is disturbed by the reaction forces of the steering wheel and by the inertia forces due to the vehicle's dynamics, forcing the driver to apply forces and moments in every direction to counteract these disturbing forces [28]. In the literature, attempts have been made to study such phenomena by using instrumented steering wheels, the only vehicle component that the driver uses for lateral vehicle control.

To develop control logics for EPS systems, sensors are needed to measure the torque on the steering wheel, to correctly set the power steering intervention. Gabrielli et al. [29] presented an instrumented steering wheel, created by cutting a steering wheel into three sectors, each one connected with the steering hub by a sensing component with strain gauges, used to measure the resultant of the forces applied to each sector. The instrumented steering wheel was used to characterize the response of healthy drivers and the response of drivers with disabilities in order to customize the EPS intervention [30]. From this paper, they found that the drivers applied a non-negligible force in a direction orthogonal to the steering plane. The instrumented steering wheel used in the cited papers did not allow measuring the force and moment components applied by each hand separately.

The authors presented an instrumented steering wheel with two six-component load cells that allow the measurement of the three forces and the three moments applied by each hand separately. The instrumented steering wheel (ISW) is described in [3], and the experimental data are available in [31]. In [32,33], the authors studied the interaction between the drivers and the steering wheel in normal driving situations and in emergency situations. They identified some control patterns in the force application on the steering wheel, that are performed unconsciously by the majority of drivers, regardless of their driving experience. In [34], a ranking method for evaluating the handling of vehicles by using the forces measured by the instrumented steering wheel was proposed.

The present paper reports an in-depth analysis of the driver-steering wheel interaction, considering the case of driving with one hand only and comparing it with the situation in which the driver acts on the steering wheel with two hands. The paper is developed as follows.

In Section 2, the instrumented steering wheel is briefly described, together with the experimental procedure. In Section 3, during an emergency manoeuvre, the forces applied by a set of drivers are analysed. The drivers hold the steering wheel with both of the hands during a kick plate manoeuvre. In Section 4, the same drivers keep only one hand on the steering wheel during the manoeuvre. In Section 5, the steering power required for counter steering is analysed.

2. Experimental Setup

In this study, an instrumented steering wheel (Smartmechanical Company s.r.l., Albano S.Alessandro (BG), Italy) was adopted, which measured the six components of force and moment applied by each hand separately on the steering wheel. Figure 1 shows the body of the steering wheel made in carbon fibre composite to guarantee the mass, inertia, and stiffness levels were very similar to a standard steering wheel, in order not to modify the dynamics of the steering system.

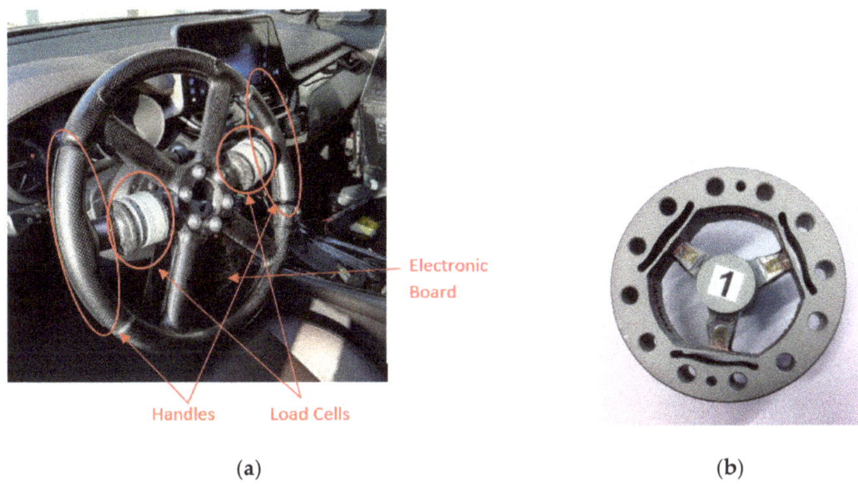

Figure 1. (**a**) Instrumented steering wheel; (**b**) six-component load cell.

Figure 1 highlights the two handles, which are the sensitive parts that the driver must grasp in order to measure the force that he/she applies to the steering wheel while driving. The handles are connected to the body of the steering wheel with two six-component load cells designed at the Politecnico di Milano (Smartmechanical Company s.r.l., Albano S.Alessandro (BG), Italy) [2] (see Figure 1).

The load cells are instrumented with twelve resistive strain gauges each, connected to form six Wheatstone half bridges. In such a way, the thermal effects are compensated because the thermal strains are measured by the strain gauges positioned on the opposite sides (ε_1 and ε_2 respectively) of the spokes. The deformation read by the Wheatstone half bridge $\varepsilon_{bridge} = \varepsilon_1 - \varepsilon_2$, the thermal effect on the bridge measurement is null. The strain gauge signals are acquired by the electronic board integrated in the steering wheel, and the scheme is shown in Figure 2. The electronic board returns the force components applied by the single hand through the relationship

$$\left\{ \begin{array}{c} \underline{F_{meas}} \\ \underline{M_{meas}} \end{array} \right\} = [C_b]\underline{\Delta V} \tag{1}$$

where $\underline{F_{meas}} = \{F_x, F_y, F_z\}^T$ and $\underline{M_{meas}} = \{M_x, M_y, M_z\}^T$ are the applied forces and moments vectors, $[Cb]$ the experimental calibration matrix [6 × 6], which links the force and moment components with the voltage outputs of the six strain gauges bridges $\underline{\Delta V} = \{\Delta V_1, \Delta V_2, \Delta V_3, \Delta V_4, \Delta V_5, \Delta V_6\}^T$. Refer to Figure 3 for the load cells reference frames.

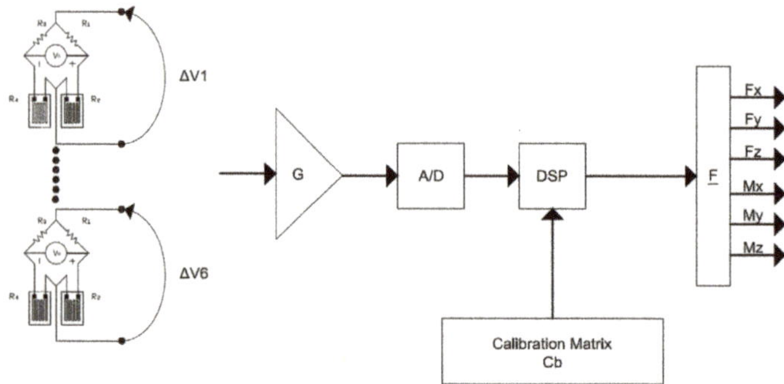

Figure 2. Electronic board of the instrumented steering wheel. G is the signal amplifier, A/D is the converter, and DSP is the digital signal processor.

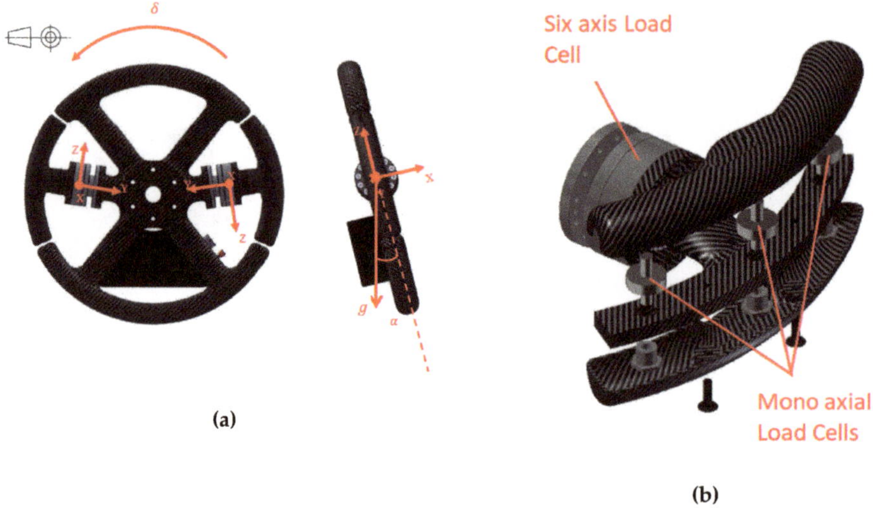

Figure 3. (**a**) Load cell reference frames. δ is the steering angle, α is the inclination angle of the steering wheel, and g is the gravity acceleration. (**b**) Mono-axial load cells in the handle to measure the grip force.

The force and moment signals, measured $\underline{F_{meas}}$, are then processed to compensate the static weight force that arises due to the inclination of the steering wheel $\underline{F_W}$, the inertial forces originated by the vehicle dynamics $\underline{F_q}$, and the steering wheel rotation $\underline{F_{SW}}$. This compensation makes the instrumented steering wheel usable on vehicles even during high speed transient manoeuvres.

$$\underline{F} = \underline{F_{meas}} - \underline{F_W} - \underline{F_q} - \underline{F_{SW}}. \tag{2}$$

The signals output from the electronic board are published on the vehicle controller area network (CAN) and saved for post-processing data analysis. The signals from the instrumented steering wheel were acquired at a frequency of 748 Hz.

Three mono-axial load cells F_i were installed inside each handle to allow measurement of the grip forces F_{grip} (see Figure 3).

$$F_{grip} = \min\left(\left|F_{x_{six-axis\ load\ cell}} - \sum_{i=1}^{3} F_i\right|, \left|\sum_{i=1}^{3} F_i\right|\right) \tag{3}$$

In addition to the instrumented steering wheel, an OxTS RT3000 (Oxford Technical Solutions Ltd., Middleton Stoney, Oxfordshire, United Kingdom) inertial platform was installed on the vehicle to measure the vehicle accelerations in three directions and the vehicle body roll, pitch, and yaw. A GPS antenna was installed on the vehicle to track the vehicle's route in the circuit. The data from the inertial platform and GPS were acquired at 100 Hz. The vehicle speed, steering angle, and yaw rate were obtained from the vehicle's CAN network at a frequency of 82 Hz. As both the vehicle data and the steering wheel data were published on the same CAN network, they were synchronous.

All the data were resampled at 100 Hz. To allow the data synchronization between the inertial platform and the vehicle CAN signals, a cross-correlation analysis on the yaw rate signal acquired by the two sensors was completed.

The manoeuvre analysed was a kick plate test (Figure 4), in which the vehicle passed at constant speed on a plate that slides sideways when the rear axle passes over it. In this way, a lateral disturbance is introduced in a safe environment and the driver is forced to counter-steer to bring the vehicle back to its original trajectory. Several tests were performed by a panel of eight drivers, who performed the manoeuvre six times keeping both hands on the steering wheel, and three times holding only one hand on the steering wheel. The eight drivers were males from 24 to 57 years old, each with a valid driving license and different driving experience.

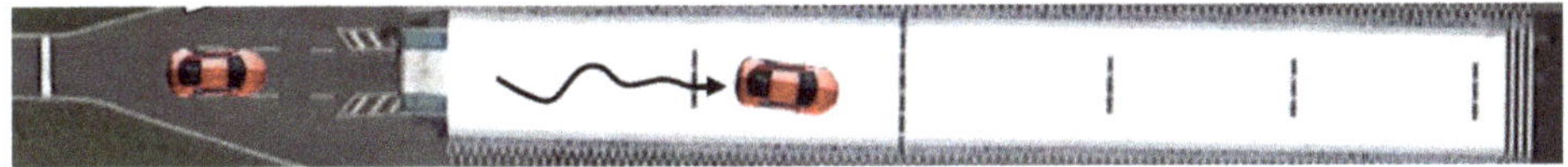

Figure 4. The kick plate test.

3. Kick Plate Disturbance while Driving with Two Hands

The first step was to analyse the manoeuvre performed by keeping both of the hands on the steering wheel, in correspondence with the two handles to allow the measurement of the forces and torques applied. Figure 5 shows that a few instants after the lateral disturbance, a torque was applied unconsciously or involuntarily by the driver, even before the steering wheel started to rotate.

In Figure 5, the lateral acceleration plot is reported, and the vehicle was disturbed laterally at instant t = 0.38s. A few moments later, at t = 0.53 s, there is a peak of torque on the steering wheel to counteract the sudden movement of the vehicle due to the kick plate input. This torque did not correspond to an effective rotation on the steering wheel, which instead began at t = 0.66 s, but was due to the driver's unconscious muscular reflex. The torque that was applied unconsciously on the steering wheel, on average on all the datasets, was about 20% in the modulus with respect to the torque applied voluntarily to bring the vehicle back to the desired path. Table 1 shows the average time delay Δt between the lateral disturbance and the peak of torque unconsciously applied (Δt1), and between the peak of torque and the effective rotation of the steering wheel (Δt2) for different drivers. The time delay between the lateral disturbance and the effective rotation of the steering wheel Δt3 = Δt1 + Δt2 is also reported.

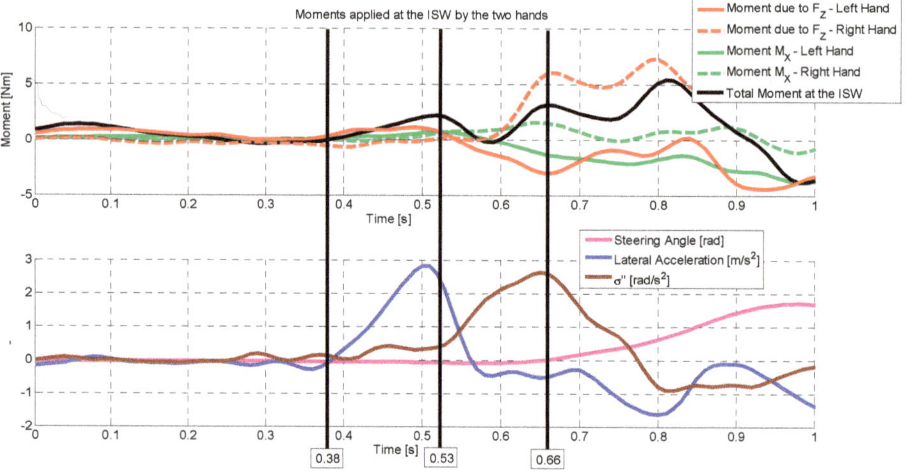

Figure 5. Moments on the steering wheel. The kick plate manoeuvre. The driver used two hands.

Table 1. Drivers with two hands on the steering wheel. Average ± Standard Deviation Values (Avg ± St. Dev.).

	Δt1 (s)	**Δt2 (s)**	**Δt3 (s)**
	Avg ± St. Dev.	Avg ± St. Dev.	Avg ± St. Dev.
Driver 1	0.138 ± 0.033	0.273 ± 0.082	0.412 ± 0.082
Driver 2	0.114 ± 0.030	0.160 ± 0.102	0.274 ± 0.102
Driver 3	0.128 ± 0.025	0.257 ± 0.099	0.385 ± 0.099
Driver 4	0.158 ± 0.015	0.360 ± 0.058	0.518 ± 0.058
Driver 5	0.120 ± 0.032	0.280 ± 0.181	0.400 ± 0.181
Driver 6	0.148 ± 0.024	0.268 ± 0.096	0.417 ± 0.096
Driver 7	0.145 ± 0.031	0.368 ± 0.082	0.513 ± 0.082
Driver 8	0.120 ± 0.141	0.250 ± 0.141	0.370 ± 0.141
Average	0.137 ± 0.030	0.282 ± 0.114	0.416 ± 0.128

Δt1: delay between the lateral disturbance and the peak of torque unconsciously applied, Δt2: delay between the peak of torque and the effective rotation of the steering wheel, and Δt3: delay between the lateral disturbance and the effective rotation of the steering wheel.

4. Kick Plate Disturbance while Driving with One Hand

The same manoeuvre considered in Section 3 was repeated while grasping the steering wheel with one hand. Figure 6 shows the forces applied to the steering wheel during this manoeuvre. The same phenomena reported in the manoeuvre performed with two hands on the steering wheel was found. At the instant t = 0.28 s, the lateral disturbance from the kick plate acted on the vehicle, and at t = 0.40 s the maximum torque peak was applied unconsciously by the driver. Again, this torque peak did not correspond to a rotation of the steering wheel, which occurred at t = 0.64 s. Even in this case, the unconsciously applied torque, on average on all the datasets, had a modulus of about 20% of the torque voluntarily applied to recover the desired trajectory.

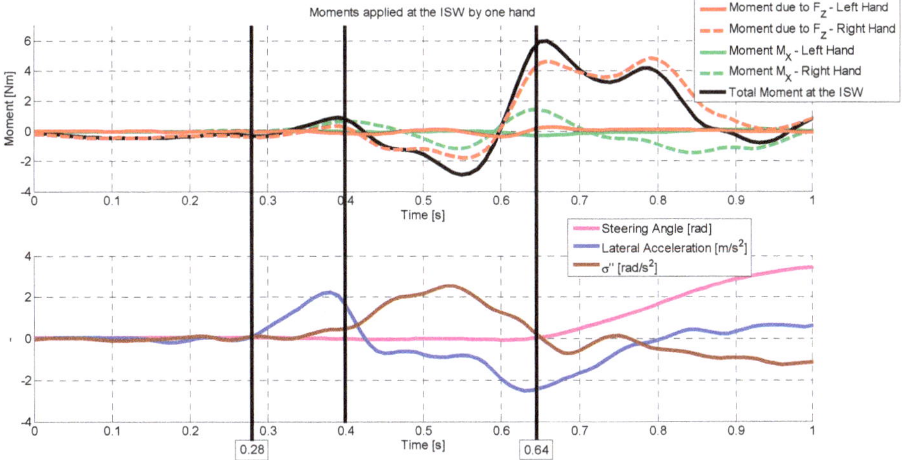

Figure 6. Moments on the steering wheel. The kick plate manoeuvre. The driver used one hand.

Referring to the panel of drivers, Table 2 shows the average durations of the described phenomena, following the same procedure described in the previous paragraph. In the case of the one-hand steering wheel operation, there was a slight extension of the delay Δt1 between the application of the lateral disturbance and the peak of the torque applied unconsciously. On the contrary, there was a slight shortening of the delay Δt2 between the peak torque and the effective rotation of the steering wheel. This shorter time can be explained by analysing the opposite torques as applied by the two hands in the time interval Δt2 shown in Figure 5. The two hands were fighting against each other in the time interval Δt2. During one-handed operation, this effect on the applied forces did not exist. The reduced time interval is also reflected in Δt3 (the delay between the lateral disturbance and effective steering rotation).

Table 2. Drivers with one hand on the steering wheel.

	Δt1 (s)	**Δt2 (s)**	**Δt3 (s)**
	Avg ± St. Dev.	Avg ± St. Dev.	Avg ± St. Dev.
Driver 1	0.150 ± 0.035	0.270 ± 0.026	0.420 ± 0.026
Driver 2	0.160 ± 0.042	0.090 ± 0.057	0.250 ± 0.057
Driver 3	0.187 ± 0.090	0.160 ± 0.069	0.347 ± 0.069
Driver 4	0.177 ± 0.015	0.193 ± 0.146	0.433 ± 0.146
Driver 5	0.137 ± 0.055	0.093 ± 0.045	0.260 ± 0.045
Driver 6	0.133 ± 0.071	0.213 ± 0.055	0.347 ± 0.055
Driver 7	0.173 ± 0.089	0.210 ± 0.060	0.383 ± 0.060
Driver 8	0.073 ± 0.035	0.300 ± 0.035	0.373 ± 0.035
Average	0.148 ± 0.051	0.208 ± 0.082	0.356 ± 0.080

Δt1: delay between the lateral disturbance and the peak of torque unconsciously applied, Δt2: delay between the peak of torque and the effective rotation of the steering wheel, and Δt3: delay between the lateral disturbance and the effective rotation of the steering wheel.

5. Steering Power Evaluation

While analysing the differences between the two types of manoeuvres (one or two hands on the steering wheel), we studied the differences in the steering power used by the drivers while counter steering. The steering power is defined as

$$P_{sw} = T_{SW} \cdot \dot{\delta} \tag{4}$$

where T_{SW} is the torque applied at the steering wheel, calculated by means of the instrumented steering wheel, and $\dot{\delta}$ is the speed of variation of the steering angle δ.

As the input of the kick plate is in random directions, the drivers reacted in two different possible ways. Some of them were able to immediately understand which steering wheel rotation direction was required to recover the desired trajectory. In other cases, they started the steering rotation in the wrong direction and then corrected the steering action. We named the first way as the correct approach, the second as the non-correct approach. In Figure 7, two examples of correct and incorrect actions are shown, together with the values of steering torque (multiplied by 10) and the steering wheel angle. When using the correct approach, the direction of the steering torque and the steering angle were immediately correct, while in the case of the incorrect approach, they were both in the wrong direction. As expected, when the incorrect action was applied, the required steering power, as defined in (4), was higher to correct the error. Given the large difference between the drivers and the large difference between the single manoeuvres, we decided to analyse the peak of steering power in the first instants in which the steering wheel started to rotate after the lateral excitation.

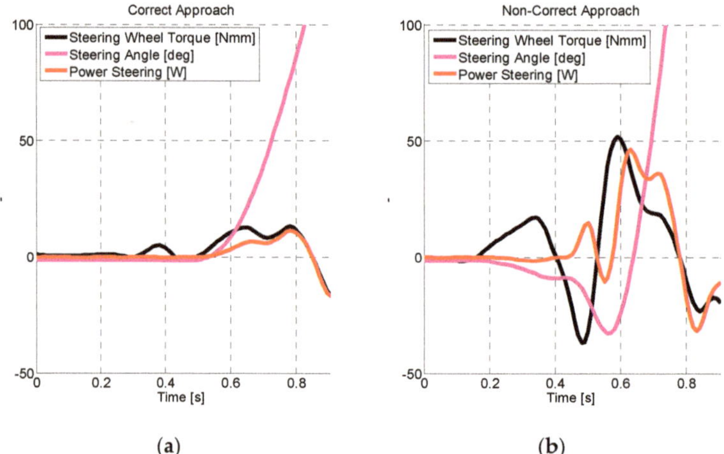

Figure 7. Different approaches in counter steering: (**a**) correct approach; (**b**) incorrect approach. The steering toque is shown multiplied by a factor 10.

Figure 8 shows the values of the steering power peak, averaged for each driver, together with its standard deviation. In the event that the driver did not perform any manoeuvres that match the description, the average value was set to zero, as well as the standard deviation value if the driver had performed only one test. Referring to Figure 8, drivers 4 and 5 performed the one-handed manoeuvres only with the incorrect approach, while drivers 6 and 8 applied only the correct approach. At the same time, driver one performed the one-hand manoeuvres one time with the correct approach and two times with the incorrect one. In these situations, the standard deviation cannot be defined. The analysis was differentiated for the case of correct and incorrect approaches since, in this second case, the applied steering power was higher. In the correct approach (Figure 8a), the power applied in the case of driving with two hands (blue circle) was, on average, lower than the power applied in the case of driving with one hand (red diamond). In the incorrect approach (Figure 8b), the power applied with two hands was higher than the power applied with one hand. Table 3 shows the peak power values averaged on all the drivers, in the case of the correct and incorrect approach, together with the standard deviations. In the correct approach, the power applied with two hands or with one hand was similar (they differed on average less than 10%). In the incorrect approach, the power applied with two hands was higher (on average more than 20%) than the power applied with one hand.

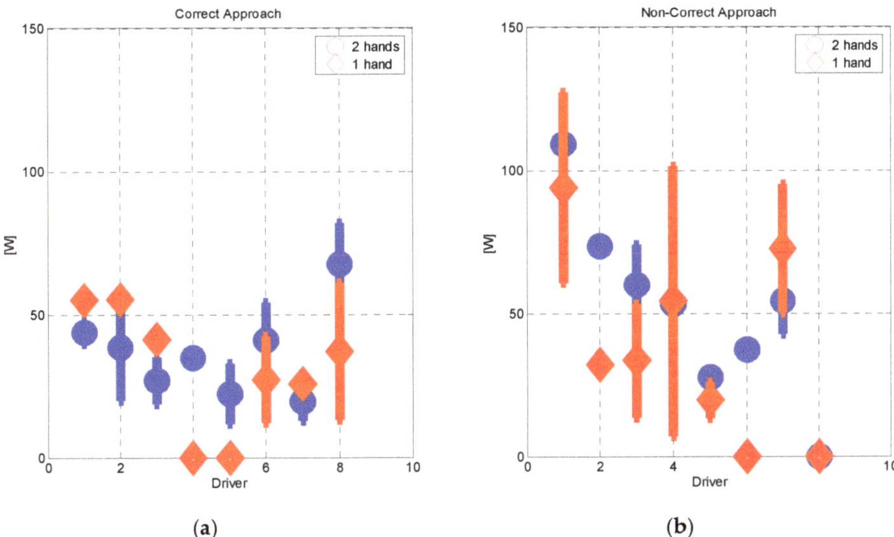

(a) (b)

Figure 8. The power required to steer for the two different approaches: (**a**) the correct approach; and (**b**) the incorrect approach.

Table 3. The power required to steer in the two manoeuvres.

		P_{sw} (W)
		Avg ± St. Dev.
Correct approach	Two hands	41.48 ± 21.76
	One hand	38.12 ± 17.19
Incorrect approach	Two hands	65.48 ± 28.67
	One hand	50.01 ± 34.98

Since the steering power is related to the steering torque T_{sw} and to the speed of variation of the steering wheel angle $\dot{\delta}$. Table 4 shows the steering torque T_{sw} and the angular speed of the steering wheel $\dot{\delta}$ averaged for all drivers in both types of approach for manoeuvring with one or two hands on the steering wheel. In the case of the correct approach, both the torque and the steering velocity were similar if applied with one single hand, or with both hands. In the case of the incorrect approach, the torque applied with two hands was higher than the torque applied with one hand. The same was found for the steering wheel angular velocity.

Table 4. The steering torque and steering angle velocity in the two manoeuvres.

		T_{sw} (Nm)	ΔT_{sw} (Nm)	$\dot{\delta}$ (rad/s)	$\Delta\dot{\delta}$ (rad/s)
		Avg ± St. Dev.	(two hands – one hand)	Avg ± St. Dev.	(two hands – one hand)
Correct approach	Two hands	4.41 ± 1.12	+13%	8.96 ± 3.22	−7%
	One hand	3.83 ± 0.67		9.62 ± 3.24	
Incorrect approach	Two hands	5.91 ± 0.94	+19%	10.80 ± 3.98	+12%
	One hand	4.81 ± 1.97		9.53 ± 3.65	

Figure 9 shows the trends of the steering torque (multiplied by 10), the steering speed, and the steering power. The peak in power always follows the peak torque by a few hundredths of a second, while the angular speed of the steering wheel continues to increase, changing its slope.

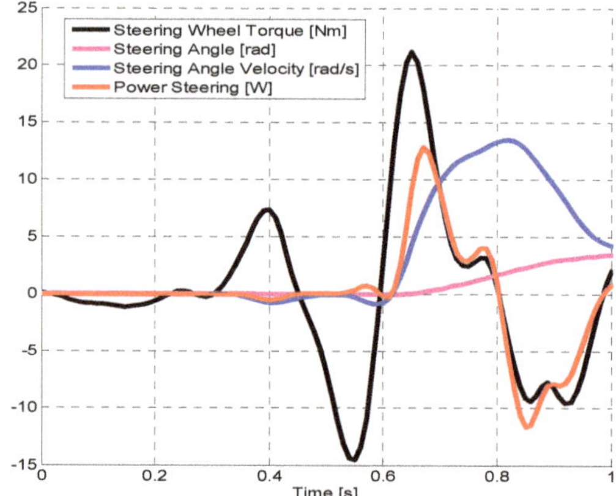

Figure 9. Power torque, steering angle velocity, and power steering.

6. Conclusions

An instrumented steering wheel that measured the forces applied to the steering wheel by the driver allowed us to study the human-machine interactions while driving using only one hand, or both hands.

The same phenomena occurred if only one hand was kept on the steering wheel and when both hands were used. After the occurrence of the lateral disturbance, a torque was applied unconsciously to the steering wheel in a very short time. This torque did not correspond to an actual rotation of the steering wheel, which occurred only afterwards. This torque peak was applied suddenly if both hands were used and was a slower phenomenon if only one hand was used. In both cases, the unconsciously applied torque has a modulus, on average, about 20% of the modulus of the torque applied voluntarily to recover the desired trajectory.

The time interval between the peak of the unconsciously applied torque and the beginning of the steering rotation presented relevant differences between the two considered driving situations. In the case of manoeuvres performed using two hands, the two hands acted by applying opposite torques to the steering wheel, and this fighting action slowed down the driving action. On the contrary, for the one-handed operation, since there was no opposite action of the two hands, the time interval was shortened. This time interval reduction was reflected in the total time interval necessary to modify the steering angle, which was lower in the case of the manoeuvre performed with just one hand.

The input of the kick plate had a random direction, and thus the drivers reacted in two different ways, either being able to immediately understand which steering wheel rotation direction was required to recover the desired trajectory (correct approach), or starting the steering rotation in the wrong direction and then correcting the steering action (incorrect approach). Following the correct approach, the drivers applied similar power if acting with one or two hands; however, in the case of the incorrect approach, the drivers applied lower steering power when manoeuvring with one hand.

The conclusions highlighted in this paper are based on data acquired from a relatively small number of drivers who performed manoeuvres by holding the steering wheel with only one hand or both hands. More accurate results could be obtained by increasing the number of drivers with different levels of driving experience and planning the test in such a way to avoid the bias due to learning the manoeuvre.

The results of this study highlight the importance of the inclusion of the neuro-muscular system (NMS) in driver models. The NSM considers both the voluntary activation of the muscles through alpha-neurons and the involuntary activation through gamma-neurons.

Author Contributions: All authors contributed to the design and implementation of the research, to the analysis of the results, and to the writing of the manuscript. All authors have read and agreed to the published version of the manuscript.

Funding: This research received no external funding.

Acknowledgments: SMARTMechanical Company s.r.l. is gratefully acknowledged for assembling the instrumented steering wheel prototype.

Conflicts of Interest: The authors declare no conflict of interest.

References

1. Pick, A.J.; Cole, D.J. A mathematical model of driver steering control including neuromuscular dynamics. *J. Dyn. Syst. Meas. Control.* **2008**, *130*, 031004. [CrossRef]
2. Wei, H.; Wu, Y.; Chen, X.; Xu, J. A Steering-Following Dynamic Model with Driver's NMS Characteristic for Human-Vehicle Shared Control. *Appl. Sci.* **2020**, *10*, 2626. [CrossRef]
3. Gobbi, M.; Comolli, F.; Hada, M.; Mastinu, G. An instrumented steering wheel for driver model development. *Mechatronics.* **2019**, *64*, 102285. [CrossRef]
4. Saito, Y.; Itoh, M.; Inagaki, T. Driver Assistance System With a Dual Control Scheme: Effectiveness of Identifying Driver Drowsiness and Preventing Lane Departure Accidents. *IEEE Trans. Human-Machine Syst.* **2016**, *46*, 660–671. [CrossRef]
5. Bonera, E.; Gadola, M.; Chindamo, D.; Morbioli, S.; Magri, P. On the Influence of Suspension Geometry on Steering Feedback. *Appl. Sci.* **2020**, *10*, 4297. [CrossRef]
6. Cai, H.; Hu, Z.; Chen, Z.; Zhu, D. A Driving Fingerprint Map Method of Driving Characteristic Representation for Driver Identification. *IEEE Access.* **2018**, *6*, 71012–71019. [CrossRef]
7. Wang, Z.; Zheng, R.; Kaizuka, T.; Shimono, K.; Nakano, K. The Effect of a Haptic Guidance Steering System on Fatigue-Related Driver Behavior. *IEEE Trans. Human-Machine Syst.* **2017**, *47*, 741–748. [CrossRef]
8. Jiang, H.; Tian, H.; Hua, Y.; Tang, B. Research on Control of Intelligent Vehicle Human-Simulated Steering System Based on HSIC. *Appl. Sci.* **2019**, *9*, 905. [CrossRef]
9. Okamoto, K.; Tsiotras, P. A Comparative Study of Data-Driven Human Driver Lateral Control Models. In Proceedings of the 2018 American Control Conference (ACC 2018), Milwaukee, WI, USA, 27–29 June 2018; pp. 3988–3993.
10. Li, A.; Jiang, H.; Zhou, J.; Zhou, X. Implementation of Human-Like Driver Model Based on Recurrent Neural Networks. *IEEE Access* **2019**, *7*, 98094–98106. [CrossRef]
11. Fang, Z.; Wang, J.; Li, P.; Xia, J. Modeling of driver's steering behavior in large-curvature path following with back propagation neural network. In Proceedings of the 38th Chinese Control Conference (CCC2019), Guangzhou, China, 27–30 July 2019; pp. 6727–6732.
12. Cheng, Z.J.; Jeng, L.W.; Li, K. Behavioral classification of drivers for driving efficiency related ADAS using artificial neural network. In Proceedings of the 2018 IEEE International Conference on Advanced Manufacturing (IEEE ICAM 2018), Yunlin, Taiwan, 16–18 November 2018; pp. 173–176.
13. Del Campo, I.; Finker, R.; Martinez, M.V.; Echanobe, J.; Doctor, F. A real-time driver identification system based on artificial neural networks and cepstral analysis. In Proceedings of the 2014 International Joint Conference on Neural Networks (IJCNN 2014), Beijing, China, 6–11 July 2014; pp. 1848–1855.
14. Martínez, M.V.; Echanobe, J.; Del Campo, I. Driver identification and impostor detection based on driving behavior signals. In Proceedings of the 19th International IEEE Conference on Intelligent Transportation Systems (IEEE ITSC 2016), Rio de Janeiro, Brazil, 1–4 November 2016; pp. 372–378.
15. Kolekar, S.; Mugge, W.; Abbink, D.A. Modeling Intradriver Steering Variability Based on Sensorimotor Control Theories. *IEEE Trans. Human-Machine Syst.* **2018**, *48*, 291–303. [CrossRef]
16. Zhu, B.; Liu, Z.; Zhao, J.; Chen, Y.; Deng, W. Driver Behavior Characteristics Identification Strategies Based on Bionic Intelligent Algorithms. *IEEE Trans. Human-Machine Syst.* **2018**, *48*, 572–581. [CrossRef]
17. You, C.; Lu, J.; Tsiotras, P. Nonlinear Driver Parameter Estimation and Driver Steering Behavior Analysis for ADAS Using Field Test Data. *IEEE Trans. Human-Machine Syst.* **2017**, *47*, 686–699. [CrossRef]
18. Gote, C.; Flad, M.; Hohmann, S. Driver characterization & driver specific trajectory planning: An inverse optimal control approach. In Proceedings of the 2014 IEEE International Conference on Systems, Man and Cybernetics (SMC2014), San Diego, California, 5–8 October 2014; pp. 3014–3021.

19. Hua, Y.; Jiang, H.; Tian, H.; Xu, X.; Chen, L. A Comparative Study of Clustering Analysis Method for Driver's Steering Intention Classification and Identification under Different Typical Conditions. *Appl. Sci.* **2017**, *7*, 1014. [CrossRef]
20. Chen, S.H.; Pan, J.S.; Lu, K. Driving behavior analysis based on vehicle OBD information and adaboost algorithms. *Lecture Notes in Engineering and Computer Science* **2015**, *1*, 102–106.
21. Pan, J.-S.; Lu, K.; Chen, S.-H.; Yan, L. Driving Behavior Analysis of Multiple Information Fusion Based on SVM. In *Lecture Notes in Computer Science*; Springer Science and Business Media LLC: Berlin/Heidelberg, Germany, 2014; Volume 8481, pp. 60–69.
22. Sathyanarayana, A.; Boyraz, P.; Purohit, Z.; Lubag, R.; Hansen, J.H.L. Driver adaptive and context aware active safety systems using CAN-bus signals. In Proceedings of the 2010 IEEE Intelligent Vehicles Symposium (IV '10), San Diego, CA, USA, 21–24 June 2010; pp. 1236–1241.
23. Van Ly, M.; Martin, S.; Trivedi, M.M. Driver classification and driving style recognition using inertial sensors. In Proceedings of the 2013 IEEE Intelligent Vehicles Symposium (IV), QLD, Australia, 23–26 June 2013; pp. 1040–1045.
24. Hallac, D.; Sharang, A.; Stahlmann, R.; Lamprecht, A.; Huber, M.; Roehder, M.; Sosič, R.; Leskovec, J. Driver identification using automobile sensor data from a single turn. In Proceedings of the 2016 IEEE 19th International Conference on Intelligent Transportation Systems (ITSC), Rio de Janeiro, Brazil, 1–4 November 2016; pp. 953–958.
25. Silva, H.; Lourenço, A.; Fred, A. In-vehicle driver recognition based on hands ECG signals. In Proceedings of the 17th International Conference on Intelligent User Interfaces (IUI 2012), Lisbon, Portugal, 14–17 February 2012; pp. 25–28.
26. Yang, C.H.; Liang, D.; Chang, C.C. A novel driver identification method using wearables. In Proceedings of the 2016 13th IEEE Annual Consumer Communications & Networking Conference (CCNC 2016), Las Vegas, NV, USA, 9–12 January 2016; pp. 1–5.
27. Schmidt, E.; Decke, R.; Rasshofer, R. Correlation between subjective driver state measures and psychophysiological and vehicular data in simulated driving. In Proceedings of the 2016 IEEE Intelligent Vehicles Symposium (IV), Gothenburg, Sweden, 19–22 June 2016; pp. 1380–1385.
28. Abe, M. *Vehicle Handling Dynamics*; Elsevier BV: Amsterdam, The Netherlands, 2015.
29. Gabrielli, F.; Pudlo, P.; Djemai, M. Instrumented steering wheel for biomechanical measurements. *Mechatronics* **2012**, *22*, 639–650. [CrossRef]
30. Schiro, J.; Gabrielli, F.; Pudlo, P.; Barbier, F.; Djemai, M. Normal/tangential force proportion during steering under simulation condition. *Comput. Methods Biomech. Biomed. Eng.* **2012**, *15*, 243–245. [CrossRef] [PubMed]
31. Gobbi, M.; Comolli, F.; Ballo, F.M.; Mastinu, G. Measurement data obtained by an instrumented steering wheel for driver model development. *Data Brief* **2020**, *30*, 105485. [CrossRef] [PubMed]
32. Comolli, F.; Ballo, F.M.; Gobbi, M.; Mastinu, G. Instrumented Steering Wheel: Accurate Experimental Characterisation of the Forces Exerted by the Driver Hands for Future Advanced Driver Assistance Systems. In Proceedings of the Volume 3: 20th International Conference on Advanced Vehicle Technologies; 15th International Conference on Design Education, Quebec City, QC, Canada, 26–29 August 2018.
33. Mastinu, G.; Gobbi, M.; Comolli, F.; Hada, M. Instrumented Steering Wheel for Accurate ADAS Development. In Proceedings of the SAE Technical Papers; SAE International, SAE WCX, Detroit, MI, USA, 9–11 April 2019; Vol. 2019-April.
34. Carrera Akutain, X.; Ono, K.; Comolli, F.; Gobbi, M.; Mastinu, G. Further understanding of steering feedback and driver behavior through the application of an instrumented steering wheel. In Proceedings of the 10th International Munich Chassis Symposium, Munich, Germany, 25–26 June 2019; pp. 481–502.

Article

Analysis of the Phenomena Causing Weave and Wobble in Two-Wheelers

Francesco Passigato [1,*], Andreas Eisele [1], Dirk Wisselmann [2], Achim Gordner [2] and Frank Diermeyer [1]

[1] Chair of Automotive Technology, Technical University of Munich, 85748 Garching bei München, Germany; andreas.eisele@tum.de (A.E.); diermeyer@ftm.mw.tum.de (F.D.)
[2] BMW Group, 80937 Munich, Germany; dwisselmann@t-online.de (D.W.); Achim.Gordner@bmw.de (A.G.)
* Correspondence: francesco.passigato@tum.de

Received: 24 August 2020; Accepted: 24 September 2020; Published: 29 September 2020

Abstract: The present work follows in the tracks of previous studies investigating the stability of motorcycles. Two principal oscillation modes of motorcycles are the well-known wobble and weave modes. The research in this field started about fifty years ago and showed how different motorcycle parameters influence the stability of the mentioned modes. However, there is sometimes a minor lack in the physical analysis of why a certain parameter influences the stability. The derived knowledge can be complemented by some mechanical momentum correlations. This work aims to ascertain, in depth, the physical phenomena that stand behind the influence of fork bending compliance on the wobble mode and behind the velocity dependence of the weave damping behaviour. After a summary of the relevant work in this field, this paper presents different rigid body simulation models with increasing complexity and discusses the related eigenvalue analysis and time behaviour. With these models, the mentioned modes are explained and the physical phenomena only partly covered by the literature are shown. Finally, the influence of the rider model on weave and wobble is presented.

Keywords: two-wheeler; vehicle dynamic modelling; wobble; weave; fork bending compliance; rider influence

1. Introduction

Motorcycle dynamics have always been an extensive research topic due to the complexity and the influence on the riders' safety. The knowledge used to develop a stable motorcycle chassis comes, to a certain extent, from experience or sensitivity analysis which sometimes lack physical and theoretical understanding of the hidden phenomena. This paper wants to clarify these phenomena, in particular with regard the understanding of the effect of the front fork bending compliance and of the weave damping behaviour with changing speed.

Two of the principal eigenmodes of a motorcycle are known as wobble and weave. The first one includes almost solely the rotation of the front assembly about the steering axis whose frequency is in the range between 7 and 10 Hz depending on the motorcycle's parameters [1]; the second one is more complex, and when affected by this mode, the motorcycle shows roll, yaw, steering-head rotation and lateral displacement [2].

The starting point for this work is an observation resulting from the comparison of experience in reality and simulation results. Using the eigenvalue analysis, the earlier motorcycle models, such as the model of Sharp [3] which only includes a tyre model but no compliances, show a stable wobble mode in the lower speed range (up to 80 km h^{-1}), and this mode becomes unstable in the higher speed range. However, the experience with a real motorcycle shows an almost opposite behaviour: lower stability (or even instability) at low speed and increasing stability at higher speed. This was demonstrated through measurements on a real motorcycle by Koch [4]. Previous research addressed this apparent incompatibility between simulation and reality, concluding that the front fork bending compliance is a key parameter when studying wobble [5–9], as discussed further in Section 2. Wisselmann [10] conducted an analysis where the motorcycle stiffnesses were progressively eliminated (i.e., set to infinity); in this case also, a remarkable variation in the motorcycle response was shown.

For weave mode, the situation is different: the well-known correlation between eigenmode-damping and speed is also reproduced by models without chassis compliances [2,5,6]. In this case, up to a certain speed, the weave mode becomes more stable with increasing speed. After a tipping point, the stability degenerates again when the speed increases. This behaviour is unexpected, as the increasing gyroscopic moments of the wheel should stabilise the motorcycle, as is the case, for example, with a gyroscope. This is a complex system to analyse because there are many possible gyroscopic moments and their combined influence is not trivial. This peculiar behaviour is further analysed in Section 4.2.

The different models that will be presented within this paper were produced using open source multibody simulation software named "MBSim" (www.mbsim-env.de) which was developed at the Institute of Applied Mechanics of the Technical University of Munich. An introduction [11] on the functionalities of this software is provided. Multibody simulation is common for vehicle dynamics, and was used, for example, by Cossalter in multiple publications [2,5,7,12]. It allows the building of the desired model through a CAD-like interface, by connecting the bodies with joints and assigning them stiffness and damping when needed. The present motorcycle model is described in Section 3.

To summarise, this work deeply investigates the influence of the front fork bending compliance on the wobble mode. Starting from the point reached by previous authors [1,5,6], Section 4.1 provides a different justification for the well-known effect of the fork bending compliance on wobble damping. Moreover, in Section 4.2 the weave mode is further analysed and a possible justification for its damping behaviour is provided. Finally Section 4.3 investigates the influence of the rider on wobbling and weaving.

2. Related Work

Vehicle dynamics is a very popular research topic which features in many books and publications. Most of the literature is focused on four-wheelers; nevertheless, there is also a huge collection of research on two-wheeled vehicles. The different relevant authors are mentioned below.

Back in 1971, Sharp [3] published a paper about the driving stability of two-wheeled vehicles. This research is considered the starting point of the modern motorcycle dynamics. In fact, thanks to a (linearised) tyre model it was possible to detect the wobble mode, which was not present in the even earlier and less complex motorcycle models by Whipple [13] (p. 195).

Cossalter [2] (p. 278) [5] and Spierings [6] presented a solution to overcome the previously mentioned incoherence between reality and rigid model simulations when analysing wobble. A fork bending compliance with a lumped stiffness was added: a revolute joint in the front fork, with its rotating axis (named twist axis [14]) perpendicular to the steering head axis. A certain stiffness and a damping are assigned to this joint. Lumped parameter models are described in [15]. This kind of modelling is also found in other sources, especially when a more complex non-rigid model is developed [16–21].

This additional fork bending results in a more realistic wobble-damping characteristic: wobble-instability at low speed and a clear stabilisation when increasing the speed. Cossalter [2] (p. 280) and Spierings [6] explain this change of characteristic with the combination of two opposed effects caused by the fork bending compliance:

1. The bending compliance itself reduces the stability.
2. The combination of wheel-spin and the rotation around the twist axis produces a gyroscopic moment about the steering axis, leading to stabilisation.

Since the gyroscopic moment is proportional to the speed, the first effect dominates at low speed, causing lower stability in this speed range. The second effect becomes dominant with increasing speed, which explains the restored stability at high speeds. Section 4.1 investigates the influence of the fork bending compliance considering another phenomenon caused by the bending compliance itself.

There are several parameters affecting a motorcycle's wobble stability; a complete list can be found in [2,3,17,21]. For this work, besides the fork bending compliance, two tyre parameters are of particular importance: the relaxation length and the cornering stiffness. Some research [3,16,17,20] has underlined that a tyre model with cornering stiffness is necessary in order to simulate the wobble mode. This also suggests an important consideration: the principal cause of the wobble is given by the tyre's response. When the wheel is subject to an outer disturbance steering the wheel itself, the tyre reacts with a side force which, thanks to the mechanical trail, produces an aligning moment around the steering axis. When returning to the equilibrium position, the wheel starts oscillating about the steering axis, thereby triggering the wobble mode [22]. Depending on the speed of the motorcycle, this oscillation will diverge (unstable behaviour) or converge to the equilibrium (stable behaviour).

The other important tyre parameter influencing wobble is the relaxation length [3,13,23]. Several works [3,16,17,20,24] investigated its effect on wobble damping, underlying how the relaxation length destabilises the wobble mode, since it generates a delay between the wheel steering and the tyre-lateral force generation. As a consequence, Sharp [3] (p. 323) clarifies that both tyre sideslip and tyre relaxation properties are fundamental for a proper representation of a motorcycle's dynamic properties. Like most of the tyre parameters, the relaxation length also depends on several factors. Pacejka et al. [25] and Sharp [1] report that the relaxation length increases with increasing speed. Moreover, a dependence on the vertical load is present. A more recent paper by Sharp [12] proposes a tyre model based on the Pacejka magic formula [26] where dependencies on both speed and vertical load are reproduced. This model and the related parameters are also used in the present work.

In summary, some minimal requisites for a correct analysis of the wobble mode can be defined. Pacejka [26] and Doria [16] suggest that the tyre properties have the greatest influence on motorcycle stability. The Pacejka magic formula [26] coupled with a first order dynamic response, as described in the previous paragraph, is necessary in order to produce reliable wobble stability analyses [13] (p. 324). This tyre model allows an accurate description of the tyre forces and moments if the frequency of the external forces is lower than 15 Hz [13]; these conditions are generally satisfied for common stability analysis, therefore the magic formula is used in almost all the latest researches based on multibody analysis [2,7,12,27–29]. In addition to the tyre modelling, the lateral compliance at the front of the machine is necessary when simulating wobble [13,30]. This can be obtained with a bending compliance of the front fork [5–9], or by adding a lumped torsional stiffness of the main frame at the steering head joint with its rotation axis perpendicular to the steering axis [12,26–28]; a combination of both stiffnesses can also be implemented, as in the present work.

Besides the wobble mode, many of the mentioned sources also analyse vehicle "weave" [2,3,5–7,17,19,21]. The weave mode can also be qualitatively represented with a motorcycle model having no flexibilities [13] (p. 195). However, the literature underlines that some additional degrees of freedom (DoF) notably influence the weave mode. In particular, Doria [17] (p. 17) and Splerings [6] (p. 28) underline that the fork bending compliance destabilises the weave mode. Taraborrelli [23] and Cossalter [7] analyse the effect of the swingarm torsional and bending compliance. Reference [23] simulated two motorcycle types with significantly different stiffness values: a super-sport motorcycle (higher stiffness), where the two swingarm compliances had almost no influence on weave, and an enduro-motorcycle, where the swingarm bending compliance slightly destabilised the weave mode, which was at the same time slightly stabilised by the torsional compliance of the swingarm. The results of [7], on the contrary, show the destabilising effect of both swingarm bending and torsional compliance; however, the influence was also small in this case.

Furthermore, the rider has proven to greatly affect the motorcycle's stability behaviour. In fact, Roe [31] points out that the stability analysis of a riderless motorcycle can be very misleading. Significant changes in the weave stability due to the rider have also been reported by Pacejka [26] (p. 535). The modelling of the rider's passive response and its influence on stability is a broad topic which will be summarised in Sections 3.3 and 4.3.

An interesting argument strictly connected to weave is the so called "cornering weave". During cornering, the in-plane eigenmodes (bounce, pitch) become increasingly coupled with the out-of-plane eigenmodes (weave, wobble), whereby the former shows degrees of freedom typical of the latter and vice versa [9,13,32]. Concerning cornering weave, the eigenvector also shows front and rear suspension travels which are not present in the straight-running weave [32]. This mode-coupling is critical because it provides a signal transmission path from road undulations to lateral motions [33]. Moreover, under cornering conditions, bounce and weave have similar eigenfrequencies in the middle to high speed range (70–120 $\mathrm{km\,h^{-1}}$) [9,24,32,34], thereby leading to mutual triggering [13,33,34]. From these considerations it should be clear that the suspension system must also be modelled when carrying out stability analysis of motorcycles under cornering conditions [13,34].

As mentioned in the introduction, the models for the present work were implemented with the "MBSim" multibody simulation software. Two interesting references in this field are the works of moreCossalter [29] and moreLot [35]. They did not use the same software, but developed two multibody-simulation tools that follow the same principle; they are both based on the symbolic calculation software "Maple" with the second one using a build-in library for multibody modelling called "MBSymba". These references underline the suitability of a multibody simulation tool for the dynamics and stability analysis of motorcycles. Another piece of software used in several works [12,27,36] is "Autosim"; in this case as well, the authors demonstrated the great advantages provided by multibody modelling when conducting both time and frequency analyses. Following this established research, this modelling and calculation method were chosen for the present work.

3. Modelling

Different models have been developed during the last 70–80 years. Their complexity is obviously related to the objectives they pursue. For example, the wobble has been simulated studying systems capable of shimmy behaviour [37] (p. 166). Limebeer [13] (p. 184) derived the equations of motion of such a system, which possesses two DoF:

1. Rotation about the steering axis (in this case perpendicular to the ground).
2. Lateral translation of the steering joint used to reproduce lateral flexibility.

The model does not take into account the gyroscopic influences. A liner model with relaxation effect is used for the tyre forces. Even with this strongly simplified configuration, some important effects are described, which also occur in motorcycle modelling. The destabilising effect of the cornering stiffness, as prescribed by Sharp [3], is reproduced. Moreover, the inversion of behaviour experienced in motorcycle simulation when using an infinite stiff front frame (Section 2) is also present. This is a remarkable result, as this model does not reproduce the gyroscopic moments, which are considered in [2,6] as responsible for the inversion of behaviour. This could suggest that also other phenomena influence the wobble stability. In Section 4.1 this hypothesis is further developed.

Limebeer [13] (p. 189) carries on his analysis by converting the shimmy model into a front fork model. In this case, the caster angle is considered for the steering axis and the gyroscopic effects are also considered. Another important feature is a revolute joint connecting the fork assembly to the ground. This simulates the motorcycle's frame torsional flexibility, which provides a lateral flexibility at wheel level, as described in Section 2. Thanks to this feature, the eigenvalue analysis of this system provides results similar to those generally obtained with motorcycle modelling. Cossalter [2] (p. 251) also produced a similar model with the aim of studying wobble; however, it lacks the mentioned lateral flexibility and transient tyre properties, thereby failing to capture the physics of the wobble mode.

The simplified front fork model is acceptable when the aim is to only reproduce wobble. When studying weave, however, a whole motorcycle model is needed. The "basic" motorcycle model is generally updated with some additional degrees of freedom in order to reproduce, for example, the frame compliance or the rider motion.

In Section 3.1 the motorcycle model used in the present work is explained.

3.1. Motorcycle Model

The model used in this work is based on [10], where a model for BMW sport motorcycles from the 1990s was developed and validated with several experimental data, especially regarding the weave and wobble modes. The kinematic structure of the rear suspension and the dataset in the present paper are based on [12]. This reference uses a model with 13 DoF:

- Six DoF related to the main frame rigid body motion: translation about x, y, z axes, roll (ρ), pitch (ϕ) and yaw (ψ).
- Two wheel spins ($\theta_{\mathrm{fw}}, \theta_{\mathrm{rw}}$).
- Two suspension travels ($z_{\mathrm{f}}, z_{\mathrm{r}}$).
- Steer rotation (δ).
- Frame torsion (α_{fr}) at the steering head joint, simulated with a revolute joint with rotation axis perpendicular to the steering axis.
- Rider lean (α_{ri}) simulated with a revolute joint between the rider's lower and upper body with rotation axis parallel to the main frame x axis.

In the present work some additional DoF are added, which are shown in Figure 1.

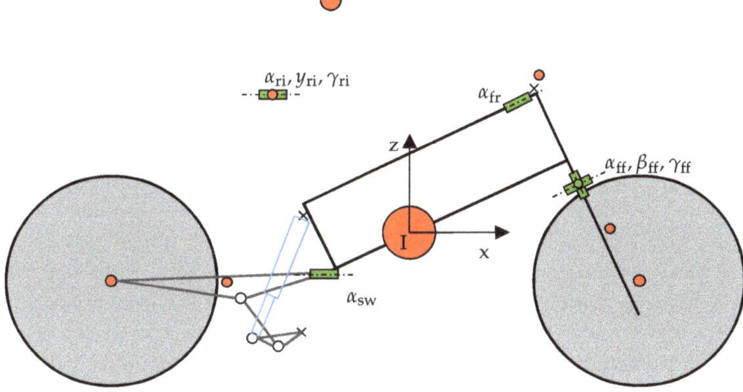

Figure 1. Motorcycle sketch with the DoF associated with the stiffnesses. The small rectangles indicate the related joints. The red points represent the bodies' centre of mass.

- Three compliances of the front fork, simulated with three revolute joints placed approximately at the half of fork length:
  - Bending about the x axis (α_{ff}); the rotation axis of the revolute joint is perpendicular to the steering axis. This compliance is particularly important for wobble, as explained in Section 2. It allows a lateral motion of the wheel along the y axis.
  - Bending about the y axis (β_{ff}); the rotation axis of the revolute joint is parallel to the front wheel axis. This flexibility allows a translation of the wheel along the x axis.
  - Torsion (γ_{ff}); the rotation axis of the revolute joint is parallel to the steering axis.
- Lateral motion of the rider's lower body with respect to the saddle (y_{ri}), simulated with a linear joint with axis parallel to the main frame y axis.
- Rider yaw rotation (γ_{ri}) simulated with a revolute joint between the rider's lower and upper body with rotation axis parallel to the main frame z axis.
- Swingarm torsion (α_{sw}), simulated with a revolute joint at the swingarm anchor point with rotation axis parallel to the main frame x axis.

The motorcycle parameters in [12] were obtained through experimental measurements or through estimates. However, some parameter variations have been used in the present work:

- The data for the rider's upper body mass and inertia tensor were taken from [12]. In [12] the mass and inertia tensor of the rider's lower body and of the motorcycle's main frame were fused together. In the present work, with the aim of allowing lateral translation of the rider's hip (considered as lower body in the present work), the mass and inertia tensor of the main frame in [12] were split between hip and main frame, while ascertaining that the combined inertia tensor and the overall centre of gravity remain unchanged. The data for hip's mass and inertia tensor were taken from [38]. According to these data, the mass and inertia tensor of the main frame were properly modified. In particular, the hip mass was subtracted from the main frame mass in [12]. Moreover, as explained above, the inertia tensor of the main frame was adapted so that the whole inertia tensor of the lower body and main frame together was equal to the main frame inertia in [12]. The rider's legs were considered fixed to the main frame, so their mass and inertia were included in it. As reported in [12] (p. 252), the rider's total mass was 72 kg.
- In the year of publication of [12] the experimental data of the front fork were not completely available. Some of them were judged by the authors of the present work as not realistic (for example, the mass of the lower fork and its lacking inertia tensor). For this reason, the front fork mass and inertia data were taken from [39], while retaining the overall geometry given in [12].

The motorcycle's parameters are reported in Appendix A.

The tyre forces and moments were reproduced with the Pacejka magic formula; further details on the tyre model are given in Section 3.4. The software for the motorcycle model is available on GitHub: https://github.com/TUMFTM/motorcycle_model.git.

3.2. Suspension

There are several types of motorcycle suspension. The simplest configuration is composed of a telescopic front suspension and a cantilever rear suspension. However, multi-link suspensions have been developed for both front and rear wheels. For the rear suspension, the multi-link solution allows a non-linear damping curve to be obtained even with a linear damper, as the kinematic function of the damper travel can be adjusted using a rocker. Multi-link at the front suspension generally increases the fork stiffness and favours the anti-dive property. For further details the reader is referred to [13] (p. 327). The motorcycle adopted in the present work uses a telescopic front suspension and a unitrack-like rear suspension (definition of [13]), where the spring-damper unit is connected to the chassis and to a rocker, while a pull rod connects the rocker to the swingarm. The kinematic functions of this rear suspension are presented in [12]. Figure 1 shows its geometry.

3.3. Rider Model

In contrast to cars, the rider's weight strongly influences the total motorcycle weight in riding conditions. Moreover, the rider has some relative motions with respect to the motorcycle. One can therefore expect the rider to have a great influence on motorcycle stability. This is indeed the case, as shown by both simulation [7,40] and experimental [41] results. In this work we only focus on the rider's passive motions. This is reasonable, as the eigenfrequencies of wobble and weave are too high for the rider to actively counteract them; therefore, it can be assumed that the rider behaves as a passive body when the motorcycle experiences wobble or weave. Under this assumption, different rider models can be developed. The most simple model has been adopted by several works [12,26–28] and includes only one DoF: the rider's upper body rotation about an axis parallel to the frame x axis. A different one-DoF model [42] reproduces the rider's rotation about the vertical axis (parallel to the frame z axis); moreover, the connection between the upper body and handlebar is included with a parallel rotational spring-damper element. A 2-DoF model can then be obtained by combining the two previously described models [40]. Finally, the lateral displacement (frame y axis) of the rider's hip can be added, thereby obtaining a 3-DoF rider model [7]. An interesting observation is made by Limebeer [13] (p. 326) about the influence of the rider's parameters: he reports that the rider's upper body parameters mainly influence the weave mode, while the lower body parameters have a greater effect on wobble. The literature also offers several examples of more complex rider models containing up to 28 DoF [43]. In this case, the rider model was used to faithfully evaluate the motorcycle race performances, so that the whole rider motion was needed, including his control action on the handlebars and his lean-in strategy. As introduced at the beginning of this paragraph, the current work pursues a very different objective, i.e., the analysis of the motorcycle eigenmodes, so that such a complex rider model is not necessary and would only complicate the interpretation of the results.

The rider model used in the present work is based on that presented in [7]. The rider is composed by two masses, representing the upper and lower body. The DoF are:

- α_{ri}: the rider's upper body lean.
- γ_{ri}: the rider's upper body yaw.
- y_{ri}: the rider's hip (lower body) lateral motion.

As previously said, the connection with the handlebar is modelled by a rotational spring-damper element which reacts to the relative rotation between the upper body and handlebar, and applies equal and opposite moments for lower body and upper body, and frame and steering head, respectively.

3.4. Tyre Model

As introduced in Section 2, the present model uses the Pacejka magic formula to describe the tyre forces. The tyre model is analogous to that reported by Sharp [12] and is suited for motorcycles, as it accounts for the lateral force generation due to the wheel roll angle (camber angle). Moreover, it also describes the longitudinal forces, thereby allowing the motorcycle to be simulated in all possible conditions. Another peculiar factor is the description of the combined longitudinal and lateral force generation. This means that the maximal tyre-lateral force is reduced when the tyre simultaneously generates a longitudinal force (and vice versa). The model also takes into account the variation of the tyre potential with changing vertical load.

The present tyre model considers the tyre width. In this way, in contrast to thin disc models, the overturning moment must not be added separately, as the lateral migration of the contact point automatically generates this overturning moment [12,28].

The magic formula needs some inputs: longitudinal slip (κ), slip angle (α), camber angle (ρ_{w}), vertical load (F_z). κ can be calculated with Equation (1).

$$\kappa = -\frac{v_{x_{\text{C}}} - \omega_y \left(r_{\text{rim}} + r_{\text{c}} \cos \rho_{\text{w}} \right)}{v_{x_{\text{C}}}}, \tag{1}$$

where $v_{x_{\text{C}}}$ is the longitudinal velocity of the tyre contact point, r_{rim} is the rim radius, r_{c} is the tyre crown radius and ω_y is the wheel rotation speed. α is calculated with a first order relaxation equation:

$$\frac{\sigma}{v_{x_{\text{C}}}} \dot{\alpha} + \alpha = \alpha_{\text{ss}}, \qquad \alpha_{\text{ss}} = -\arctan\left(\frac{v_{y_{\text{C}}}}{v_{x_{\text{C}}}} \right), \tag{2}$$

where α_{ss} is the slip angle in steady-state conditions, while $v_{y_{\text{C}}}$ is the lateral velocity of the tyre contact point. This means that the lateral force due to the slip angle is generated with some delay depending on the relaxation length σ. Its value is not constant and depends on both longitudinal speed and vertical load; these dependencies are taken into account in the present model. The longitudinal force and the lateral force due to the camber angle are assumed to build up instantaneously, so no relaxation equation is used.

At this point, a geometrical model of the tyre contact point and a related reference frame are needed in order to calculate the velocity components $v_{y_{\text{C}}}, v_{x_{\text{C}}}$ at the contact point itself and the vertical load F_z. Firstly, the orientation of the reference frame at the contact point has to be obtained. It should satisfy the following requirements: z axis being perpendicular to the road; x axis being in the wheel plane and parallel to the road. Such a reference frame can be derived with two rotations starting from the reference frame of the wheel carrier, indicated in the following with the subscript Wc. The intermediate reference frame W is obtained as follows:

$$\begin{aligned} {}_{\mathrm{W}}e_x &= \frac{{}_{\mathrm{Wc}}e_y \times {}_{\mathrm{I}}e_z}{|{}_{\mathrm{Wc}}e_y \times {}_{\mathrm{I}}e_z|}, \\ {}_{\mathrm{W}}e_y &= {}_{\mathrm{Wc}}e_y, \\ {}_{\mathrm{W}}e_z &= {}_{\mathrm{W}}e_x \times {}_{\mathrm{W}}e_y, \end{aligned} \tag{3}$$

where ${}_{\mathrm{ref}}e_i$ is the unit vector of the ith axis of the reference frame named "ref", and ${}_{\mathrm{I}}e_z = [0,0,1]^{\mathrm{T}}$. The frame W has the x axis parallel to the ground. In order to get C, another transformation is needed, which allows the z axis to be perpendicular to the ground:

$$\begin{aligned} {}_{\mathrm{C}}e_x &= {}_{\mathrm{W}}e_x, \\ {}_{\mathrm{C}}e_z &= {}_{\mathrm{I}}e_z, \\ {}_{\mathrm{C}}e_y &= {}_{\mathrm{C}}e_z \times {}_{\mathrm{C}}e_x. \end{aligned} \tag{4}$$

The unit vectors in Equation (3) define the rotation matrix ${}_{\mathrm{I}}\mathbf{S}_{\mathrm{W}} = [{}_{\mathrm{W}}e_x, {}_{\mathrm{W}}e_y, {}_{\mathrm{W}}e_z]$ from frame W to the inertial frame I. Similarly, the rotation matrix from frame C to I is defined as ${}_{\mathrm{I}}\mathbf{S}_{\mathrm{C}} = [{}_{\mathrm{C}}e_x, {}_{\mathrm{C}}e_y, {}_{\mathrm{C}}e_z]$. Now the longitudinal and lateral velocities $v_{x_{\mathrm{C}}}, v_{y_{\mathrm{C}}}$ can be calculated in the following steps:

$$\begin{aligned} {}_{\mathrm{W}}v_{\mathrm{Wc}} &= {}_{\mathrm{I}}\mathbf{S}_{\mathrm{W}}^{\mathrm{T}}\, {}_{\mathrm{I}}v_{\mathrm{Wc}}, \\ {}_{\mathrm{W}}v_{\mathrm{C}} &= {}_{\mathrm{W}}v_{\mathrm{Wc}} + {}_{\mathrm{W}}\omega_{\mathrm{Wc}} \times [0,\, -r_{\mathrm{c}} \sin\rho_{\mathrm{W}},\, -r_{\mathrm{rim}} - r_{\mathrm{c}} \cos\rho_{\mathrm{W}}]^{\mathrm{T}}, \\ {}_{\mathrm{I}}v_{\mathrm{C}} &= {}_{\mathrm{I}}\mathbf{S}_{\mathrm{W}}\, {}_{\mathrm{W}}v_{\mathrm{C}}, \\ {}_{\mathrm{C}}v_{\mathrm{C}} &= {}_{\mathrm{I}}\mathbf{S}_{\mathrm{C}}^{\mathrm{T}}\, {}_{\mathrm{I}}v_{\mathrm{C}}. \end{aligned} \tag{5}$$

${}_{\mathrm{B}}v_{\mathrm{A}}$ is the absolute velocity vector of the generic point A in the generic frame B. The vector ${}_{\mathrm{C}}v_{\mathrm{C}}$ represents the velocity of the contact point in frame C; its first and second components provide $v_{x_{\mathrm{C}}}, v_{y_{\mathrm{C}}}$.

The last remaining input for the magic formula is the vertical load. This is composed by a constant $F_{\mathrm{z,stat}}$ and a varying part $F_{\mathrm{z,dyn}}$. The constant one is obtained in static equilibrium conditions. The second one depends on motorcycle trim and can be calculated using the tyre carcass compression Δz_{tyre} from the nominal state.

$$\Delta z_{\mathrm{tyre}} = -r_{\mathrm{rim}} \cos\rho_{\mathrm{W}} - r_{\mathrm{c}} + {}_{\mathrm{I}}z0_{\mathrm{C}} - {}_{\mathrm{I}}z_{\mathrm{Wc}}. \tag{6}$$

${}_{\mathrm{I}}z0_{\mathrm{C}} > 0$ is the vertical distance under static equilibrium conditions between the origin of the inertial frame I and the contact point. ${}_{\mathrm{I}}z_{\mathrm{Wc}} > 0$ is the instantaneous vertical distance between the origin of the inertial frame I and the wheel carrier Wc. Finally, the tyre vertical load F_z is calculated as follows:

$$F_z = F_{\mathrm{z,stat}} + F_{\mathrm{z,dyn}} = F_{\mathrm{z,stat}} - c_z\, \Delta z_{\mathrm{tyre}} - d_z\, v_{z_{\mathrm{C}}}. \tag{7}$$

c_z, d_z are the tyre vertical stiffness and damping, respectively. It is worth pointing out that Δz_{tyre} is negative when the carcass is compressed; therefore, the carcass compression leads, through Equation (7), to an increased vertical load.

All the inputs for the magic formula are now available. The equations of the magic formula are taken from [12], so they are not repeated here. The outputs are: longitudinal force, lateral force and aligning moment. These outputs are applied in the wheel carrier and not at the contact point as in [12]. This choice was made because in MBSim the forces must be applied to a body, which would require the definition of a body in the contact point, which in the actual model structure is not an easy task. Shifting the tyre forces in the wheel carrier requires applying additional moments; this is, however, easily done, as all the lever arms are known or can be derived with geometric reasoning. The equivalence between this force system and the one with the forces applied at the contact point has been verified on a simplified model with a single wheel. The results are equivalent, thereby confirming the correctness of the choice made.

3.5. Validation

As explained in Section 3.1, the motorcycle model in the present work is based on the data of [12]. The obvious solution is then to validate the present model against this reference by Sharp. In order to do so, the front fork data were set to the values in [12]; moreover, the additional DoF were "switched off." This was possible thanks to a functionality developed for this purpose in MBSim, which allows the selection of the DoF to be considered. Reference [12] reports the simulation results of a steady-state cornering condition with fixed roll angle and at different speeds. Several variables are then shown under this condition; particularly important for steady-state cornering are: steering torque, front and rear tyre-lateral forces and aligning moments. The same condition has been tested in the present work and the variables compared. The accordance is very good, with deviations less than 5 % for all values. This is shown Figure 2, where F_{y_f}, F_{y_r} are the front and rear tyre-lateral forces, while M_{z_f}, M_{z_r} are the front and rear tyre aligning moments and T_δ is the steer torque.

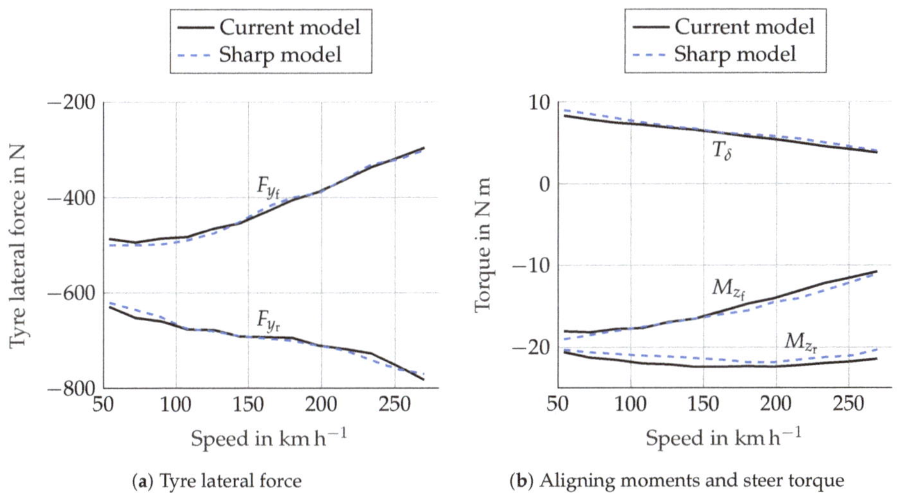

(**a**) Tyre lateral force (**b**) Aligning moments and steer torque

Figure 2. Model validation with steady-state analysis: comparison between the current model and the model of Sharp for a steady-state cornering maneuver. F_{y_f}, F_{y_r} are the front and rear tyre-lateral forces, while M_{z_f}, M_{z_r} are the front and rear tyre aligning moments and T_δ is the steer torque.

As further validation, the present model was also compared with that in [26] (p. 508) by Pacejka. This model renounces some DoF with respect to [12]: the suspensions are not modelled. Moreover, the frame torsional stiffness is not present, while the front fork has bending compliance. Using the same data of [26] and with the same DoF, a comparison of the eigenvalue analysis has been carried out. Additionally, in this case the agreement of the results is very good, with the curves of the wobble and weave eigenfrequencies and damping as a function of the speed almost overlapping, as shown in Figure 3. In this figure, the frequency is expressed in rad s^{-1} in order to facilitate the comparison with the results of [26], where this unit of measurement is used. In the subsequent paragraphs, however, the frequency will be expressed in Hz. Additionally, the selection of the speed range in Figure 3 was made according to [26].

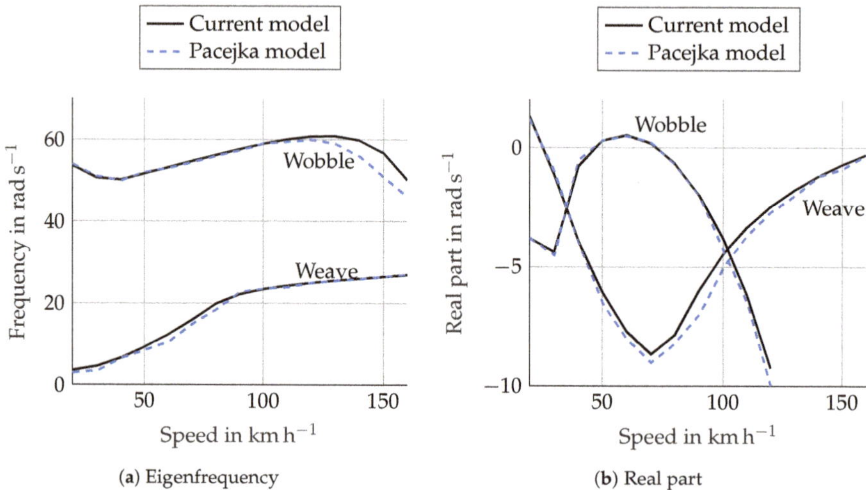

(**a**) Eigenfrequency (**b**) Real part

Figure 3. Model validation with eigenvalue analysis: comparison between the current model (parametrised with the same data of the Pacejka model) and the model of Pacejka.

4. Results

This section shows the eigenvalue analysis and the time response for the presented model. Starting from these plots, some physical relations will be derived and will help to understand the important phenomena influencing the behaviour of the wobble and weave eigenmodes as a function of vehicle speed.

4.1. Wobble

Section 2 describes how the front fork lateral flexibility (bending-compliance about the x axis) strongly influences the stability of wobble. This difference is shown in Figure 4. In order to isolate the influence of the fork compliance, the curve relative to the rigid fork was obtained with a 9-DoF model, wherein the DoF used to represent the motorcycle's stiffnesses, as described in Section 3.1, have been eliminated. Moreover, the rider's DoF are not present. In the curve relative to the flexible fork, only the DoF α_{ff} has been added, so that the DoF are 10 in total. With the parameter set used in the present work and in the speed range considered, the wobble mode remains well damped. This is due to the relatively high value of the steering damper. In the following part of this section, the reason for such a drastic behaviour change between rigid and flexible fork is addressed. In particular, a possible cause is presented which partially differs from the explanation provided by Cossalter [2] (p. 280) and Spierings [6].

A key concept described in Section 2 is the correlation between tyre properties and wobble. In particular, a tyre model with cornering stiffness is necessary to reproduce wobble [3,16,17,20]. Therefore, the tyre and specifically the lateral force due to the slip angle is a fundamental factor when considering wobble stability. However, what is the reason for the behaviour inversion in Figure 4 and how can it be attributed to the tyre response? Equation (2) shows how the slip angle depends on the lateral velocity of the wheel contact point v_{y_C}. The fork lateral flexibility allows some lateral motion of the wheel because of the rotation about the revolute joint axis (fork x axis), thereby influencing v_{y_C}.

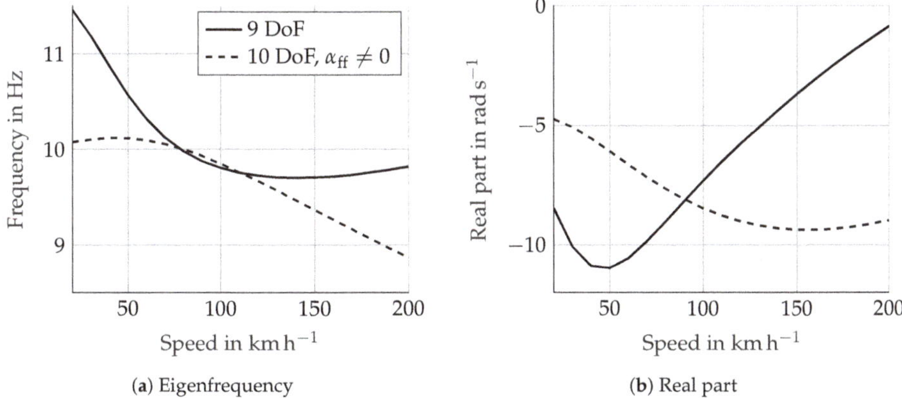

(**a**) Eigenfrequency (**b**) Real part

Figure 4. Influence of front fork flexibility on wobble.

This concept is shown schematically in Figure 5. The variation of v_{y_C} due to the fork lateral flexibility can be expressed as:

$$\Delta v_{y_C} = h\,\dot{\alpha}_{ff}. \tag{8}$$

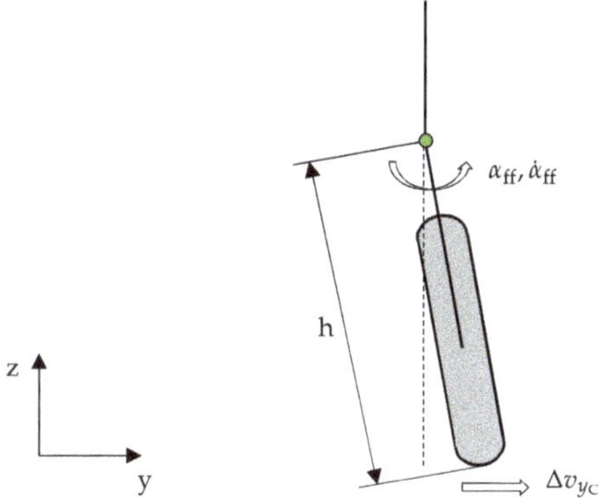

Figure 5. Variation of the lateral velocity at the contact point due to lateral fork flexibility.

To prove the effect of Δv_{y_C}, it is compensated by subtracting its value from v_{y_C} in Equation (2). Figure 6 shows the corresponding eigenvalue analysis with the dots-marked curve. The compensation of Δv_{y_C} remarkably stabilises wobble. Moreover the curve shows the same behaviour for the 9-DoF model. At this point a conclusion can be drawn: the fork lateral flexibility influences the wobble stability through its effect on the lateral velocity v_{y_C} of the contact point.

In a subsequent step, it is of particular interest to observe whether Δv_{y_C} increases or reduces the value of v_{y_C}. Figure 7 shows a time simulation at 50 km h^{-1} with a steer torque impulse as excitation. The full line represents v_{y_C} obtained from Equation (5); the dashed line was obtained with $v_{y_C} - \Delta v_{y_C}$, whereby Δv_{y_C} was derived from Equation (8). The subtraction of Δv_{y_C} increases the value of v_{y_C}: Δv_{y_C}, which is normally included in v_{y_C}, thereby reduces the amplitude of v_{y_C}. The tyre-lateral force is proportional to the slip angle, which is proportional to v_{y_C} (Equation (2)). This force tends to oppose the wheel motion due to the wobble oscillation. Therefore, the fork flexibility introduces a Δv_{y_C}, which reduces the tyre-lateral force. This explains why the curve with $\Delta v_{y_C} = 0$ in Figure 6 is significantly more stable than the 10-DoF case with $\Delta v_{y_C} \neq 0$.

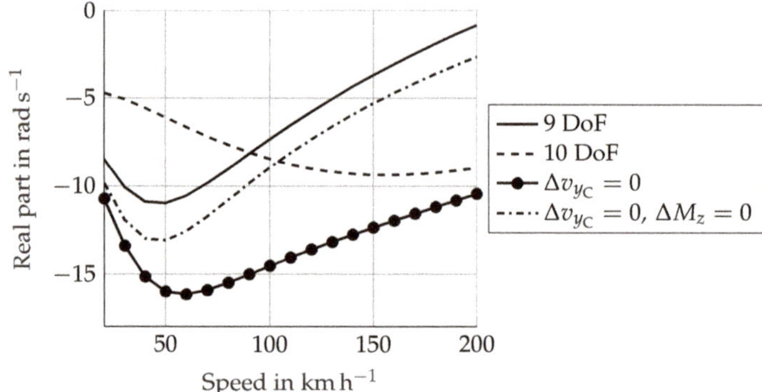

Figure 6. Effect of compensating Δv_{y_C} on wobble damping. In the dotted curve only Δv_{y_C} is compensated. In the dash-dotted curve both Δv_{y_C} and ΔM_z (with Equation (9)) are compensated.

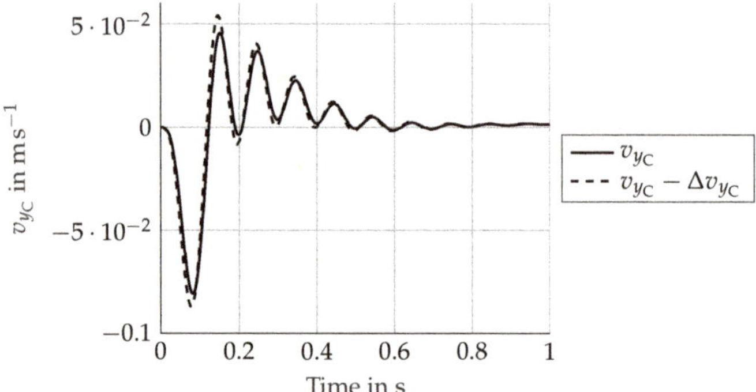

Figure 7. Variation of the lateral velocity of the contact point v_{y_C} due to front fork flexibility. Time simulation at 50 km h^{-1} and with a steer torque impulse as excitation.

Even if the curve with compensated Δv_{y_C} in Figure 6 assumes the same qualitative behaviour as the 9-DoF model, the damping values are still significantly different. In order to get a curve similar to that of the 9-DoF model, the gyroscopic moment introduced by fork bending must be compensated. The moment acts about the steering axis and can be calculated with Equation (9) (this formula differs from the complete Euler equations (compare to Equation (10)), whose left-hand side defines the whole gyroscopic effects. Equation (9) is used here because its simple structure allows one to compensate the gyroscopic effects about the steering axis as external moments in the used multibody simulation software), as shown by Cossalter [5]. This formula is also well-known in the literature, as it is present in the book on vehicle dynamics by Mitschke [44]. The symbol Δ in Equation (9) indicates that the formula only refers to the gyroscopic moment due to the fork lateral flexibility.

$$\Delta M_z = \dot{\alpha}_{\mathrm{ff}}\, \omega_{y_{\mathrm{fw}}}\, I_{yy_{\mathrm{fw}}}, \tag{9}$$

where $\omega_{y_{\mathrm{fw}}}$ is the front wheel rotation speed and $I_{yy_{\mathrm{fw}}}$ is its polar moment of inertia. The dash-dotted curve in Figure 6 shows the effect of compensating ΔM_z. As expected, the damping curve gets very close to the 9-DoF model. The remaining offset can be attributed to other secondary phenomena, which have not yet been completely identified.

In summary, the influence of the fork bending compliance on wobble can be explained by considering two main phenomena:

1. Gyroscopic moment;

2. Variation of the lateral speed of the wheel contact point.

Both of them originate through the rotation α_{ff} and its time derivatives.

When observing the curve of the 10-DoF case in Figure 6, it is evident that it is less stable than the 9-DoF case only up to about 80 km h^{-1}. Above this speed, the higher stability of the 10-DoF model compared to the 9-DoF model can be justified with the same consideration made by Cossalter [5]: the gyroscopic moment ΔM_z about the steering axis stabilises the wobble with increasing speed.

At this point another question may be raised: Which phenomenon is responsible for the decreasing wobble damping shown by the 9-DoF model and the model with compensated Δv_{y_C} compared to the 10-DoF one? To answer this question, a model containing only the front fork and front wheel has been used. With this model two cases are studied: two DoF characterised by the rigid fork and three DoF with a flexible fork; the first model correlates with the 9-DoF model, and the second with the 10-DoF model.

Figure 8a shows the damping curves for these two models. Different variations were made: the wheel inertia tensor $\boldsymbol{I}_{fw}$ was set to 1% of the original value (to eliminate the gyroscopic moments) and Δv_{y_C} has been compensated as in Figure 6. Some important conclusions can be drawn:

- The 2-DoF model did not show a decrease in damping with increasing speed, as did the 9-DoF model.
- The damping of the 3-DoF model grew more rapidly than the one of the 10-DoF model.
- The case $\boldsymbol{I}_{fw} = 1\,\%$, $\Delta v_{y_C} = 0$ is very close to the curve of the 2-DoF model, as was expected by to theoretical observations. In fact, the 2-DoF model had no gyroscopic moments about the steering axis, as no rotation about the fork x axis was present. In Figure 6 the curve with $\Delta v_{y_C} = 0$ shows a remarkable offset compared to the 9-DoF curve because of the gyroscopic moments. With $\boldsymbol{I}_{fw}$ set to 1 % they almost vanish, resulting in a smaller offset.

The difference in behaviour between the 2 and 9-DoF models can be justified by the additional roll motion of the 9-DoF case. This motion produces additional gyroscopic moments which apparently destabilise the wobble with increasing speed. These moments are also clearly present in the 10-DoF model. However, in this case they are compensated by the gyroscopic moment produced by the fork flexibility. The opposition of these two effects is underlined by the saturation shown in the 10-DoF model at about 150 km h^{-1}, which is not present in the 3-DoF model.

In summary, the wobble damping behaviour of a real motorcycle is determined by three principal mechanisms. The front fork flexibility produces a lateral movement of the front wheel, which can be represented with a rotation α_{ff}. This reduces the tyre-lateral force and thus decreases the wobble damping at low speed compared to a rigid fork. When the speed increases, the gyroscopic moment caused by the rotation α_{ff} increases in magnitude and stabilises the wobble, thereby making the 10 DoF model more stable than the 9-DoF model. The third mechanism is related to the gyroscopic moments caused by the motorcycle's roll motion. They destabilise the wobble with increasing speed, thereby explaining why the wobble damping in the 9-DoF model decreases with increasing speed. This does not happen in the 10-DoF model because the stabilising gyroscopic moment caused by α_{ff} outweighs the destabilising effect.

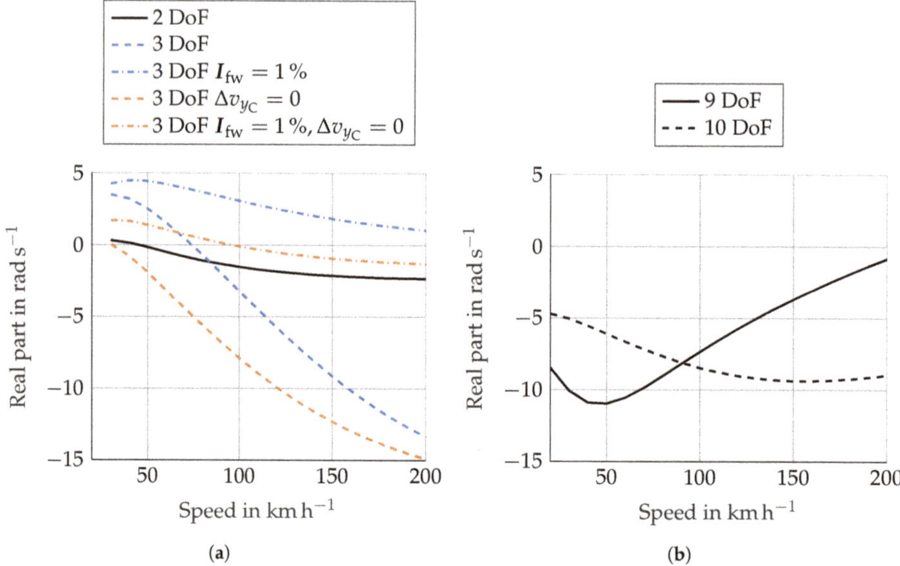

Figure 8. (**a**) Wobble damping for the 2 and 3-DoF models with variation of the tensor I_{fw} and compensation of Δv_{y_C}. The curve with $I_{fw} = 1\,\%$ for the 2-DoF model is identical to that with $I_{fw} = 100\,\%$ so it was not plotted. (**b**) Figure 4b copied to facilitate the comparison.

4.2. Weave

The weave mode was already captured in its essence by the very first motorcycle model by Whipple [13] (p. 195). Moreover, there is no evidence of parameters which produce a behaviour inversion of the damping curve, as can be observed for the wobble mode. This already suggests that the weave mode reproduces in some way the "basic" motorcycle behaviour. This idea was also proposed by Schröter [45] (p. 28), who describes the weave mode as a degenerated dynamic stabilisation process involving steering, roll and yaw oscillations. This can be shown with a simple time simulation where the upright riding motorcycle is excited with a lateral force impulse applied to the frame. Assuming that the wobble is well damped, the motorcycle reacts to this excitation with a weave oscillation, which may diverge or stabilise depending on the motorcycle's speed and parameters.

Figure 9 shows the frequency and damping of the weave mode as a function of the speed. The effect of the rider is analysed in Section 4.3, so this figure refers to the case with a rigid rider; the other DoF are all present. In order to represent the saturation shown by the damping curve at high speed, the speed range differs from the one in Figure 4. The behaviour of the weave damping shows some interesting characteristics: below 70 km h^{-1} weave is stabilised with increasing speed; above it is destabilised. However, above 220 km h^{-1} the weave damping shows a plateau. This particular behaviour suggests that the weave mode changes with speed. Moreover, the speed dependency could be explained by a certain influence of the gyroscopic effects, which are also speed dependent. The gyroscopic effects are obtained with the whole left-hand side of the Euler equations, which is shown in Equation (10) for the front wheel.

$$I_{fw}\dot{\omega}_{fw} + \omega_{fw} \times (I_{fw}\omega_{fw}) = M_{ext}, \tag{10}$$

where ω_{fw} is the rotational velocity vector of the front wheel and I_{fw} is the front wheel inertia tensor.

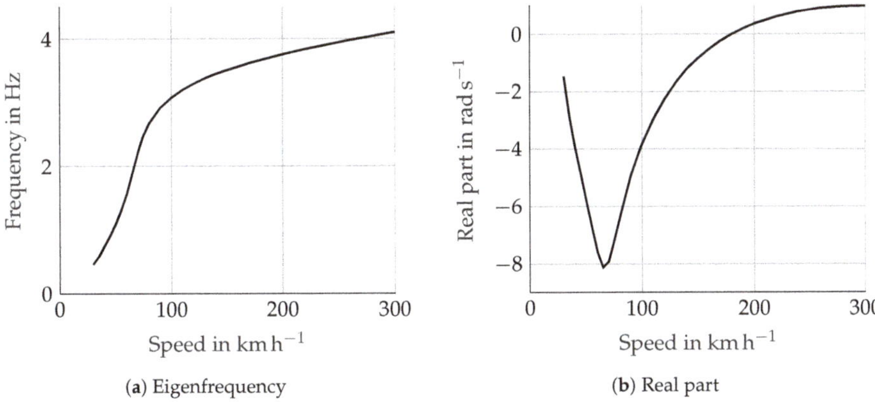

(**a**) Eigenfrequency (**b**) Real part

Figure 9. Weave eigenfrequency and damping as a speed function.

Figure 10 further illustrates this concept. In this figure, the weave eigenvector is shown with the so-called compass plot. The arrows are named phasors and represent how the motorcycle's DoF are involved in the weave oscillation. An arrow's length is the magnitude of the motion of the related DoF. In order to maintain the information about the absolute amplitude of the phasors, no normalisation is present. The phasor's angle shows the phase. In the following paragraphs the relative angle between the phasors is called relative phase.

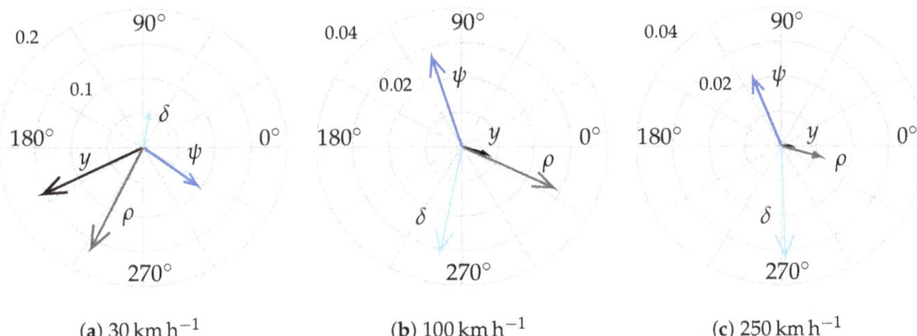

(**a**) 30 km h^{-1} (**b**) 100 km h^{-1} (**c**) 250 km h^{-1}

Figure 10. Compass plot of the weave eigenvector at different speeds. δ is the steering angle, ψ is the yaw angle, ρ is the roll angle and y is the lateral displacement.

At low speed, (Figure 10a) the roll (ρ) and lateral displacement (y) show big amplitudes. The latter is due to the small gyroscopic effects, so that if weave is excited (for example, with a lateral impulse), the motorcycle describes a "slalom" at low frequency (see Figure 9) involving a significant lateral displacement. This can be described as a low-frequency "self-stabilising" motion, which the literature already considers as weave.

As the speed increases (Figure 10b), the relative phase between the motions significantly changes, whereby the lateral displacement is no longer important. The other DoF show a similar amplitude. Finally, when the speed increases further (Figure 10c), the relative phase does not change remarkably. The most important observation, however, is the reduction of the roll (ρ) amplitude with increasing speed, which can be seen when comparing the related phasor in Figure 10b,c. This can also be observed in the time simulation. If the motorcycle is excited by a lateral impulse at high speed, the resulting weave oscillation does not show a significant roll motion, while the front wheel and the frame rotate around the steering and vertical axis almost in opposition of phase, as shown by the vectors δ, ψ in Figure 10c. The reduction in roll motion with increasing speed is also shown by a free rolling wheel. This phenomenon can be explained considering the gyroscopic effects which increase with speed, thereby preventing the wheel (or the motorcycle) capsizing.

Which are the possible causes of the peculiar damping behaviour shown in Figure 9b? Some useful knowledge can be derived by an eigenvalue analysis with reduced wheel inertia. The correlation to Figure 10 is shown in Figure 11, for which the total wheel inertia tensors $\boldsymbol{I}_{\text{fw}}, \boldsymbol{I}_{\text{rw}}$ was set to 10 % of the original tensors. The first consideration resulting from the comparison is that in Figure 11 the relative phase between the phasors no longer changes with speed, as it happens in Figure 10. The plot at 30 km h^{-1} is not present, as the weave mode is no longer available at this speed. It is important to point out that, because of this massive inertia reduction, the gyroscopic effects almost disappear. Two main observations can be made:

1. The missing weave mode at low speed can be attributed to the (almost) missing gyroscopic effects. The time simulation at 30 km h^{-1} with a lateral force impulse shows that the motorcycle does not react to the excitation with a weave oscillation, but it capsizes. With the original inertia values the gyroscopic effects are present, despite the low speed. They prevent the motorcycle from capsizing immediately after the excitation and producing the low-frequency "self-stabilising" motion.
2. The gyroscopic effects are also the main cause for the change in the weave eigenvector shown in Figure 10. In fact, when they are very small, as in Figure 11, the relative phase between the eigenvector components is no longer speed dependent.

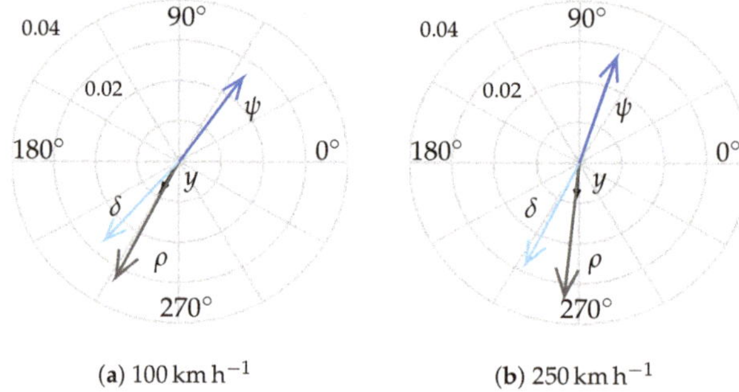

(**a**) 100 km h^{-1} (**b**) 250 km h^{-1}

Figure 11. Compass plot of the weave eigenvector at different speeds with $\boldsymbol{I}_{\text{fw}}, \boldsymbol{I}_{\text{rw}} = 10\,\%$.

The correlation with Figure 9 is shown in Figure 12. The speed range starts now from 80 km h^{-1}, as for the following reasoning only this range is necessary. Moreover, the weave with reduced wheel inertia is not present up to about 50 km h^{-1} and in the range 50–80 km h^{-1} it is extremely well damped (real part > 25 rad s^{-1}). Similarly to the full inertia case, the damping continues to decrease with increasing speed in the small inertia case shown in Figure 12b. As the gyroscopic effects are almost missing, the reason for this behaviour can be determined by investigating in the tyre response. The tyre-lateral force is proportional to the slip angle which is in turn proportional to the ratio between v_{y_C} and v_{x_C} (Equation (2)). The authors verified that v_{y_C} mainly depends on the kinematic steering angle (see [2] for its definition), which is proportional to the steering angle. Figure 11 shows that above a certain speed (100 km h^{-1}) the amplitude of steering motion in the weave eigenvector remains fairly constant. The numerator of α_{ss} in Equation (2) is therefore almost constant with speed. The denominator of α_{ss} is given by v_{x_C}, which clearly increases with speed. As a result, α_{ss} decreases with increasing speed and so does the tyre-lateral force. The tyre-lateral forces tend to bring the motorcycle back to the equilibrium position, thereby stabilising the weave. The decreasing tyre force is therefore responsible for the decreasing weave damping and the hyperbolic behaviour of the dashed curve in Figure 12b.

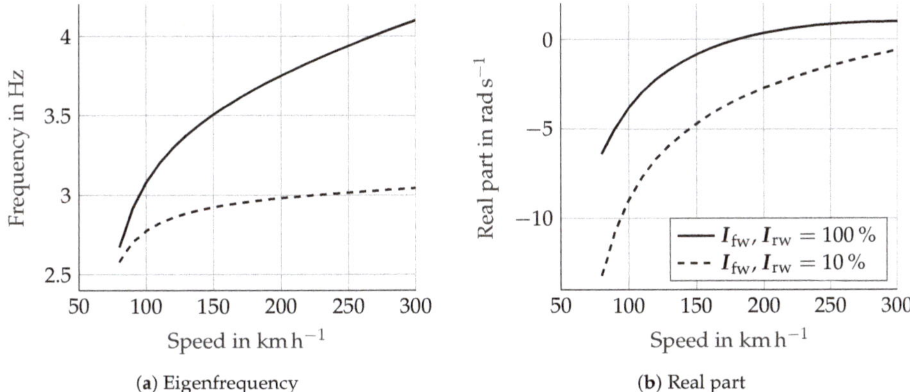

(**a**) Eigenfrequency (**b**) Real part

Figure 12. Comparison between the eigenvalue analysis with original wheel inertia tensors and reduced inertia tensors.

Considering again the curve with the original wheel inertia in Figure 12, the first observation is the smaller weave damping. This can be explained by taking into account the additional impact of the gyroscopic effects. The gyroscopic effects of the front wheel about the steering axis are in counterphase with respect to the tyre moments, as Figure 13 shows, therefore they work against them. For this reason, the weave damping is lower compared to the reduced inertia case. The gyroscopic effects also justify the slower damping decrease in the case with original wheel inertia (see Figure 12). In fact, due to the progressively changing eigenvector (Figure 10), i.e., the decreasing roll motion amplitude, the gyroscopic effects increase underproportionally with the speed. For example, in a time simulation with lateral force excitation, the first peak of the front wheel gyroscopic effects about the z axis increase from 100 to 150 $\mathrm{km\,h^{-1}}$ by 37.5%, while they only increase by 14.5% from 250 to 300 $\mathrm{km\,h^{-1}}$. As a consequence, they also underproportionally reduce the tyre forces with increasing speed. This, combined with the decreasing tyre forces, leads to the mentioned saturation above 220 $\mathrm{km\,h^{-1}}$, which is not present in the reduced inertia case.

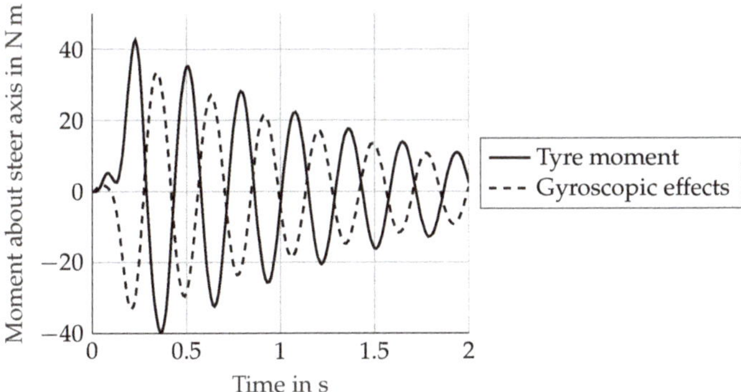

Figure 13. Front wheel gyroscopic effects and tyre moments about the steering axis. The time simulation was obtained with a lateral force excitation on the motorcycle frame at the speed of 150 $\mathrm{km\,h^{-1}}$.

The last characteristic to be observed is the increase in damping shown below 80 $\mathrm{km\,h^{-1}}$ in Figure 9b. The authors' belief is that in this lower speed range the weave mode can still be seen as the already mentioned low-speed "self-stabilising" motion. When increasing the speed, this motion progressively changes to the "real" weave, thereby producing the observed increase in damping.

Figure 14 illustrates this fact. Following the three compass plots with increasing speed, one can see that the phase and amplitude of the phasors progressively change; i.e., the weave changes from the "self-stabilising" motion to the classical weave. In fact, above 100 $km\,h^{-1}$ the relative angle between the phasors no longer changes significantly, as the comparison of Figure 10b,c shows.

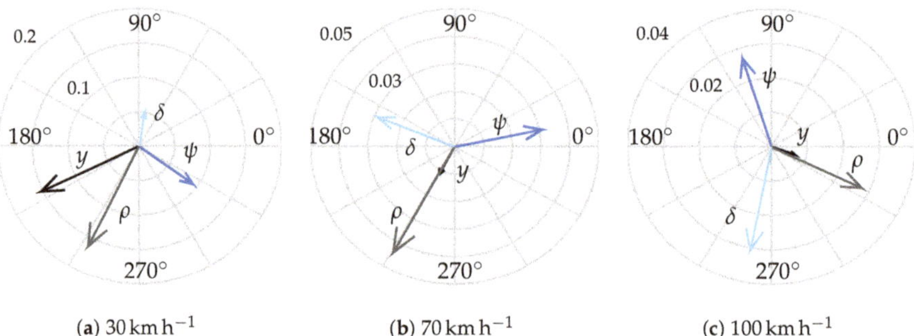

(a) 30 $km\,h^{-1}$ (b) 70 $km\,h^{-1}$ (c) 100 $km\,h^{-1}$

Figure 14. Compass plot of the weave eigenvector at different speeds.

4.3. Rider Influence

As explained in Section 3.3, the literature offers different examples of rider modelling. These references often analyse the influences of the rider on weave and wobble stability. For example, reference [42] investigates the effect of the rider yaw combined with the connection to the handlebar. Reference [7] proposes the eigenvalue analysis with the 3-DoF rider also used in the present work. In order to obtain a better overview of the single influences, this section aims to briefly summarise the essential differences between the three different DoF models used for rider modelling. This is shown in Figure 15. The effect of the single DoF can be summarised as follows:

- The rider lean α_{ri} stabilises the weave and destabilises the wobble at high speed. The increased stability of weave is a reasonable result because during the weave oscillation the rider lean is almost in counterphase to the motorcycle's roll [41], thereby damping out this motion.
- The rider yaw γ_{ri} combined with the connection to the handlebar destabilises the weave, massively stabilises the wobble and also increases its eigenfrequency. This is in accordance with [42] and is a reasonable result, as the connection acts in a similar way to a steering damper, which also stabilises the wobble while destabilising the weave.
- The rider's hip lateral motion y_{ri} hardly affects the eigenmodes, as also reported by [31]. The only remarkable influence is the increase in the wobble frequency at high speed.

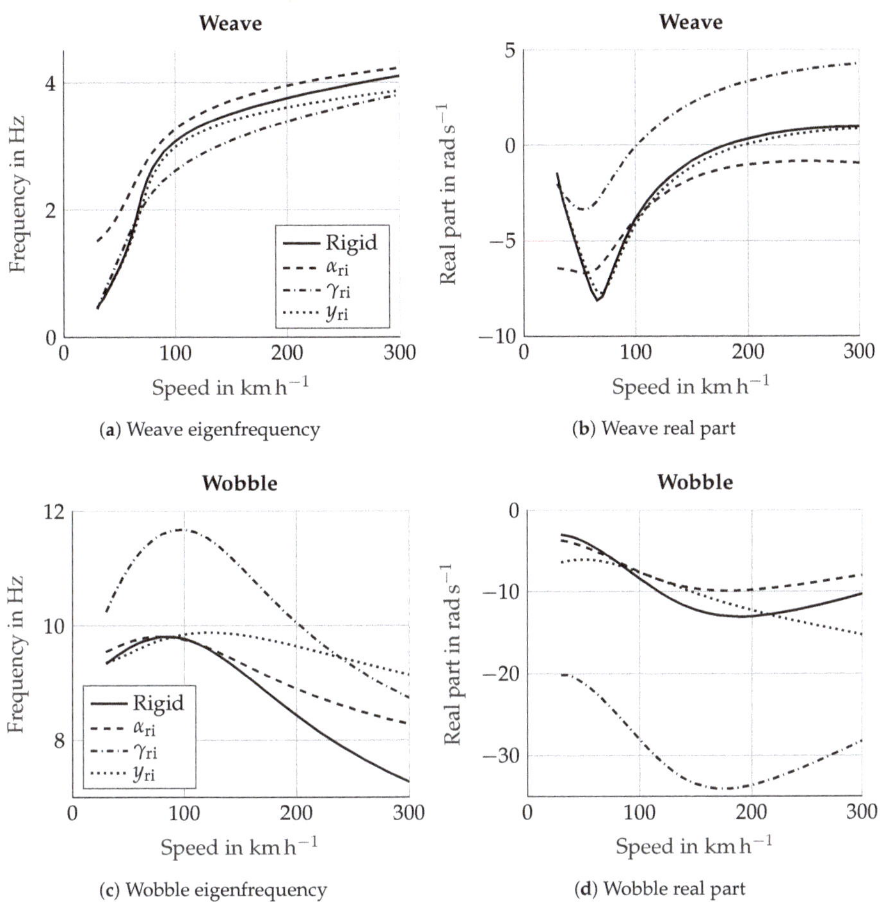

(**a**) Weave eigenfrequency

(**b**) Weave real part

(**c**) Wobble eigenfrequency

(**d**) Wobble real part

Figure 15. The rffects of the single rider's DoF on the weave and wobble modes.

Figure 16 shows the effect of the whole rider model, containing the three DoF $\alpha_{ri}, y_{ri}, \gamma_{ri}$. In the weave the effect of α_{ri} overcomes the others, thereby causing the already mentioned stabilisation at high speed. When concerning wobble, the effect of γ_{ri} and the connection with the handlebar are dominant, leading to an increase in both frequency and damping.

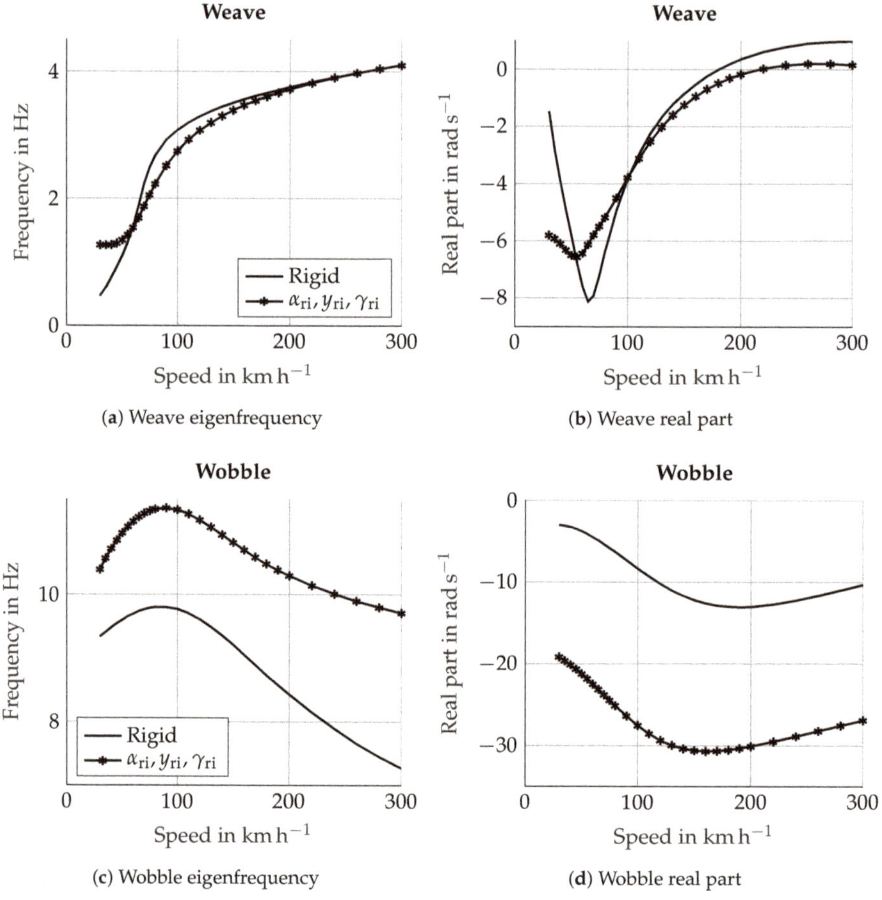

(**a**) Weave eigenfrequency

(**b**) Weave real part

(**c**) Wobble eigenfrequency

(**d**) Wobble real part

Figure 16. Effect of the whole rider model (3-DoF) on the weave and wobble modes.

5. Conclusions and Outlook

The present work further analysed the physical phenomena behind the wobble and weave eigenmodes. In particular, the effect of the front fork bending compliance has been discussed and a possible justification for the well-known weave damping behaviour with increasing speed is provided. Finally, the effect of the single rider's DoF on the weave and wobble mode has been shown.

The first aspect has been addressed before by Cossalter [5] and Spierings [6]. They noticed that the modelling of the fork bending compliance allows very similar results to the real driving experience to be obtained: the wobble mode is unstable at low speed and stabilises with increasing speed. Without this parameter, the simulation results are not realistic. Cossalter [5] justifies this behaviour considering the superposition of two effects, both caused by the fork bending compliance: this compliance alone is destabilising but it also produces a gyroscopic moment about the steering axis that stabilises the wobble. The first effect dominates at low speeds, while the second prevails at high speed. The present work provides an additional perspective for explaining the effect of the fork bending compliance, though maintaining the validity of the previous results [5]. The additional insight is given by the tyre behaviour. It was demonstrated that the lateral motion of the wheel contact point caused by the fork bending compliance reduces the lateral component of the contact point velocity

v_{yC}. The tyre-lateral force is proportional to this velocity component through the slip angle. Reducing v_{yC} also reduces the tyre force. This causes a reduction in the wobble damping at below 80 km h^{-1}. At higher speeds, the effect of the gyroscopic moments introduced by the fork flexibility leads to increasing wobble damping with increasing speed, as explained by Cossalter [5].

The weave damping behaviour is well-known and involves a progressive stabilisation up to about 80 km h^{-1} (with the present parameters), then the damping decreases with increasing speed until it reaches a saturation above 220 km h^{-1}. This peculiar behaviour can now be explained. The weave eigenvector changes with speed. In the lower speed range this corresponds to a change from a low frequency self-stabilising motion involving a lot of frame lateral displacement to a weave oscillation where the lateral motion is no longer significant. The eigenvector change is supposed to be the main cause of the damping increase below 80 km h^{-1}. Above this lower speed range, the weave eigenvector does not change in the relative phase between phasors, but the roll component becomes progressively smaller. This influences the gyroscopic effects, which underproportionally increase with speed, thereby causing the plateau above 220 km h^{-1}.

The multibody model used in this work was provided with a functionality to select the different DoF. This gives the chance to investigate the separated influence of the single DoF used in the rider model. In particular, the rider lean stabilises the weave and slightly destabilises the wobble; the rider yaw plus the connection with the handlebar destabilises the weave and remarkably stabilises the wobble. The whole 3-DoF rider model produces the same effect of the rider lean as regards weave and the same effect of rider yaw as regards wobble.

The contributions of the present paper to general motorcycle dynamics knowledge can be summarised as follows. A review of the literature on motorcycle dynamics and stability behaviour; this knowledge has been interpreted in order to derive some minimal prerequisites to the motorcycle model, with the aim of conducting a stability analysis. After that, the effect of the tyre response on the wobble damping was analysed, thereby leading to considerations that fuse together with and partially complete the theory exposed by Cossalter [5] and Spierings [6]. Moreover, a possible justification for the well-known weave damping behaviour was given, which was not found in the literature. Finally, the influence of the rider model on stability, which was already studied in previous works, has been summarised, thereby facilitating the interpretation of the effect of the single rider's DoF.

Further development of the present paper could involve the inclusion of a "flexible body" to faithfully reproduce the different frame flexibilities without using lumped stiffnesses.

Author Contributions: F.P. is the leading author who wrote the whole paper. The project and problem formulation were conceptualised by F.P. and D.W.; these authors also built the motorcycle model and produced the results presented in this paper. D.W., A.E., F.D. and A.G. revised the paper critically for important intellectual content. Conceptualisation, F.P. and D.W.; formal analysis, F.P. and D.W.; methodology, F.P. and D.W.; supervision, A.E., D.W., A.G. and F.D.; validation, F.P. and D.W.; visualization, F.P.; writing—original draft, F.P.; writing—review and editing, A.E., D.W., A.G. and F.D. All authors have read and agreed to the published version of the manuscript.

Funding: This research received no external funding.

Acknowledgments: The motorcycle model is completely based on the multibody simulation software "MBSim". The author would like to thank Martin Förg (University of Landshut), who is one of the developers of "MBSim", for the support and suggestions during the building phase of the model. Several structural choices, such as the possibility to select the model DoF, were possible thanks to his contribution.

Conflicts of Interest: The authors declare no conflict of interest.

Nomenclature/Notation

Symbol	Unit	Description
α	rad	Tyre slip angle
α_{ff}	rad	Front fork bending about fork x axis
α_{fr}	rad	Frame torsion at steering head joint
α_{ri}	rad	Rider's upper body lean angle
α_{sw}	rad	Swingarm torsion
β_{ff}	rad	Front fork bending about fork y axis
γ_{ff}	rad	Front fork torsion
γ_{ri}	rad	Rider's upper body yaw rotation
δ	rad	Steer rotation
ΔM_z	N m	Gyroscopic moment due to fork flexibility
Δv_{y_C}	m s^{-1}	Variation of the lateral velocity of the wheel contact point
Δz_{tyre}	m	Tyre carcass compression
θ_{fw}	rad	Front wheel spin angle
θ_{rw}	rad	Rear wheel spin angle
κ	-	Longitudinal slip
ρ	rad	Main frame roll angle
ρ_{w}	rad	Wheel camber angle
σ	m	Relaxation length
ϕ	rad	Main frame pitch angle
ψ	rad	Main frame yaw angle
${}_{\text{B}}\boldsymbol{\omega}_{\text{A}}$	rad s^{-1}	Rotational velocity vector of the generic point A in the generic frame B
$\boldsymbol{\omega}_{\text{fw}}$	rad s^{-1}	Front wheel rotational velocity vector
ω_y	rad s^{-1}	Wheel rotational speed
c_z	N m^{-1}	Tyre carcass vertical stiffness
d_z	N s m^{-1}	Tyre carcass vertical damping
${}_{\text{ref}}\boldsymbol{e}_i$	-	Unit vector of the *i*th axis of the "ref" reference frame
$F_{y_{\text{f}}}$	N	Front tyre-lateral force
$F_{y_{\text{r}}}$	N	Rear tyre-lateral force
F_z	N	Tyre vertical load
$\boldsymbol{I}_{\text{fw}}$	kg m^2	Front wheel inertia tensor
$\boldsymbol{I}_{\text{rw}}$	kg m^2	Rear wheel inertia tensor
M_δ	N m	Steer torque
$M_{z_{\text{f}}}$	N m	Front tyre aligning moment
$M_{z_{\text{r}}}$	N m	Rear tyre aligning moment
${}_{\text{B}}\boldsymbol{S}_{\text{A}}$	-	Transformation matrix from system A to system B
${}_{\text{B}}\boldsymbol{v}_{\text{A}}$	m s^{-1}	Absolute velocity vector of the generic point A in the generic frame B
v_{x_C}	m s^{-1}	Longitudinal speed of the wheel contact point
v_{y_C}	m s^{-1}	Lateral speed of the wheel contact point
v_{z_C}	m s^{-1}	Vertical speed of the wheel contact point
y_{ri}	m	Rider's lower body lateral motion
z_{f}	m	Front suspension travel
z_{r}	m	Rear suspension travel

Appendix A

Appendix A.1. Geometric Parameters

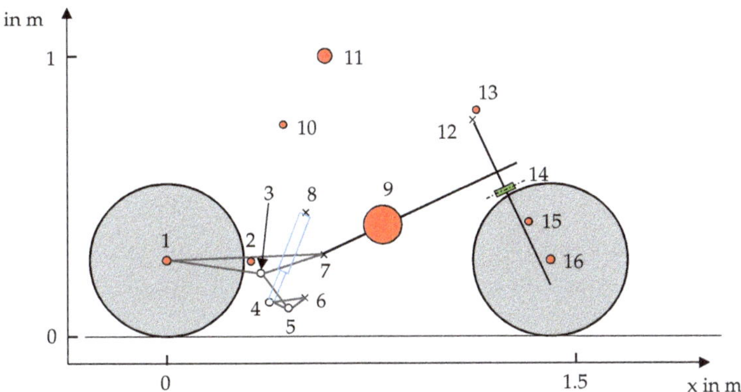

Figure A1. Scheme for the geometric parameters.

Table A1. Dimensions.

Point	x in m	z in m
1	0	0.297
2	0.196	0.3113
3	0.3722	0.2748
4	0.4443	0.1782
5	0.4946	0.1522
6	0.539	0.1878
7	0.549	0.3608
8	0.487	0.4888
9	0.6779	0.4724
10	0.364	0.8438
11	0.415	1.14
12	1.173	0.749
13	1.1803	0.7818
14	1.3036	0.5209
15	1.3125	0.4357
16	1.41	0.282

Table A2. Wheel geometry.

Parameter	Symbol	Value in m
Front wheel radius	r_{fw}	0.282
Front wheel crown radius	$r_{\mathrm{c\,fw}}$	0.06
Front wheel rim radius	$r_{\mathrm{rim\,fw}}$	0.222
Rear wheel radius	r_{rw}	0.297
Rear wheel crown radius	$r_{\mathrm{c\,rw}}$	0.095
Rear wheel rim radius	$r_{\mathrm{rim\,rw}}$	0.202

Appendix A.2. Tyre Coefficients

Table A3. Tyre vertical stiffness and damping.

Parameter	Symbol	Value
Front tyre vertical stiffness	$c_{z\,\mathrm{fw}}$	130,000 N m^{-1}
Front tyre vertical damping	$d_{z\,\mathrm{fw}}$	800 N s m^{-1}
Rear tyre vertical stiffness	$c_{z\,\mathrm{rw}}$	141,000 N m^{-1}
Rear tyre vertical damping	$d_{z\,\mathrm{rw}}$	800 N s m^{-1}

The other tyre coefficients used in the present work were taken from [12].

Appendix A.3. Stiffness and Damping

Table A4. Lumped stiffness and damping values. The DoF refer to Figure 1.

DoF	Stiffness	Damping
Frame torsion α_{fr}	163,000 N m rad^{-1}	100 N m s rad^{-1}
Front fork bending α_{ff}	55,300 N m rad^{-1}	20 N m s rad^{-1}
Front fork bending β_{ff}	55,300 N m rad^{-1}	20 N m s rad^{-1}
Front fork torsion γ_{ff}	16,000 N m rad^{-1}	20 N m s rad^{-1}
Swingarm torsion α_{sw}	63,700 N m rad^{-1}	20 N m s rad^{-1}
Rider rotation about x axis α_{ri}	380 N m rad^{-1}	34 N m s rad^{-1}
Rider rotation about z axis γ_{ri}	75.8 N m rad^{-1}	4.79 N m s rad^{-1}
Rider arms	1053.4 N m rad^{-1}	19.28 N m s rad^{-1}
Rider translation along y axis y_{ri}	38,416 N m rad^{-1}	981 N s m^{-1}

Appendix A.4. Suspensions

Table A5. Suspension's parameters.

Parameter	Value
Front spring stiffness	17,000 N m^{-1}
Front spring unloaded length	0.35 m
Front spring preload	876.231 N
Front compression damping force at 2 m s^{-1}	203 N
Front rebound damping force at 2 m s^{-1}	865.8 N
Rear spring stiffness	58,570 N m^{-1}
Rear spring unloaded length	0.3435 m
Rear spring preload	248.87 N
Rear compression damping force at 1 m s^{-1}	9600 N
Rear rebound damping force at 1 m s^{-1}	13,700 N

Appendix A.5. Mass and Inertia

Table A6. Mass parameters.

Parameter	Symbol	Value
Frame mass	m_{fr}	153.716 kg
Rider lower body mass	$m_{ri\,lo}$	11.414 kg
Rider upper body mass	$m_{ri\,ub}$	33.68 kg
Upper fork mass	$m_{ff\,up}$	9.3 kg
Lower fork mass	$m_{ff\,lo}$	6.5 kg
Front wheel mass	m_{fw}	11.9 kg
Swingarm mass	m_{sw}	8 kg
Rear wheel mass	m_{rw}	14.7 kg

Table A7. Inertia parameters.

Parameter	Symbol	Value
Frame inertia	$\boldsymbol{I}_{fr}$	$\begin{bmatrix} 9.2774 & 0 & 2.267 \\ 0 & 19.0193 & 0 \\ 2.267 & 0 & 13.6702 \end{bmatrix}$ kg m^2
Rider lower body inertia	$\boldsymbol{I}_{ri\,lo}$	$\begin{bmatrix} 0.1163 & 0 & -0.0055 \\ 0 & 0.0942 & 0 \\ -0.0055 & 0 & 0.1036 \end{bmatrix}$ kg m^2
Rider upper body inertia	$\boldsymbol{I}_{ri\,ub}$	$\begin{bmatrix} 1.4280 & 0 & -0.443 \\ 0 & 1.347 & 0 \\ -0.443 & 0 & 0.916 \end{bmatrix}$ kg m^2
Upper fork inertia (z axis ≡ steer axis)	$\boldsymbol{I}_{ff\,up}$	$\begin{bmatrix} 0.5 & 0 & 0 \\ 0 & 0.4 & 0 \\ 0 & 0 & 0.13 \end{bmatrix}$ kg m^2
Lower fork inertia (z axis ≡ steer axis)	$\boldsymbol{I}_{ff\,lo}$	$\begin{bmatrix} 0.3 & 0 & 0 \\ 0 & 0.25 & 0 \\ 0 & 0 & 0.09 \end{bmatrix}$ kg m^2
Front wheel inertia	$\boldsymbol{I}_{fw}$	$\begin{bmatrix} 0.27 & 0 & 0 \\ 0 & 0.484 & 0 \\ 0 & 0 & 0.27 \end{bmatrix}$ kg m^2
Swingarm inertia	$\boldsymbol{I}_{sw}$	$\begin{bmatrix} 0.02 & 0 & 0 \\ 0 & 0.259 & 0 \\ 0 & 0 & 0.259 \end{bmatrix}$ kg m^2
Rear wheel inertia	$\boldsymbol{I}_{rw}$	$\begin{bmatrix} 0.383 & 0 & 0 \\ 0 & 0.638 & 0 \\ 0 & 0 & 0.383 \end{bmatrix}$ kg m^2

References

1. Sharp, R.S. Stability, Control and Steering Responses of Motorcycles. *Veh. Syst. Dyn.* **2001**, *35*, 291–318. [CrossRef]
2. Cossalter, V. *Motorcycle Dynamics*, 2nd ed.; Lulu Press: Morrisville, NC, USA, 2006.
3. Sharp, R.S. The stability and control of motorcycles. *J. Mech. Eng. Sci.* **1971**, *13*, 316–329. [CrossRef]
4. Koch, H. *Zusammenhänge und Einflüsse von Lebensalter, Fahrerfahrung und Motorleistung auf die Unfallverwicklung von Motorradfahranfängern*; VDI-Berichte: Düsseldorf, Germany, 1987.
5. Cossalter, V.; Lot, R.; Massaro, M. The influence of frame compliance and rider mobility on the scooter stability. *Veh. Syst. Dyn.* **2007**, *45*, 313–326. [CrossRef]
6. Splerings, P.T.J. The Effects of Lateral Front Fork Flexibility on the Vibrational Modes of Straight-Running Single-Track Vehicles. *Veh. Syst. Dyn.* **1981**, *10*, 21–35. [CrossRef]
7. Cossalter, V.; Lot, R.; Massaro, M. An advanced multibody code for handling and stability analysis of motorcycles. *Meccanica* **2011**, *46*, 943–958. [CrossRef]

8. Koenen, C.; Pacejka, H.B. The Influence of Frame Elasticity and Simple Rider Body Dynamics on Free Vibrations of Motorcycles in Curves. *Veh. Syst. Dyn.* **1981**, *10*, 70–73. [CrossRef]
9. Cossalter, V.; Doria, A.; Mitolo, L. *Inertial and Modal Properties of Racing Motorcycles*; SAE Technical Papers; SAE International: Warrendale, PA, USA, 2002.
10. Wisselmann, D. Motorrad-Fahrdynamik-Simulation. Modellbildung, Validierung und Anwendung. Ph.D. Thesis, VDI-Berichte, Dachau, Germany, 1992.
11. Schindler, T.; Förg, M.; Friedrich, M.; Schneider, M.; Esefeld, B.; Huber, R.; Zandler, R.; Ulbrich, H. Analysing Dynamical Phenomenons: Introduction to MBSim. In Proceedings of the 1st Joint International Conference on Multibody System Dynamics, Lappeenranta, Finland, 25–27 May 2010.
12. Sharp, R.S.; Evangelou, S.A.; Limebeer, D.J.N. Advances in the Modelling of Motorcycle Dynamics. *Multibody Syst. Dyn.* **2004**, *12*, 251–283. [CrossRef]
13. Limebeer, D.J.N.; Massaro, M. *Dynamics and Optimal Control of Road Vehicles*; Oxford University Press: Oxford, UK, 2018.
14. Cossalter, V.; Doria, A.; Massaro, M.; Taraborrelli, L. Experimental and numerical investigation on the motorcycle front frame flexibility and its effect on stability. *Mech. Syst. Signal Process.* **2015**, *60–61*, 452–471. [CrossRef]
15. Meirovitch, L. *Elements of Vibration Analysis*; Mc Grow Hill: New York, NY, USA, 1975.
16. Doria, A.; Favaron, V.; Taraborrelli, L.; Roa, S. Parametric analysis of the stability of a bicycle taking into account geometrical, mass and compliance properties. *Int. J. Veh. Des.* **2017**, *75*, 91–123. [CrossRef]
17. Doria, A.; Roa, S. On the influence of tyre and structural properties on the stability of bicycles. *Veh. Syst. Dyn.* **2018**, *56*, 947–966. [CrossRef]
18. Klinger, F.; Nusime, J.; Edelmann, J.; Plöchl, M. Wobble of a racing bicycle with a rider hands on and hands off the handlebar. *Veh. Syst. Dyn.* **2014**, *52*, 51–68. [CrossRef]
19. Sharp, R.S.; Alstead, C.J. The Influence of Structural Flexibilities on the Straight-running Stability of Motorcycles. *Veh. Syst. Dyn.* **1980**, *9*, 327–357. [CrossRef]
20. Plöchl, M.; Edelmann, J.; Angrosch, B.; Ott, C. On the wobble mode of a bicycle. *Veh. Syst. Dyn.* **2012**, *50*, 415–429. [CrossRef]
21. Roa, S.; Doria, A.; Muñoz, L. Optimization of the Bicycle Weave and Wobble Modes. *Am. Soc. Mech. Eng.* **2018**, *3*. [CrossRef]
22. Diermeyer, F.; Eisele, A. Lecture notes of the course Motorradtechnik. 2018, Unpublished manuscript.
23. Taraborrelli, L.; Favaron, V.; Doria, A. The effect of swingarm stiffness on motorcycle stability: experimental measurements and numerical simulations. *Int. J. Veh. Syst. Model. Test.* **2017**, *12*, 240. [CrossRef]
24. Weir, D.H.; Zeller, J.W. Experimental investigation of the transient behaviour of motorcycles. *SAE Trans.* **1979**, *88*, 962–978.
25. De Vries, E.; Pacejka, H.B. Motorcycle Tyre Measurements and Models. *Veh. Syst. Dyn.* **1998**, *29*, 280–298. [CrossRef]
26. Pacejka, H.B.; Besselink, I. *Tire and Vehicle Dynamics*, 3rd ed.; Butterworth-Heinemann: Oxford, UK, 2012.
27. Evangelou, S.A.; Limebeer, D.J.N.; Sharp, R.S.; Smith, M.C. Mechanical steering compensators for high-performance motorcycles. *J. Appl. Mech. Trans.* **2007**, *74*, 332–346. [CrossRef]
28. Sharp, R.S.; Limebeer, D.J.N. A Motorcycle Model for Stability and Control Analysis. *Multibody Syst. Dyn.* **2001**, *6*, 123–142. [CrossRef]
29. Cossalter, V.; Lot, R.; Maggio, F. A Multibody Code for Motorcycle Handling and Stability Analysis with Validation and Examples of Application. In *SAE International400 Commonwealth Drive*; SAE Technical Paper Series; SAE International: Warrendale, PA, USA, 2003. [CrossRef]
30. Limebeer, D.J.N.; Sharma, A. Burst Oscillations in the Accelerating Bicycle. *J. Appl. Mech. Trans.* **2010**, *77*, 061012. [CrossRef]
31. Roe, G.E.; Thorpe, T.E. A solution of the low-speed wheel flutter instability in motorcycles. *J. Mech. Eng. Sci.* **1976**, *18*, 57–65. [CrossRef]
32. Cossalter, V.; Lot, R.; Maggio, F. The Modal Analysis of a Motorcycle in Straight Running and on a Curve. *Meccanica* **2004**, *39*, 1–16. [CrossRef]
33. Limebeer, D.J.N.; Sharp, R.S.; Evangelou, S.A. Motorcycle Steering Oscillations due to Road Profiling. *J. Appl. Mech. Trans.* **2002**, *69*, 724–739. [CrossRef]

34. Sharp, R.S. The Influence of the Suspension System on Motorcycle Weave-mode Oscillations. *Veh. Syst. Dyn.* **1976**, *5*, 147–154. [CrossRef]
35. Lot, R. A Motorcycle Tire Model for Dynamic Simulations: Theoretical and Experimental Aspects. *Meccanica* **2004**, *39*, 207–220. [CrossRef]
36. Evangelou, S.A.; Limebeer, D.J.N.; Tomas-Rodriguez, M. Suppression of Burst Oscillation in Racing Motorcycle. *J. Appl. Mech.* **2013**, *80*, 011003. [CrossRef]
37. Karnopp, D. *Vehicle Dynamics, Stability, and Control*, 2nd ed.; CRC Press Taylor and Francis: Boca Raton, FL, USA, 2013.
38. Robbins, D.H. *Anthropometry of Motor Vehicle Occupants: Final Report October 1980–December 1983*; University of Michigan Transportation Research Inst: Ann Arbor, MI, USA, 1983; Volume 83-53-2.
39. Wunram, K.; Eckstein, L.; Rettweiler, P. *Potenzial Aktiver Fahrwerke zur Erhöhung der Fahrsicherheit von Motorrädern: Bericht zum Forschungsprojekt FE 82.328/2007*; Berichte der Bundesanstalt für Strassenwesen, Fahrzeugtechnik; Wirtschaftsverl. NW Verl. für Neue Wiss: Bremerhaven, Germany, 2011; Volume 81.
40. Sharp, R.S.; Limebeer, D.J.N. On steering wobble oscillations of motorcycles. *Proc. Inst. Mech. Eng. Part C J. Mech. Eng. Sci.* **2004**, *218*, 1449–1456. [CrossRef]
41. Doria, A.; Formentini, M.; Tognazzo, M. Experimental and numerical analysis of rider motion in weave conditions. *Veh. Syst. Dyn.* **2012**, *50*, 1247–1260. [CrossRef]
42. Cossalter, V.; Doria, A.; Lot, R.; Massaro, M. The effect of rider's passive steering impedance on motorcycle stability: Identification and analysis. *Meccanica* **2011**, *46*, 279–292. [CrossRef]
43. Sequenzia, G.; Oliveri, S.M.; Fatuzzo, G.; Calì, M. An advanced multibody model for evaluating rider's influence on motorcycle dynamics. *Proc. Inst. Mech. Eng. Part K J. Multi-Body Dyn.* **2015**, *229*, 193–207. [CrossRef]
44. Mitschke, M. *Dynamik der Kraftfahrzeuge*; Springer: Berlin/Heidelberg, Germany, 1972.
45. Schröter, K.G. Brake Steer Torque Optimized Corner Braking of Motorcycles. Ph.D. Thesis, Technische Universität Darmstadt, Darmstadt, Germany, 2015.

Article

Grousers Effect in Tracked Vehicle Multibody Dynamics with Deformable Terrain Contact Model

Francesco Mocera *,†, Aurelio Somà † and Andrea Nicolini †

Department of Mechanical and Aerospace Engineering—Politecnico di Torino, Corso duca degli Abruzzi 24, 10129 Torino, Italy; aurelio.soma@polito.it (A.S.); andrea.nicolini@polito.it (A.N.)
* Correspondence: francesco.mocera@polito.it
† These authors contributed equally to this work.

Received: 4 August 2020; Accepted: 27 August 2020; Published: 21 September 2020

Abstract: In this work, a multibody model of a small size farming tracked vehicle is shown. Detailed models of each track were coupled with the rigid body model of the vehicle. To describe the interaction between the track and the ground in case of deformable soil, custom defined forces were applied on each link of the track model. Their definition derived from deformable soil mechanics equations implemented with a specifically designed routine within the multibody code. According to the proposed model, it is assumed that the main terrain deformation is concentrated around the vehicle tracks elements. The custom defined forces included also the effects of the track grousers which strongly affect the traction availability for the vehicle. A passive soil failure model was considered to describe the terrain behaviour subjected to the grousers action. A so developed model in a multibody code can investigate vehicle performance and limit operating conditions related to the vehicle and soil characteristics. In this work, particular attention was focused on the results in terms of traction force, slip and sinkage on different types of terrain. Tests performed in the multibody environment show how the proposed model is able to obtain tractive performance similar to equivalent analytical solutions and how the grousers improve the availability of tractive force for certain type of soil characteristics.

Keywords: deformable soil; tracked vehicle; multibody; contact model

1. Introduction

Numerical simulations are very useful tools to evaluate complex mechanical systems behaviour during their early design stage as well as during product development. Multibody models are widely used to predict vehicle kinematics and dynamics [1], giving the ability to identify/predict safety critical working conditions to be avoided [2,3]. Focusing on wheeled or tracked vehicles, machine-soil contact modelling is relevant to assess its performance. Several mathematical models have been proposed in the literature to characterize wheel or track-terrain interaction [4,5].

A comprehensive contact model is a particularly complex task to be accomplished. The identification of the terrain mechanical characteristics is strongly affected by its composition and humidity content. A terrain can be described as an elasto-plastic medium which exhibits a strong non-linearity in the force-deformation characteristic. The behaviour of different types of soil has been characterized by means of several empirical parameters for terrain cohesion, internal friction angle and stiffness [6–8]. Several numerical models have been proposed in the literature to study and simulate wheel-soil interaction [9–11] as well as the experimental test rigs to characterize them [12]. Similarly, empirical methods, discrete elements and finite elements methods as well as reduced order models have been proposed to model the terrain-tyre and terrain track interaction response when in contact with machines wheels or tracks modelled as rigid or flexible bodies [13–15]. Two mathematical models are usually considered to represent the track(or wheel)-soil interaction—the Bekker's model and the

Janosi-Hanamoto's law. Bekker's equation relates the normal pressure exerted on the ground with the vertical sinkage of a circular or rectangular plate used for the soil characterization procedure through a non-linear elastic relationship. When modelling the sinkage behaviour of the tracked machine also the soil damping contribution should be considered.Janosi-Hanamoto equation relates shear stress to the corresponding shear displacement of the soil. The difference between the actual shear stress and the maximum one admissible by the terrain strongly affects the availability of tractive force to be developed by the tracked vehicle. Both the models depend on empirical parameters to be derived through specific terrain characterization procedures [16]. A novel soil-tyre model has been recently proposed in Reference [17] integrating soil mechanics behaviour with a soft soil-elastic tyre interaction model. The tyre and soil deformations are computed iteratively solving the equilibrium problem as implemented by Sharaf et al. in Reference [18] to model an off-road vehicle. On the other hand, different procedures have been adopted to model a deformable terrain in multibody (MTB) codes. Rubinstein and Hitron [19] modelled the terrain shape with rigid bodies whereas in other MTB codes discrete elements method (DEM), finite element or mesh free methods [20] have been proposed. A springs-damper-oil sand slots connected by spherical joints terrain modelling approach was proposed by Frimpong and Thiruvengadam [21,22]. Soil modelling through triangular meshes or discrete elements was also considered in the Adams Tracked Vehicle Toolkit (ATV) [20,23,24]. In other numerical environments, soil behaviour under load has also been investigated thanks to Finite Element Models (FEMs) [25] or Semi-empirical Contact Models (SCMs) [26]. There are two different types of tracks classified as rigid or flexible. Rubber tracks, reinforced with a steel core, are usually classified as flexible whereas tracks made of steel links are classified as rigid. The latter are composed of rigid links [27] usually connected in multibody models with rotational joints [28] or bushing elements [29]. Bushing elements can also damp the system to limit numerical errors [30]. In fact, simple rotational joints without damping can lead to high vibration phenomena of the track links [31].

In this paper a MTB model of a tracked machine with a deformable terrain model was developed, also thanks to the authors' research group previous experience in MTB contact [32] and vehicle modelling [33]. The main assumption of the model considers the terrain deformation located in a close area around the vehicle tracks. Thus, the soil deformation can be kinematically correlated to the track motion. The soil mechanics equations were implemented on each rigid body used to model the rubber tracks. Each link is connected to each contiguous one by means of bushing elements aimed to replicate a realistic track elastic characteristic. Soil mechanics equations were implemented through a specifically designed routine which solved only the equations for the track links in contact with the terrain. The algorithm defined at each iteration the forces applied to each track link in contact with the terrain. Moreover, the effects of grousers in the overall tractive performance of the machine were investigated. To model grousers effects, soil mechanics equations related to passive terrain failure [34,35] were considered and machine draw-bar pull was evaluated for different types of soil.

2. Soil Mechanics

In this section, the soil mechanics equations implemented in the multibody environment are summarized. The soil behaviour when compressed in normal direction can be described with the Bekker equation (Equation (1)), also known as the pressure-sinkage relationship [6].

$$p = \left(\frac{k_c}{b} + k_\phi\right) \cdot y^n = k_{eq} \cdot y^n, \tag{1}$$

with y representing the sinkage coordinate while k_c, k_ϕ are empirical coefficients describing the cohesion and the internal friction angle of the considered type of soil. n is the sinkage exponent and b the reference dimension of a circular or rectangular plate in contact with the terrain. A different formulation of Bekker law was proposed by Reece in Reference [16]. Wong et al. [11] formulated a pressure-sinkage relationship describing also the soil response during an unloading cycle or when pressure removal occurs. The soil shear stress-shear displacement relationship in the longitudinal direction of the track can be described trough the Janosi-Hanamoto law [36] (Equation (2))

$$\tau = \tau_{max} \cdot \left(1 - e^{-j/K}\right), \tag{2}$$

where j is the shear displacement of the terrain and K is the shear deformation parameter which represents the shear displacement required for the development of the maximum shear stress. The maximum shear stress for a certain type of terrain is expressed by Mohr-Coulomb equation (Equation (3)) and depends on soil cohesion c and internal friction angle ϕ.

$$\tau_{max} = c + \sigma_n \cdot \tan\phi, \tag{3}$$

with σ_n bring the normal pressure (p in Bekker law). Values of the above mentioned empirical parameters k_{eq},c,ϕ and n can be found in the literature [37] for several terrain types. Also numerical methods for their real time determination have been developed [38]. These iterative algorithms allow to know some variable terrain characteristics during tracked machine motion. Focusing on tracked or wheeled vehicles on deformable terrain, the ϕ values for rubber-soil metal-soil contact are very similar to the soil internal friction angle. Adhesion values between rubber and soil or metal and soil differ from cohesion values [39]. If a shearing action is exerted on the soil by a track or a wheel, shear displacement of the terrain occurs. The shear displacement j is equivalent to the relative motion between track and soil (slip). As discussed in the following sections, the traction force developed by a tracked machine coincides with the integral of Equation (2) on the track area in contact with the soil. If no slip occurs no drawbar pull is developed. As shown in Figure 1 Janosi-Hanamoto law differs from experimental results in particular for compact terrains and high normal pressure [40]. Janosi-Hanamoto law expressed in Equation (2) is monotone increasing, whereas shear stress-shear displacement curves for dense frictional-cohesive soils exhibits a maximum [41]. To model a maximum in the shear stress—shear displacement behaviour other formulations are available in the literature [31].

$$\tau = \tau_{max} K_r \cdot \left[1 + \left(\frac{1}{K_r \cdot (1 - e^{-1})} - 1\right) \cdot e^{(1-j/K_W)}\right] \cdot \left(1 - e^{-j/K_W}\right), \tag{4}$$

where K_r represents the ratio between the residual shear stress and its maximum admissible value while K_w represents the shear displacement at which the maximum shear stress occurs.

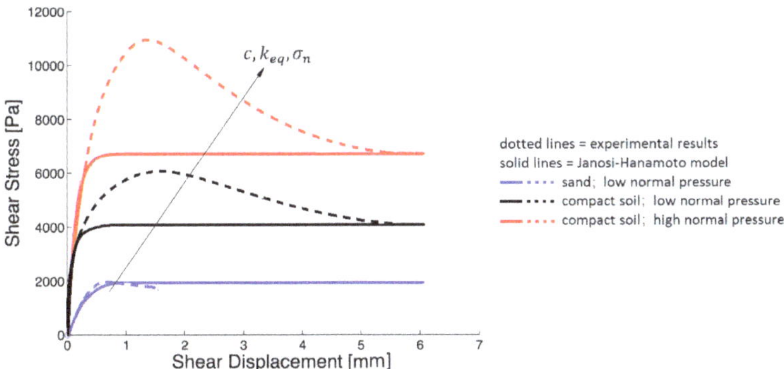

Figure 1. Janosi-Hanamoto model and experimental results, for different types of soil. Experimental data usually exhibit a peak in the traction capabilities for certain values of the shear displacement.

Equation (2) was implemented in the MTB model developed in this paper and also in other similar MTB models available in the literature [19]. As discussed, the application of this equation can be justified for loose soils and low normal pressure values. In the present MTB model Equation (2) was also used for compact agricultural terrain due to the limited normal pressure on the soil exerted by the small size tracked vehicle. Focusing on the interaction of a cutting surface as a blade or a grouser with the soil, passive earth failure must be considered. In fact, the terrain in front of the cutting blade or behind the grouser can be brought in state of passive failure, due to the applied force [37]. According to Mohr criterion the passive failure of a soil prism occurs when the principal stress σ_p on its vertical sides is reached (Figure 2).

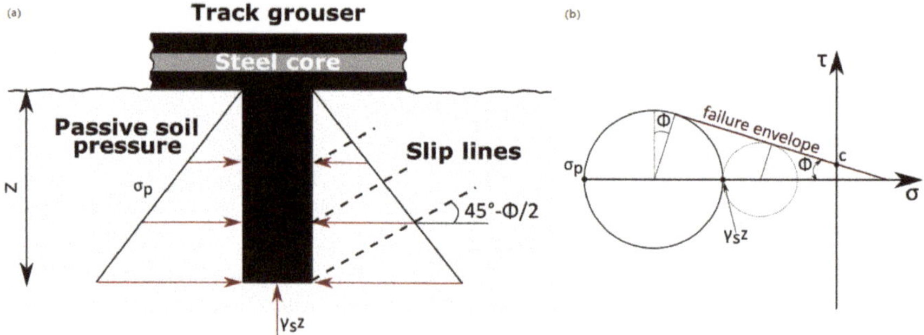

Figure 2. (**a**) Soil prism under passive compression; (**b**) Mohr circle.

Since the major principal stress is horizontal, failure of the soil prism takes place on slip lines sloped at $(45° - \phi/2)$.

Integrating the principal passive earth pressure σ_p on the grouser vertical area (Figure 3) the horizontal force acting on the lug can be determined (Equation (5)). It is the contribution to the vehicle traction force offered by a single grouser sunk into the soil.

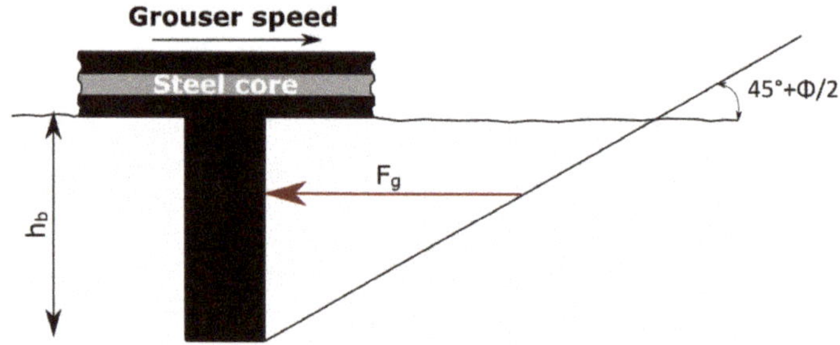

Figure 3. Interaction of a grouser with the terrain.

$$F_g = b\int_0^l \sigma_p dz = b\int_0^{h_b}\left(\gamma_s z N_\phi + 2c\sqrt{N_\phi}\right)dz = b\left(\frac{1}{2}\gamma_s h_b^2 N_\phi + 2ch_b\sqrt{N_\phi}\right), \tag{5}$$

in which γ_s is the specific weight of the soil, h_b is the grouser height sunk into the soil and N_ϕ is equal to $\tan^2(45° + \phi/2)$. If there is a normal pressure q acting on the terrain surface behind the grouser the resultant force is:

$$F_g = b\left(\frac{1}{2}\gamma_s h_b^2 N_\phi + qh_b N_\phi + 2ch_b\sqrt{N_\phi}\right). \tag{6}$$

It has to be pointed out that the passive earth pressure approach should be integrated also considering the existence of friction and adhesion on grouser vertical sides. Furthermore because of the existence of shear stress at grouser-soil interface the normal pressure considered is no more a principal stress and the vertical surface of a grouser can be inclined with respect to the horizontal plane of the terrain. Reference [34] proposed a mathematical solution to correctly evaluate the force applied by a grouser taking into account its inclination and the presence of shear stress on its sides. Bekker proposed an equation based on an elastic stress distribution into the soil in order to consider the traction force developed by the vertical shear areas of a grouser. In Reference [42], Reece proposed a different equation based on static equilibrium to consider the same contribution. He formulated an equation of the total traction force given by a grouser considering the latter as a cutting surface and referring to a failure state of the soil.

3. Methods: Multibody Modelling

The MTB model of the tracked vehicle was built into the MSC ADAMS® (Newport Beach, CA, USA) software. Modelling of the tracks and the tensioning system was combined with a simplified body representing the main machine geometries. However, the mass and moments of inertia were estimated accordingly to the vehicle characteristics of a real machine for farming applications taken as a reference (see Table 1). Each track was made of 40 rigid bodies (links). The sprocket teeth transmit the motion by inserting into the links holes. The lateral movement of the track was bound with respect to the idle wheel by the central rail formed by each track segment (Figure 4).

Table 1. Main characteristics of the tracked vehicle.

Vehicle Properties	
Mass	750 kg
Length	1500 mm
Width	700 mm
Height	1130 mm
Track width	180 mm
Top speed	4 km/h

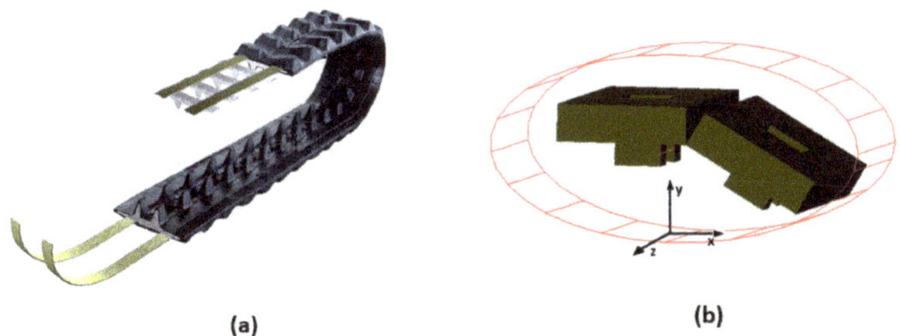

Figure 4. (**a**) Track structure, (**b**) links connected by a bushing element.

Small size tracked machines are usually equipped with rubber tracks. They have a steel core which gives a greater tensile strength. To obtain a track model with behaviour and properties similar to a real rubber track it was necessary to properly model the joints between consecutive links. In this work a bushing element was considered (Figure 4b). Three-dimensional bushing joints allowed to transmit elastic and damping forces and torques between two consecutive links. The required stiffness and damping coefficients were defined applying specific loads and comparing the virtual track behaviour with a reasonable response of a real one. In particular, a static test as the one shown in Figure 5 was simulated to see if the tensioning force and the proposed parameters for the bushing elements were coherent with the deflection of a track with a mass of 20 kg placed in between the idle wheel and the sprocket. Stiffness values are listed in Table 2.

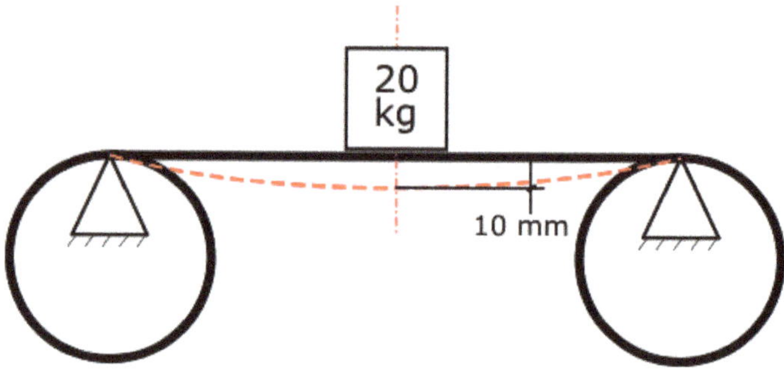

Figure 5. Deflection test.

Table 2. Bushing parameters.

	x	y	z
Translational Stiffness N/mm	10^4	10^4	10^4
Rotational Stiffness Nmm/deg	10^4	10^4	200

Negligible displacements and angular misalignments in x and y directions between two consecutive links were obtained using high translational and rotational stiffness values along those directions. Comparing the deformed shape of a real rubber track with the deformed shape of its MTB model, rotational stiffness in z direction was set. Track vibrations were limited by setting appropriate rotational damping in z direction, also considering rubber damping effect [43]. Idle wheels, driven wheels and oscillating plates with road-wheels were properly modelled (Figure 6). Contact forces were defined between each track link and each wheel. Low penetration depth and friction coefficient related to rubber-steel contact were defined. A damping value of 5.0 Ns/mm and a stiffness value of 100 N/mm were considered for the contact model of these elements. The driving wheel teeth insertion into links holes allowed torque transmission thanks to the contact force definition. In Figure 6 the position of the machine centre of mass (CM) is highlighted. Its position, as well as the overall moments of inertia of the vehicle are coherent with the vehicle characteristics shown before in Table 1.

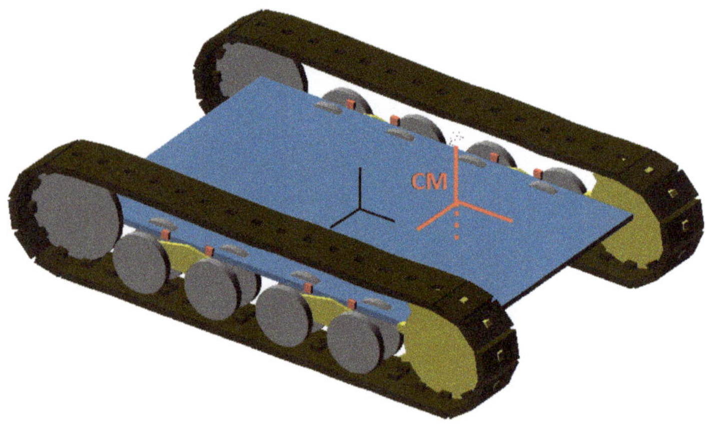

Figure 6. Complete model of the farming vehicle.

The track tensioning system was modelled so that the simultaneous rotation and translation of the idle wheel could be allowed (Figure 7). A translational joint was applied between a cubic pivot (dummy) body and the main body of the machine. The idle wheel was connected to the pivot body by means of a rotational joint. Considering typical tensioning force values for rubber tracks of small size machines, F force value (Figure 7) was set equal to 5100 N in such a way that the static track deformation would satisfy the test shown in Figure 5.

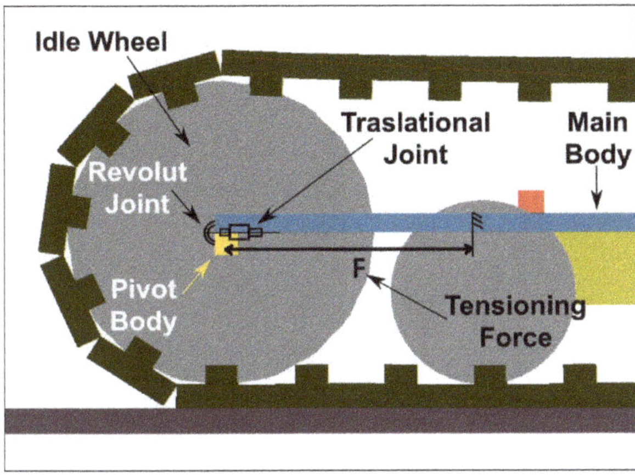

Figure 7. Tensioning system.

3.1. Deformable Ground Contact Model

The mathematical model of deformable soil was integrated into the MTB environment to correctly simulate the contact and the behaviour of the track on agricultural terrain. An efficient track-terrain contact model was developed implementing soil mechanics equations within routines defining different force elements applied to each rigid link of the track. Resulting forces acting on a link depend on ground characteristics, although they do not affect soil body deformation. The following routines description refers to smooth tracks (without grousers). Bekker equation Equation (1) was implemented in the routine related to sinkage behaviour of track links. The soil was modelled as a rigid body which may present irregularities such as peaks or valleys. Equation (1) multiplied for the link contact area was implemented in a non-linear Single-component-force (S-force) applied to the center of mass of each link and moving together with it (Figure 8). Thus each track link correctly sank into the ground body. The routine was defined to solve the Bekker force equation only when a contact between the soil body and the link was detected. The distance between the free surface of the soil and the contact base of a track link (y in Equation (1)) must be known at each time step of the solution. The terrain was modelled in the MTB environment as a rigid body with the desired shape to replicate the real undeformed one. To solve the contact equations, the information of the undeformed shape of the terrain was taken into account using a particular modelling strategy. A set of small and with low mass spheres forced to move with each track links but free to follow normally the terrain profile was used. The spheres CM position was evaluated at each time step, thus the information of the terrain geometry was available to the developed routine.

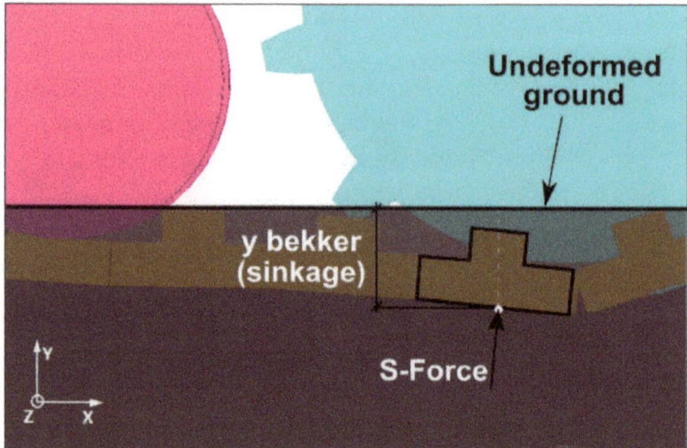

Figure 8. Bekker force acting on a single-track link.

The sinkage value y was constantly given as the difference between the base of the track link and the lowest contact point of each sphere (function of the sphere CM position and its radius). In other words, the marker centred in the sphere CM was one side of the non-linear spring defined by Equation (1) and applied to the track link surface. As a result, the tracks sinkage and the machine position depended on the terrain characteristics and shape (Figure 9).

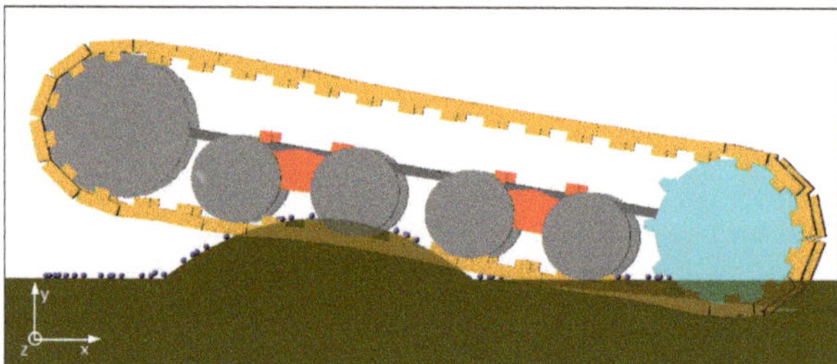

Figure 9. Auxiliary spheres following the undeformed shape of the terrain profile.

To be sure to not interfere with the track link movement, but to correctly sense the terrain shape, a frictionless contact model was implemented between each sphere and the ground body. Each sphere had to be vertically aligned with the corresponding link. For this reason, two motion laws were imposed. The first (between each sphere and the corresponding link), had a constant null displacement in z direction, and no constraints in the other directions. In the second motion law (between each sphere and the terrain) the sphere speed value in horizontal direction was set to be equal to the corresponding track link speed in the same direction. Soil damping was considered into account using typical values of terrain damping coefficients [44] and modifying Bekker equation as follows.

$$F = A\left(\frac{k_c}{b} + k_\phi\right) y^n + Ac\dot{y} = Ak_{eq}y^n + Ac\dot{y}. \tag{7}$$

The complete algorithm can be summarized as in Figure 10 in which F_B and F_t are the Bekker force and the traction force respectively.

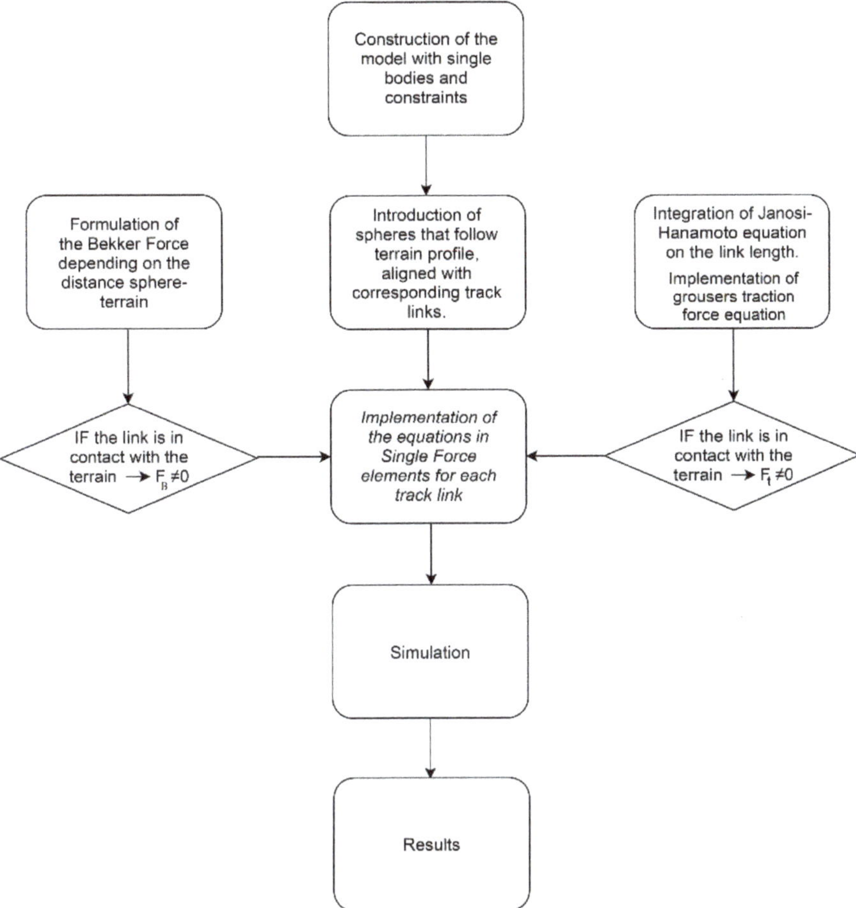

Figure 10. Complete multibody (MTB) code scheme

The contribution of each track portion in terms of traction force was obtained employing Janosi-Hanamoto equation in a specific routine inserted in another Single-component-force (S-force) defined for each track link. The internal shear stress – shear displacement law is reported below.

$$\tau = \tau_{max} \cdot \left(1 - e^{-j/K}\right) = (c + p(x)\tan\phi) \cdot \left(1 - e^{-j/K}\right) \tag{8}$$

As explained in Section 2, Equation (8) was implemented both for loose soil and compact agricultural terrain. Shear displacement j can be defined as the product ix, being

$$i = \frac{v_t - v}{v_t} = \frac{v_j}{v_t}. \tag{9}$$

i is also known as slip coefficient, v_t is the theoretical speed of the vehicle, v is the actual one and v_j is the relative speed between track portion and soil. In other words, v_j is the absolute link speed in x direction: when no slip occurs the track link in contact with the terrain is still. In this approach, the x coordinate is correctly identified as in Figure 11 and the theoretical speed is defined by Equation (10)

$$v_t = \omega_{sprocket} \cdot \left(r_{sprocket} + s_{track} \right). \quad (10)$$

$\omega_{sprocket}$ and $r_{sprocket}$ are the angular speed and the radius of the driven wheel, respectively, and s_{track} is the track thickness.

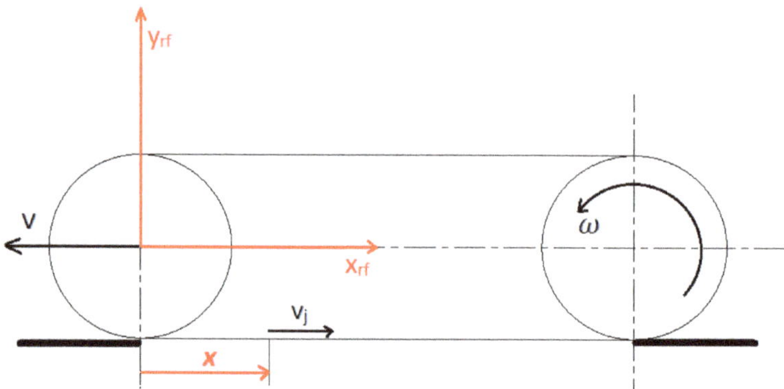

Figure 11. Coordinate system for traction force equation integration.

The syntax of the traction force was written into a specific routine which solved it only if contact between soil body and links was detected. The traction force expression was obtained integrating Equation (8) along the length of each track link, for the portion of the track in contact with the soil.

$$F = l \int_{b_1}^{b_2} (c + p(y) \tan\phi) \cdot \left(1 - e^{-ix/K}\right) dx = A\,(c + p(y)\tan\phi) \left[1 + \frac{K}{ib}\left(e^{-ib_1/K} - e^{-ib_2/K}\right)\right], \quad (11)$$

where A was the link base area, b was the link length and b_1 and b_2 were the limits of integration for a single link (Figure 12). They were moving limits of integration because the x coordinate was referred to the moving reference frame centred on the front idle wheel. Each link varied its position with respect to the idle wheel during vehicle motion as shown in Figure 11. The traction force routine was implemented into each link S-force element (Figure 12). Traction force sign was always opposite to the sign of the corresponding link speed (slip speed) coherently to the machine movement. The product between the normal pressure and the link contact area (term $Ap(y)$ in Equation (11)) coincided with the Bekker force previously implemented. Within the traction force formulation, the term $Ap(y)$ was replaced by the Bekker force definition. In this way the traction force evaluation depended on the current normal pressure under the track link (influenced by the terrain shape) and the approximation of the normal pressure with the overall average one was avoided. To sum up, soil mechanics equations were implemented only within force elements applied to the track links. No other equations related to soil stress and deformation were computed during simulations. However, tracks sinkage and developed drawbar pull were always related to deformable terrain characteristics and shape by mean of moving spheres constrained to each link as shown in Figure 12, where the sphere A is the sphere corresponding to the highlighted link.

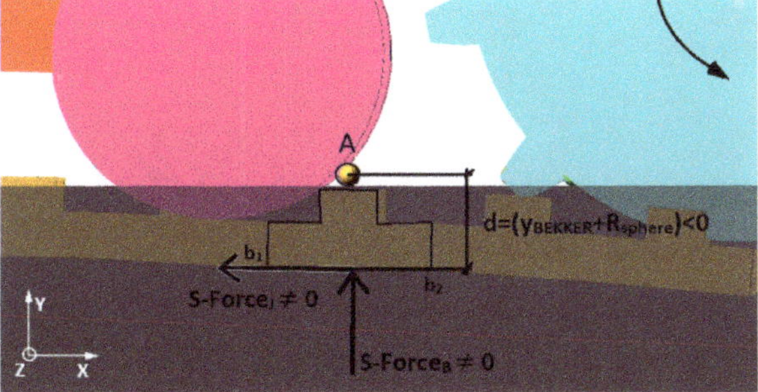

Figure 12. Forces acting on the single link.

3.2. Grousers Modelling

In order to evaluate the drawbar pull developed by a tracked machine the contribution in terms of traction force offered by grousers must be considered. The traction force equation presented in the previous section, and implemented for each track link, considered only the shear stress between the track flat surface and the soil. Focusing on rubber tracks with lugs, the cutting action of grousers sunk into the soil must be considered. In fact, grousers generate a greater traction effort contribution. This contribution varies with soil characteristics and grousers number and dimensions. In the presented model grousers as large as the track and 25 mm height were considered, one for each link.

For the sake of simplicity, no new rigid bodies were introduced into the MTB model and the link shape was not modified to include the grousers. The virtual grouser position is shown in Figure 13. Traction and Bekker force equations referred to a single grouser were properly related to its size, even if grouser body was not geometrically modelled. Fg in Figure 14 indicates the traction force developed by a grouser. The sinkage behaviour of the whole track changes with the presence of grousers because they are the first parts coming into contact with the soil. No new force elements were introduced but Bekker equation referred to the grouser base was introduced in the force formulation previously written for each link. If the base of the grouser sinks into the ground but the latter does not touch the link base, the track portion receives an upward force equal to:

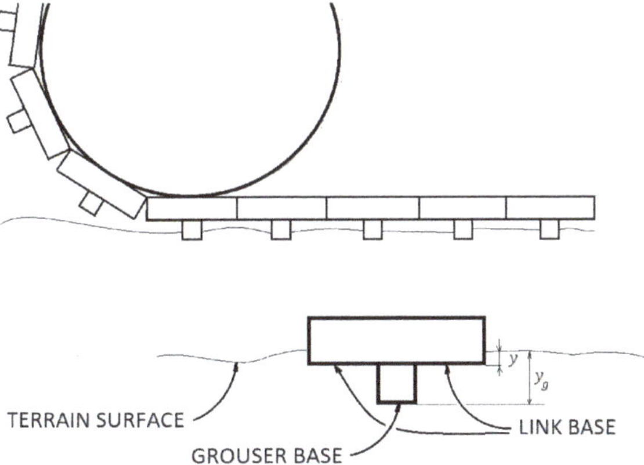

Figure 13. Virtual grouser position.

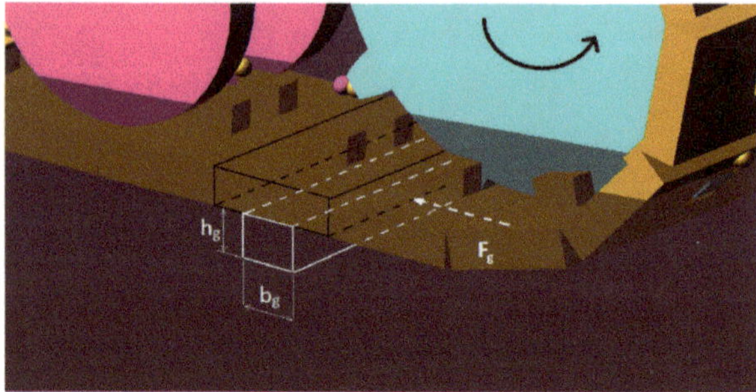

Figure 14. Grousers shape and position.

$$F_B = A_g\left(\frac{k_c}{b} + k_\phi\right) \cdot y_g^n + A_g c\dot{y}_g = A_g k_{eq} y_g^n + A_g c\dot{y}_g, \tag{12}$$

in which A_g is the grouser base area (Figure 13). If also the link base sinks into the soil the track portion receives an upward force equal to:

$$F_B = A_{link} k_{eq} y^n + A_{link} c\dot{y} + A_g k_{eq} y_g^n + A_g c y_g^n, \tag{13}$$

in which A_{link} is the link base area (Figure 13); y_g differs from y by the grouser height. Also, for the traction force developed by grousers no new force elements were introduced into the MTB model. The expression of the new force contribution was integrated into the existent traction force routine implemented for each track link. The resultant equation was a parametric equation in which grousers number and dimensions could be changed. The traction force offered by a grouser is related to the passive soil pressure and expressed by Equation (6), where q is different from zero only if the link base touches the soil. In fact, it represents the surcharge acting on the terrain surface portion under the grouser. For this reason, it can be set equal to the Bekker pressure acting on the link base. The grouser height sunk into the soil h_b depends on soil characteristics and Bekker equation. To sum up, the whole traction force equation implemented on each link force element depended on two contribution: the shear stress between link and/or grouser base and soil, and the grousers cutting action. If the whole grouser sank into the terrain and the latter touches the link base, the traction force expression for the link is reported in Equation (14).

$$F_t = A\left(c + p(y)\tan\phi\right) \cdot \left[1 + \frac{K}{ib}\left(e^{-ib_1/K} - e^{-ib_2/K}\right)\right] + F_g\left(1 - e^{-i\bar{x}/K}\right)_{q \neq 0}, \tag{14}$$

where A is the total link base area (which includes the grouser base area) and y_g is reasonably approximated with y values. If only part of the grouser sank into the soil, the traction force expression for the link which the grouser was attached to was expressed as in Equation (15).

$$F_t = A_g\left(c + p(y_g)\tan\phi\right) \cdot \left[1 + \frac{K}{ib}\left(e^{-ib_1/K} - e^{-ib_2/K}\right)\right] + F_g\left(1 - e^{-i\bar{x}/K}\right)_{q=0}. \tag{15}$$

The grouser traction force F_g was multiplied by the term $\left(1 - e^{-i\bar{x}/K}\right)$, in which $\bar{x}$ is the x coordinate value of the link CM marker. In fact, if no slip (i) was observed, grouser traction force contribution was null because no compressive stress was exerted on the soil portion behind the grouser. On the contrary, if slip value increased the grouser traction force contribution value rapidly tended to F_g. In fact, the initial phase of traction force generation in compressible terrain is soil compression by grousers in the x direction, which divides the soil under a track into separate blocks. This compression increases at least to the transition point, when a block is sheared off and starts sliding along the channel formed by the preceding grouser [45]. Furthermore, passive soil pressure is related to a state of incipient plastic flow of the soil. The following Table 3 resumes the nomenclature adopted to explain the construction of the routine implemented into the MTB code.

The algorithm of the routine related to the traction and Bekker forces for a single-track link with grouser is summarized in Figure 15.

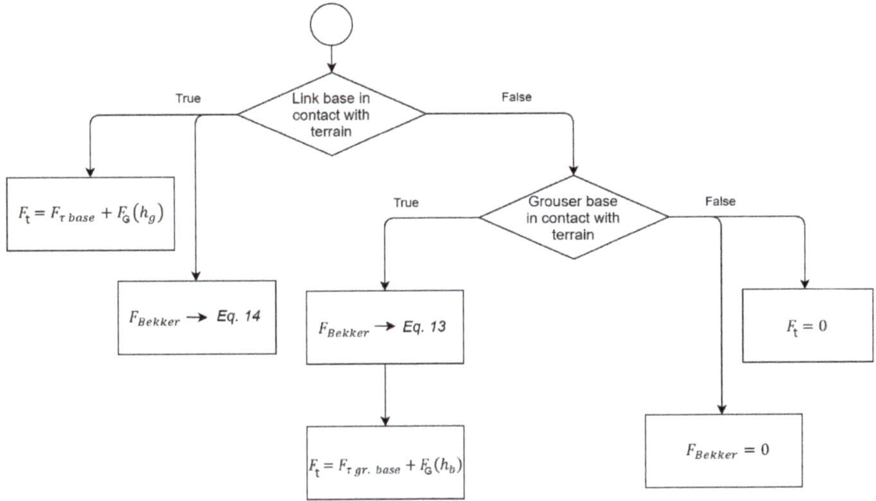

Figure 15. Evaluation strategy for the forces acting on each track link with grouser.

Table 3. Grousers MTB routine nomenclature.

Symbol	
h_g	Grouser height
h_b	Grouser height sunk into the soil
$F_{\tau gr,base}$	Traction force related to Janosi-Hanamoto law (first term of Equation (15)) acting on the grouser base
$F_{\tau base}$	Traction force related to Janosi-Hanamoto law (first term of Equation (14)) acting on the total link base area (grouser base + link base)
$F_G(h_b)$	Traction force developed by a grouser, equal to $F_g\left(1 - e^{-i\tilde{x}/K}\right)$
F_t	Total traction force developed by a track link.

4. Simulations and Results

In this section, performance of the tracked vehicle on deformable soil are discussed. Before implementing the moving ground contact model for deformable terrains, the MTB model of the vehicle was validated with tests performed assigning simple contact elements with coulomb friction between each track link and the soil body [43].

4.1. Smooth Tracks

After this first stage the mathematical model of deformable soil described before was implemented into the MTB environment. Results related to the sinkage and the traction force of the tracks without grousers are reported. Changing accordingly the values of the constants k_{eq}, c, K and ϕ previously explained two different types of terrain (sand and agricultural soil) were simulated.

4.1.1. Normal Pressure Distribution and Track Sinkage

The same test was performed on sand and compact ground. Selecting the same location on the ground body (rigid), it was possible to evaluate different track deformation and sinkage for the two types of soil. The highest sinkage reached on sand was equal to 50 mm, whereas the maximum sinkage value on compact agricultural terrain was equal to 8 mm, because of the higher equivalent stiffness. Dividing the Bekker force values acting on each track link by its base area the pressure distribution both for sand and compact soil was obtained (Figure 16). A linear pressure distribution was also plotted to simply show the influence of the CM position and modifying the simplified assumption of constant pressure under tracks proposed in Reference [37].

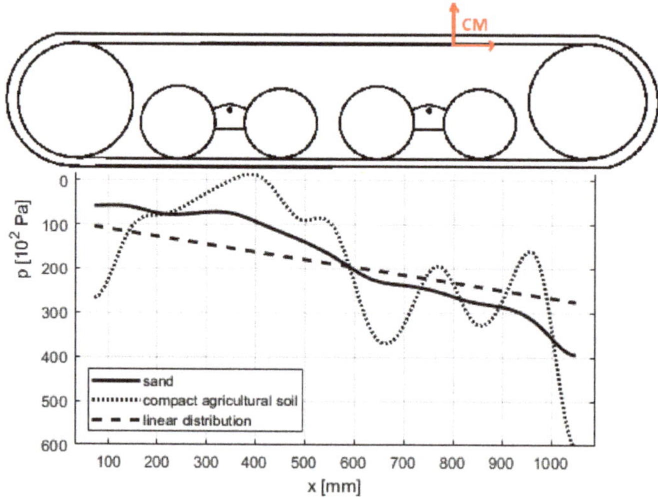

Figure 16. Pressure distribution for the track-terrain contact.

As shown in Reference [46], local rise of the pressure distribution is obtained at road-wheel locations. Moreover, pressure maxima values were grater for a compact terrain, as reported by Reference [37]. The mean value of normal pressure increased along x coordinate according to CM position. Local pressure peaks were also read under each grouser base, due to the grater sinkage of the latter with respect to the corresponding link base.

4.1.2. Traction Force-Slip Curves

Traction force-slip curves evaluated within the MTB code for smooth tracks are reported in Figure 17. Also, analytical curves evaluated both for sandy and compact terrain are presented and compared with MTB results. The only contribution to the traction effort derives by the shearing action between tracks and soil. To measure traction force and slip into the MTB environment, a spring force element was applied between the rear part of the vehicle and a fixed reference frame and the same increasing torque ramp was applied to the sprockets. The traction force as a function of time (linearly increasing with machine displacement) was read from the spring element. Slip as a function of time was evaluated as $i = 1 - v/v_t$. The two entities were then reported on the same plot (Figure 17).

The maximum traction force value reached on compact agricultural soil was greater than the traction force value on sand, mainly due to higher ϕ and c values.

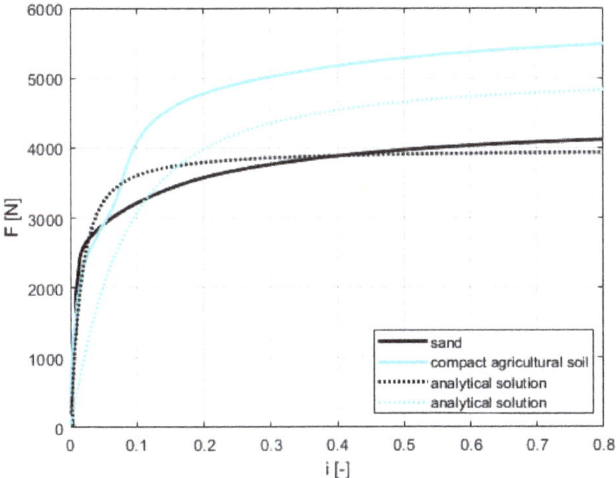

Figure 17. Traction force-slip curves.

4.2. Tracks with Grousers

Results in term of drawbar pull and sinkage for complete tracks model with grousers are reported. Comparison with results obtained with smooth tracks is also pointed out.

4.2.1. Sinkage

On soft sandy terrain 12 mm height grousers completely sank into the soil, due to ground characteristics and machine weight (Figure 18a).

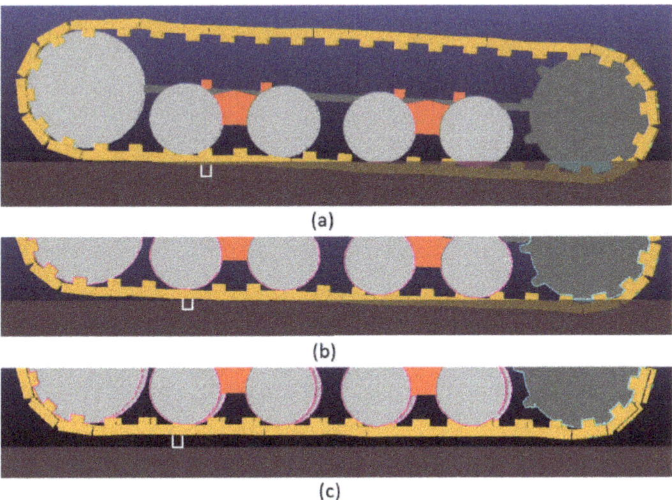

Figure 18. Tracks sinkage with 12 mm high grousers on sand (**a**), 25 mm high grousers on sand (**b**), and 25 mm high grousers on compact terrain (**c**). In each figure, only one virtual grouser is represented on each image to figure out the track sinkage.

With 25 mm height grousers the global vehicle sinkage was slightly smaller (Figure 18b), according to Bekker force acting on grousers base area. On compact agricultural soil grousers partially sank into the ground, but the latter did not come into contact with track links base (Figure 18c). Figure 18a–c highlight just one grouser for the sake of clarity. However as discussed before, simulations take into account the effects from all the grousers of the tracks.

4.2.2. Traction Force-Slip Curves

The grousers action contribution to the traction effort is reported and discussed. The procedure to obtain traction force-slip curves in MTB model was the same reported in Section 4.1.2. Traction force developed by tracks with grousers was investigated for two different types of soil: sand and compact agricultural terrain. The effect of grousers height was also highlighted. The maximum traction force value was reached on sand with 25 mm height grousers (Figure 19). It was about 250% of the value reached with smooth tracks (on sand). Comparing the results obtained for different grouser height on sand (12 mm and 25 mm), it is clear that the variation of traction force maximum values was not proportional to grousers height variation. The mean term of Equation (6) reported below is proportional both to grouser height sunk into the soil (h_b) and surcharge q. Moreover, it is about 100 times greater than other two terms. The height of the grouser part sunk into the soil (h_b) on sand was almost equal to the total grouser height, both for 25 mm and 12 mm grousers height (Figure 18). On the contrary pressure q exerted by links bases was greater for 12 mm height grousers on sand, due to the greater global sinkage. For this reason, maximum traction effort reached with 12 mm height grousers was about 80% of the value referred to 25 mm height grousers (Figure 19).

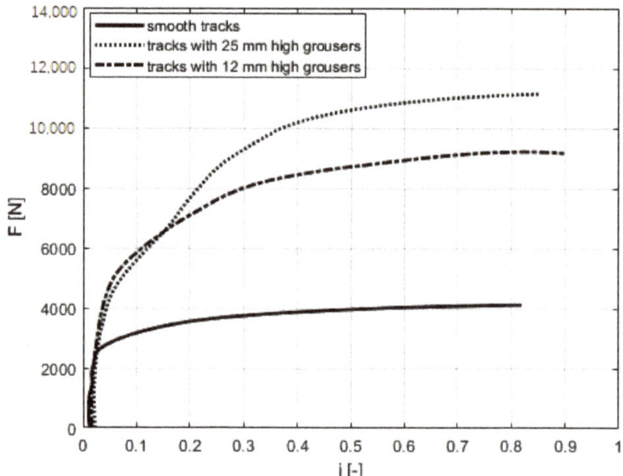

Figure 19. Traction force-slip curves on sand.

On compact agricultural soil the maximum value of the traction force was smaller than the value reached with smooth grousers (Figure 20). Grousers contribution in terms of traction effort is limited, because soil stiffness does not allow them to sink. Shearing action between tracks and terrain had the main role in drawbar pull development, but the sum of the grousers shear base areas is smaller than the total area of smooth tracks.

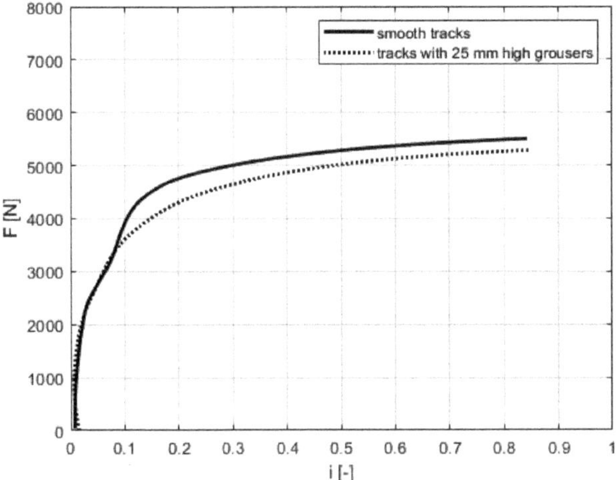

Figure 20. Traction force-slip curves on soil.

5. Discussion and Conclusions

A multibody model of a tracked vehicle for off-road application is presented. A specifically designed algorithm to model track-deformable terrain interaction was developed. Bekker and Janosi-Hanamoto soil mechanics equations were implemented to properly simulate both sinkage and traction behaviour of the tracked machine. Moreover, a passive earth failure model was adopted to also consider the contribution of grousers in terms of drawbar pull. The two flexible tracks were modelled as groups of rigid links constrained by bushing elements able to replicate the elastic behaviour of the track. On each track link, Single component Force elements were applied and related to the soil mechanics equation solved only for those links in contact with the terrain. The use of small rigid spheres kinematically constrained to the link motion, allowed to have at each integration step the local shape of the terrain body in contact with their corresponding link. A simple terrain modelling procedure was adopted to obtain indicative results about tracked vehicles behaviour on deformable soil, in favour of computational times. Vehicle sinkage, developed traction force and pressure distribution under tracks on different type of soil were highlighted in the presented results. The type of terrain was varied into the MTB environment changing the soil characteristic coefficients. In particular, traction force-slip curves were compared with the analytical ones and good agreement is shown. Pressure peaks under road-wheels were correctly evaluated and the actual pressure under each links contributed to the overall traction force computation. Drawbar pull developed by smooth tracks on different type of soil was compared with drawbar pull developed by tracks with grousers. The influence of grousers height and contact surface amplitude was also analysed highlighting the higher impact in traction force on soil when compared with smooth links. As expected, a much smaller impact of the grousers on compact terrain model was highlighted due to the limited sinkage.

Author Contributions: Conceptualization, F.M., A.S. and A.N.; methodology, F.M., A.S. and A.N.; software, F.M. and A.N.; validation, F.M., A.S. and A.N.; formal analysis, F.M. and A.N.; investigation, F.M., A.S. and A.N.; data curation, F.M. and A.N.; writing—original draft preparation, F.M. and A.N.; writing—review and editing, F.M., A.S. and A.N.; visualization, F.M., A.S. and A.N.; supervision, A.S.; project administration, A.S. All authors have read and agreed to the published version of the manuscript.

Funding: This research received no external funding

Conflicts of Interest: The authors declare no conflict of interest.

References

1. Blundell, M.; Harty, D. *The Multibody Systems Approach to Vehicle Dynamics*, 2nd ed.; Butterworth-Heinemann: Oxford, UK, 2014.
2. Pascuzzi, S. A multibody approach applied to the study of driver injuries due to a narrow-track wheeled tractor rollover. *J. Agric. Eng.* **2015**, *46*, 105–114. [CrossRef]
3. Melzi, S.; Sabbioni, E.; Vignati, M.; Cutini, M.; Brambilla, M.; Bisaglia, C.; Cavallo, E. Multibody model of fruit harvesting trucks: Comparison with experimental data and rollover analysis. *J. Agric. Eng.* **2015**, *49*, 92–99. [CrossRef]
4. Yong, R.N. Track-soil interaction. *J. Terramech.* **1984**, *21*, 133–152. [CrossRef]
5. Pacejka, H.B. *Tyre and Vehicle Dynamics*; Butterworth-Heinemann: Oxford, UK, 2002.
6. Bekker, M.G. *Theory of Land Locomotion*; University of Michigan Press: Ann Arbor, MI, USA, 1962.
7. Wong, J.Y.; Preston-Thomas, J. On the characterization of the shear stress-displacement relationship of terrain. *J. Terramech.* **1982**, *19*, 225–234. [CrossRef]
8. Wong, J.Y. Data processing methodology in the characterization of the mechanical properties of terrain. *J. Terramech.* **1980**, *17*, 13–41. [CrossRef]
9. Taheri, S.; Sandu, C.; Taheri, S.; Pinto, E.; Gorsich, D. A technical survey on terramechanics models for tire–terrain interaction used in modeling and simulation of wheeled vehicles. *J. Terramech.* **2014**, *57*, 1–22. [CrossRef]
10. Wong, J.Y. Some recent developments in the computer-aided methods for design evaluation of off-road vehicles. *Proc. Symp. Transp. Syst.* **1990**, *28*, 269–288.
11. Wong, J.Y. On the study of wheel-soil interaction. *J. Terramech.* **1984**, *21*, 117–131. [CrossRef]
12. Senatore, C.; Iagnemma, K. Analysis of stress distributions under lightweight wheeled vehicles. *J. Terramech.* **2013**, *51*, 1–17. [CrossRef]
13. Shabana, A.A. ANCF Tire Assembly Model for Multibody System Applications. *J. Comput. Nonlinear Dyn.* **2015**, *10*, 024504. [CrossRef]
14. Patel, M.; Orzechowski, G.; Tian, Q.; Shabana, A.A. A New Multibody System Approach for Tire Modeling Using ANCF Finite Elements. *Proc. Inst. Mech. Eng. Part K J. Multi Body Dyn.* **2016**, *230*, 69–84. [CrossRef]
15. Sivo, S.; Stio, A.; Mocera, F.; Somà, A. A study of a rover wheel for Martian explorations, based on a flexible multibody approach. *Proc. Inst. Mech. Eng. Part J. Multi Body Dyn.* **2019**, *234*, 306–321. [CrossRef]
16. Reece, A.R. Principles of soil-vehicle mechanics. *Proc. Inst. Mech. Eng. Automob. Div.* **1965**, *108*, 45–66. [CrossRef]
17. Harnisch, C.; Lach, B.; Jakobs, R.; Troulis, M.; Nehls, O. A new tyre–soil interaction model for vehicle simulation on deformable ground. *Veh. Syst. Dyn.* **2005**, *3*, 384–394. [CrossRef]
18. Sharaf, A.M.; Rahnejat, H.; King, P.D. Analysis of handling characteristics of all-wheel-drive off-road vehicles. *Int. J. Heavy Veh. Syst.* **2008**, *15*, 89–106. [CrossRef]
19. Rubinstein, D.; Hitron, R. A detailed multibody model for dynamic simulation of off-road tracked vehicles. *J. Terramech.* **2004**, *41*, 163–173. [CrossRef]
20. Madsen, J., Heyn, T.; Negrut, D. *Methods for Tracked Vehicle System Modelling and Simulation*; Technical Report 2010–01; Department of Mechanical Engineering, University of Wisconsin: Madison, WI, USA, 2010.
21. Frimpong, S.; Thiruvengadam, M. Contact and joint forces modeling and simulation of crawler-formation interactions. *J. Powder Metall. Min.* **2015**, *4*, 1–14. [CrossRef]
22. Frimpong, S.; Thiruvengadam, M. Rigid multi-body kinematics of shovel crawler-formation interactions. *Int. J. Mining Reclam. Environ.* **2016**, *30*, 347–369. [CrossRef]
23. Baik, D.K. *Systems Modeling and Simulation: Theory and Applications*; Springer: Jeju Island, Korea, 2005; pp. 553–558.
24. Matej, J. Tracked mechanism simulation of mobile machine in MSC.ADAMS/View. *Res. Agric. Eng.* **2010**, *56*, 1–7. [CrossRef]
25. Liu, C.H.; Wong, J.Y.; Mang, H.A. Large strain finite element analysis of sand: Model, algorithm and application to numerical simulation of tire-sand interaction. *Comput. Struct.* **1999**, *74*, 253–265. [CrossRef]
26. Krenn, R.; Hirzinger, G. Simulation of rover locomotion on sandy terrain-modeling, verification and validation. In Proceedings of the 10th ESA Workshop on Advanced Space Technologies for Robotics and Automation—ASTRA 2008, Noordwijk, The Netherlands, 11–13 November 2018.

27. Gao, Y.; Wong, J.Y. The development and validation of a computer aided method for design evaluation of tracked vehicles with rigid links. *Proc. Inst. Mech. Eng. Part D J. Automob. Eng.* **1994**, *208*, 207–215. [CrossRef]
28. Nakanishi, T.; Shabana, A.A. Contact forces in the non-linear dynamic analysis of tracked vehicles. *Int. J. Numer. Methods Eng.* **1994**, *37*, 1251–1275. [CrossRef]
29. Omar, M.A. Modular multibody formulation for simulating off-road tracked vehicles. *Stud. Eng. Technol.* **2014**, *1*, 77–100. [CrossRef]
30. Ryu, H.S.; Bae, D.S.; Choi, J.H.; Shabana, A.A. A compliant track link model for high-speed, high-mobility tracked vehicles. *Int. J. Numer. Methods Eng.* **2000**, *48*, 1481–1502. [CrossRef]
31. Bando, K.; Yoshida, K.; Hori, K. The development of the rubber track for small size bulldozers. *SAE Trans. Sect. 2 J. Commer. Veh.* **1991**, *100*, 339–347.
32. Bosso, N.; Spiryagin, M.; Gugliotta, A.; Somà, A. *Mechatronic Modeling of Real-Time Wheel-Rail Contact*; Springer: Berlin/Heidelberg, Germany, 2013.
33. Nicolini, A.; Mocera, F.; Somà, A. Multibody simulation of a tracked vehicle with deformable ground contact model. *Proc. Inst. Mech. Eng. Part K J. Multi Body Dyn.* **2019**, *233*, 152–162. [CrossRef]
34. Hettiaratchi, D.R.P.; Reece, A.R. The calculation of passive soil resistance. *Geotecnique* **1974**, *24*, 289–310. [CrossRef]
35. Wong, J.Y.; Garber, M.; Preston-Thomas, J. Theoretical prediction and experimental substantiation of the ground pressure distribution and tractive performance of tracked vehicles. *Proc. Inst. Mech. Eng. Part D J. Automob. Eng.* **1984**, *198*, 265–285. [CrossRef]
36. Janosi, Z.J.; Hanamoto, B. The analytical determination of drawbar pull as a function of slip for tracked vehicles in deformable soils. In Proceedings of the 1st International Conference on the Mechanics of Soil Systems, Torino, Italy, 12–16 June 1961.
37. Wong, J.Y. *Terramechanics and Off-Road Vehicles Engineering*, 2nd ed.; Butterworth-Heinemann: Oxford, UK, 2010.
38. Hutangkabodee, S.; Zweiri, Y.H.; Seneviratne, L.D.; Althoefer, K. Validation of soil parameter identification for track-terrain interaction dynamics. In Proceedings of the 2007 IEEE/RSJ International Conference on Intelligent Robots and Systems, San Diego, CA, USA, 29 October–22 November 2007.
39. Neal, M.S. Friction and adhesion between soil and rubber. *J. Agric. Eng. Res.* **1966**, *11*, 108–112. [CrossRef]
40. Jayakumar, P.; Melanz, D.; MacLennan, J.; Gorsich, D.; Senatore, C.; Iagnemma, K. Scalability of classical terramechanics models for lightweight vehicle applications incorporating stochastic modeling and uncertainty propagation. *J. Terramech.* **2014**, *54*, 37–57. [CrossRef]
41. Grecenko, A. Binomic slip-thrust equation for tractors in predominantly frictional soil. *J. Terramech.* **1967**, *4*, 37–54. [CrossRef]
42. Reece, A.R. The fundamental equation of earthmoving mechanics. In *Proceedings of the Institution of Mechanical Engineers, Conference Proceedings*; Sage: London, UK, 1964; Volume 179, pp. 16–22.
43. Mocera, F.; Nicolini, A. Multibody simulation of a small size farming tracked vehicle. *Procedia Struct. Integr.* **2018**, *8*, 118–125. [CrossRef]
44. Rubinstein, D.; Coppock, J.L. A detailed single-link track model for multi-body dynamic simulation of crawlers. *J. Terramech.* **2007**, *44*, 355–364. [CrossRef]
45. Grecenko, A. A Re-examined principles of thrust generation by a track on soft ground. *J. Terramech.* **2006**, *44*, 123–131. [CrossRef]
46. Keller, T.; Arvidsson, J. A model for prediction of vertical stress distribution near the soil surface below rubber-tracked undercarriage systems fitted on agricultural vehicles. *Soil Tillage Res.* **2016**, *155*, 116–123. [CrossRef]

Article

Concept and Preliminary Simulations of a Driver-Aid System for Transport Tasks of Articulated Vehicles with a Hydrostatic Steering System

Marian J. Łopatka * and Arkadiusz Rubiec

Faculty of Mechanical Engineering, Military University of Technology, Kaliskiego 2, 00-908 Warsaw, Poland; arkadiusz.rubiec@wat.edu.pl
* Correspondence: marian.lopatka@wat.edu.pl

Received: 8 July 2020; Accepted: 17 August 2020; Published: 19 August 2020

Abstract: Heavy-wheeled vehicles with articulated hydraulic steering systems are widely used in construction, road building, forestry, and agriculture, as transport units and tool-carriers because they have many unique advantages that are not available in car steering systems, based on the Ackermann principle, such as—high cross-country mobility, excellent maneuverability, and high payload and lift capacity, due to heavy axles components. One problem that limits their speed of operation and use efficiency is that they have poor directional stability. During straight movement, articulated tractors' deviate from a straight line and permanent driver correction is required. This limits the vehicles' speed and productivity. In this study, we describe a driver-aid system concept that would improve the directional stability of articulated vehicles. Designing such a system demands a comprehensive knowledge of the reasons for the snaking phenomenon and driver behaviors. The results of our articulated vehicle directional stability investigation are presented. On this basis, we developed models of articulated vehicles with hydraulic steering systems and driver interaction. We next added the stabilizing system to the model. A simulation demonstrated the possibility of directional stability improvement.

Keywords: articulated vehicle; hydraulic steering system; directional stability

---

1. Introduction

Wheel machines with an articulated frame due to rigid axles, large tires, and hydraulic steering systems have many unique features, such as high maneuverability, high cross-country mobility, high payload capacity, and high durability and reliability, with low manufacturing costs, as a result of the simplicity of construction and the serial, mass production of components. For these reasons, they are widely used with high efficiency in agriculture like heavy tractors; in forestry, as tree-harvesters and forwarders for transport tasks; in mining sites as transportation vehicles, and on construction and mining sites, as loaders for realization load and carry tasks, and as dumpers for hauling and transportation tasks. According to Petraska et al. [1], transportation productivity mainly depends on average speed. Lopatka et al. [2] indicated that this speed for articulated vehicles is mainly limited by the snaking phenomenon—low direction stability of machine movement. According to Meng et al. [3], the average speed of an articulated hauler on construction sites is about 25 km/h, with a top speed of 54 km/h. According to previous studies [2,4,5], low directional stability demands higher effort from drivers, increasing their concentration levels; additionally, cases of dynamic loads on bearings in joints and in the hydraulic cylinder are a serious problem. The robotization of such equipment is a main development trend, and according to Alshaer et al. [6], the dynamics of vehicle steering systems are critical for path planning for articulated vehicles. Wider use of articulated vehicles in transport applications and increasing their productivity require improvements to their average speed, by limiting the snaking phenomenon at high speeds. The effective realization of transport tasks requires duplication of road speed and achieving about 60 km/h, during driving in good road conditions.

The low lateral stability of articulated machines is a broadly known problem; for this reason, the international standard [7] was developed. It describes the method of dynamic testing of vehicle stability when moving at high speeds. Lateral stability of articulated vehicles is the subject of many studies. A general model of heavy articulated vehicles with hydraulic steering systems was presented [8]. It enabled investigating the influence of geometrical parameters on directional and roll-over lateral stability, and the influence of torsional joint stiffness on the front and rear parts of vehicles and their inertia. Additionally, the lateral and radial stiffness of tires were considered.

Crolla et al. [9] described the handling behavior of articulated vehicles riding on smooth and rough surfaces at high speeds. The model used had three degrees of freedom (DOF) and was linear. Steering actuators were modeled as torsional springs and dampers at the articulation joints and were not actively controlled during riding. The main assumption for this investigation was that low directional stability is caused by mutual front and rear frame oscillation, due to low hydraulic cylinder stiffness. For the studied model, whose weight was 3000 kg, the authors found that with higher speed (12 m/s), folding instability could occur in addition to oscillation. The authors indicated that the most important parameter for stability is oil volume in the hydraulic steering system. Crolla et al. [10] added a model of a hydraulic steering system with linear directional valves. The authors showed the influence of cylinder dimensions and leakage in the hydraulic circuit and additional friction in the articulated joint, on the possibility of mutual frame oscillation damping and lateral vehicle stability improvements. The advanced model revealed that an oversteer mode is also possible. The authors concluded that the interaction between oversteering response and driver action is, in practice, the most likely cause for weaving oscillations, called the snaking phenomenon.

A linear model of an articulated vehicle with a rotary proportional valve, corresponding to the popular technical hydraulic feedback solution was presented [11]. The authors analyzed the understeering or oversteering behavior of articulated tractors with hydraulic steering systems, in the same manner as in vehicles with steering systems based on the Ackermann principle, according to Wong [12].

Azad et al. [13] described the lateral stability of an articulated forestry skidder witch pulling logs. The authors investigated the impact of hydraulic cylinder stiffness, play in the steering system, and locking the front and rear differentials on the stability of such systems. They assumed that frame oscillation is caused by low cylinder stiffness of up to 1°. The used models of vehicle and hydraulic steering system were linear.

A similar assumption was used in other studies [14,15], but the authors noted that for handling the characteristic of articulated vehicles with hydraulic steering systems, one of the most important features is front and rear frame inertia, which affects yaw acceleration and real yaw steer angle. For better results, they used a multi-body model of vehicle build in the MSC.ADAMS software and added a Fiala tire model. The experimental investigation of a scale model of articulated vehicles was prepared, according to this assumption [16].

According to the Standard PN-EN 12643:2000 [17], letting go of the steering wheel or the steering lever should not result in a deviation from rectilinear vehicle travel, coupled with crossing the boundaries of the standard 125% of vehicle width, for a distance less than 20 m. Such requirements were adopted as the technical stability criterion for the investigation of rectilinear travel of an articulated 10-ton loader [18]. The authors used the multi-body model of a vehicle with a locked hydraulic steering system, consisting of a hydraulic cylinder and hoses. They used a single obstacle as disturbance, acting only on the right front wheel. The results of the tests showed that the flexibility of the hydraulic pipelines and flexibility of large-size tires have a strong influence on the total (equivalent) torsional stiffness of the steering system, and consequently, on the snaking of vehicles of this class.

The necessity of using more advanced models of articulated vehicles and a hydraulic steering system was confirmed [3]. Widely used linear models show relatively low accuracy and are not suitable for solving stability problems. The authors proposed using a 3D multi-body vehicle model, together with a nonlinear hydraulic steering model, considering the elastic modulus.

The influence of nonlinearity of flow area on hydraulic steering system characteristics was reported [19]. The researchers showed the influence of dead zones on the rotary valve and the nonlinearity of flow areas, as a function of the relative angular position between the spool and the sleeve in hydrostatic steering units. When modeling the dynamic behavior of hydrostatic steering systems, special attention should be paid to dead zones, which delay the systems' actions, increase errors, and change system responsiveness.

Daher et al. [20] investigated the yaw stability of an articulated five-ton loader model with active steering control, during a steady-state cornering maneuver and a single-lane change maneuver. The efficiency of such a system on a slippery (snowy) surface during a single-lane change maneuver and J-turn maneuver with a speed of 20 km/h was investigated [21]. The proposed active system used a lateral acceleration and a frame-steer angle sensor. An electrohydraulic proportional valve was used for the hydraulic steering system control. The proposed system was intended for increasing the safety level on a slippery surface during maneuvering at high speed.

2. Experimental Identification Research

All work discussed above assumed that poor directional stability is caused by low torsional stiffness of the hydraulic system and a low damping coefficient. As a result, the snacking phenomenon and frame oscillation can occur. According to previous studies [15,18], the frame, which has limited stiffness and tire flexibility due to the hydraulic cylinder, can oscillate by up to ±2°, with a period of one to two seconds and an oscillation damping ratio of about 0.05–0.1.

To verify this assumption, an experimental investigation on a 20-ton articulated loader was conducted [22]. For this purpose, the pressure in hydraulic actuators was measured. As a forced signal, a fast steering wheel revolution rate was used (about 1 rps). During testing, the loader was placed on a hard surface. The pressure pulsation (Figure 1) showed that fast steering wheel action could generate oscillation, but its period was much smaller, at about 0.4 sec. (with a frequency of about 2.5 Hz) and a damping ratio of about 0.3. The maximal actuator displacement was below 0.1 mm and the angular oscillation did not exceed 0.1°. This was more than 10 times less than that in theoretical assumptions.

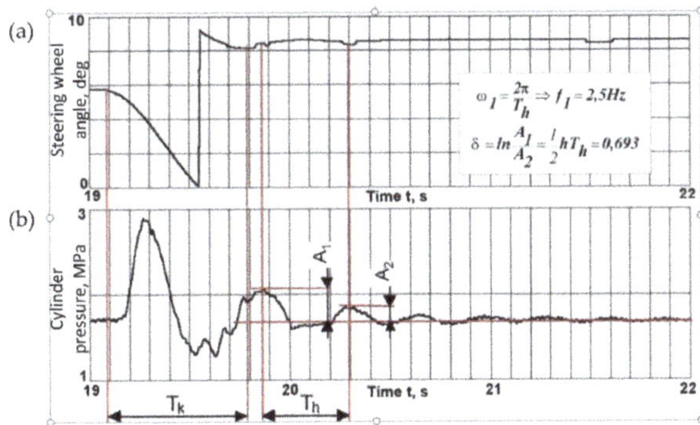

Figure 1. Pressure pulsation in hydraulic actuator of the steering system as a response to steering wheel forcing action—(**a**) forcing signal and (**b**) response signal. T_S—time of steering wheel turn, and T_{FD}—period of articulated frame oscillation.

Recognition of the snaking phenomenon required a vehicle stability investigation at top speed on a straight-line strip and, 125% of vehicle width over tire [7,17]. We investigated 20-ton loaders with a hydraulic steering system with mechanical feedback (5.5 revolution per full-frame turn), and with a hydraulic steering unit (hydraulic feedback) working in a load-sensing system (LS) with a hydraulic amplifier (3.2 revolutions per full frame turn). The results (Figure 2a) showed that during stable movement, lateral displacement exhibited sinusoidal change and could reach 20 cm. In a system with a higher gain (Figure 2b), drivers with lower experience could lose stability, and displacement would constantly grow, causing the driver to stop the vehicle.

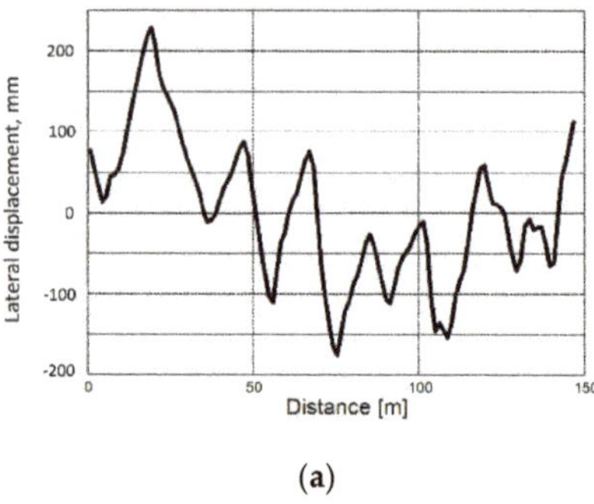

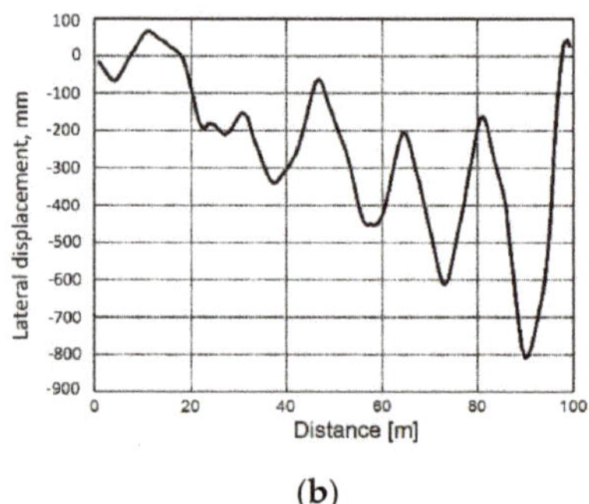

(a) (b)

Figure 2. Example of driver lateral displacement during straight-line riding at 30 km/h speed—(**a**) hydraulic steering system with mechanical feedback and (**b**) hydraulic steering system with load-sensing system (LS) and hydraulic amplifier.

When riding, the length of the hydraulic actuators continuously changed (Figure 3) due to the driver constantly acting on the steering wheel. The steering wheel's rotation as the driver reacted to the on-vehicle's location, strongly depended on speed, driver experience, and type of hydraulic steering system, but typically during stable movements, steering wheel rotation did not exceed 20°, with an average frequency of 0.4 Hz [23]. According to Chaczaturow et al. [24], driver action caused driver lateral displacement or driver angular deviation from the strip axle. During straight-line car driving, angular deviation did not exceed 0.5° and the reason for driver action was mainly lateral driver displacement. Previous research [2] confirmed this and showed that driver perception and sensing were limited. During driving on a straight strip, humans do not sense a lateral displacement error of up to 0.1 m and a yaw deviation of up to 0.5°.

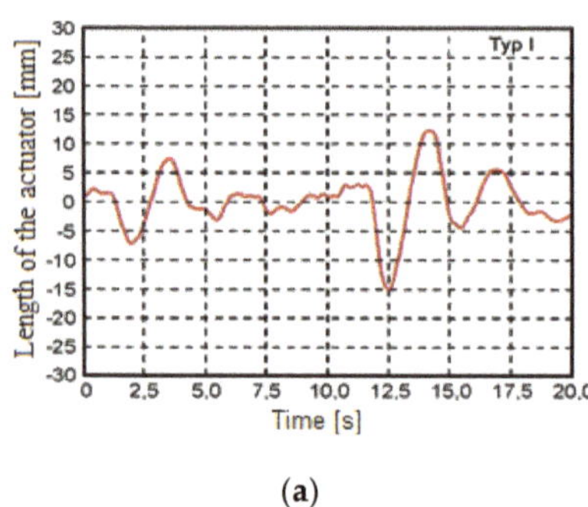

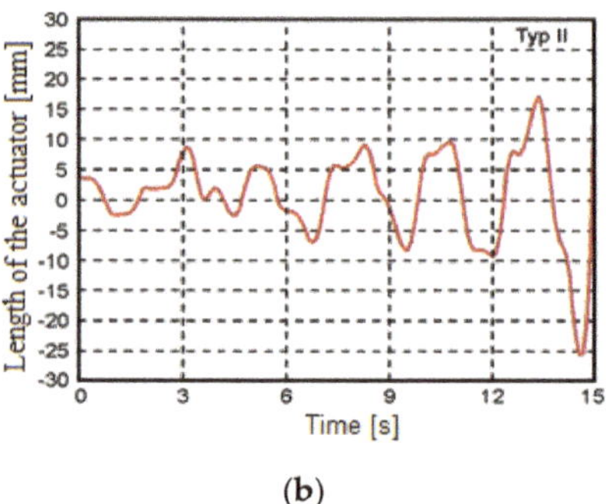

(a) (b)

Figure 3. Example of hydraulic actuator displacement during straight-line riding at 30 km/h speed—(**a**) hydraulic steering system with mechanical feedback and (**b**) hydraulic steering system with LS and hydraulic amplifier.

The values of the human sensing level of angular deviation during articulated vehicle riding permanently exceeded a frequency of about 0.4 Hz (Figure 4a). Deviations of the front frames achieved ±1°. Rear frame achieved a bigger deviation—of up to ±1.7°, due to a lower rear tire pressure and lateral stiffness. Frame turn in articulated joints did not typically exceed 1° (Figure 4b). As in articulated wheel loaders, the driver seat is typically placed above an articulated joint and connected to the rear frame, these deviations trigger driver action and activate the snacking phenomenon.

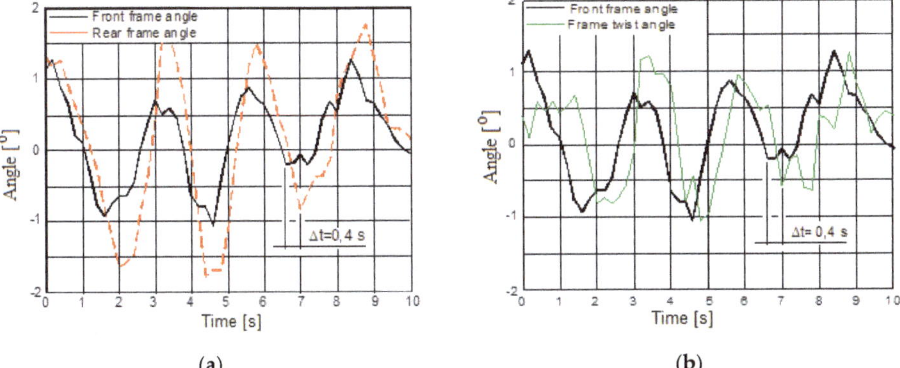

(a) (b)

Figure 4. Angular frame deviation during straight-line drive (**a**) and frame turn and angular deviation of front frame (**b**).

This phenomenon did not exist in vehicles with a steering system, based on the Ackermann principle. The design and work characteristics (positive caster angle) of such a system enable acting stabilizing torque [12], which straightens the leading wheel position. In articulated vehicles with hydraulic steering systems, stabilizing moments do not exist. All errors must be eliminated by the drivers, which have limited ability as their acting and feeling accuracy is about 0.5°.

Initial deviation is caused by limited driver accuracy during maneuvering and the driver sensing only one frame position; at the same time the second frame can be twisted in an articulated joint, which leads to yaw. After the driver closes directional valves in the hydraulic actuator chambers, residual pressure tension of up to 1 MPa can remain, due to friction in the actuator sealing. During riding, this pressure acts on pistons and causes additional frame turn, due to vibration.

According to Figure 4, the leading vehicle front-frame deviation preceded the rear frame yaw. Hence, the loader vehicle drivers led the part deviation with a time lag. This delayed reaction and increased lateral vehicle displacement and angular deviation. Moreover, hydraulic steering systems introduced an additional time lag, as a result of the mechanical and hydraulic characteristics of the system [2]. In hydraulic systems with mechanical feedback, this delay amounted to 0.45 s and in systems with LS and a hydraulic amplifier, it could amount to 0.35 s. Other authors [22] showed that to limit internal leakage, the rotary valve of the steering unit had a positive overlay, which caused a 2° dead zone and disturbed the steering system action. Finally, in articulated loaders, the driver reaction was permanently delayed and this caused the necessity of a permanently high level of activity from the driver.

In modified articulated loaders, the driver seat was in the front of the vehicle and was connected to the front frame. Thus according to Figure 4, the drivers felt a deviation earlier because the rear frame deviation was delayed and their reaction to the leading frame's position could be quicker and provided better directional stability. To improve vehicle reaction, the steering system should be quicker, without LS, and a hydraulic amplifier. To verify this theory, study of a modified loader was conducted.

For the test, a direct hydraulic steering system without LS and amplification was used and the driver was moved to the front of the vehicle to heighten his sensing of yaw (Figure 5). The time lag, which is caused in such hydraulic systems through a pressure transmission in the pipeline, was about 0.05 s. During riding, the average steering wheel rotation was similar (a bit smaller), compared to other hydraulic systems, and its values were about 15°–20°. Driver and front frame yaw deviation were on the same level, at about 1°. However, an increase in low driver activity duration was observed (Figure 5b);—points 1 to 2 and 3 to 4, and the interval between interventions increased from 2.5 s to about 4 s. During low driver activity, steering wheel rotation was about 5°, and was probably caused by driver vibration and inertia during riding. Considering the dead zone in the rotary valve, these steering wheel rotations had no significant influence on the steering system operation. This was confirmed by the hydraulic actuator length measurements. This meant that better yaw sensing and a shorter time lag in the steering system could significantly improve the stability of articulated vehicles and enable a higher driving speed. However, the main problem was the absence of mechanisms that straighten the frame like in vehicles with an Ackermann-based steering system principle. Necessity of continuous correction of frame twist by the driver was the main limitation of the articulated vehicle speed. During riding with a higher speed, the driver sensing and acting ability could be too low and the vehicle could lose stability. In order to increase the articulated vehicle speed, it was necessary to introduce a driver-aid system correcting the frame twist. This should lower the driver effort and reduce driver activity to levels observed in vehicles with a steering system, based on the Ackermann principle.

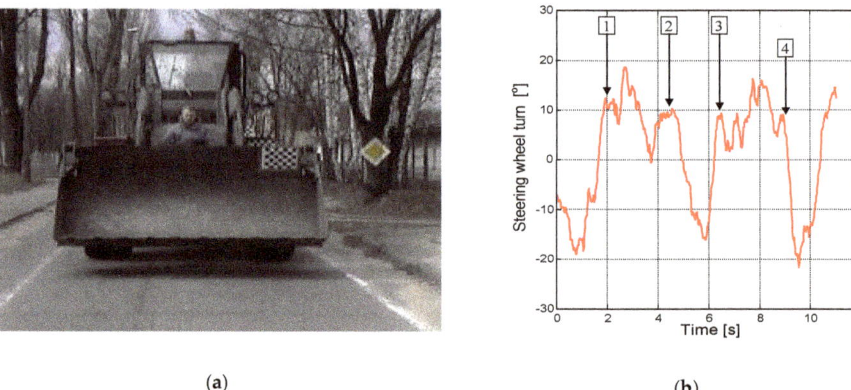

(a) (b)

Figure 5. View of the modified loader with driver before front axle and direct hydraulic steering system during straight-line test (**a**), and driver reactions during modified loader straight-line test at a speed 28 km/h (**b**).

3. Concept of Driver-Aid System

Vehicles (e.g., cars) with Ackermann-based steering systems due to their geometry and characteristics (positive caster angle), have the ability to stabilize vehicle direction. In contrast, the geometry of the steering system in articulated vehicles, makes the driver responsible for stabilizing the steering system. The driver's perception limits the accuracy of vehicle heading, the directional stability of movement, and the top permissible speed. To improve the directional stability of the articulated tractors, the limited sensitivity of the human brain should be supported by the driver-aid system. Since the driver sensitivity of yaw deviation is limited to about 0.5° and this is the main cause of poor directional stability of articulated vehicles, the driver-aid system should straighten the frame and eliminate small frame twist that lead to growing deviations.

For this reason, the proposed system consists of a steering wheel rotation, a frame articulation sensor, and an additional hydraulic circuit for straightening the frame (Figure 6a). For the driver-aid system, we assumed that such a system should act only when the main steering system controlled by the driver was not operating (maneuver was not conducted), and that its flow should be as small as possible and not disturb the main steering system operations, in case of driver-aid system failure. Moreover, its energy consumption should be low and jerk accelerations should not be generated. Hence, the driver-aid system switched on an additional pump to straighten the frame when the steering system wheel sensor indicated that the frame twist was below 1°. It was assumed that such a situation corresponded to straight-line riding. The system should stop when the frame articulation sensor indicated frame straightening. Due to safety reasons, the flow in the driver-aid system was assumed to be 10% of the main steering system flow, considering the maximal steering wheel angular speed was 100 rpm, according to [25,26]. In the analyzed case, it was 10 dm^3/min.

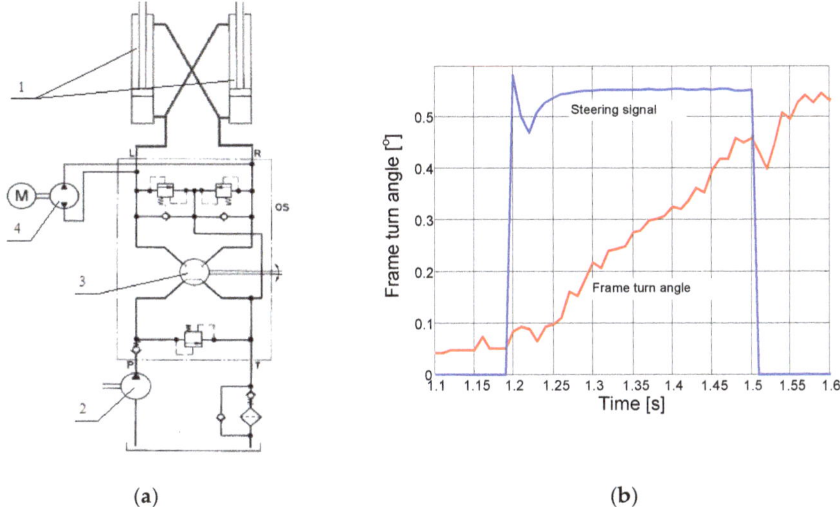

(a) (b)

Figure 6. Driver aid system—(**a**) diagram of hydraulic steering system with additional driver-aid system: 1—steering hydraulic actuator, 2—main pump, 3—hydraulic steering unit, and 4—additional pump with electric drive; and (**b**) frame turns as results of additional gear pump operation.

To verify the operation capability, the possibility of such system actions was investigated, considering a high moment of inertia of the rotated front and rear parts of the vehicle. For this reason, an additional hydraulic system powered with a 24V electric motor, 2.2 kW power, and 10 dm^3/min pump flow was installed on 20-ton wheel loader, and the time domain characteristics were determined. The results (Figure 6b) showed that during 0.3 s of operation, the pump achieved almost nominal flow at 8 dm^3/min, and the frame turned about 0.5°, a value comparable to the initial frame twist. This was a favorable result because it indicated a high speed of turn corrections almost without any time lag, and with a low power consumption at a small pump flow rate, and indicated the possibility to operate such systems.

4. Method

4.1. Vehicle Model

For evaluating the operational efficiency of such a system and estimating its main parameters, a 3D multi-body model simulation of an articulated vehicle as a steered vehicle was developed in the Adams 2014.0.1 (MSC Software Corporation, Newport Beach, CA, USA) software A collaboration co-simulation model of a driver and hydraulic steering system was developed in the Easy5 2015.0.1 Version 9.1.1 (MSC Software Corporation, Newport Beach, CA, USA) software. The assumed principle of the vehicle model is shown in Figure 7 and its parameters are presented in Table 1.

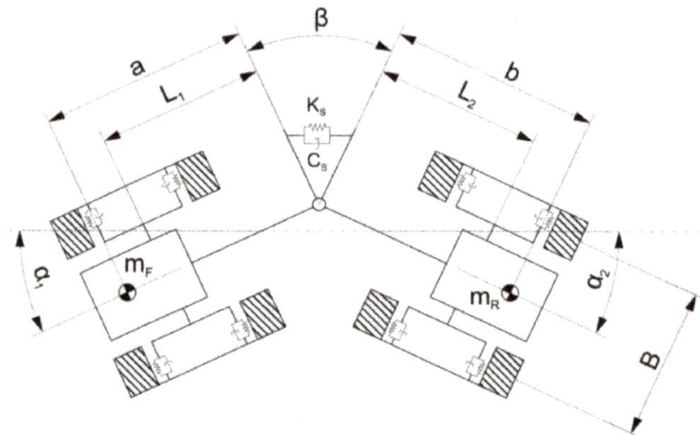

Figure 7. Assumption for model of the articulated vehicle as a steered vehicle.

Table 1. Parameters of the Articulated Vehicle Model.

Parameter	Value
M_F—front frame mass	8500 kg
M_R—rear frame mass	9500 kg
M_W—tire mass	250 kg
D_T—tire diameter	1.62 m
B_T—tire width	0.60 m
B—track width	2.08 m
L_F—front axle distance from joint	1.57 m
L_R—rear axle distance from joint	1.57 m
A—front gravity center distance	1.70 m
B—rear gravity center distance	2.00 m
H_F—height of front gravity center	1.25 m
J_F—front frame moment of inertia	42,000 kgm^2
J_R—rear frame moment of inertia	56,000 kgm^2
k_s—torsional stiffness coefficient	112×10^5 Nm/rd
c_s—torsional damping coefficient	1×10^5 Nms^2/rd^2
K_F—front tire lateral stiffness	6.1×10^5 N/m
K_R- rear tire lateral stiffness	6.1×10^5 N/m
C_F—front tire damping coefficient	5×10^4 Ns^2/m^2
C_R—rear tire damping coefficient	5×10^4 Ns^2/m^2
K_{FY}—front tire radial stiffness	700 kN/m
K_{RY}—rear tire radial stiffness	700 kN/m
H_O—height of rear axle joint	0.80 m
H_R—height of rear gravity center	1.35 m
A—active area of hydraulic actuator	160 cm^2
p_F—front tire pressure	280 kPa
p_R—rear tire pressure	280 kPa

4.2. Driver Model

Modeling human driver behavior is a well-known problem in automotive research. Many different models are used, including linear transfer function models, non-linear models, adaptive controllers, neural networks, and fuzzy logic controllers for modeling complex task performance, including acceleration, braking, and steering, with a driver in the loop. These models can be used to simulate human drivers when testing new vehicle designs and technology to verify developed solutions or concepts [27,28]. According to previous studies [13,29], low order models are often sufficient for many control tasks and are more reliable and easier to apply, because model parameters are relatively easy to identify and determine. The essential requirements of a model should include a time delay due to human processing and acting ability [28,30]. The time delay is strongly connected with driver concentration level, and depend, for example, on vehicle velocity and vehicle and strip width.

Vehicle steering models can use one or more feedback signals as input to the driver model including—lateral displacement, lateral velocity, lateral acceleration, yaw angle, yaw rate, yaw acceleration, roll angle, roll rate, roll acceleration, and other parameters, such as sound and vibration [13,24,26]. Driver behavior emerges from visual cues, motion effects, and tactile feedback.

During straight-line driving and vehicle stability investigation [24,31], the main signals for driver behaviors were lateral displacement and yaw angle (Figure 8)—and models with these inputs were sufficient. Considering the results from investigating the nonlinearity, reflecting time lag and limited driver sensing were introduced (Figure 9). The assumption for the model was that the driver always observed a point on the road. The observable point lay on the required vehicle trajectory at a distance of L from the driver. Distance L depends on vehicle velocity and curvature of trajectory. Any lateral or angular driver inclinations might cause the driver's reaction.

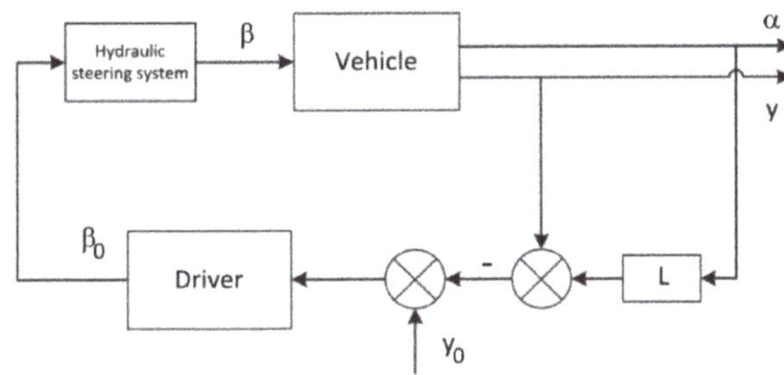

Figure 8. Diagram of steering control loop with driver input as lateral displacement and yaw angle.

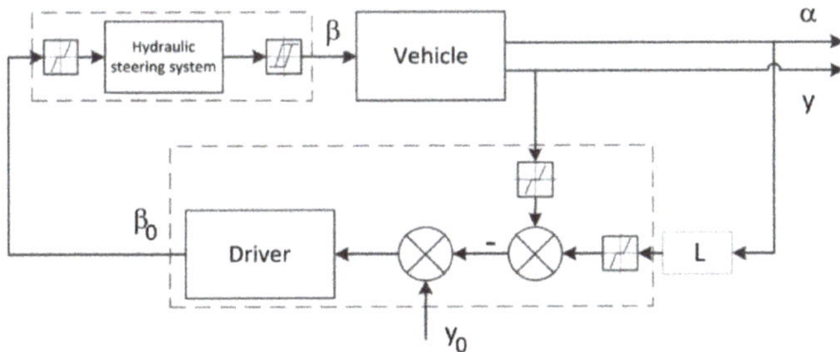

Figure 9. Steering control loop with non-linearity in driver behavior.

Typically, driver behaviors are described as an anticipative model [24,31,32]. To obtain more credible results, both models were used. Due to the lateral driver displacement being below the driver perception level, according to our assumption, the anticipative driver behavior was described as (based only on angular displacement):

$$\beta_D(s) = K\left\langle\left\{\alpha(s)\left[\left(\frac{K_D(T_1 s+1)}{T_2 s+1}\right) - \alpha_o(s)\right]\right\}\left(\frac{K_I}{s} + K_P\right)\right\rangle \tag{1}$$

where, $\beta_D(s)$—driver output signal; $\alpha(s)$—driver input signal (yaw angle); $\alpha_o(s) = 0$—desired yaw angle; K—system gain; K_D—derivative gain; K_I—integral gain; and K_P—proportional gain. The driver model parameters after tuning are shown in Table 2. In the inert model, the driver behavior was described as:

$$\beta_D(s) = (\alpha(s) - \alpha_0(s))\left[K\left(\frac{\omega^2(T_1' s+1)}{s^2+2hs+\omega^2}\right)e^{-\tau s}\right] \tag{2}$$

The model parameters after tuning are shown in Table 3. The whole model developed in the Easy5 2015.0.1 Version 9.1.1 (MSC Software Corporation, Newport Beach, CA, USA) software of a vehicle control system is shown in Figure 10.

Table 2. Parameter of Anticipative Driver Model.

Parameter	Driver on the Front of the Vehicle Next to Front Axle	Driver Above Articulated Joint Connected with Rear Frame
K—system gain	3.5	4.2
K_D—derivative gain	1.3	0.7
K_I—integral gain	0	0
K_P—proportional gain	1.2	0.6
T_1—derivative time	0.45 s	0.79 s
T_2—damping time	0.02 s	$T_2 = 0.02$ s

Table 3. Parameter of the Inert Driver Model.

Parameter	Driver on the Front of the Vehicle Next to Front Axle	Driver Above Articulated Joint Connected with Rear Frame
K—system gain	1	0.67
T_1 -derivative time	0.33 s	0.545 s
ω—natural driver frequency	10 1/s	10 1/s
h—dumping coefficient	2	$h = 2$
τ—driver delay time	0.25 s	0.05 s

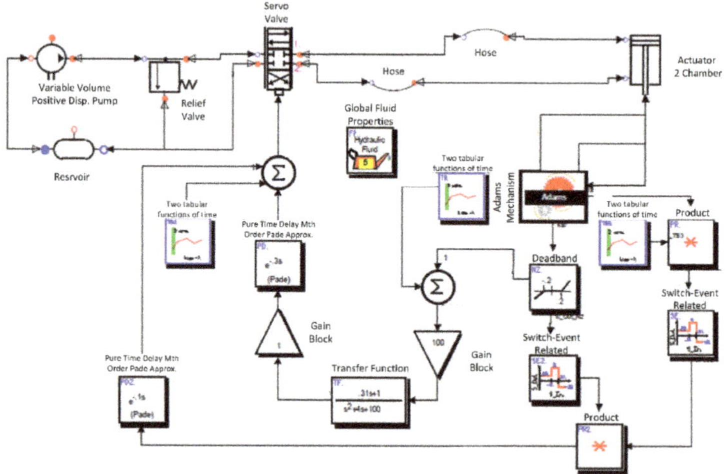

Figure 10. Diagram of vehicle control system developed in the MSC.Easy5 software.

5. Results and Discussion

The goal of the simulation was to select the best driver model, then to verify the acting efficiency of the proposed driver-aid system and finally to investigate the possibility of increasing vehicle velocity. For these reasons, the simulation of straight-line riding with a speed of 30 km/h was made. To compare simulation results with experimental results, a simulation was conducted for both anticipative and inert driver models. Apart from the two driver positions, sitting at the front and above the articulated joint were taken into consideration. In the beginning, the linear model of drivers without a dead zone of sensing was used. For both driver models (anticipative and inert), the snacking phenomenon did not occur. The system was stable without any oscillation. After adding dead zones of driver perception, oscillation appeared in both models. This confirmed that limited driver perception was crucial for the occurrence of the snaking phenomenon. For these reasons, the non-linear driver model was used for the following simulation.

The conducted simulations confirmed the high quality of the used models and the coherence of the results. For the anticipative driver model, results for sitting at the front of the vehicle are shown together with long low driver activity periods (Figure 11a). The length of the hydraulic actuator changed by about ±15 mm. The lateral axle displacement reached 70 mm and the driver yaw angular deviation was about 0.65° (Figure 11b,c, respectively). The oscillation period was about 5.2 s. In the case of the driver sitting above the joint and connected with the rear frame, there was no low driver activity period (Figure 11d) and the length of the actuator changed by up to ±50 mm. The driver was constantly active and the oscillation period was about 4 s. This meant that driver effort was much higher. Despite this lateral axle, displacement reached 120 mm and the driver yaw angular deviation was about 1.45° (Figure 11e,f, respectively).

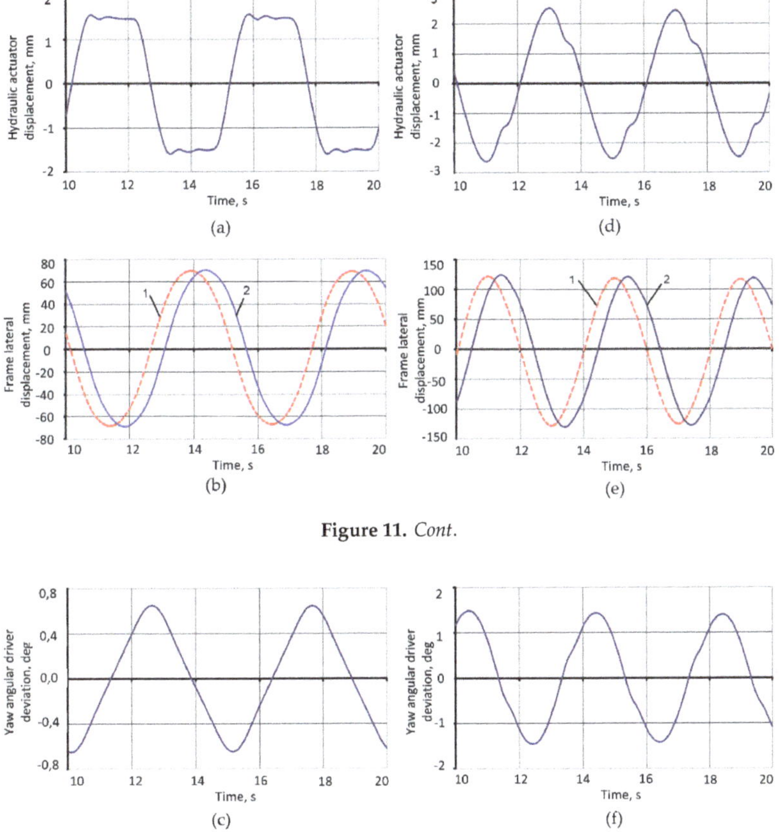

Figure 11. *Cont.*

Figure 11. Example of simulation for anticipative driver siting before front axle of vehicle—(**a**) hydraulic actuator displacement, (**b**) lateral displacement of front—1 and rear—2 axle, (**c**) yaw angular driver deviation, (**d**) hydraulic actuator displacement, (**e**) lateral displacement of front—1 and rear—2 axle, and (**f**) yaw angular driver deviation.

The simulation performed for the same assumption but with an inert driver model produced slightly different results. For a driver model sitting at the front of the vehicle, similar long low driver activity periods arose (Figure 12a), but the oscillation period was shorter, at about 4.5 s, and the hydraulic actuator length changed by about ±12 mm. The lateral axle displacement reached 51 mm and driver yaw angular deviation was about 0.52° (Figure 12b,c, respectively). In the case of the driver sitting above the joint and connected with the rear frame, there was no low driver activity period (Figure 12d) but the length of the hydraulic actuator changed by only up to ±10 mm. The oscillation period was shorter, at about 3.4 s. Despite this, the lateral axle displacement reached 170 mm and driver yaw angular deviation was about 2.3° (Figure 12e,f, respectively). Comparisons of experimental and simulated results showed that the inert model produced much better results. The period of oscillation and lateral displacement was reflected better. It was particularly visible in the simulation with the driver siting over the articulating joint and connected with the rear frame of vehicle. For these reasons, the inert model was selected as a better driver model for future simulations.

As the driver was sitting at the front of the vehicle in transport applications, such as articulated dump trucks and forwarders, and given the possibility of the speed increasing, it was crucial for these applications that the following simulation for higher vehicle velocity was made for this configuration. The results showed (Table 4) that the system could be stable up to 60 km/h but the driver must adapt their behavior. The period of oscillation caused by the snaking phenomenon for varied vehicle velocity was at similar level and the driver had to introduce corrections to the steering system at intervals of approximately 2 s. This indicated a demand for high driver concentration level and heavy physical effort during driving. The proposed driver-aid system should reduce this effort. To evaluate the efficiency of the driver-aid system, its operation was simulated. As an assumption for the simulation, straight-line driving was taken, with the driver triggering the aid system after introducing correction to the steering system. In automated mode, triggering should occur when the frame turn was below 0.5° and driver activity (steering wheel rate) was low. The aim of the aid system action was to prevent articulated frame twist, if it occurred, but the flow in the aid system was limited to 10% of the main steering system. The results of the simulation are shown in Figure 13. After 0.3 s, the frame twist was reduced to the demanded level and the yaw angle of the driver and vehicle was below 0.1°. The angular deviation did not increase and the lateral displacement increased very slowly, reaching the permissible 0.1 m (driver perception limit) in about 10 s. During this period, no driver action was needed. The vehicle movement was stable to aid the driver system's operation and driver effort was significantly lowered—the interval between driver action increased by four to five times, meaning that lateral displacement was a major factor of driver reaction. Thus, the system's behavior was similar to cars' steering systems, and it allowed reaching higher speeds of vehicle motion with a lower level of driver effort and concentration.

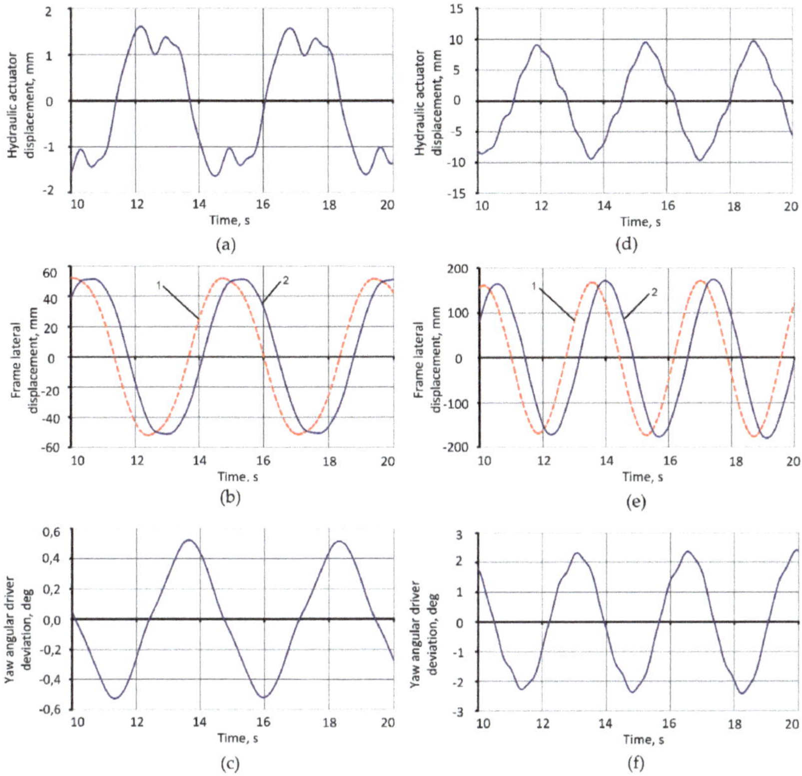

Figure 12. Example of simulation for inert driver siting before front axle of vehicle—(**a**) hydraulic actuator displacement, (**b**) lateral displacement of front—1 axle and rear—2 axles, (**c**) yaw angular driver deviation, (**d**) hydraulic actuator displacement, (**e**) lateral displacement of front—1 axle and rear—2 axles, and (**f**) yaw angular driver deviation.

Table 4. Simulation results of high-speed driving with an inert driver model and a driver at the front of the vehicle.

Vehicle Velocity (km/h)	Derivative Time (s)	Actuator Displacement (mm)	Oscillation Period (s)
30	0.33	1.5	4.40
33	0.36	1.65	4.40
36	0.38	1.4	4.22
39	0.41	1.1	4.26
42	0.44	1.1	4.33
45	0.47	1.1	4.30
48	0.50	1.0	5.12
51	0.51	1.0	4.55
54	0.52	0.8	6.40
57	0.53	0.8	5.43
60	054	0.7	6.10

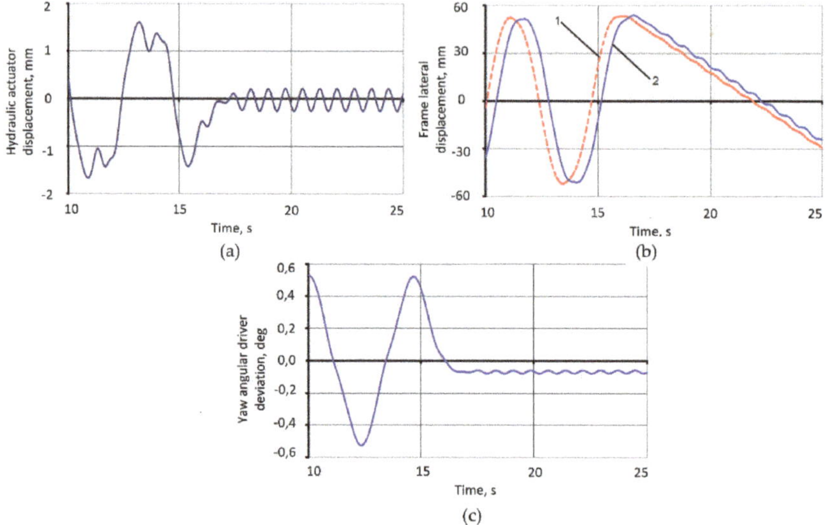

Figure 13. Example of simulation for inert driver model, siting at the front of the vehicle—(**a**) hydraulic actuator displacement, (**b**) lateral displacement of front (1) and rear (2) axle, and (**c**) yaw angular driver deviation.

6. Conclusions

The conducted research demonstrated that the low directional stability of articulated vehicles with hydraulic steering systems was mainly caused by limited driver perception in sensing angular yaw and the absence of mechanisms that straightened the twisted frame and prevented increasing lateral and angular deviation. In car steering systems, this task was performed by special steering knuckle geometry, which generate straightening torque. Hence, yaw was eliminated automatically and drivers had to mainly heave to react to lateral displacement. In the case of articulated vehicles, in contrast to cars, angular deviations were dominant due to the limited possibility to detect and eliminate them. Unnoticed yaw could reach 0.5° and unnoticed lateral displacement was 0.1 m. The conducted tests showed that a hydraulic steering system, especially one operating in an LS system, could introduce a time lag of up to 0.4 s. This lowered the overall steering system efficiency and the vehicle's directional stability. The driver's location significantly influenced movement stability. Driver placement at the front of the vehicle accelerated sensing of lateral and angular deviations from the path.

During the simulation, inert and anticipative driver models were used. The best results and higher simulation accuracy were obtained with the inert model.

Simulations showed that limitation of the snaking phenomenon during movement of articulated vehicles with the driver-aid system was possible. The principles of their operation were verified using simulations. A limited flow rate in the driver-aid system did not cause lateral jerk and acceleration, and high dynamic reactions.

The unique developed model of an articulated vehicle as a steered vehicle, exhibited full usefulness for such simulations and could be used in future work on developing the electronic control and the sensor systems necessary for stabilizing articulating vehicles.

The proposed system could increase the productivity of articulated equipment, due to a significant reduction in driver activity. It lowered the necessary driver effort and concentration level. It, thus, enabled the development of higher travel speed during operations. As was mentioned above, a vehicle's velocity is crucial for its productivity and operation efficiency. The proposed system could also be used to improve the stability of robotized articulated vehicles—one of the main development direction of modern mining equipment. In automated or robotized equipment, it is strongly recommended to measure the frame twist and connect sensors of deviation to the front frame of vehicle.

This paper only presented the concept and preliminary simulations of the new system. The simulations showed that it is possible to increase directional stability of the articulated vehicles. The results should be verified using wider simulations with different scenarios (on various types of terrain and with different slopes). Before the implementation of such a system on vehicles, risk assessments should be conducted and the implementation possibility of additional safety systems for the driver should be taken into consideration.

Author Contributions: Conceptualization: M.J.Ł.; Methodology: M.J.Ł.; Software: M.J.Ł. and A.R.; Validation: A.R., Formal analysis: M.J.Ł. and A.R.; Investigation: M.J.Ł. and A.R.; Resources: M.J.Ł.; Data curation: A.R.; Writing—original draft preparation: M.J.Ł.; Writing—review and editing: M.J.Ł. and A.R.; Visualization: M.J.Ł. and A.R.; Supervision: M.J.Ł.; Project administration: M.J.Ł. and A.R.; Funding acquisition: M.J.Ł. All authors have read and agreed to the published version of the manuscript.

Funding: This research was funded by Military University of Technology under project number 22—758 "Badania konstrukcji maszyn i innowacyjnych technologii wytwarzania oraz systemów Bezzałogowych Platform Lądowych".

Conflicts of Interest: The authors declare no conflict of interest.

References

1. Petraska, A.; Cizuniene, K.; Jarasuniene, A.; Maruschak, P.; Prentkovskis, O. Algorithm for the assessment of heavyweight and oversize cargo transportation routes. *J. Bus. Econ. Manag.* **2017**, *18*, 1098–1114. [CrossRef]
2. Lopatka, M.J.; Muszynski, T. Research of the snaking phenomenon to improve directional stability of remote controlled articulated wheel tool-carrier. In Proceedings of the 20th International Symposium on Automation and Robootics Construction (ISARC), Technische Universiteit Eindhoven (TU/e), Eindhoven, The Netherlands, 21–25 September 2003; pp. 95–101.
3. Meng, Q.; Gao, L.; Xie, H.; Dou, F. Analysis of the dynamic modeling method of articulated vehicles. *J. Eng. Sci. Technol. Rev.* **2017**, *10*, 18–27. [CrossRef]
4. Dzyura, V.O.; Maruschak, P.O.; Zakiev, I.M.; Sorochak, A.P. Analysis of inner surface roughness parameters of load-carrying and support elements of mechanical systems. *Int. J. Eng. Trans. B Appl.* **2017**, *30*, 1170–1175.
5. Maruschak, P.O.; Panin, S.V.; Zakiev, I.M.; Poltaranin, M.A.; Sotnikov, A.I. Scale levels of damage to the raceway of a spherical roller bering. *Eng. Fail. Anal.* **2016**, *59*, 69–78. [CrossRef]
6. Alshaer, B.J.; Darabseh, T.T.; Alhanouti, M.A. Path planning, modeling and simulation of an autonomous articulated heavy construction machine performing a loading cycle. *Appl. Math. Model.* **2013**, *37*, 5315–5325. [CrossRef]
7. ISO. *Standard ISO 5010:2007-06 (E): Earth-Moving Machinery—Rubber-Tyred Machines—Steering Requirements*; ISO: Geneva, Switzerland, 2007.
8. Malinowskij, E.J.; Gajcgori, M.M. *Dinamika Samochodnych Maszin s Szarnirnoj Ramoj*; Maszinostrojenie: Moscow, Russia, 1974. (In Russian)
9. Crolla, D.A.; Horton, D.N.L. The steering behaviour of articulated body steer vehicles. In Proceedings of the I. Mech. E. Conference on Road Vehicle Handling, Loughborough, UK, 22–24 May 1983; Mechanical Engineering Publications: London, UK, 1983.
10. Horton, D.N.L.; Crolla, D.A. Theoretical analysis of the steering behaviour of articulated frame steer vehicles. *Veh. Syst. Dyn.* **1986**, *15*, 211–234. [CrossRef]
11. He, Y.; Khajepour, A.; McPhee, J.; Wang, X. Dynamic modelling and stability analysis of articulated frame steer vehicles. *Int. J. Heavy Veh. Syst.* **2005**, *12*, 28–59. [CrossRef]
12. Wong, J.Y. *Theory of Ground Vehicles*; John Wiley & Sons, Inc.: Toronto, ON, Canada, 1998.
13. Azad, N.L.; McPhee, J.; Khajepour, A. Off-road lateral stability analysis of an articulated steer vehicle with a rear-mounted load. *Int. J. Veh. Syst. Model. Test.* **2005**, *1*, 106–130. [CrossRef]
14. Rehnberg, A. Vehicle Dynamic Analysis of Wheel Loader with Suspended Axles. Ph.D. Thesis, KTH, School of Engineering Sciences (SCI), Aeronautical and Vehicle Engineering, Vehicle Dynamics, Stockholm, Sweden, 2008.

15. Rehnberg, A.; Drugge, L.; Stensson Trigell, A. Snaking stability of articulated frame steer vehicles with axle suspension. *Int. J. Heavy Veh. Syst.* **2010**, *167*, 119–138. [CrossRef]
16. Rehnberg, A.; Edrén, J.; Eriksson, M.; Drugge, L.; Stensson Trigell, A. Scale model investigation of the snaking and folding stability of an articulated frame steer vehicle. *Int. J. Veh. Syst. Model. Test.* **2011**, *6*, 126–144. [CrossRef]
17. ISO. *Standard PN-EN 12643:2000: Earthmoving Machines, Rubber Wheeled Machines: Requirements Concerning the Steering System*; ISO: Geneva, Switzerland, 2000.
18. Dudzinski, P.; Skurjat, A. Directional dynamics problems of an articulated frame steer wheeled vehicles. *J. KONES Powertr. Transp.* **2012**, *19*, 89–98. [CrossRef]
19. Zardin, B.; Borghi, M.; Gherardini, F.; Zanasi, N. Modeling and simulation of a hydrostatic steering system for agricultural tractors. *Energies* **2018**, *11*, 230. [CrossRef]
20. Daher, N.; Ivantysynova, M. A virtual yaw rate sensor for articulated vehicles featuring novel electro-hydraulic steer-by-wire technology. *Control Eng. Pract.* **2014**, *30*, 45–54. [CrossRef]
21. Daher, N.; Ivantysynova, M. Yaw stability control of articulated frame off-highway vehicles via displacement controlled steer-by-wire. *Control Eng. Pract.* **2015**, *45*, 46–53. [CrossRef]
22. Lopatka, M.J. *Investigation of Snaking Phenomenon Limitations in Articulated Combat Equipment*; Grant Report KBN nr 0T00B01921; Military University of Technology: Warsaw, Poland, 2004.
23. Lopatka, M.J.; Muszynski, T.; Przychodzien, T. Research of high speed articulated wheel tool-carrier steering systems. In Proceedings of the 9th European Regional Conference of The International Society of Terrain-Vehicle Systems (ISTVS), Harper Adams University College, Newport, UK, 8–11 September 2003; pp. 261–272.
24. Chaczaturow, A.A.; Afanasjew, W.L.; Wasiliew, W.S.; Goldin, G.W.; Dodonow, B.M.; Żigariew, W.P.; Kołcow, W.I.; Jurik, W.S.; Jakobliew, E.I. *Dinamika Sistemy Doroga-Szina-Awtomobil-Woditiel*; Maszinostrojenie: Moscow, Russia, 1976. (In Russian)
25. Danfoss. *Hydrostatic Steering Components: Hydrostatic and Hydromechanical Steering Systems HK.20.B1.02*; Danfoss: Nordborg, Denmark, 2019.
26. Eaton. *Steering Control Units and Torque Generators: Char-Lynn Power Steering No 11-872*; Eaton Corporation: Dublin, Ireland, 2002.
27. Delice, I.I.; Ertugrul, S. Inteligent modeling of human driver: A survey. In Proceedings of the IEEE 2007 Intelligent Vehicles Symposium, Istanbul, Turkey, 13–15 August 2007; pp. 648–651.
28. MacAdam, C.C. Understanding and modeling the human driver. *Veh. Syst. Dyn.* **2003**, *40*, 101–134. [CrossRef]
29. Brito Palma, L.; Vieira Coito, F.; Sousa Gil, P. Low orders models for human controller-mouse interface. In Proceedings of the IEEE 2012 16th International Conference on Intelligent Engineering Systems (INES), Lisbon, Portugal, 13–15 June 2012; pp. 515–520.
30. Kesting, A.; Treiber, M. How reaction time, update time, and adaptation time influence the stability of traffic flow. *Comput. Aided Civil Infrastruct. Eng.* **2008**, *23*, 125–137. [CrossRef]
31. Andrzejewski, R.; Awrejcewicz, J. *Nonlinear Dynamics of a Wheeled Vehicle*; Springer: New York, NY, USA, 2005.
32. Chen, X.; Li, R.; Xie, W.; Shi, Q. Stabilization of traffic flow based on multi-anticipative intelligent driver model. In Proceedings of the 12th International IEEE Conference on Intelligent Transportation Systems 2009, St. Louis, MO, USA, 4–7 October 2009.

Article

Theoretical Design of a Novel Vibration Energy Absorbing Mechanism for Cables

Zhen Qin [1,2,†], Yu-Ting Wu [1,†], Aihua Huang [2], Sung-Ki Lyu [1,*] and John W. Sutherland [2]

1 School of Mechanical and Aerospace Engineering, ReCAPT, 501, Gyeongsang National University, Jinju-daero, Jinju-si, Gyeongsangnam-do 52828, Korea; musicboy163@naver.com (Z.Q.); didawyt@163.com (Y.-T.W.)

2 Laboratory for Sustainable Manufacturing, Environmental and Ecological Engineering, Purdue University, 500 Central Drive, West Lafayette, IN 47907, USA; huan1363@purdue.edu (A.H.); jwsuther@purdue.edu (J.W.S.)

* Correspondence: sklyu@gnu.ac.kr; Tel.: +82-55-772-1632

† These authors contributed equally to this paper.

Received: 29 June 2020; Accepted: 29 July 2020; Published: 31 July 2020

Abstract: A novel design of a vibration energy absorbing mechanism (VEAM) that is based on multi-physics (magnetic spring, hydraulic system, structural dynamics, etc.) for cable vibration is proposed. The minimum working force of the hydraulic cylinder has been exploited in this design in order to combine a non-linear stiffness vibration isolation module that is composed of permanent magnetic springs with hydraulic viscous vibration damping modules. In response to different environmental vibration impacts, VEAM can automatically switch the vibration control modes without an electronic mechanism. Additionally, the non-contact design effectively reduces the wear that is induced by the reciprocating motion of the small amplitude of the hydraulic viscous dampers. The proposed mechanism is explained and a theoretical model is established. The transmissibility of the two modules at a single degree of freedom is derived using the harmonic balance method. After that, a series of variable control numerical simulations were performed for each important parameter. Empirical rules for designing the system were created by comparing the influence of each parameter on the vibration isolation performance of the entire system.

Keywords: vibration energy control; multi-physics mechanism theory; non-linear stiffness; isolation; hydraulic system; magnetic spring

1. Introduction

Cables are widely used in construction, cable cars, traction, and other fields due to their extremely high tensile strength, long-distance capabilities, and low cost. This article will conduct research that is based on the cable-stayed bridge, which is the most representative and technically demanding application of cables, and illustrate a novel vibration energy absorbing mechanism that effectively responds to common vibration that is caused by environmental impacts with the automatic switching modes of the non-stiffness vibration isolators with the hydraulic viscous dampers.

In recent years, the cable-stayed bridge design has gained popularity as the design of choice for the construction of long-span bridges. This can be attributed to its simple and beautiful design, and its ability to span longer distances at relatively low construction costs [1]. From 2016 to 2019, a total of 20 cable stayed bridges with spans over 500 m were built around the world. Currently, the longest span cable-stayed bridge in the world is the Russky Bridge built over the Eastern Bosphorous in Russia, which has a span of 1104 m. Although the span of this bridge is quite short when compared to suspension bridges, the trend in construction seems to be moving towards an increasing number and spans of cable-stayed bridges. The cable-stayed bridge is mainly composed of bridge pylons, cables, and decks. The cables connect the decks to the pylons directly, so that the pylons carry the entire load. Like all structures, cable-stayed bridges are also subjected to vibration resulting from such exogenous factors, such as wind, rain, temperature change, and the movement of pedestrians and vehicles. These vibrations are transmitted to pylons and decks by the cables [2]. If the excitation frequency is close to a natural frequency of the bridge system, resonance will occur. Resonance is regarded as one of the main causes of bridge damage leading to fatigue and shortened life [3–7]. Therefore, a great amount of effort has been undertaken to develop methods for avoiding/mitigating the effects of vibration on cable-stayed bridge structures. Increasing the damping between the cable and the bridge deck is one of the key methods used to improve the robustness of cable-stayed bridges by preventing the occurrence of resonance.

Various types of vibration control systems are used in a wide variety of fields, such as automotive, aerospace, and structural engineering. According to their working mechanisms, vibration control systems for cable-stayed bridges can be divided into: (1) high-damping rubber (HDR) dampers, (2) viscous dampers, (3) magneto-rheological (MR) dampers, and (4) friction dampers.

Rubber dampers are the oldest type of damping systems used in cable-stayed bridges. These dampers dissipate vibration energy via hysteretic losses. Nakamura et al. [8] proposed an external high-damping rubber (HDR) damper for cable-stayed bridges and experimentally evaluated its performance against dynamic vibration. Cu et al. [9,10] formulated a complex eigen value problem in order to address the different damping effects of a single HDR damper and two attached HDR dampers under different installation conditions and material loss factors. Their work provides a strong theoretical basis for the design of HDR dampers. The damping performance of the HDR damper is highly dependent on the damping characteristics of its material, e.g., rubber. However, rubber materials have a limited energy consumption performance and they are easily affected by environmental factors, such as temperature.

The viscous damper category is mainly comprised of hydraulic cylinder type dampers that dissipate the vibration energy through the friction between the hydraulic fluid and their orifice(s). The maximum damping ratio of a viscous damper has a relationship with the relative distance between the support and the damper. This was proposed by Pacheco et al. [11], who also derived an optimal damping constant for viscous dampers with low modes of vibration. A later study by Krenk [12] optimized the modal damping values of the lowest five modes of vibration. Sawka [13] experimentally studied the total pressures of viscous dampers as a function of time after excitation by a mass impact. This work showed that viscous dampers show better damping performance against large dynamic vibrations, but they are not sensitive enough to small amplitude vibrations. Tomski and Kukla et al. [14] confirmed these results through numerical simulations. Viscous dampers are susceptible to reduction in load capacity due to load misalignment. In order to explore this further, Gamez-Montero et al. [15] performed a buckling analysis on hydraulic cylinders, while Niu et al. [16] proposed an oil damper with variable stiffness (ODVS). In their designs, the orifice size of the variable overflow valve can be changed to achieve a variable damping ratio according to different hydraulic loads. Viscous dampers are also susceptible to leakage due to thermal deformation and aging of the seal ring due to frequent reciprocating movements that are mostly associated with low amplitude vibrations. To overcome the problem of leakage, Zhan et al. [17] proposed the design of a third-generation hydraulic cylinder with a special lip shape at the piston periphery that varies the clearance between the piston and the cylinder according to the change in hydraulic pressure. A common method for evaluating the isolation performance of dampers is to calculate their transmissibility. Tao et al. [18] established a mathematical model for calculating the damped power transmissibility of a vibration isolation system. Guo et al. [19] analyzed a single degree of freedom (SDOF) displacement vibration isolation system (DVIS) and a forced vibration isolation system (FVIS) with a nonlinear viscous damper; simulations of system transmissibility were obtained. It has been noted that the non-linear viscous damper can significantly improve the damping performance across a wide range of frequencies. However, the problem of wear has not yet been solved due to frequent reciprocating movements.

Magneto-rheological (MR) fluid is a force transmission fluid that can be precisely controlled while using a magnetic field. Therefore, it is widely used in controllable damping systems. Wu et al. [20] proposed a tuned mass damper magneto-rheological (TMD-MR) design, which showed good damping performance in all modes during experimental verification. Macháček et al. [21] incorporated an elastic metal bellows in the spring-MR damper, which significantly improved the vibration absorption over the entire frequency range. MR dampers require the employment of an active feedback control system in order to operate properly. The controller senses the input vibration parameters and accordingly modulates the electrical signal that influences the MR fluid characteristics. This requires the use of complex electronic equipment and a constant supply of electrical power. These requirements increase the system complexity and costs related to installation, operation, and maintenance. They may also increase the chances of system failure due to a variety of causes including electrical failure.

Friction dampers use friction to dissipate vibration energy. Using a mathematical model, Nguyen et al. [22] analyzed the damping performance of two friction dampers on a stay cable and proposed the best design and installation method for friction dampers. However, friction dampers have some limitations, such as their inability to dampen low amplitude vibrations and the use of consumable friction materials that need to be replaced regularly, which serves to increase the system's operating and maintenance costs.

With the development of industrial technology, the requirements for high vibration isolation performance are also increasing [23–25]. It is apparent from the preceding discussion that there is a need for the development of a vibration damping system. Moreover, a vibration damping system is needed that can handle both high and low amplitude vibrations without requiring complex hardware, continuous power supply, and frequent maintenance. One approach that offers promise is non-linear damping technology that has attracted a great amount of attention from researchers in recent years [26–28]. The existing literature shows that the natural frequency of an isolation system can be decreased by introducing a non-linear stiffness [29–31]. This means that the introduction of non-linear stiffness is an effective method for broadening the isolation bandwidth of a damping system. Carrella et al. [32] proposed a quasi-zero-stiffness (QZS) vibration isolation system with non-linear stiffness consisting of a positive stiffness spring and two negative stiffness springs. The high-static-low-dynamic-stiffness (HSLDS) vibration isolation characteristic of their system was confirmed by calculating its transmissibility. Numerous innovative QZS designs have been proposed and applied in various fields due to the desirable performance of QZS vibration isolation [33–36]. Permanent magnets (PM), when placed in an opposing repulsive manner, can generate a non-linear stiffness. This characteristic of PMs has resulted in their wide usage in the design of QZS vibration isolators in recent years. A large number of studies have shown that vibration isolators that are composed of PMs have better vibration isolation performance than linear vibration isolators [37–39]. Yan et al. [40,41] proposed the design of a vari-stiffness nonlinear isolator (VSNI) and a bi-state nonlinear vibration isolator (BS-NVI). In their designs, the nonlinear stiffness is controlled through the manipulation of the magnetic fields. This means that the isolation performance of the system is controlled by changing the distance and included angle between the PMs. Thus, the transmissibility models established were verified through experiments.

Matsumoto et al., through field observations and wind tunnel tests, conducted in-depth evaluation and analysis of the influence of rain and wind induced vibrations on a cable bridge [42–46]. Ma et al. [47] and Du et al. [48] used wind tunnel tests and digital imaging techniques to study the dynamic characteristics of cable-stayed bridges. These studies confirmed that the amplitude of rain, wind, and traffic induced vibrations at the installation position of the damping system is usually very small. Large amplitude vibrations only occur under special circumstances and are, thus, less frequent. Hence, cable-stayed bridges can greatly benefit from the design of a vibration damping system that does not require external power supply and can handle small amplitude vibrations with minimal wear while also being able to cater to larger amplitude vibrations.

Consequently, this paper proposes a hybrid damping system design that combines a PM based non-linear stiffness vibration isolator and a hydraulic viscous damper. In the proposed design, when the amplitude of vibration in the cable is less than a preset threshold value, the non-linear stiffness isolator damps the vibration transmitted to the deck and the viscous damper is not actuated. This will help to increase the overall system life, as it protects the viscous damper from thermal deformation and excessive wear of the seal ring due to its high frequency reciprocating motion that is associated with low amplitude vibrations. When the vibration amplitude becomes greater than the threshold value, the PMs travel until there is a minimum relative distance between them and then the viscous dampers are actuated to further dissipate the vibration energy. Thus, the presented system, which we have called a vibration energy absorbing mechanism (VEAM), is developed to meet three objectives: (1) reduce the wear due to inefficient working conditions of the viscous dampers used in the system. (2) Provide a non-contact link between the viscous damper and the cable so that shear damage to the damper due to multi-directional cable motion can be avoided. (3) Achieve damping of a wide range of vibration amplitudes without using complex electronic components, such as sensors and controllers. The detailed design of the VEAM and validation through simulations are presented in this paper.

2. Mechanism of the VEAM

Figure 1 presents the 3D model of VEAM. The cable passing through the center is supported on six sides by the developed multi-physic dampers, which are in turn connected to the bridge deck. Traditional viscous damper systems usually adopt a three damper arrangement [16]. In contrast, the VEAM adopts an arrangement of three opposing pairs of dampers resulting in a six-directional cable support system. This has been done to better counter the randomness of stay cable vibration. Figure 2a shows the internal configuration of the VEAM. Because the system is symmetrically constructed using six identical modules, we will only use one of those modules to illustrate the design principles. The complete system is fitted to the cable via a central hex piece that is divided into two parts which are assembled using long through bolts. The six damper assemblies are attached to this hex piece, one on each side. The damper assemblies consist mainly of a ball-joint connected to the hex piece, a PM based non-linear stiffness isolator, a viscous damper, and a rotary mount to allow the rotation of the damper assembly about an axis perpendicular to the damper's axis of travel. The vibration of the stay cable moves the magnets, which dampen it due to non-linear stiffness. The vibration force is transmitted to the viscous damper through repulsion between the PMs placed with like poles facing each other. If the force magnitude is greater than the minimum force required to actuate the viscous damper, then the damper starts moving and dissipates the vibration energy through viscous drag.

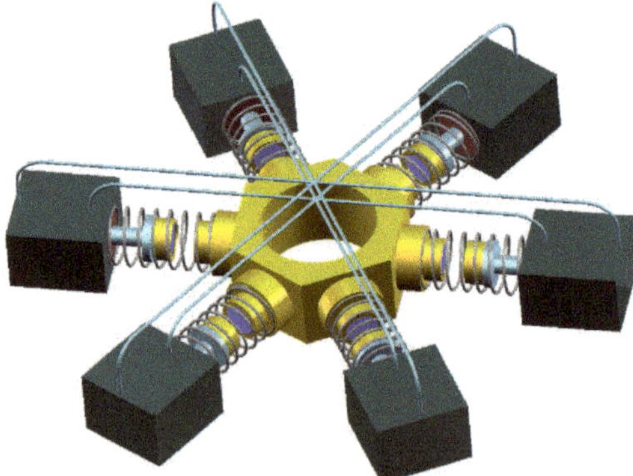

Figure 1. Three-dimensional (3D) model of the vibration energy absorbing mechanism (VEAM).

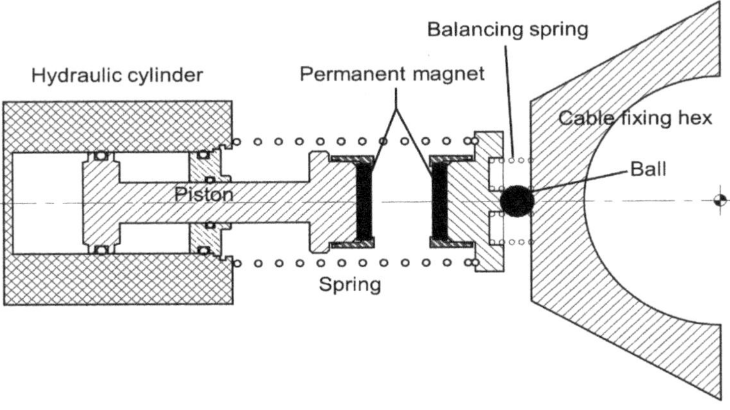

(**a**) Overall cross-section

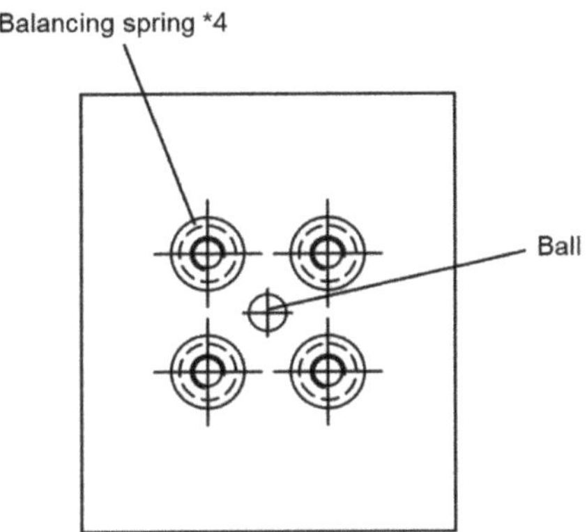

(**b**) Ball-spring balancing module

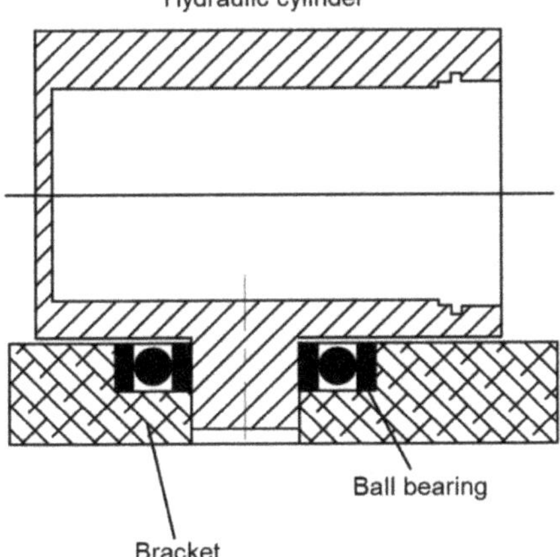

(**c**) Cylinder rotation module

Figure 2. Internal configuration of 1/6 piece of VEAM.

The thrust force applied to the piston should be aligned with the cylinder axis and should have no radial component in order to keep the viscous damper moving smoothly with minimal amount of wear. In the proposed system, the source of this thrust force is the PM based isolator attached to the end of the hydraulic piston. Thus, the necessary conditions for the smooth operation of the hydraulic cylinder can only be guaranteed if the force applied to the PM assembly has no radial component. However, in reality, the direction of vibration is not always aligned with the center axis of the damper assembly. Therefore, a compliant attachment between the damper assembly and the cable is required to ensure that only axial forces are transmitted to the damper assembly. In the proposed design, a ball joint is added between the hex and the PM side of the damper assembly in order to allow for the cable fixing hex to change angles with respect to the damper assembly. This joint, as shown in Figure 2b, consists of a ball at the center with four retaining springs around it to prevent excessive movement and damage to the joint. At the other end of the damper assembly, the cylinder of the viscous damper is connected to the system frame using a ball bearing based rotary joint shown in Figure 2c that allows for rotation of the damper assembly around an axis perpendicular to the cylinder axis. The addition of these two joints ensures that the two PMs are always coaxial and only axial loads are applied to the viscous damper. Furthermore, to provide structural damping while keeping the PMs in parallel [49,50], the design also includes a spring connecting the hex side PM and the viscous damper cylinder.

It can be seen from Figure 3a that when the hex moves due to cable movement, the six damper assemblies rotate to different angles to compensate for this displacement. However, the PMs and viscous damper always remain aligned with each other. The rotation angle of each damper assembly can be derived while using the geometric relationship between the key elements. Figure 3b shows the simplified schematic diagram of this scenario. In this figure, points A, B, C, D, E, and F represent the rotating centers of the bearings between the hydraulic cylinders and the system structure, while d and $\angle\theta$ represent the magnitude and angle (acute angle with respect to the horizontal) of displacement of the center point of the cable fixing hex. Circles with black solid hatching represent the joint balls. G is a ball in the initial (default) position and G′ is the final position of ball G after the displacement.

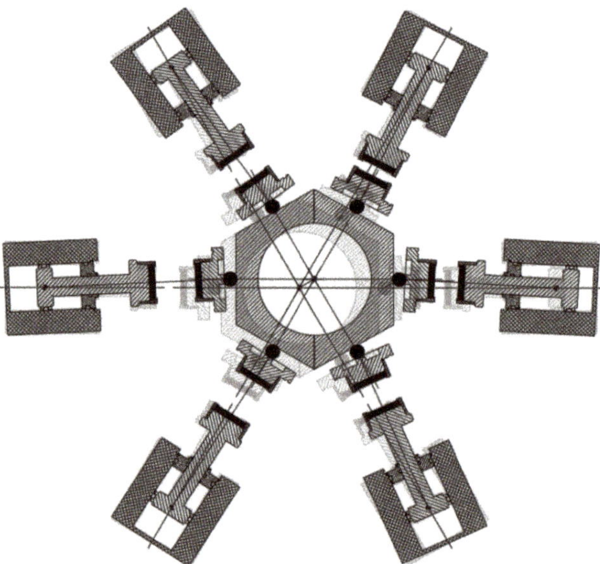

(**a**) Comparison of system positions before and after displacement

Figure 3. *Cont.*

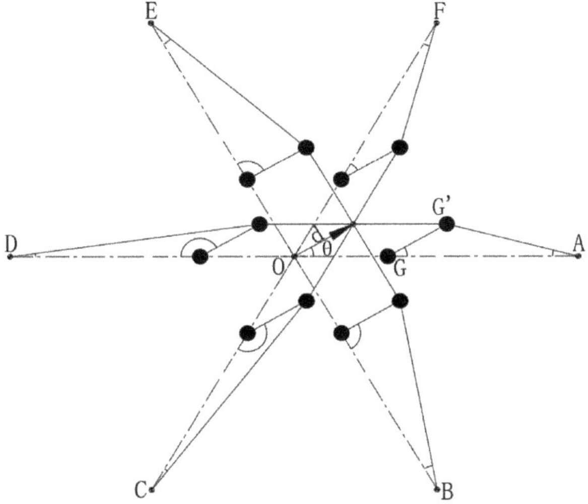

(**b**) Simplified schematic of the movement

Figure 3. Kinematic response of the VEAM to cable displacement.

It can be observed that the displacement GG′ is equal to d and the angle ∠G′GA is equal to ∠θ. The distances from the cable fixing hex center to the ball and to the rotating center of the bearing at the initial position are known. Thus, the distance between the ball at position G and the bearing rotating center at the initial position is GA = OA − OG.

According to the law of cosines, G′A can be deduced as:

$$\mathrm{G'A} = GA^2 + d^2 - 2{\cdot}GA{\cdot}d{\cdot}\cos\theta \tag{1}$$

Here, we assume that the radial deformation of the spring is zero, which means that it can only perform telescopic movement along the axial direction. Through the law of sines, the rotation angle of the hydraulic cylinder can be expressed as:

$$\angle\mathrm{A} = \arcsin\left(\frac{d{\cdot}\sin\theta}{G'A}\right) \tag{2}$$

From the geometric relationship, it can be seen that the rotation angle of the cable fixing hex side magnet around the ball center and the rotation angle of the hydraulic cylinder are alternate interior angles. Thus, combining Equations (1) and (2), the ∠A can be rewritten as:

$$\angle\mathrm{A} = \arcsin\left(\frac{d{\cdot}\sin\theta}{GA^2 + d^2 - 2{\cdot}GA{\cdot}d{\cdot}\cos\theta}\right) \tag{3}$$

The rotation angles of the other hydraulic cylinders can also be calculated with Equations (4)–(8):

$$\angle\mathrm{B} = \arcsin\left(\frac{d{\cdot}\sin\left(\theta + \frac{\pi}{3}\right)}{GA^2 + d^2 - 2{\cdot}GA{\cdot}d{\cdot}\cos\left(\theta + \frac{\pi}{3}\right)}\right) \tag{4}$$

$$\angle\mathrm{C} = \arcsin\left(\frac{d{\cdot}\sin\left(\theta + \frac{2\pi}{3}\right)}{GA^2 + d^2 - 2{\cdot}GA{\cdot}d{\cdot}\cos\left(\theta + \frac{2\pi}{3}\right)}\right) \tag{5}$$

$$\angle D = \arcsin\left(\frac{d \cdot \sin(\pi - \theta)}{GA^2 + d^2 - 2 \cdot GA \cdot d \cdot \cos(\pi - \theta)}\right) \quad (6)$$

$$\angle E = \arcsin\left(\frac{d \cdot \sin\left(\frac{2\pi}{3} - \theta\right)}{GA^2 + d^2 - 2 \cdot GA \cdot d \cdot \cos\left(\frac{2\pi}{3} - \theta\right)}\right) \quad (7)$$

$$\angle F = \arcsin\left(\frac{d \cdot \sin\left(\frac{\pi}{3} - \theta\right)}{GA^2 + d^2 - 2 \cdot GA \cdot d \cdot \cos\left(\frac{\pi}{3} - \theta\right)}\right) \quad (8)$$

A conventional viscous damper forms a hydraulic circuit by connecting the two hydraulic chambers cut off by the piston by opening an orifice in the piston itself. In this case, mainly the friction between the hydraulic oil and the orifice provides the damping effect. However, because the orifice is inside the hydraulic cylinder, it is difficult to change the orifice size after assembly. In order to facilitate the adjustment of the damping ratio, this proposed design utilizes an external circuit design, as shown in Figure 4. The double triangle marks in the figure indicate the orifice regulating valves, while LL, LR, RL, and RR represent the hydraulic chambers divided by the pistons. When the cable fixing hex is in the default position, the left and right subsystems of the VEAM are in a balanced state and the pistons on both sides are positioned in the middle of their respective hydraulic cylinders with equal pressures on both sides of the pistons. The LL chamber is connected to the RR chamber and the LR chamber is connected to the RL chamber. The purpose of this inter-connection is to ensure the continuity of the damping process. Assuming that the repulsive force between the PMs is large enough, when the cable fixing hex moves leftwards the magnet pushes the left side piston, reducing the LL chamber volume and increasing its internal pressure. Therefore, the hydraulic oil in the LL chamber flows through the orifice and the piping to the RR chamber, which is at a relatively lower pressure. At the same time, the pressure in the LR chamber is reduced, which sucks in the hydraulic oil from the RL chamber. This creates a pressure difference between the RL and RR chambers that causes the right side piston to move out. This mechanism ensures that when the cable fixing hex moves, the pistons on both sides simultaneously move in the same direction. In this way, when the hex moves to the maximum amplitude in one direction and then moves in the opposite direction, the two PMs on the opposite side will be close enough to generate a repulsive force form the movement onset, thus preventing the occurrence of sudden cable movement or the jumping phenomenon. Because the damper assemblies are designed to move, the hydraulic lines used in the prototype have been coiled to reduce the stress at coupling points. Conventional viscous dampers usually use internal or external springs to restore the piston to its default position. Because of the design of the cable-stayed bridge, the cables are always in tension and so always tend to return to their default position. The VEAM design exploits this characteristic and uses the cable movement to automatically restore the hydraulic dampers to their default positions, thus doing away with the need for an additional spring.

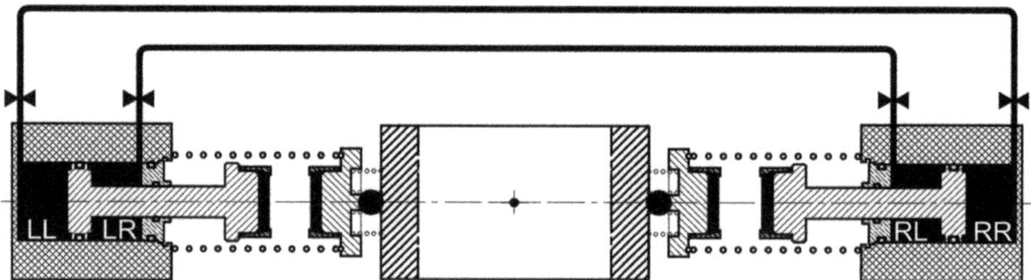

Figure 4. Schematic diagram of the hydraulic connection between opposing cylinders.

3. Theoretical Modeling of the VEAM

The VEAM is composed of six identical damper assemblies, where each pair of opposing damper assemblies is connected through a hydraulic circuit to form a sub-system. Thus, each sub-system is composed of two magnetic non-linear stiffness modules and two hydraulic viscous damping modules. The isolation and damping performance of the VEAM can be evaluated by calculating its transmissibility [51]. The harmonic balance method (HBM) is an efficient method for modeling periodic problems and it is widely used to derive transmissibility [41]. Therefore, in this paper, HBM is used to establish theoretical models for the non-linear stiffness and the viscous damping modules. In this section, a mathematical model for the mechanism of a subsystem of VEAM in a single degree of freedom is established.

3.1. Automatic Switching Between Non-Linear Stiffness and Viscous Damping Modules

Figure 5a presents a simplified single degree of freedom theoretical model of a subsystem of the VEAM along the direction of its central axis. The thick line in the middle indicates the cable, with its two ends fixed between the deck and the pylon under a tensile force (T) acting at each end. The total length of the cable is L. The VEAM is fixed at a distance L1 from the deck. The cable fixing hex is connected to the cable and the damping system structure is fixed to the deck. In the figure, y_c represents the excitation of the stay cable and y_d represents the absolute displacement of the deck. As we are only considering the single degree of freedom case, the role of the ball joint and the rotary joint of the hydraulic cylinder is not considered. Hence, VEAM is simplified as a model in which the magnetic force that is generated by the PMs is connected in series with the hydraulic cylinder and then in parallel with the spring.

The minimum working pressure of the hydraulic cylinder is mainly caused by the friction of the piston seal, which is mostly dependent on the manufacturing and assembly accuracies, etc. In recent years, in order to improve the efficiency of hydraulic cylinders, researchers have tried to minimize this minimum working pressure through various methods, but static friction that is caused by seals is inevitable. With the development of industrial technology, it is not difficult to maintain the minimum working pressure at a fixed value by controlling the precision of machining and seal manufacturing. This minimum working pressure can be easily determined through experiments. The VEAM realizes automatic switching between the non-linear vibration isolator and the hydraulic viscous damper based on the minimum working pressure of the hydraulic cylinder.

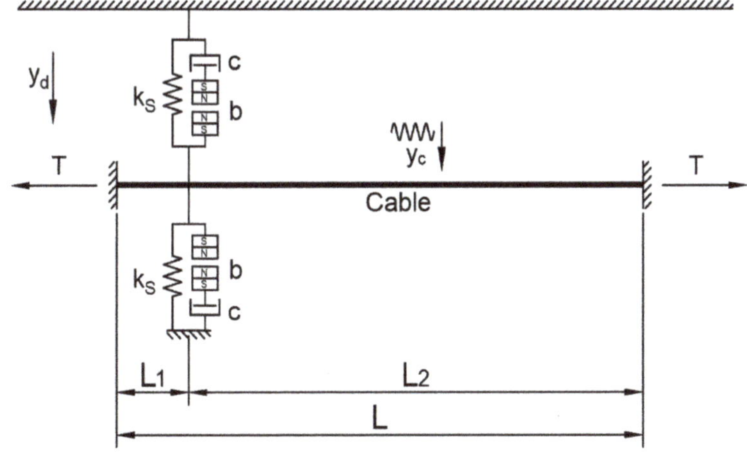

(**a**) Schematic of the sub-system mechanism

Figure 5. *Cont.*

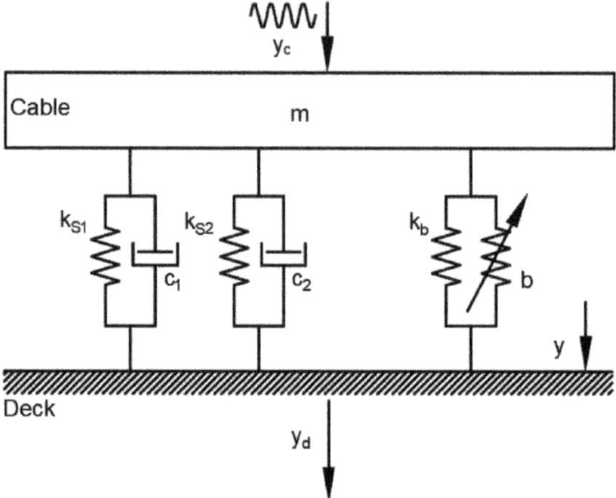

(b) Mechanism of the nonlinear vibration isolator module

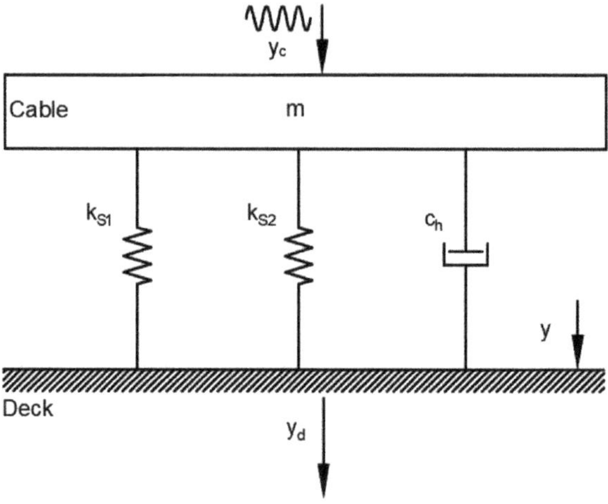

(c) Mechanism of the hydraulic viscous damper module

Figure 5. Simplified theoretical model of the VEAM.

For each subsystem, it can be observed from Figures 2 and 4 that one PM is fixed on the piston rod and the other is connected to the vibration excitation side, such that like the poles of the two PMs are opposite to each other. When a large amplitude vibration occurs at the excitation side, the distance between the PMs becomes small and the repulsive force increases. This repulsive force is transmitted to the piston rod as a thrust force (F_m). When this thrust force is less than the minimum working force (F_f) of the hydraulic cylinder only the non-linear vibration isolator works. Conversely, only the hydraulic viscous damper works when the thrust force is greater than the minimum working force of the hydraulic cylinder. Yamakawa et al. [52] proposed that the repulsive force between two PMs is inversely proportional to the nth power of the distance between them. The two magnets that are discussed in this article are identical and, therefore, satisfy the conditions for the application of this relationship. Thus, the repulsive force between the two PMs can be taken as,

$$F_m = a(l - y)^{-n} \tag{9}$$

where, a (Unit: Nm^{-n}) is the equivalence coefficient of the repulsive force, n denotes the characteristic index, l (Unit: m) represents the preset distance between the PMs, and y (Unit: m) denotes the relative displacement between the cable and ground. When only the magnetic non-linear stiffness module works, the inequality $F_m \leq F_f$ exists. Substituting Equation (9) into this inequality, we get the relationship between the preset distance of the two PMs and the excitation amplitude:

$$l \leq y + \sqrt[n]{\frac{a}{F_f}} \tag{10}$$

3.2. Nonlinear Stiffness Isolator Model

In the VEAM sub-system, there is a set of mutually exclusive PMs at each end of the cable fixing hex, these two sets of PMs form a magnetic spring. The force exerted on the cable fixing hex by the repulsion of the magnetic spring can be expressed as the following Duffing type cubic polynomial equation [40,41].

$$F_{ms} = k_b y + b y^3 \tag{11}$$

where, F_{ms} represents the resultant force provided by the magnetic spring and k_b and b represent the equivalent linear and non-linear stiffness coefficients, respectively.

When $F_m \leq F_f$, the hydraulic cylinder does not work, and the PM at the end of the hydraulic piston rod can be assumed as grounded. The theoretical model of the non-linear stiffness isolator module of the VEAM can be established as a system composed of; (1) the mass (m), (2) the linear stiffness of the coil spring (k_{S1}, k_{S2}), (3) the structural damping from the two coil springs on each side of the sub-system (which is simplified as Rayleigh damping in this study) (c_1, c_2), and (4) the equivalent linear and non-linear stiffness due to the magnetic spring (k_b), as shown in Figure 5b. The linear stiffness coefficient of the VEAM subsystem can be integrated as $k_l = k_{S1} + k_{S2} + k_b$.

To simplify the calculations, the vibration excitation from the stay cable is calculated as the following sine function:

$$y_c = Y_c \sin(\omega t + \varphi) \tag{12}$$

where, Y_c is the vibration amplitude, ϖ is the angular frequency, t is the time, and φ is the phase shift. Thus, the displacement of the VEAM subsystem (y), which is the relative displacement between the stay cable and the deck, can be expressed as:

$$y = Y \sin(\omega t) \tag{13}$$

When the VEAM sub-system receives the vibration transmitted from the stay cable and enters a steady state, the equation of motion of the sub-system according to the HBM can be written as the following Duffing type equation [53,54]:

$$m\ddot{y} + 2c\dot{y} + k_l y + b y^3 = -m\ddot{y}_c \tag{14}$$

Because the same springs are used on both sides of the sub-system, we assume that their structural damping coefficients are equal, i.e., $c = c_1 = c_2$.

Substituting Equations (12) and (13) into Equation (14) and deriving the time, the following equation is obtained.

$$-mY\omega^2 \sin(\omega t) + 2cY\omega \cos(\omega t) + k_l Y \sin(\omega t) + bY^3 \sin^3(\omega t) = mY_c\omega^2[\sin(\omega t)\cos\varphi + \cos(\omega t)\sin\varphi] \tag{15}$$

The cubic term in this equation can be broken down as: $\sin^3(\omega t) = [3\sin(\omega t) - \sin(3\omega t)]/4$. The non-harmonic high frequency term that, in general, does not simply decay with time can be omitted and the constant terms with $\sin(\omega t)$ and $\cos(\omega t)$ can be integrated as:

$$-mY\omega^2 + k_l Y + \frac{3}{4}bY^3 = mY_c\omega^2 \cos\varphi \tag{16}$$

and

$$2cY\omega = mY_c\omega^2 \sin\varphi \tag{17}$$

The absolute displacement of the deck is the sum of the vibration excitation of the stay cable and the relative displacement between the stay cable and the deck:

$$y_g = y + y_c = Y\sin(\omega t) + Y_c \sin(\omega t + \varphi) \tag{18}$$

Therefore, the displacement transmissibility of the VEAM sub-system can be sorted as:

$$T = \frac{y_g}{y_c} = \sqrt{\frac{(Y + Y_c\cos\varphi)^2 + (Y_c\sin\varphi)^2}{Y_c^2}} = \sqrt{\frac{Y^2 + 2YY_c\cos\varphi + Y_c^2}{Y_c^2}} \tag{19}$$

Substituting Equation (16) into Equation (19), we get:

$$T = \sqrt{\frac{Y^2}{Y_c^2} + 1 + \frac{2Y}{Y_c}\left(\frac{-mY\omega^2 + k_l Y + \frac{3}{4}bY^3}{mY_c\omega^2}\right)} \tag{20}$$

3.3. Hydraulic Viscous Damper Model

When F_f is less than F_m, the PM pushes the piston and the hydraulic viscous damper begins to work. The damping due to the hydraulic cylinder is much greater than that due to the structural damping of the coil springs. Therefore, the structural damping is neglected in the theoretical model. The VEAM sub-system can be simplified as an SDOF model, as shown in Figure 5c. The figure graphically shows that the damping of the hydraulic cylinder (c_h) is connected in parallel with the stiffness of the two coil springs ($k_s = k_{S1} + k_{S2}$). Thus, the equation of motion of the VEAM sub-system can be expressed as:

$$m\ddot{y} + c_h\dot{y} + k_s y = -m\ddot{y}_c \tag{21}$$

The derivation process to get the displacement transmissibility is similar to that described in Section 3.2. Thus, the equation is obtained, as follows:

$$T = \sqrt{\frac{Y^2}{Y_c^2} + 1 + \frac{2Y}{Y_c}\left(\frac{-mY\omega^2 + k_s Y}{mY_c\omega^2}\right)} \tag{22}$$

4. Numerical Simulation

VEAM is a multi-physics system composed of a non-linear permanent magnet vibration isolator module and a hydraulic viscous damping module. The vibration isolation or damping performance of the VEAM may be greatly influenced by the different design factors. It is necessary to explore the influence of each design factor on system performance in order to design the VEAM system, so that it meets the highest requirements for cable-stayed bridges. In this section, we use the control variates method to change each of the key parameter values to observe and discuss its impact on the overall system performance. Table 1 lists the default parameters used in the simulation. In each simulation, one of these values is changed, while the rest remain constant. The simulations were coded using Python 3.7 and processed on a workstation with a 3.6 GHz Intel Xeon processor and 16GB RAM. The non-linear equations were solved using the newton_krylov method in the scipyoptimize module [55].

Section 4.1 discusses the relationship between the required distance between PMs and the starting force of the hydraulic cylinder with respect to different magnetic factors. Section 4.2 shows the simulation and discussions addressing the characteristics of the non-linear stiffness magnetic isolation module when the parameters of the coil spring and magnet are varied. In Section 4.3, the characteristics of hydraulic viscous damper under different design parameters are reported.

Table 1. Default parameters for simulation.

Item	Value
Stiffness of the springs, $k_{s1} = k_{s2}$ $\left(Nm^{-1}\right)$	2000
Stiffness of the linear item of magnetic spring, k_b $\left(Nm^{-1}\right)$	3500
Stiffness of the non-linear item of magnetic spring, b $\left(Nm^{-3}\right)$	8.1×10^6
Mass, m (kg)	0.5
Structural damping ratio of the springs, $\zeta_{s1} = \zeta_{s2}$	0.02
Damping ratio of the hydraulic cylinder module, ζ_h	0.2
Amplitude of vibration excitation, Y_c (m)	0.0005

4.1. Preset Distance of Magnet

In Section 3.1, we mentioned that the automatic switching of the non-linear stiffness vibration isolator and hydraulic viscous damper is mainly determined by the distance between the magnets (l) and the minimum working force (F_f) of the hydraulic cylinder. However, different permanent magnets exhibit different magnetic repulsive force characteristics. We must understand the effect of the hydraulic cylinder working force on the distance between PMs under different PM parameters in order to design a VEAM that can switch modes when the input vibration amplitude is above the specified value. In order to evaluate this, we set the mode switching amplitude y as 0.002 m in this simulation.

Figure 6 presents the relationship between the designed distance between the PMs and the minimum working force of the hydraulic cylinder according to Equation (10). We performed simulations that are based on three kinds of PMs with different magnetic repulsive force characteristics. The solid line represents the curve obtained using the PMs with a = 0.4 and n = 1.0, where a and n are consistent with Equation (9). The dotted line and the dash-dot-dash line represent the curves for when a = 0.6, n = 1.1, and a = 0.8, n = 1.2, respectively. Because the magnet calculation formula involves the power function calculation that may get an infinite value, the ordinate is only taken to an effective distance of 0.05 m.

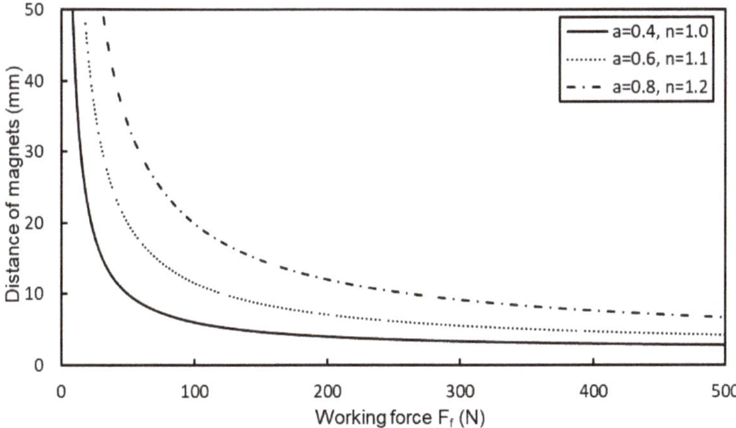

Figure 6. Graph showing the relationship between the preset distance between Permanent magnets (PMs) and the working force of the hydraulic cylinder.

It can be observed that the three curves show an exponential downward trend as the minimum working force required by the hydraulic cylinder increases, which also reflects the characteristics of the magnet repulsive force. We can also observe that, for the curve with a larger magnetic repulsion factor, the allowable distance between PMs is also greater. Assuming that the minimum working force of the hydraulic cylinder is 200 N, the curve obtained using the magnet repulsive force factor of a = 0.4 and n = 1.0 shows a distance of 0.004 m, which is very close to the preset amplitude value. The reciprocating motion that is caused by vibration is likely to cause the PMs to collide with each other, which may cause damage to the magnet or other components of the VEAM. During the design stage, the allowable distance between PMs should be greater than the preset amplitude to ensure the safety of the system. At the same minimum working force, the allowable distance on the dash-dot-dash curve with a strong magnetic repulsive force is 0.012 m, which is two times higher than the solid curve. Even so, when considering the limitations of geometry and safety factor in actual application, the allowable distance between PMs provided by the simple structural Neodymium magnets (Nd2Fe14B) cannot fully guarantee the safety of the system. However, this work only provides a basis for the design optimization of the VEAM. Based on these methods, we can devise a safer system using stronger magnets in the future by possibly utilizing some sort of motion limiting mechanism or collision prevention buffers between the magnets.

A Simulink model is established using the Simscape module available in Matlab 2019, which is a special numerical simulation software for solving multi-physical problems, in order to verify the automatic switching function between the nonlinear vibration isolator and the hydraulic viscous damper modules. Figure 7 shows the Simulink model of the entire VEAM subsystem. Table 1 shows the default parameter values used in this simulation. The nonlinear stiffness isolator module, hydraulic viscous damper module, and the entire VEAM subsystem are simulated separately in order to discuss whether the developed model is valid. In the simulation of the nonlinear stiffness isolator module, the input amplitude and frequency are set at 0.001 m and 1 Hz. In the simulations of hydraulic viscous damper module and the entire VEAM subsystem, the input amplitude and frequency are set at 0.005 m and 1 Hz, respectively. In order to reduce the calculation time, the symmetrically connected hydraulic cylinders are simplified as general viscous damping and the simulation is performed using ode23t solver. No errors or warnings were reported during the entire simulation processes, which shows that each mechanism can operate normally, and that the nonlinear vibration isolator and hydraulic viscous damper can be switched without any problems. Figure 8 shows the simulation results for the acceleration and the corresponding Fourier spectrum of each mechanism. As shown in Figure 8a, due to the nonlinear preloading force, the acceleration is relatively large at the beginning, and it tends to gradually reduce to a steady state. It can be seen from the corresponding Fourier spectrum (Figure 8b) that the acceleration presents a non-linear decreasing curve over the entire bandwidth. When the nonlinear stiffness isolation module is removed and only the hydraulic damping module is operating, the output becomes a familiar damped simple harmonic motion form, this simulation result is shown in Figure 8c,d. It can be seen from Figure 8e that the high-frequency harmonics appear at the beginning of the bandwidth and their period cannot be clearly distinguished when the entire VEAM is in a state where the nonlinear stiffness isolation module and the hydraulic damping module operate together. This may be attributed to the nonlinear dynamics of the magnetic force that is generated by the permanent magnets. Figure 8f plots the corresponding Fourier spectrum implies the visible chaotic phenomenon in the continuous band.

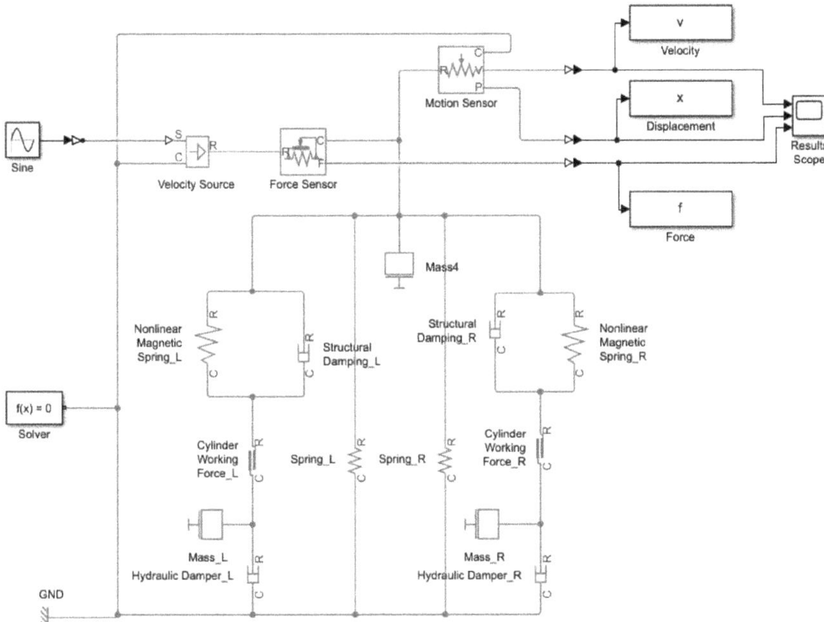

Figure 7. Simscape model of the VEAM.

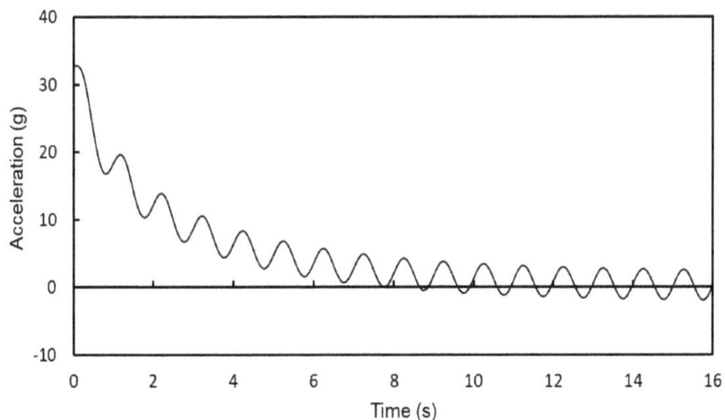

(**a**) Simulation results for the acceleration of the nonlinear vibration isolator module

Figure 8. *Cont.*

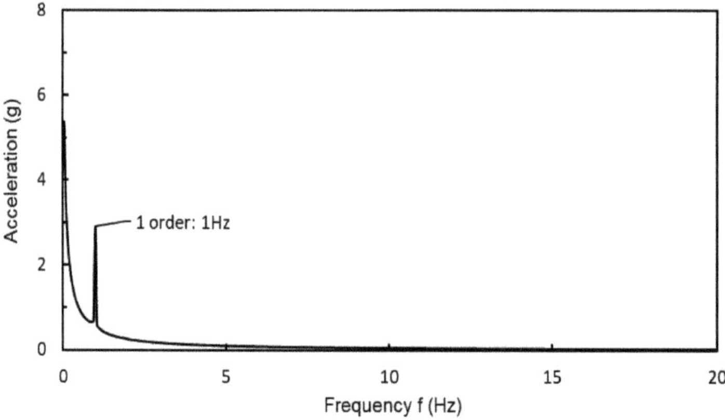

(**b**) The corresponding Fourier spectrum of nonlinear vibration isolator module

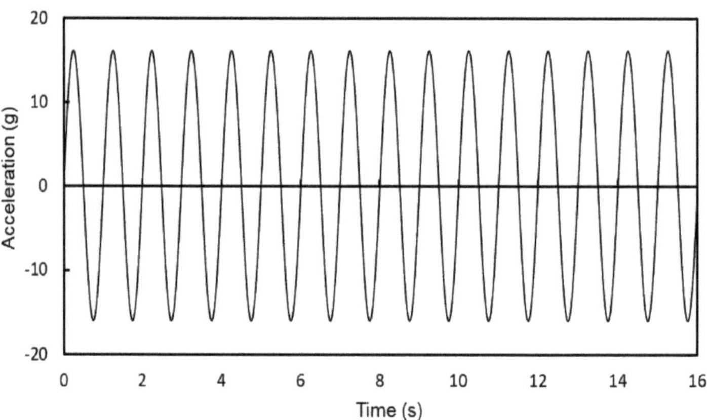

(**c**) Simulation results for the acceleration of the hydraulic viscous damper module

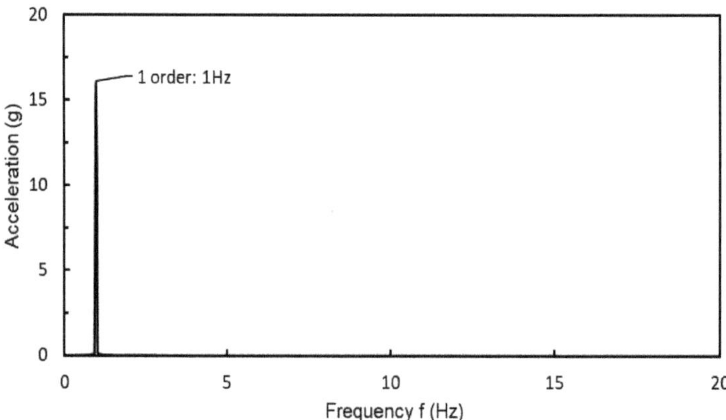

(**d**) The corresponding Fourier spectrum of hydraulic viscous damper module

Figure 8. *Cont.*

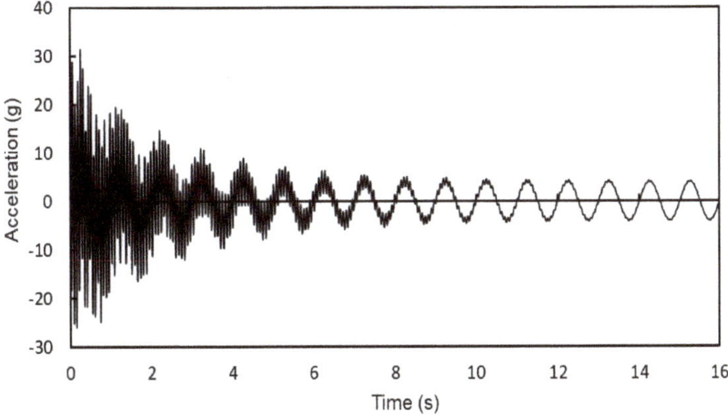

(e) Simulation results for the acceleration of the mechanism sub-system

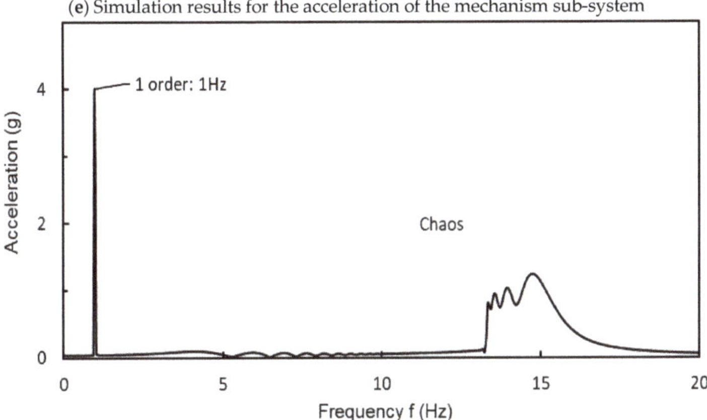

(f) The corresponding Fourier spectrum of the mechanism sub-system

Figure 8. Simulation results for the acceleration and the corresponding Fourier spectrum of the VEAM.

4.2. Non-Linear Stiffness Isolator

When the magnetic repulsive force is less than the minimum working force of the hydraulic cylinder, only the non-linear stiffness vibration isolator composed of the magnetic springs works. In this section, we will individually alter the key parameters of the coil springs and PMs used in the non-linear vibration isolator in order to observe the influence of the parameters k_s, k_b, b, and ζ on the vibration isolation performance of the system and summarize the design experience.

In the first case, the linear stiffness of the coil springs is set as 1000 Nm^{-1}, 2000 Nm^{-1}, and 3000 Nm^{-1}, while the other parameters are kept at the default values that are shown in Table 1, and the displacement transmissibility of the non-linear stiffness isolator module is observed. The simulation results are shown in Figure 9. It can be observed that the simulated transmissibility curve moves to the high frequency region with the increase in linear stiffness of the spring. This is accompanied by an increase in the peak value of transmissibility. When the stiffness is 1000 Nm^{-1}, the maximum displacement transmissibility of 3.9 occurs at a frequency of 16.5 Hz. When the stiffness is 3000 Nm^{-1}, the maximum displacement transmissibility of 4.5 occurs at a frequency of 21.7 Hz. This shows that increasing the linear stiffness of the spring in the VEAM can effectively reduce the displacement transmissibility in the low frequency region. This means that the vibration isolation performance of the VEAM at a tuning frequency can be optimized using springs with higher stiffness values.

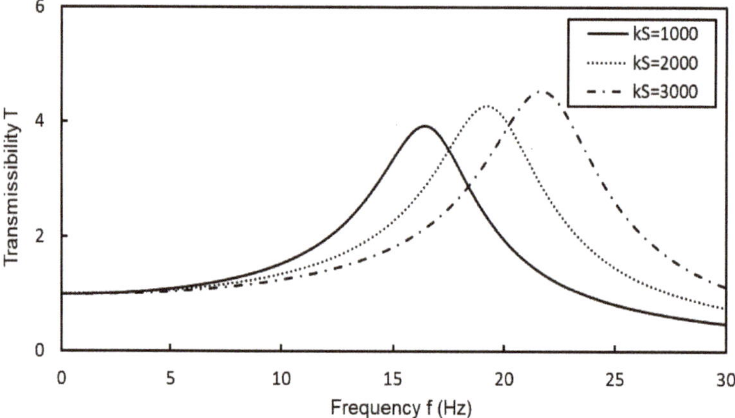

Figure 9. Transmissibility of the non-linear stiffness isolator module of VEAM with varying values of k_s.

The results of the second case are presented in Figure 10, which shows the effect of different equivalent linear stiffness obtained using different magnetic springs on the displacement transmissibility of the VEAM. The solid, dotted, and dash-dot-dash curves represent the magnetic spring equivalent linear stiffness values of 2000 Nm^{-1}, 3500 Nm^{-1}, and 5000 Nm^{-1}, respectively. It can be seen that the shifting of the transmissibility curve is similar to that observed while varying the linear stiffness of the spring; i.e., as the equivalent linear stiffness increases the curve shifts to the high frequency region. At frequencies below 18.5 Hz, the VEAM sub-system with a k_b value of 5000 Nm^{-1} has a lower transmissibility than that of the sub-system with a k_b value of 2000 Nm^{-1}. At 18.5 Hz, the transmissibility in both cases is equal with a value of 3.2. Beyond this frequency, the transmissibility with a k_b value of 5000 Nm^{-1} becomes higher than that with a k_b value of 2000 Nm^{-1}. Therefore, according to these results the VEAM sub-system with larger k_b value will have better vibration isolation performance in the low frequency range, but it will not work satisfactorily in the high frequency range.

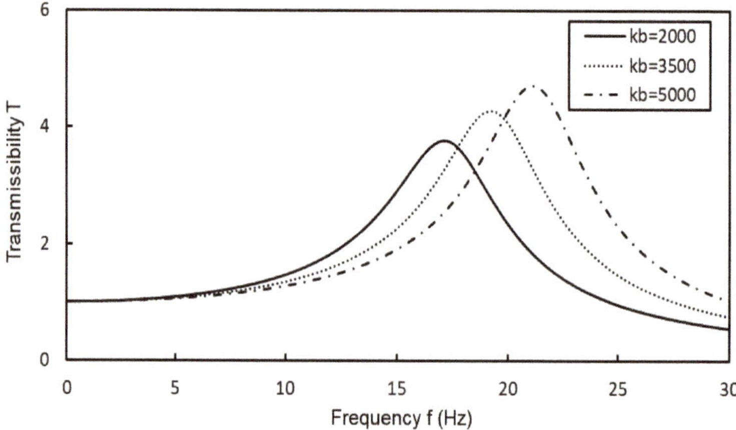

Figure 10. Transmissibility of the non-linear stiffness isolator module of VEAM with varying values of k_b.

b is the only non-linear stiffness parameter in the VEAM system, so it is necessary to consider its influence on the vibration isolation performance of the whole sub-system. The solid, dotted, and dash-dot-dash curves in Figure 11 indicate the simulated transmissibility of the VEAM sub-system with respect to b values of 0 (no non-linear stiffness component in the system), default value (8.1×10^6), and 10×10^9 (an ideal value chosen in order to clearly see the effect of increasing the value), respectively. It can be observed that, as the value of b increases, the peak value of the transmissibility decreases and shifts towards the high-frequency region. However, the curve appears to bend to the right and, as the value of b gets larger, the amount of bending also becomes greater (as shown in the dash-do-dash curve). A similar sort of transmissibility curve bending towards the high frequency region has also been reported in other works [40,41]. We speculate that the excessive curve bending caused by the large b value will cause multiple transmissibility values to exist in a part of the frequency range, which means that there will be a multiple states phenomenon. This may cause the dynamics of the system to become complicated with the change in amplitude and frequency of excitation and chaotic motion may occur, as shown in Figure 8b.

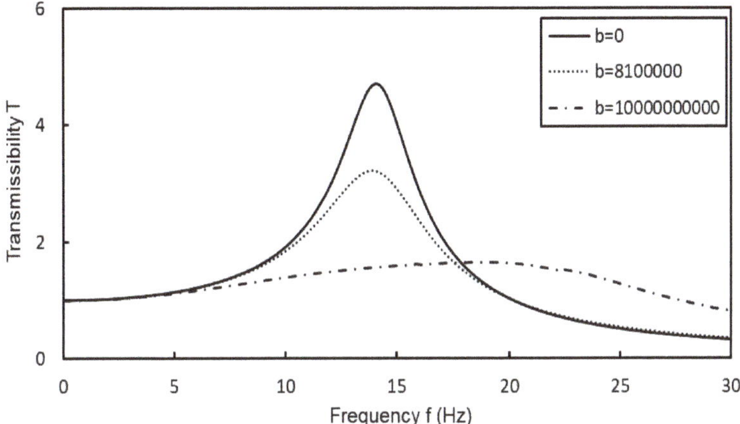

Figure 11. Transmissibility of the non-linear stiffness isolator module of VEAM with varying values of b.

In the non-linear stiffness isolator module consisting of a magnetic spring, there is no additional damping, except for the structural damping of the springs. We simulated the transmissibility of the VEAM sub-system with spring structural damping ratios of 0.02, 0.04, and 0.06. Figure 12 shows the results of these simulations. It can be observed that, the larger the structural damping ratio, the smaller the transmissibility of the VEAM sub-system. It is also apparent that the structural damping has almost no effect on the change of the frequency domain of the peak transmissibility value. When the frequency is 13.0 Hz, the maximum transmissibility of the sub-system with a spring structural damping ratio of 0.06 is 65.2% lower than that with a structural damping ratio of 0.02. This result shows that structural damping has a considerable effect on the overall vibration damping characteristic of the system.

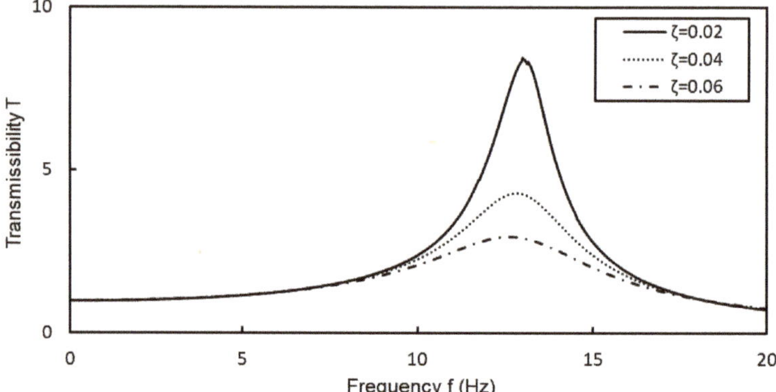

Figure 12. Transmissibility of the non-linear stiffness isolator module of VEAM with varying values of ζ.

4.3. Hydraulic Damper

Under normal conditions, the stay cable vibration has a small amplitude. However, under special environmental impacts, such as earthquakes and typhoons, the amplitude of the stay cable vibration can become large. The VEAM incorporates hydraulic viscous dampers in its design to dampen such large amplitude vibrations by exploiting their high damping performance. This section explores the effects of damping and spring stiffness on the performance of this damper module. In this section, the excitation amplitude Y_c is set as 0.005 m.

In practical applications, the damping ratio of the hydraulic damper used in VEAM can be adjusted by changing the size of the orifice in the flow path. To make performance comparisons, we set the damping ratios of the hydraulic cylinders at 0.2, 0.6, and 0.8. Figure 13 shows the simulated transmissibility results of the VEAM sub-system. It can be seen that the higher the damping ratio of the hydraulic cylinder the lower the peak value of the transmissibility. However, although the curve with the lowest damping ratio of 0.2 has a higher transmissibility in the low-frequency region (region of amplification) within a characteristic frequency, the frequency region after this frequency, which is called the "isolation region", shows better vibration isolation performance.

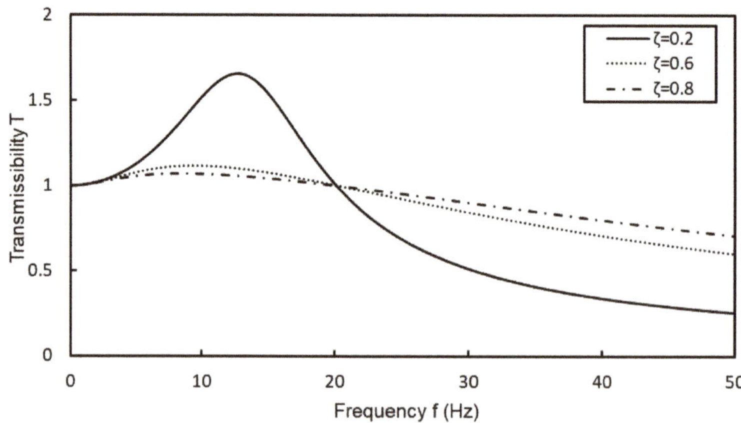

Figure 13. Transmissibility of the hydraulic viscous damper module of VEAM with varying values of ζ.

Figure 14 shows a comparison of the simulated transmissibility with varying values of k_s. Similar to Figure 9, the frequency domain of the peak wave of transmissibility curve becomes wider as the value of k_s increases, and the maximum transmissibility occurs at a higher frequency. However, the difference between the two figures is that the peak value of the transmissibility is not affected by the change in the value of k_s. This is because, when the hydraulic shock absorber is in operation, the magnetic spring stops working, and the equivalent stiffness of the entire system no longer changes due to the nonlinear stiffness. Thus, for the damping module, the spring with a lower k_s value shows better isolation performance in the high frequency region.

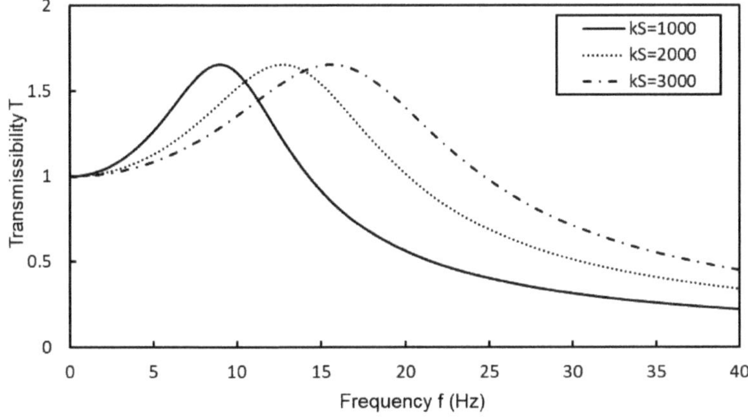

Figure 14. Transmissibility of the hydraulic viscous damper module of VEAM with varying k_s.

5. Conclusions

This study has proposed a novel damper design for cable stayed bridges that can respond to different vibration amplitudes that are caused by different environmental phenomena. It is a hybrid system that combines non-linear stiffness vibration isolation modules and hydraulic viscous damper modules. The VEAM proposes a novel way to utilize the minimum working force of the hydraulic cylinders, which is an unavoidable characteristic that can be leveraged to combine the two modules together to form a hybrid system that is able to automatically switch the module in operation based on the input vibration amplitude. This paper presented the mechanism design and theoretical model of the VEAM. Multiple numerical case studies were performed in order to observe the influence of various factors on the vibration isolation and damping performance of the entire system. The key findings of this research are as follows:

1. VEAM mechanism, through the combined action of hydraulic cylinder bearings, ball joints, and springs, can maintain the alignment of PMs at all times, and will not cause vibration force transmission instability due to tangential shift.
2. The external hydraulic circuit connecting the opposing damper pairs can maintain motion synchrony to avoid the jumping phenomenon, thus preventing system damage. This method can also easily make adjustments of the orifice regulating valves to achieve a variable damping ratio. Through the Simscape simulation of the entire VEAM subsystem, the design theory of the automatic switching mechanism was tested. The results showed that, when the whole system is working, due to the non-linear stiffness, chaotic output is observed in the high frequency bandwidth. The implications of this behavior need to be verified in future experiments with the real system.
3. We can use strong repulsive magnets to ensure the maintenance of a safe distance between the PM pairs. The use of magnets with stronger linear stiffness will reduce the transmissibility of the nonlinear stiffness module in the low frequency region, but it will increase the transmissibility in the high frequency region. However, magnets with stronger nonlinear stiffness can effectively

reduce the peak of transmissibility. In the design stage, according to the needs of the actual situation, the selection of PMs of appropriate magnetic strength can ensure a high performing safe design, while also ensuring a safe distance. However, during prototype manufacturing, we found that, due to space constraints, it is difficult to only use the magnetic force between the two PMs to actuate the hydraulic cylinder. Accordingly, it may be necessary to use crash pads between the magnets to achieve rigid transmission of force. The magnetic resistance of the surrounding environment and design involving stronger PMs will be considered in future studies.

4. In the low-frequency region, the greater the spring linear stiffness, the lower will be the transmissibility of both the non-linear stiffness vibration isolation module and the hydraulic damper module, i.e., there will be better vibration isolation performance. However, in the high-frequency region, the stiffness of the spring will affect the vibration isolation performance when only the nonlinear stiffness vibration isolator is in operation. Therefore, the common effect of both modules must be considered when selecting the spring stiffness.
5. The structural damping of the spring will undoubtedly provide good isolation performance when the non-linear stiffness module is working. However, the viscous damping of the hydraulic cylinder is different. Excessive viscous damping will affect the vibration isolation performance of the VEAM and increase the possibility of structural damage. On the other hand, small viscous damping has a large area of the isolation region, which can improve the poor vibration isolation that is caused by the shift towards the high frequency region of the transmissibility caused by excessive spring stiffness.

Author Contributions: Z.Q. and Y.-T.W. contributed equally. Conceptualization, Z.Q. and Y.-T.W.; methodology, A.H.; software, Z.Q. and Y.-T.W.; validation, Z.Q., Y.-T.W. and A.H.; formal analysis, S.K.L.; investigation, Y.-T.W.; resources, S.-K.L.; data curation, Z.Q., Y.-T.W. and A.H.; writing—original draft preparation, Z.Q. and Y.-T.W.; writing—review and editing, A.H., S.-K.L. and J.W.S.; visualization, Z.Q.; supervision, Y.-T.W.; project administration, S.-K.L.; funding acquisition, S.-K.L. All authors have read and agreed to the published version of the manuscript.

Funding: This work was supported by the Regional Leading Research Center of NRF and MOCIE (NRF-2019R1A5A808320112).

Conflicts of Interest: The authors declared that they have no conflict of interest to this work.

References

1. Chen, W.-L.; Gao, D.; Laima, S.; Li, H. A Field Investigation on Vortex-Induced Vibrations of Stay Cables in a Cable-Stayed Bridge. *Appl. Sci.* **2019**, *9*, 4556. [CrossRef]
2. Xu, Y.; Zeng, Z.; Cui, C.; Zeng, S. Practical Design Method of Yielding Steel Dampers in Concrete Cable-Stayed Bridges. *Appl. Sci.* **2019**, *9*, 2857. [CrossRef]
3. Qin, Z.; Wu, Y.T.; Eizad, A.; Lee, K.H.; Lyu, S.K. Design and evaluation of two-stage planetary gearbox for special-purpose industrial machinery. *J. Mech. Sci. Technol.* **2019**, *33*, 5943–5950. [CrossRef]
4. Qin, Z.; Zhang, Q.; Wu, Y.T.; Eizad, A.; Lyu, S.K. Experimentally Validated Geometry Modification Simulation for Improving Noise Performance of CVT Gearbox for Vehicles. *Int. J. Precis. Eng. Manuf.* **2019**, *20*, 1969–1977. [CrossRef]
5. Qin, Z.; Wu, Y.T.; Eizad, A.; Jeon, N.S.; Kim, D.S.; Lyu, S.K. A Study on Simulation Based Validation of Optimized Design of High Precision Rotating Unit for Processing Machinery. *Int. J. Precis. Eng. Manuf.* **2019**, *20*, 1601–1609. [CrossRef]
6. Qin, Z.; Son, H.I.; Lyu, S.K. Design of anti-vibration mounting for 140A class alternator for vehicles. *J. Mech. Sci. Technol.* **2018**, *32*, 5233–5239. [CrossRef]
7. Qin, Z.; Wu, Y.T.; Lyu, S.K. A Review of Recent Advances in Design Optimization of Gearbox. *Int. J. Precis. Eng. Manuf.* **2018**, *19*, 1753–1762. [CrossRef]
8. Nakamura, A.; Kasuga, A.; Arai, H. The effects of mechanical dampers on stay cables with high-damping rubber. *Constr. Build. Mater.* **1998**, *12*, 115–123. [CrossRef]
9. Cu, V.H.; Han, B. High-damping rubber damper for taut cable vibration reduction. *Aust. J. Struct. Eng.* **2015**, *16*, 283–291. [CrossRef]

10. Cu, V.H.; Han, B.; Wang, F. Damping of a taut cable with two attached high damping rubber dampers. *Struct. Eng. Mech.* **2015**, *55*, 1261–1278. [CrossRef]
11. Pacheco, B.M.; Fujino, Y.; Sulekh, A. Estimation curve for modal damping in stay cables with viscous damper. *J. Struct. Eng.* **1993**, *119*, 1961–1979. [CrossRef]
12. Krenk, S. Vibrations of a taut cable with an external damper. *J. Appl. Mech. Trans. ASME* **2000**, *67*, 772–776. [CrossRef]
13. Sawka, B. *Experiments on dynamic response of hydraulic props. Report of GIG Chief Mining Institute*; Główny Instytut Górnictwa (GIG): Katowice, Poland, 1973.
14. Tomski, L.; Kukla, S. Dynamical response of bar-fluid-shell system simulating hydraulic cylinder subjected to arbitrary axial excitation. *J. Sound Vib.* **1984**, *92*, 273–284. [CrossRef]
15. Gamez-Montero, P.J.; Salazar, E.; Castilla, R.; Freire, J.; Khamashta, M.; Codina, E. Misalignment effects on the load capacity of a hydraulic cylinder. *Int. J. Mech. Sci.* **2009**, *51*, 105–113. [CrossRef]
16. Niu, J.; Ding, Y.; Shi, Y.; Li, Z. Oil damper with variable stiffness for the seismic mitigation of cable-stayed bridge in transverse direction. *Soil Dyn. Earthq. Eng.* **2019**, *125*, 105719. [CrossRef]
17. Zhan, C.; Deng, J.; Chen, K. Research on low-friction and high-response hydraulic cylinder with variable clearance. *Jixie Gongcheng Xuebao* **2015**, *51*, 161–167. [CrossRef]
18. Tao, J.; Mak, C.M. Effect of viscous damping on power transmissibility for the vibration isolation of building services equipment. *Appl. Acoust.* **2006**, *67*, 733–742. [CrossRef]
19. Guo, P.F.; Lang, Z.Q.; Peng, Z.K. Analysis and design of the force and displacement transmissibility of nonlinear viscous damper based vibration isolation systems. *Nonlinear Dyn.* **2012**, *67*, 2671–2687. [CrossRef]
20. Wu, W.J.; Cai, C.S. Theoretical exploration of a taut cable and a TMD system. *Eng. Struct.* **2007**, *29*, 962–972. [CrossRef]
21. Macháček, O.; Kubík, M.; Strecker, Z.; Roupec, J.; Mazůrek, I. Design of a frictionless magnetorheological damper with a high dynamic force range. *Adv. Mech. Eng.* **2019**, *11*, 1–8. [CrossRef]
22. Nguyen, D.T.; Vo, D.H. A Study on Combination of Two Friction Dampers to Control Stayed-Cable Vibration Under Considering its Bending Stiffness. In *CIGOS 2019, Innovation for Sustainable Infrastructure 2020*; Springer: Singapore, 2019; pp. 87–92.
23. Huang, X.; Liu, X.; Sun, J.; Zhang, Z.; Hua, H. Effect of the system imperfections on the dynamic response of a high-static-low-dynamic stiffness vibration isolator. *Nonlinear Dyn.* **2014**, *76*, 1157–1167. [CrossRef]
24. Han, L.; Wei, H.; Wang, F. Study on the Vibration Isolation Performance of Composite Subgrade Structure in Seasonal Frozen Regions. *Appl. Sci.* **2020**, *10*, 3597. [CrossRef]
25. Hao, Z.; Cao, Q.; Wiercigroch, M. Nonlinear dynamics of the quasi-zero-stiffness SD oscillator based upon the local and global bifurcation analyses. *Nonlinear Dyn.* **2017**, *87*, 987–1014. [CrossRef]
26. Ding, H.; Ji, J.; Chen, L. Nonlinear vibration isolation for fluid-conveying pipes using quasi-zero stiffness characteristics. *Mech. Syst. Signal Process.* **2019**, *121*, 675–688. [CrossRef]
27. Wang, X.; Zhou, J.; Xu, D.; Ouyang, H.; Duan, Y. Force transmissibility of a two-stage vibration isolation system with quasi-zero stiffness. *Nonlinear Dyn.* **2017**, *87*, 633–646. [CrossRef]
28. Liu, C.; Yu, K. A high-static–low-dynamic-stiffness vibration isolator with the auxiliary system. *Nonlinear Dyn.* **2018**, *94*, 1549–1567. [CrossRef]
29. Virgin, L.N.; Santillan, S.T.; Plaut, R.H. Vibration isolation using extreme geometric nonlinearity. *J. Sound Vib.* **2008**, *315*, 721–731. [CrossRef]
30. Wang, Y.; Li, S.; Neild, S.A.; Jiang, J.Z. Comparison of the dynamic performance of nonlinear one and two degree-of-freedom vibration isolators with quasi-zero stiffness. *Nonlinear Dyn.* **2017**, *88*, 635–654. [CrossRef]
31. Sun, M.; Dong, Z.; Song, G.; Sun, X.; Liu, W. A Vibration Isolation System Using the Negative Stiffness Corrector Formed by Cam-Roller Mechanisms with Quadratic Polynomial Trajectory. *Appl. Sci.* **2020**, *10*, 3573. [CrossRef]
32. Carrella, A.; Brennan, M.J.; Waters, T.P. Static analysis of a passive vibration isolator with quasi-zero-stiffness characteristic. *J. Sound Vib.* **2007**, *301*, 678–689. [CrossRef]
33. Lan, C.C.; Yang, S.A.; Wu, Y. Design and experiment of a compact quasi-zero-stiffness isolator capable of a wide range of loads. *J. Sound Vib.* **2014**, *333*, 4843–4858. [CrossRef]
34. Araki, Y.; Asai, T.; Kimura, K.; Maezawa, K.; Masui, T. Nonlinear vibration isolator with adjustable restoring force. *J. Sound Vib.* **2013**, *332*, 6063–6077. [CrossRef]

35. Zhou, J.; Wang, X.; Xu, D.; Bishop, S. Nonlinear dynamic characteristics of a quasi-zero stiffness vibration isolator with cam-roller-spring mechanisms. *J. Sound Vib.* **2015**, *346*, 53–69. [CrossRef]
36. Zhou, J.; Xu, D.; Bishop, S. A torsion quasi-zero stiffness vibration isolator. *J. Sound Vib.* **2015**, *338*, 121–133. [CrossRef]
37. Zhou, N.; Liu, K. A tunable high-static-low-dynamic stiffness vibration isolator. *J. Sound Vib.* **2010**, *329*, 1254–1273. [CrossRef]
38. Wu, W.; Chen, X.; Shan, Y. Analysis and experiment of a vibration isolator using a novel magnetic spring with negative stiffness. *J. Sound Vib.* **2014**, *333*, 2958–2970. [CrossRef]
39. Zheng, Y.; Zhang, X.; Luo, Y.; Yan, B.; Ma, C. Design and experiment of a high-static-low-dynamic stiffness isolator using a negative stiffness magnetic spring. *J. Sound Vib.* **2016**, *360*, 31–52. [CrossRef]
40. Yan, B.; Ma, H.; Zhao, C.; Wu, C.; Wang, K.; Wang, P. A vari-stiffness nonlinear isolator with magnetic effects: Theoretical modeling and experimental verification. *Int. J. Mech. Sci.* **2018**, *148*, 745–755. [CrossRef]
41. Yan, B.; Ma, H.; Jian, B.; Wang, K.; Wu, C. Nonlinear dynamics analysis of a bi-state nonlinear vibration isolator with symmetric permanent magnets. *Nonlinear Dyn.* **2019**, *97*, 2499–2519. [CrossRef]
42. Matsumoto, M.; Shiraishi, N.; Kitazawa, M.; Knisely, C.; Shirato, H.; Kim, Y.; Tsujii, M. Aerodynamic behavior of inclined circular cylinders-cable aerodynamics. *J. Wind Eng. Ind. Aerodyn.* **1990**, *33*, 63–72. [CrossRef]
43. Matsumoto, M.; Saitoh, T.; Kitazawa, M.; Shirato, H.; Nishizaki, T. Response characteristics of rain-wind induced vibration of stay-cables of cable-stayed bridges. *J. Wind Eng. Ind. Aerodyn.* **1995**, *57*, 323–333. [CrossRef]
44. Matsumoto, M.; Daito, Y.; Kanamura, T.; Shigemura, Y.; Sakuma, S.; Ishizaki, H. Wind-induced vibration of cables of cable-stayed bridges. *J. Wind Eng. Ind. Aerodyn.* **1998**, *74*, 1015–1027. [CrossRef]
45. Matsumoto, M.; Yagi, T.; Shigemura, Y.; Tsushima, D. Vortex-induced cable vibration of cable-stayed bridges at high reduced wind velocity. *J. Wind Eng. Ind. Aerodyn.* **2001**, *89*, 633–647. [CrossRef]
46. Matsumoto, M.; Shiraishi, N.; Shirato, H. Rain-wind induced vibration of cables of cable-stayed bridges. *J. Wind Eng. Ind. Aerodyn.* **1992**, *43*, 2011–2022. [CrossRef]
47. Ma, C.M.; Duan, Q.S.; Liao, H.L. Experimental investigation on aerodynamic behavior of a long span cable-stayed bridge under construction. *KSCE J. Civ. Eng.* **2017**, *22*, 2492–2501. [CrossRef]
48. Du, W.; Lei, D.; Bai, P.; Zhu, F.; Huang, Z. Dynamic measurement of stay-cable force using digital image techniques. *Meas. J. Int. Meas. Confed.* **2020**, *151*, 107211. [CrossRef]
49. Barton, M.A.; Kuroda, K. Ultralow frequency oscillator using a pendulum with crossed suspension wires. *Rev. Sci. Instrum.* **1994**, *65*, 3775–3779. [CrossRef]
50. Barton, M.A.; Kanda, N.; Kuroda, K. A low-frequency vibration isolation table using multiple crossed-wire suspensions. *Rev. Sci. Instrum.* **1996**, *67*, 3994–3999. [CrossRef]
51. Carrella, A.; Brennan, M.J.; Waters, T.P.; Lopes, V., Jr. Force and displacement transmissibility of a nonlinear isolator with high-static-low-dynamic-stiffness. *Int. J. Mech. Sci.* **2012**, *55*, 22–29. [CrossRef]
52. Yamakawa, I.; Takeda, S.; Kojima, H. Behavior of a New Type Dynamic Vibration Absorber Consisting of Three Permanent Magnets. *Bull. JSME* **1977**, *20*, 947–954. [CrossRef]
53. He, J.; Cai, J. Dynamic Analysis of Modified Duffing System via Intermittent External Force and Its Application. *Appl. Sci.* **2019**, *9*, 4683. [CrossRef]
54. Fathi, M.F.; Bakhshinejad, A.; Baghaie, A.; D'Souza, R.M. Dynamic Denoising and Gappy Data Reconstruction Based on Dynamic Mode Decomposition and Discrete Cosine Transform. *Appl. Sci.* **2018**, *8*, 1515. [CrossRef]
55. Knoll, D.A.; Keyes, D.E. Jacobian-free Newton–Krylov methods: A survey of approaches and applications. *J. Comput. Phys.* **2004**, *193*, 357–397. [CrossRef]

Article

Modeling and NVH Analysis of a Full Engine Dynamic Model with Valve Train System

Xu Zheng, Xuan Luo, Yi Qiu * and Zhiyong Hao

College of Energy Engineering, Zhejiang University, Hangzhou 310027, China; zhengxu@zju.edu.cn (X.Z.); 3090100538@zju.edu.cn (X.L.); haozy@zju.edu.cn (Z.H.)

* Correspondence: yiqiu@zju.edu.cn; Tel.: +86-057-87953286

Received: 6 July 2020; Accepted: 23 July 2020; Published: 27 July 2020

Abstract: The valve train system is an important source of vibration and noise in an engine. An in-depth study on the dynamic model of the valve train is helpful in understanding the dynamic characteristics of the valve train and improving the prediction accuracy of vibration and noise. In the traditional approaches of the dynamic analyses, the simulations of the valve train system and the engine are carried out separately. The disadvantages of these uncoupled approaches are that the impact of the cylinder head deformation to the valve train and the support and constraints of the valve train on the cylinder head are not taken into consideration. In this study, a full engine dynamic model coupled with a valve train system is established and a dynamic simulation and noise vibration harshness (NVH) analysis are carried out. In the coupled approach, the valve train system is simulated simultaneously with the engine, and the complexity of the model has been greatly increased. Compared with the uncoupled approach, more detailed dynamic results of the valve train can be presented, and the subsequent predictions of vibration and noise can also be more accurate. The acoustic results show that the difference from the experimental sound power level is reduced from 1.8 dB(A) to 0.9 dB(A) after applying the coupled approach.

Keywords: valve train; dynamics; coupled approach; NVH

---

1. Introduction

The valve train system is one of the important sources of vibration and noise in an internal combustion engine. For an ideal valve train system, on the one hand, the valve should be opened and closed as quickly as possible to improve the ventilation efficiency; on the other hand, each component should be operated smoothly to avoid the occurrence of jump, bounce, and overload, causing failures or noise vibration harshness (NVH) problems. The valve train dynamic model is used to simulate the motions and forces of the valve, spring, rocker, camshaft, hydraulic lash adjuster (HLA), and other parts during the actual operation of the engine. Therefore, the establishment of a comprehensive and accurate dynamic model is of great significance for studying dynamic problems and predicting the vibration and noise.

The earliest valve train dynamic model is a single degree of freedom (DoF) model proposed by W. M. Dudley [1]. All the valve train parts, including the valve, washer, lock, and part of the spring, are simplified as an equivalent mass. The model is too simple to predict the valve motion accurately. N. S. Eiss established a valve train dynamic model with two DoFs, and demonstrated how the parameters of the model were selected to give the minimum vibration amplitude [2]. Since the 1980s, it has become easier to solve complex multivariate dynamic equations with the assistance of the modern computer. Models with distributed masses take elastic deformations into consideration, and more factors can be studied. C. Chan formulated a 6-DoF dynamic model to predict the forces and the motions of the end-pivot rocker-arm cam system components [3]. S. Seidlitz built a valve train dynamic model with 21 DoFs to simulate the single valve train system of an overhead valve (OHV) engine [4]. K. Nagaya studied the jump and bounce phenomena of a driven valve system based on a multiple-DoF dynamic model [5]. A. P. Pisano and F. Freudenstein considered the valve spring as a continuous flexible body. The valve spring was modeled as a distributed parameter system governed by the damped wave equation [6]. W. Lassaad et al. investigated the nonlinear dynamic behavior of a cam mechanism with an oscillating roller follower in the presence of defects by a multiple-DoF lumped-mass model with two nonlinear hertzian contacts [7]. J. Guo et al. proposed a mode matching method (MMM) for valve train components in a multiple-DoF dynamic model. The amount of lumped masses for each flexible component was determined based on its natural frequencies and the considered frequency range [8]. C. J. Zhou et al. proposed an enhanced flexible dynamic model for a valve train with clearance and multi-directional deformations and verified the model with experiments. They investigated the effects of valve clearance and cam rotation speed on the contact force, acceleration, and dynamic transmission error [9].

Based on various multiple-DoF valve train dynamic models, each body and joint in the system has been modeled and analyzed in detail. However, all the above approaches just consider the valve train system as an independent system, ignoring the interaction between the valve train system and other parts of the engine, which will affect the accuracy of the dynamic modelling. In this paper, the dynamic models of the valve train system and other parts of the engine are combined into one model, forming a completed coupled engine dynamic model with the valve train system. In this coupled approach, not only the internal interactions of the valve train system, but also the external interactions between the valve train system and other parts of the engine are taken into consideration. For a valve train system in the form of end-pivot rocker arm with roller finger follower, the external interactions act not only on valves, springs, and bearings, but also on HLAs. Based on the full engine model coupled with valve train system, the dynamic simulation is performed, and the vibration and noise characteristics are predicted.

2. Theory

The dynamic model is based on the floating frame of reference (FFoR) formulation [10]. For large-scale displacement, the Lagrangian formulation is employed. The small elastic deformations are considered relative to a local frame of reference, while this local frame of reference undergoes large motions with respect to a fixed global frame of reference [11]. Each discrete mass of the model is governed by Newton–Euler equation of motion. The analysis is performed by the implicit integration method (Newmark) in the time domain [12].

Considering the complexity of the engine model, it needs to be greatly simplified, as shown in Figure 1. Besides the parametric and lumped-mass models, the finite element method (FEM) is commonly used for simplifications. An accurate NVH analysis requires that the element size should not be too small. The lower limit of the wavelength is calculated according to the upper limit of the analysis frequency, and then the density of the element is determined. Due to the complexity of the engine structure of the components, there are hundreds of thousands of elements in the engine FEM model. In order to reduce the DoFs and modes of the matrices to an acceptable level for calculation, component mode synthesis (CMS) is applied in the condensation of the FEM models [13]. Only a few main modes of substructures are retained in the condensed FEM models and participate in the calculations of the equations of motion.

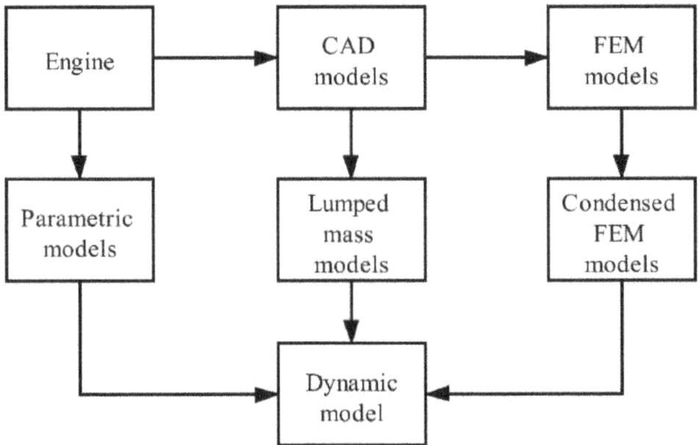

Figure 1. Simplification process of the engine model.

The schematic representation of the FFoR is shown in Figure 2. $e^T = (e_1, e_2, e_3)$ and $b^T = (b_1, b_2, b_3)$ denote the global coordinate system and the local coordinate system, respectively.

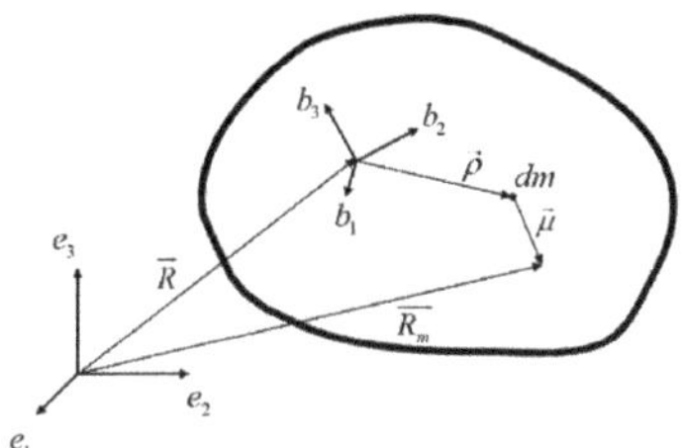

Figure 2. Schematic representation of the floating frame of reference (FFoR) approach.

$\vec{R}$ denotes the global body position vector, $\vec{\rho}$ denotes the global coordinate system initial position vector, and $\vec{\mu}$ denotes the elastic displacement of the node dm. The reference position vector can be written as:

$$\vec{R}_m = \vec{R} + \vec{\rho} + \vec{\mu} = e^T \hat{R} + b^T \hat{\rho} + b^T \hat{\mu}, \tag{1}$$

where $\hat{R}$, $\hat{\rho}$, $\hat{\mu}$ are the component matrices of $\vec{R}$, $\vec{\rho}$, $\vec{\mu}$.

The elastic displacement expanded in modal coordinates can be written as:

$$\vec{\mu} = b^T \Phi \varepsilon, \tag{2}$$

where Φ is the eigenvector; ε is the modal participation factor.

The Lagrange equation is defined as:

$$L = T - U, \tag{3}$$

where T is the kinetic energy; U is the potential energy.

$$T = \frac{1}{2}\int \left|\vec{v}_m\right| \cdot \left|\vec{v}_m\right| dm = \frac{1}{2}\int \left|\frac{d\vec{R}_m}{dt}\right| \cdot \left|\frac{d\vec{R}_m}{dt}\right| dm, \tag{4}$$

$$U = \frac{1}{2}\varepsilon^T \Lambda \varepsilon, \tag{5}$$

where $\vec{v}_m$ is the velocity vector of dm; Λ is the diagonal matrix of the natural frequencies, defined as $\Lambda = diag\begin{bmatrix} \lambda_1^2 & \lambda_2^2 & \cdots \end{bmatrix}$.

The basic dynamic equations in the FFoR can be obtained by substituting Equations (4) and (5) into the Lagrange equation:

$$\frac{d}{dt}\frac{\partial L}{\partial \dot{q}} - \frac{\partial L}{\partial q} = Q_q, \tag{6}$$

where Q_q denotes generalized force, and q denotes generalized coordinate.

Based on the Newton–Euler method, the governing equation of motion in the multi-body system can be expressed as:

$$M{\cdot}\ddot{q} + C{\cdot}\dot{q} + K{\cdot}q = f_e + f_l + f_c, \tag{7}$$

where M, C, and K denote mass, damping, and stiffness matrix; q denotes the vector of the system generalized coordinates; f_e denotes the excitation loads; f_l denotes the external loads; f_c denotes the constraint forces.

Both the uncoupled and coupled approach adopted the classical Newton–Euler equation, but the difference is the consideration of the valve train system. In the uncoupled approach, the forces to the engine from the valve train are considered as external loads in the term f_l. In the coupled approach, the components of the valve train system are merged in the matrices and vectors at the left-hand side of the equation.

The condensation approach is based on the Craig–Bampton method, which is an improved fixed-interface CMS method [14]. The main modes of each substructure are retained, and the DoFs of the substructures are reduced. The omitted internal displacements r_i are determined by:

$$\begin{bmatrix} r_i \\ r_e \end{bmatrix} = \begin{bmatrix} B & \Phi \\ I & 0 \end{bmatrix}\begin{bmatrix} r_e \\ \varepsilon \end{bmatrix}, \tag{8}$$

where r_e is the external displacements, B is the static modes, Φ is the component modes (eigenvectors), and ε is the modal participation factor.

3. Modeling and Analysis

3.1. Engine Basic Parameters

The engine is a four-cylinder turbocharged gasoline engine, the layout of the valve train is a double overhead camshaft (DOHC), with an end-pivot rocker arm with a roller finger follower and hydraulic lash adjuster, which is in the most common form in passenger cars. The basic specification of the engine is shown in Table 1.

Table 1. Specification of the engine.

Engine type	Four cylinders four strokes gasoline engine
Intake type	Turbocharged
Displacement	1798 cc
Bore	82.5 mm
Stroke	84.2 mm
Max power	120 kW@5500 rpm
Max torque	250 Nm@1500–4500 rpm
Balance	Double balancing shaft
Valve train	DOHC End-pivot rocker arm (roller finger follower) HLA

3.2. FEM Models

Based on three-dimensional digital models of the engine main components provided by the manufacturer, the FEM models of crankshaft, block, cylinder head, cylinder head cover, timing cover, oil pan, intake manifold, and exhaust system are established, as shown in Figure 3. Given the fact that the crankshaft has a regular and symmetrical structure, the first order hexahedral elements are adopted in the FEM model of the crankshaft. However, other components with many fillets, bolt holes, and oil channels are complex; therefore, second order tetrahedral elements are adopted in the FEM models of the rest components. The parameters of the FEM models are shown in Table 2.

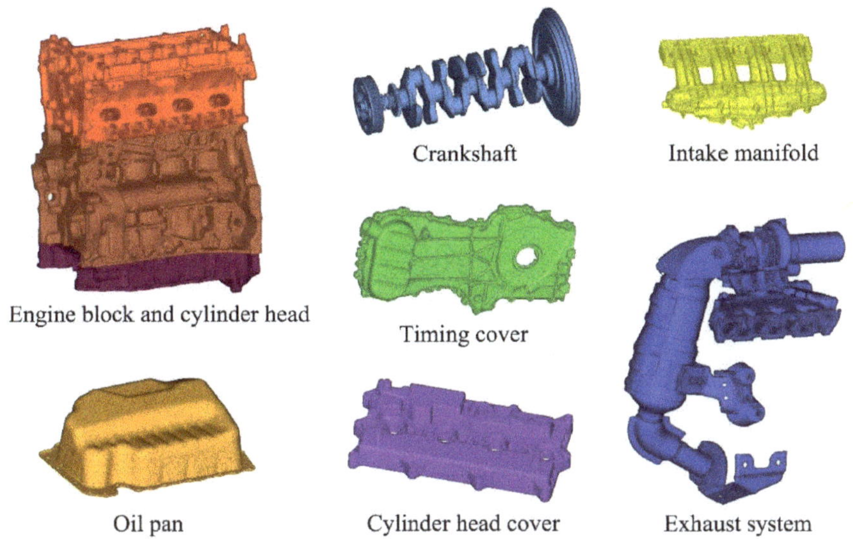

Figure 3. Finite element method (FEM) models of the components.

Table 2. Specification of the engine components.

	Elements	Material	Density (g/cm^3)	Young's Modulus (MPa)	Poisson's Ratio
Crankshaft	93,196	Steel	7.85	210,000	0.30
Oil pan	9043				
Block	269,359	Cast iron	7.30	110,000	0.30
Cylinder head	139,218	Aluminum alloy	2.73	76,550	0.34
Timing cover	91,996				
Cylinder head cover	64,609	Polypropylene	1.40	3150	0.35
Intake manifold	166,391	Polyamide	1.50	1800	0.30
Exhaust system	98,330	Cast iron	7.80	191,000	0.30

The natural frequencies and mode shapes of the main components are measured by the experimental free modal analysis through the Lanczos algorithm and the multi-input single-output (MISO) hammering method [15]. As shown in Figure 4, the components are hung on flexible ropes, a hammer of type PCB 086C03 is used to excite the structures, and an acceleration sensor of type PCB 356A16 is glued to the structures to acquire the vibration signals. The hammer and acceleration sensor are connected to the LMS SCADAS data acquisition front-end, and the signal acquisition and analysis software is LMS Test.Lab.

Figure 4. Modal experiments of components.

The first five modal frequencies of the simulations and measurements are listed in Table 3. Comparing with the experimental results of the components, the errors of the simulation results are within plus or minus 6%, and the mode shape of each component is consistent with the corresponding experimental results. The accuracies of the FEM models of main components are verified, and the FEM models can be used in the following simulations.

Table 3. Modal frequencies of components: comparison of the results from the simulation and measurement.

Components	Order	Simulation/Hz	Measurement/Hz	Error
Engine block (up) and cylinder head	1	545.5	543.6	0.3%
	2	987.3	1042.0	−5.2%
	3	1368.2	1435.9	−4.7%
	4	1455.3	1488.6	−2.2%
	5	1635.9	1673.8	−2.3%
Engine block (low)	1	327.5	309.8	5.7%
	2	793.3	790.9	0.3%
	3	940.5	969.9	−3.0%
	4	1182.8	1145.3	3.3%
	5	1461.8	1385.9	5.5%
Crankshaft	1	406.5	309.6	4.1%
	2	553.5	529.8	4.4%
	3	907.9	881.0	3.1%
	4	950.9	920.9	3.3%
	5	961.6	923.3	4.2%

Table 3. *Cont.*

Components	Order	Simulation/Hz	Measurement/Hz	Error
Cylinder head cover	1	128.5	133.2	−3.5%
	2	204.1	201.7	1.2%
	3	348.6	346.6	0.6%
	4	386.6	401.1	3.6%
	5	439.6	435.8	0.9%
Timing cover	1	197.6	194.3	1.7%
	2	255.6	250.6	2.0%
	3	539.4	531.5	1.5%
	4	692.3	678.9	2.0%
	5	812.9	804.1	1.1%
Oil pan	1	193.4	197.7	−2.2%
	2	323.1	329.5	−1.7%
	3	502.7	490.9	2.4%
	4	572.0	578.1	−1.1%
	5	775.8	807.4	−3.9%

3.3. Multi-Body Dynamic Model

Based on the actual structures and connections of the engine, a simplified multi-body dynamic model is established, as shown in Figure 5. The excitation of the model comes from the pressure in each cylinder. The pressure forces act downward on the pistons and upward on the cylinder head and valves (omitted in Figure 5). The moving parts of the engine include pistons, conrods, crankshaft, valve train, and balance shafts. The forces from the moving parts act on the structures through joints, including liners, bearings, and seats. The structure bodies of the engine include a cylinder block, cylinder head, cylinder head cover, intake manifold, exhaust system, timing cover, and oil pan.

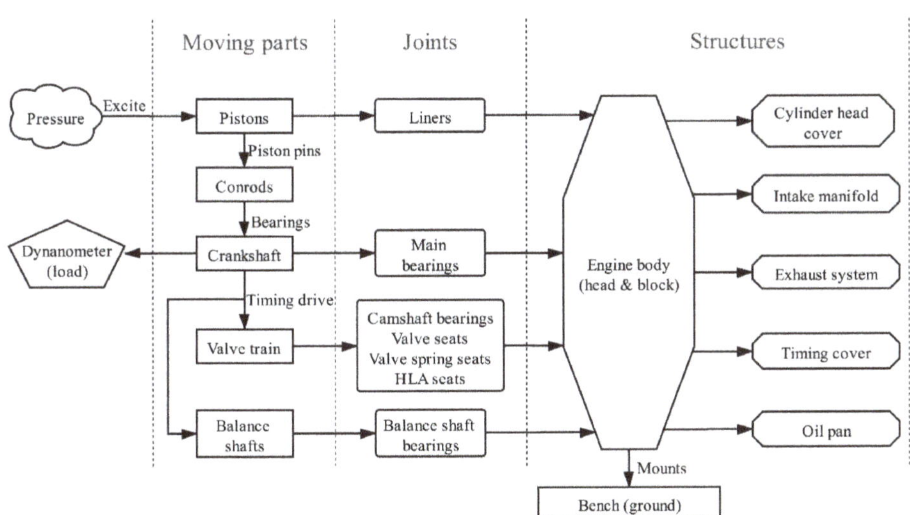

Figure 5. Schematic diagram of the engine model structure.

The conrod is simplified as a beam element, and the mass is distributed to the big end and the small end as lumped mass points. The piston–liner interaction relationship is simplified as a joint with a spring and a damper, as shown in Figure 6. The piston with a piston ring and piston pin is simplified to a mass point which is connected to the points on the liner thrust side (TS) and anti-thrust side (ATS). Each point-to-point connection consists of a linear spring and a damper.

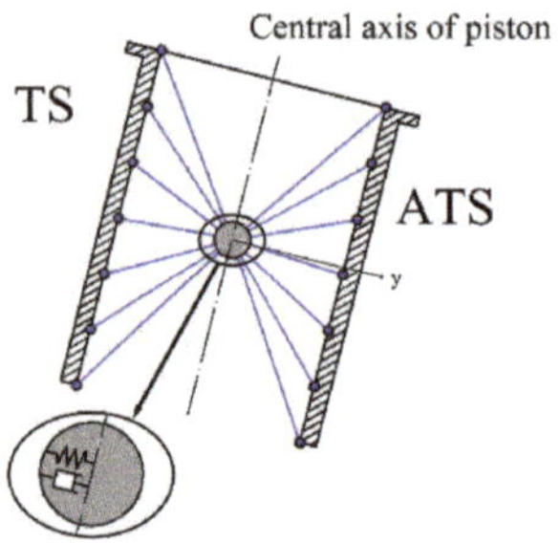

Figure 6. Simplified piston–liner interaction model.

Since the piston's second-order motion will cause a shift in the piston's center of mass, the force in the y direction of the piston can be calculated by the following formula:

$$F_p = -k\Delta u_y - c\Delta \dot{u}_y, \tag{9}$$

where k is the equivalent spring stiffness, c is the equivalent damping, Δu_y is the piston's displacement in the y direction, and $\Delta \dot{u}_y$ is the component of the piston velocity projected in the y direction.

The interaction relationship between the conrod big end and the crank pin is simplified as a point-to-point joint with a non-linear spring and a non-linear damper.

In local coordinates, the bearing force can be calculated as:

$$F_b = k_0 \left(\frac{k_b}{k_0}\right)^{\frac{|\Delta u_{ij}|}{u_0}} \Delta u_{ij} + c_0 \left(\frac{c_b}{c_0}\right)^{\frac{|\Delta u_{ij}|}{u_0}} \Delta \dot{u}_{ij}, \tag{10}$$

where Δu_{ij} is the displacement vector from the center of the conrod big end to the center of the crank pin, k_0 and c_0 are the equivalent stiffness and damping when $|\Delta u_{ij}| = 0$, k_b and c_b are the equivalent stiffness and damping corresponding to Δu_{ij}, and u_0 is the bearing radial clearance.

The joint connecting the main bearing and the main journal is very important in transmitting the load from the crankshaft to the engine block, which is significant to the vibration and acoustic calculation. Therefore, the nonlinear spring and damper (NONL) model is used for the main joint, which is more precise and detailed.

Considering that the oil film only transmits pressure but not tension, only a part of the oil film is compressed when offset at the center of the main journal. As shown in Figure 7, the compressed oil film accounts for approximately 1/3 of the entire bearing (120 degrees).

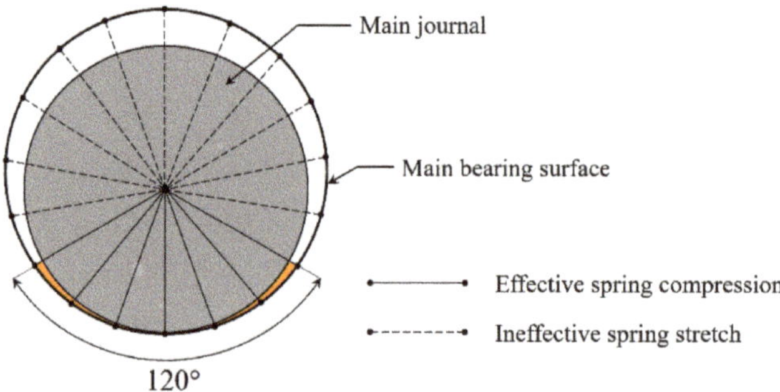

Figure 7. Simplified main journal–main bearing interaction model.

Each line in Figure 7 represents a set of spring and damper. Considering the oil leakage at both ends of the main bearing, the oil film pressure should be parabolic in the axial direction. The stiffness of each spring is defined as:

$$k_{MB} = \frac{F_{\max MB}}{\mu_{0MB} \cdot r} \cdot A, \quad (11)$$

where $F_{\max MB}$ is the maximum main bearing pressure, which can be obtained through the theoretical calculation according to the pressure curve in the cylinder; μ_{0MB} is the radius clearance of the main bearing; r is the number of all the effective compression springs in the bearing; A is the correction factor, $A = 1.5 \sim 3.0$.

The loads of the multi-body dynamic model come from the in-cylinder pressure, the piston slap, and the valve train system.

In-cylinder mixed gas combustion produces high pressure periodically. The pressure is imposed to the piston downward and to the bottom of the cylinder head upward. The measured in-cylinder pressure at 2500 rpm under the wide open throttle (WOT) condition is shown in Figure 8. Under this working condition, the engine outputs the maximum torque at a medium rotation speed, which is relatively frequently used when accelerating. The piston reaches the top dead center (TDC) on the compression stroke when the crank angle is at zero degrees.

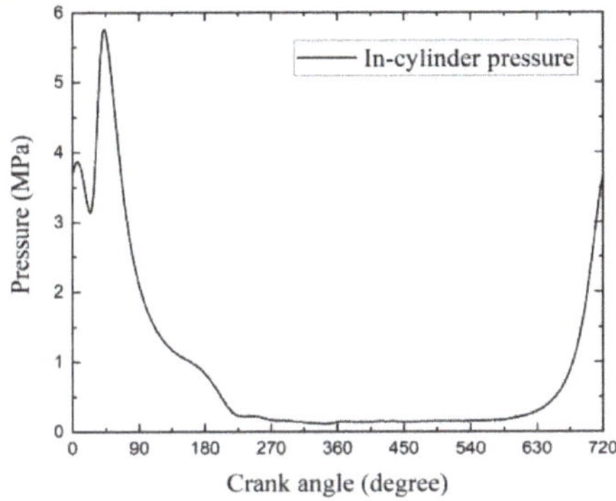

Figure 8. In-cylinder pressure at 2500 rpm under the wide open throttle (WOT) condition.

The load of the in-cylinder pressure is transmitted to the crankshaft through the piston-linkage mechanism, as shown in Figure 9. F_P is the downward force of the in-cylinder pressure on the piston. The conrod force F_C transmitting along the conrod is divided into the radial component F_R and the tangential component F_T.

$$\begin{aligned} F_R &= F_C \cos(\alpha + \beta) = F_P \frac{\cos(\alpha+\beta)}{\cos\beta} \\ F_T &= F_C \sin(\alpha + \beta) = F_P \frac{\sin(\alpha+\beta)}{\cos\beta} \end{aligned} \quad (12)$$

where α is the angle between F_R and F_P, and β is the angle between F_C and F_P.

Due to the existence of the cylinder clearance, the piston has a lateral translation and a rotation around the piston pin during the stroke. The piston lateral force at different height points on the liner TS and ATS is calculated based on the piston dynamic model, then the piston slap excitation to the cylinder liner is obtained. For example, the load of cylinder 1# piston-liner contact force is calculated and shown in Figure 10.

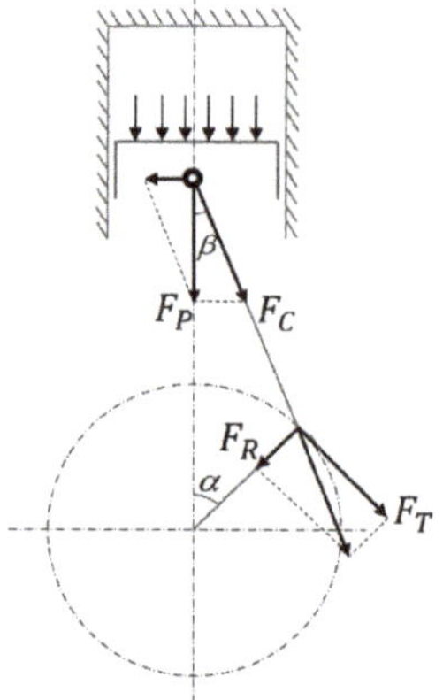

Figure 9. The force transmitting through the conrod.

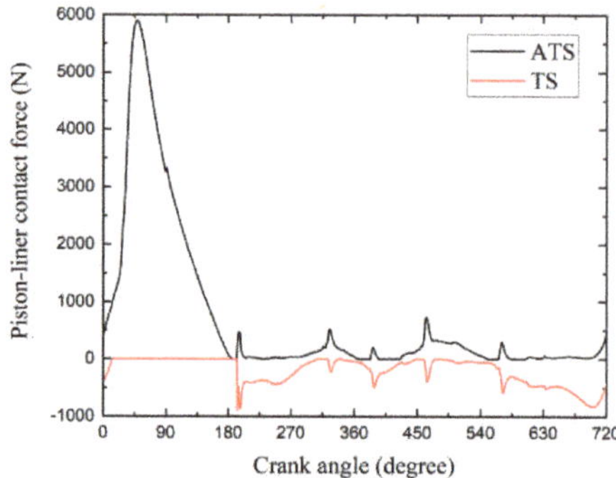

Figure 10. The load of cylinder 1# piston-liner contact force.

3.4. Valve Train System Model

In the uncoupled approach, the dynamic calculation of the valve train system and the dynamic calculation of the engine are carried out separately. The valve train dynamic model is built and calculated independently from the engine; the force results of the joints between the valve train and the engine are loaded onto the engine dynamic model. The disadvantages of the approach are that the impact of the cylinder head deformation to the valve train and the support and constraints of the valve train on the cylinder head are not taken into consideration. The crank speed fluctuation is also neglected or pre-set. In fact, there are important interactions between the valve train and the cylinder head which have significant impacts on the precision of vibration calculation and noise prediction.

In the proposed coupled approach, the dynamic models of the valve train and the engine are combined into one model, forming a full engine dynamic model with the valve train. In this model, the dynamic process between the valve train and the engine is fully analyzed, and the engine dynamic results including the valve train are obtained. The challenge of the coupled approach is the reduced efficiency and solver stability due to the increased complexity of the model. The schematic diagram of differences between the uncoupled and coupled approach is shown in Figure 11.

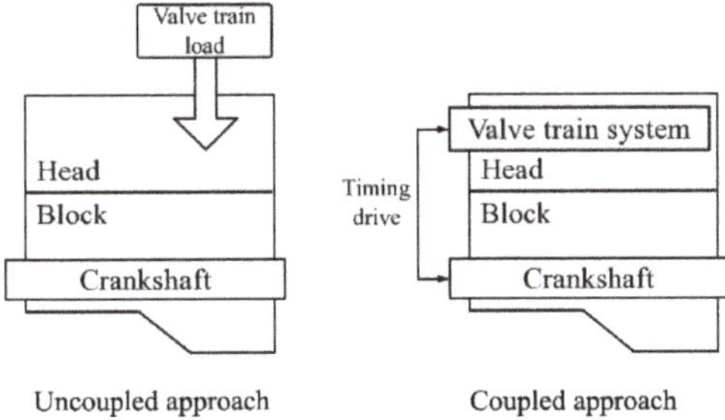

Figure 11. Differences between the uncoupled and coupled approach.

The valve train system model consists of two camshafts with variable valve timing (VVT) phasers, eight single intake valve trains, and eight single exhaust valve trains. The valve train type of the engine is the end-pivot rocker arm overhead camshaft (OHC), as shown in Figure 12. The topology of the valve train system is shown in Figure 13.

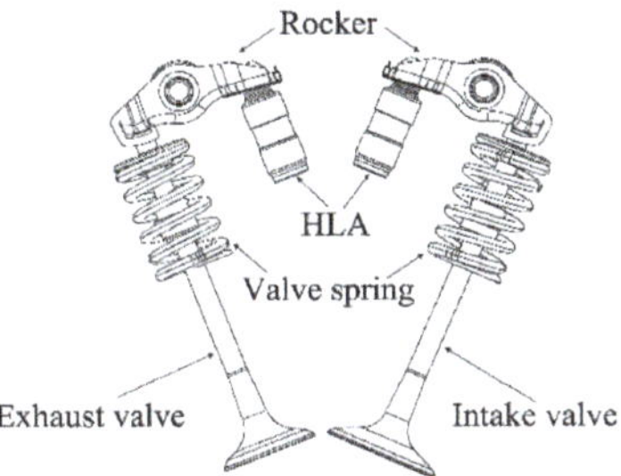

Figure 12. The structure of the intake and exhaust single valve train (SVT) system.

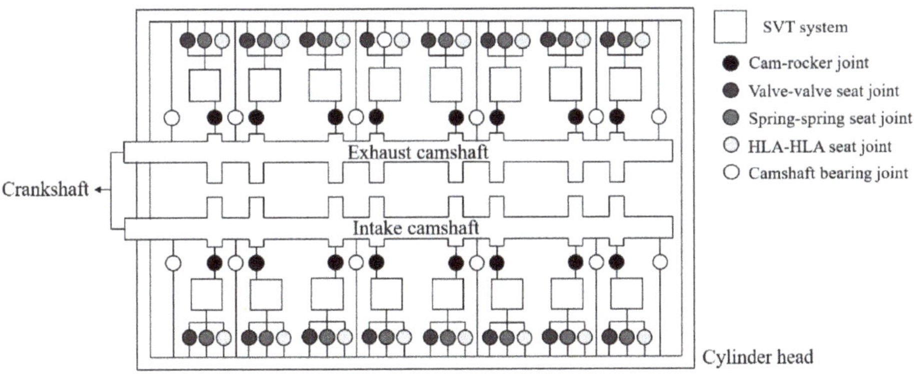

Figure 13. The topology of the valve train system.

Each single valve train (SVT) system model consists of a rocker, a hydraulic lash adjuster (HLA), a valve, and a valve spring, as shown in Figure 14. In each SVT system model, in addition to the joint connected to the camshaft, there are three joints connected to the cylinder head: valve–valve seat joint; valve spring–valve spring seat joint; and HLA–HLA seat joint. Each camshaft is connected to the cylinder head by six radial bearings and two thrust bearings (omitted in Figure 13). The front ends of the two camshafts are connected to the front end of the crankshaft, and the camshafts rotate at a speed of 1/2 of the crankshaft.

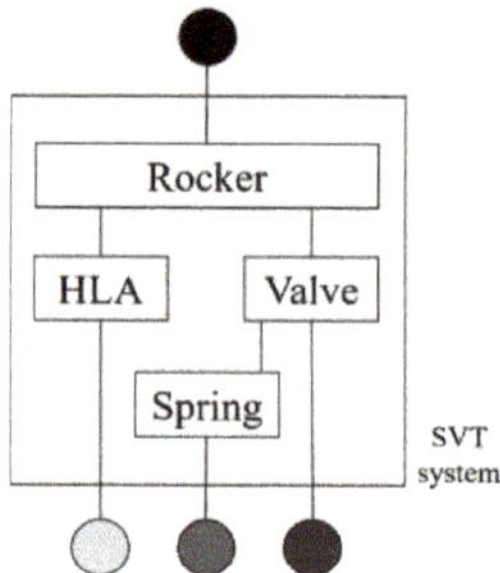

Figure 14. The topology of the SVT system.

The governing equations of motion are obtained for each element of the lumped masses through the Newton–Euler method and solved together in a time marching scheme [16]. In order to improve the accuracy of the model, the structure of the valve and the valve spring can be split into several discrete mass elements respectively, and the elements are connected in series by springs and dampers. Both the spring and the valve are limited to one-dimensional motion along the valve axis. The mass and stiffness of each discrete element of the spring are defined as [17]:

$$\begin{aligned} m_i &= 2\pi\rho A\Delta N_i r_i \\ \kappa_i &= \frac{I_t G_t \cos(\theta)}{2\pi {r_i}^3 \Delta N_i} \end{aligned} \tag{13}$$

where m is the mass, ρ is the density, A is the cross section area, ΔN is the discretized coil number, r is the radius of the discretized coil, κ is the stiffness, I is the torsional moment of inertia, G is the shear modulus, and θ is the pitch angle.

The rocker of the valve train is the roller finger follower type, which is mostly used in modern passenger cars, and has both rotational and translational DoFs. The overhead cam pushes the roller in the middle of the rocker, with the HLA as the pivot point, driving the valve to open and close. The deformation of the rocker has a great effect on the valve dynamic results. However, in uncoupled approaches, the components are generally simplified as rigid bodies connected by springs with constant stiffness, although the stiffness of the rocker arm varies continuously during the valve event [18]. In order to accurately simulate the rocker stiffness, the condensed rocker FEM model is adopted as shown in Figure 15. The rotational and translational DoFs of three nodes of the pivot point, the roller center, and the center of contact circle are retained.

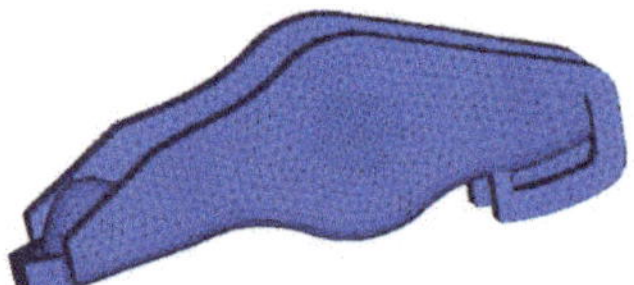

Figure 15. FEM model of the rocker.

A typical HLA, as shown in Figure 16, is used to maintain zero valve clearance, containing a plunger, outer housing, supply chamber, check valve, return spring, and high-pressure chamber (HPC). The lubricant flowing into the HPC is controlled by the check valve at the bottom of the plunger. The check valve is normally closed. After a certain amount of lubricant is lost, the check valve opens to refill the HPC. Due to the extension of the plunger, the valve clearance decreases. When the engine valve enters the lift phase, the check valve is closed, and the pressure in the oil chamber keeps the plunger in place and maintains an effective length to support the rocker [19].

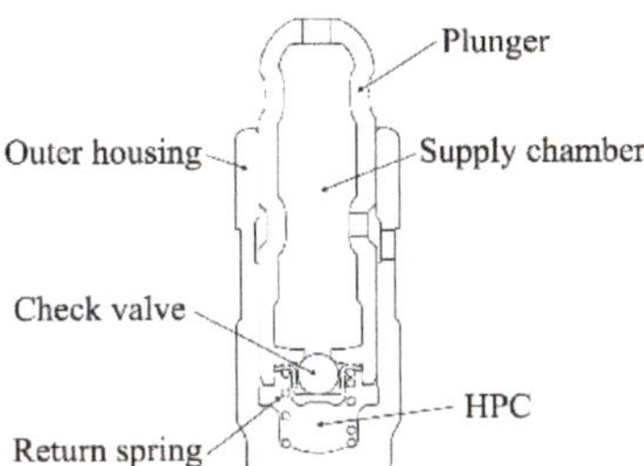

Figure 16. The structure of the hydraulic lash adjuster (HLA).

The lubricant pressure in the HPC is governed by the first order differential equation:

$$\frac{dP(t)}{dt} = \frac{\beta}{V(t)}\left(-\frac{dV(t)}{dt} + Q_{cv}(t) + Q_l(t)\right), \tag{14}$$

where β is the bulk modulus, V is the HPC volume, Q_{cv} is the volumetric flow through the check valve, and Q_l is the leakage through the clearance between the plunger and the outer housing.

The volumetric flow through the check valve can be obtained by:

$$Q_{cv} = \pm C_d S \sqrt{\frac{2|\Delta p|}{\rho}}, \tag{15}$$

where C_d is the discharge coefficient, S is cross-sectional area of the lubricant flow, ρ is the lubricant density, and Δp is the pressure difference between the HPC and the supply chamber.

The leakage through the clearance between the plunger and the outer housing can be obtained by:

$$Q_l = \pm \frac{\pi d \delta^3}{12\mu} \frac{|\Delta p|}{L}, \tag{16}$$

where d is the plunger diameter, δ is the radial clearance between the plunger and the outer housing, μ is the dynamic viscosity of the lubricant, and L is the leakage length.

3.5. Acoustic Model

Besides the classic methods of FEM and BEM (boundary element method), the Trefftz methods are applied in acoustic prediction. The wave-based technique (WBT) is an indirect and frameless Trefftz method [20]. Unlike the traditional element-based methods, WBT uses wave functions that exactly satisfy the governing dynamic equations but do not necessarily satisfy the boundary conditions [21]. The steady-state pressure p at position r is governed by the homogeneous Helmholtz equation.

$$\nabla^2 p(r) + k^2 p(r) = 0, \tag{17}$$

where ∇^2 is the Laplace operator, $k = \omega/c$ is the acoustic wave number, and c is the speed of sound.

As shown in Figure 17, the positions of sound level meters in the acoustic prediction model are arranged around the engine following ISO 3744-2010 [19], which is consistent with that in the measurements.

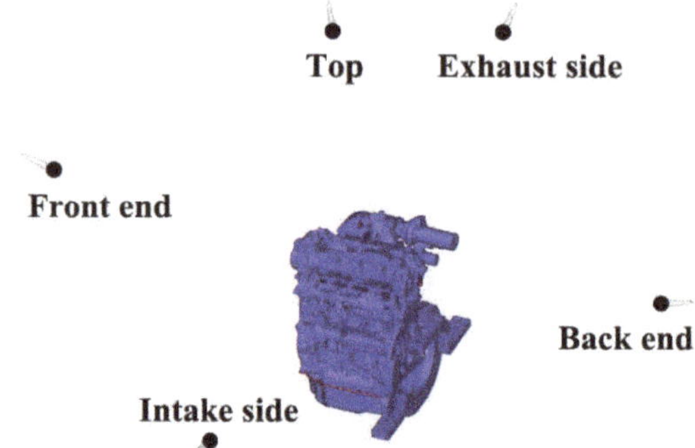

Figure 17. The positions of sound level meters in the acoustic prediction model.

4. Experimental Setup

Noise and vibration verification experiments of the engine are conducted with an AVL electric dynamometer of type 204/8 SL in a semi-anechoic room. The sound pressure level of background noise is below 25 dB(A) in the semi-anechoic room. As shown in Figure 18, five noise measuring points are arranged around the engine, following ISO 3744-2010. The distance between the microphones and the corresponding end faces of the engine is set to 1 m. Three vibration measuring points are arranged on the surface of the intake side of the cylinder head, as shown in Figure 19. All the microphones of type PCB 378B02 and acceleration sensors of type PCB 356A16 are connected to the LMS SCADAS data acquisition front-end after calibration. The signal acquisition and analysis software is LMS Test.Lab. The sampling frequency is 20,480 Hz, and the sampling time is 5 s. Measure twice and take the average value for each working condition. During the experiments, the water temperature was guaranteed to keep in a normal state, and the intake, exhaust, and coolant pipes were all properly led out, so the influences of external factors on the experiments were minimized.

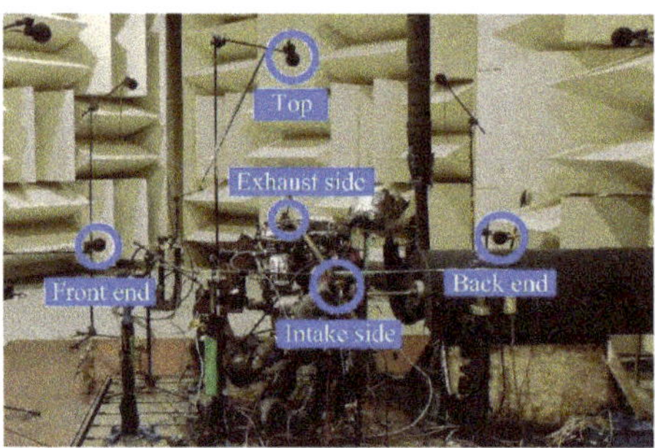

Figure 18. The positions of the microphones in the noise experiments.

Figure 19. The positions of the acceleration sensors in the vibration experiments.

The sound pressure in pascals is measured. The A-weighted sound pressure level is expressed in decibels:

$$L_{pA} = 10\lg\frac{p^2}{p_0^2}, \tag{18}$$

where the reference value $p_0 = 2 \times 10^{-5}$ Pa.

The A-weighted sound power level is expressed in decibels:

$$L_{WA} = 10\lg\left[\frac{1}{N}\sum_{i=1}^{N} 10^{0.1L_{pAi}}\right] + 10\lg\left(\frac{S_1}{S_0}\right), \tag{19}$$

where N is the number of measuring points, L_{pAi} is the A-weighted sound pressure level at the ith measuring point, S_1 is the area of the measuring surface, and $S_0 = 1\ \text{m}^2$.

5. Results and Discussion

5.1. Valve Train Dynamic Analysis

It is a common approach to consider the forces from the valve train to the engine body as external loads, which can be called an uncoupled approach. Based on the uncoupled approaches, previous researchers performed dynamic simulations [9,22]. In this section, the dynamic analysis results using the coupled approach are compared with those in the uncoupled approach under the 2500 r/min WOT condition. The crank angles of 0 degree in the results are the moments of TDC on the compression stroke.

Figure 20 shows the exhaust valve displacement and velocity comparisons between the two approaches. It can be found from Figure 20a that the moments of valve opening, maximum lift, and closing are basically the same in the two approaches. Due to the vibration of the cylinder head, the displacement should fluctuate slightly after the closure of the exhaust valve. In the uncoupled approach, the valve is completely stationary when the valve is closed, since the valve train dynamic model is independent of the cylinder head, while the slight displacement fluctuation can be captured accurately in the coupled approach. A slight fluctuation in the valve velocity curves in the coupled approach in Figure 20b can also be observed. The zooming in views at the opening and closing sections of the exhaust valve velocity are shown in Figure 20c,d. Since the rotation speed fluctuation and the disturbance from the cylinder head are not taken into consideration in the uncoupled approach, the valve velocity curve is relatively smooth and idealized except for the moment of opening. In the coupled approach, there are some fluctuations throughout the valve opening section. In the valve closing section, the velocity curve in the coupled approach is also not as smooth as that in the uncoupled approach.

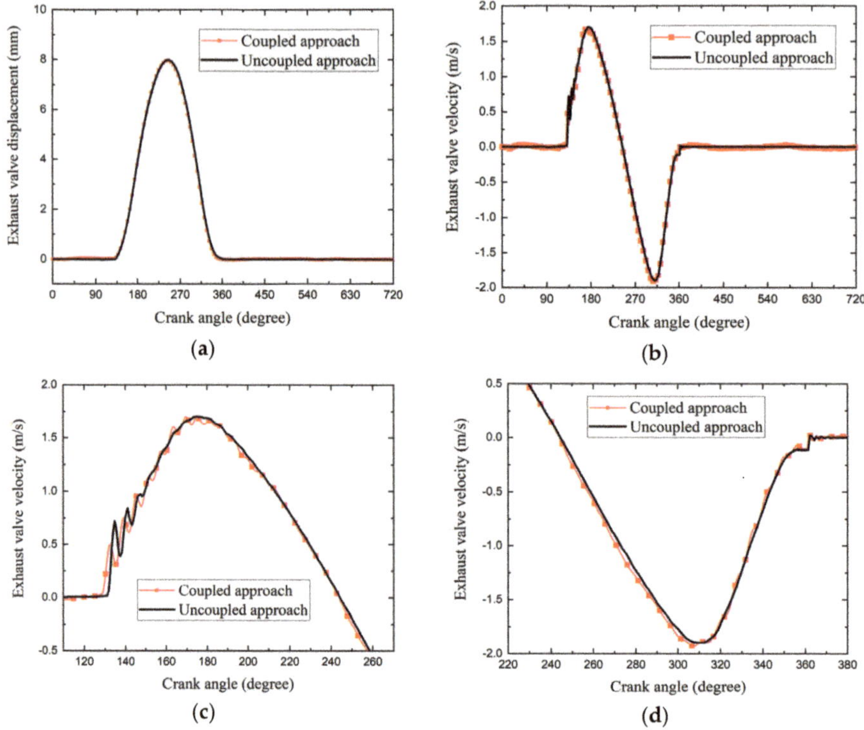

Figure 20. The exhaust valve displacement and velocity comparisons between the two approaches: (**a**) exhaust valve displacement; (**b**) exhaust valve velocity; (**c**) exhaust valve velocity (opening section); (**d**) exhaust valve velocity (closing section).

As shown in Figure 21, compared with the uncoupled approach, the acceleration curve of the coupled approach has a more obvious fluctuation, especially in the positive acceleration section. When the valve is seating, there is an acceleration peak of 4474 m/s^2 in the coupled approach and 5050 m/s^2 in the uncoupled approach.

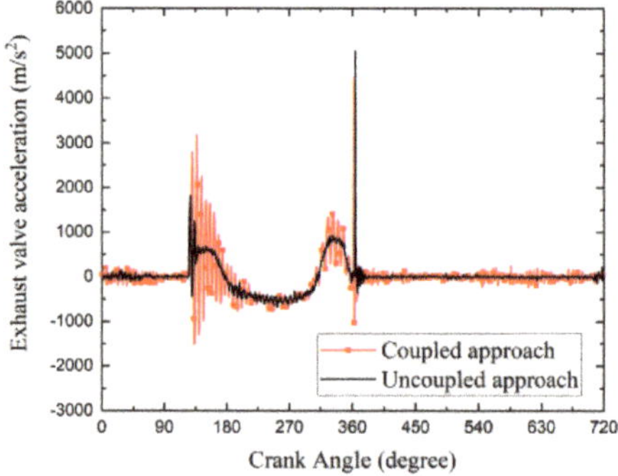

Figure 21. The exhaust valve acceleration comparison between the two approaches.

The impact force from the engine valve is an important source of excitation from the valve train system. The exhaust valve seating force comparison is shown in Figure 22. When the valve is closed, the valve seat force is positively related to the in-cylinder pressure; therefore, the seat force curves of the two approaches are basically the same. When the valve is seated at the crank angle of 385 degrees, the peak force of the uncoupled approach is 412 N, while that of the coupled approach is 292 N. This shows that the flexible cylinder head model in the proposed approach may reduce the overestimated results of the valve seating force in the uncoupled approach.

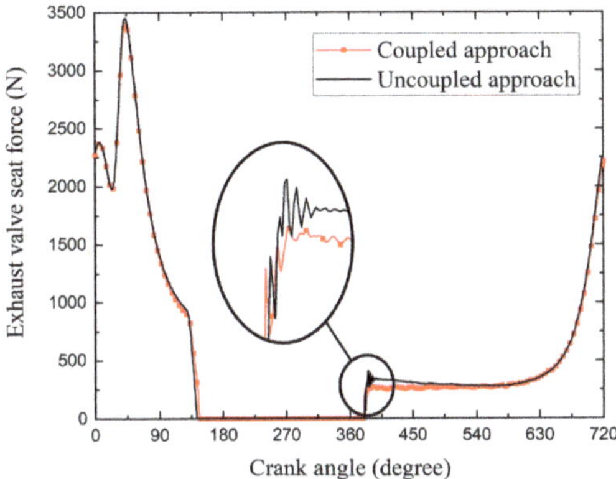

Figure 22. The exhaust valve seating force comparison between the two approaches.

The camshaft bearing force is another important source of excitation to the engine. The comparison of the camshaft bearing force in the Z-direction is shown in Figure 23. When the valve is opening, the peak force of the uncoupled approach is 2586 N, while that of the coupled approach is 3385 N. In the coupled approach, the camshaft bearing not only bears the force of the cam to open the valve, but also reflects the impact of the cylinder head vibration. Therefore, in the coupled approach, there are more fluctuations in the bearing force curve even when the valve is closed, compared with those in the uncoupled approach.

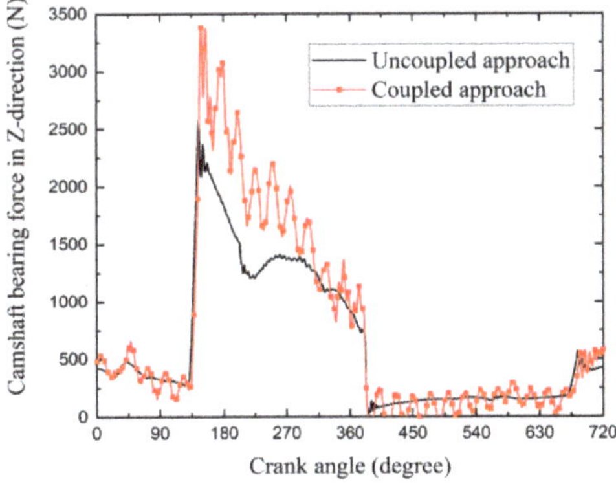

Figure 23. The camshaft bearing force in the Z-direction comparison between the two approaches.

When the valve is opening, the HLA as a pivot bears the force of the valve opening. The two curves have similar trends, as shown in Figure 24. Due to the HLA seat force of the coupled approach reflecting the vibration of the cylinder head, there are more fluctuations in the curve. The peak HLA seat force of the uncoupled approach is 651 N, while that of the coupled approach is 659 N.

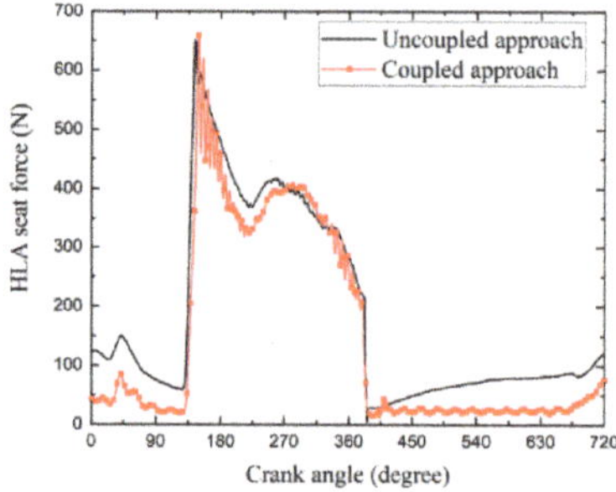

Figure 24. The HLA seat force comparison between the two approaches.

The force of the valve spring seat force is proportional to the valve lift. The preloads of the spring are 240 N. The trend of the two curves are basically the same as that shown in Figure 25. The curve of the coupled approach fluctuates more than that of the uncoupled approach, especially when the valves are opening and closing.

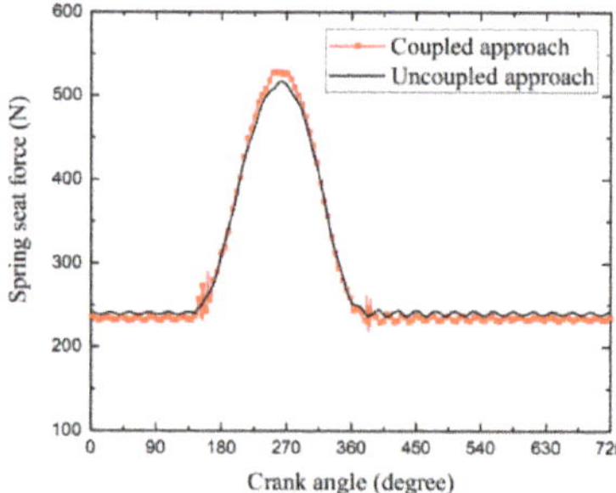

Figure 25. The spring force comparison between the two approaches.

Therefore, the above comparisons have verified that the results obtained by the proposed coupled approach are more reasonable than the uncoupled approach.

5.2. Vibration and Noise Results Analysis

In the experiment, the engine rotational speed is 2500 r/min, thus the fundamental frequency is 41.7 Hz. There are obvious peaks at the second (83.3 Hz), fourth (166.7 Hz), and sixth order (250.0 Hz) frequencies in both the vibration and sound spectrums. As shown in Figure 26, the 1/3 octaves of the measuring point 2 acceleration on the cylinder head vertical to the surface in two simulation approaches are compared with the measured value in the experiment. In the uncoupled approach, the amplitude of the 80 Hz center frequency band is 4.68 m/s^2, which is much larger than the experimental value 1.43 m/s^2. The amplitude at the second-order frequency of 83.3 Hz is significantly overestimated using the uncoupled approach. In the other frequency bands, the amplitude difference between the uncoupled approach and the coupled approach is relatively small, and there is not much difference with the experimental value. The amplitudes of the low frequency bands have a great influence on the overall value; therefore, the vibration acceleration is severely overestimated in the uncoupled approach.

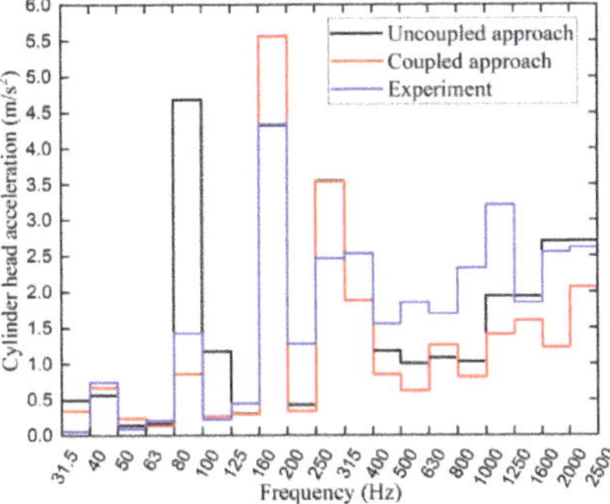

Figure 26. The 1/3 octave of measuring point 2 acceleration comparison between two approaches.

The vibration normal velocity level comparison colormaps between the two simulation approaches at the second order frequency of 83.3 Hz and the fourth order frequency of 166.7 Hz are shown in Figures 27 and 28, respectively. In order to show the velocity results of the cylinder head and the block more clearly, the intake manifold and the exhaust system are hidden in the colormaps, while in fact they are also involved in the calculation of vibration and noise.

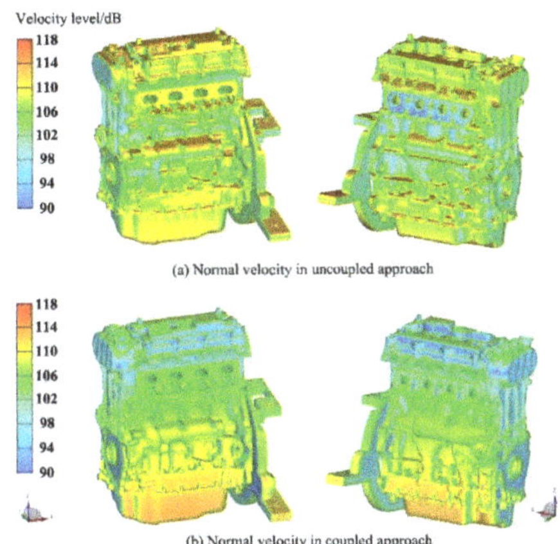

Figure 27. The colormaps of the vibration normal velocity level comparison between the two approaches at the frequency of 83.3 Hz.

The velocity levels at the second order frequency of 83.3 Hz are where the two approaches differ most. As shown in Figure 27, the normal velocity levels of the upper parts of the engine including the cylinder head and the cylinder head cover in the uncoupled approach are larger than those in the coupled approach. This is consistent with the previous conclusion that the uncoupled approach overestimates the second-order vibration. The velocity levels at the fourth order frequency of 166.7 Hz are relatively close between the two approaches, as shown in Figure 28. In the coupled approach, the vibration velocities of the cylinder head and the block are slightly higher than those in the uncoupled approach, which is consistent with the previous 1/3 octave acceleration results on the cylinder head.

The comparison of the acoustic results shows similar trends, as shown in Figure 29. The sound pressure level at the top measuring point at the second order frequency is 63.6 dB(A) in the uncoupled approach, which it is significantly overestimated compared with the experimental value 52.1 dB(A). The coupled approach value is 56.6 dB(A), which is much closer to the experimental value. The coupled approach has a higher accuracy at the fourth order frequency. The coupled approach value is 59.5 dB(A), and the difference from the experimental value 60.4 dB(A) is only 0.9 dB(A). The uncoupled approach value is 54.8 dB(A), and the difference from the experimental value is 5.6 dB(A). At the sixth order frequency of 250 Hz, the coupled approach value also has a better precision. The measured value is 52.4 dB(A), and the value of the coupled approach is 57.7 dB(A), while the value of the uncoupled approach is 61.1 dB(A). The coupled approach shows a better accuracy on noise prediction in the lower frequency bands, which has a great impact on the overall sound pressure level value.

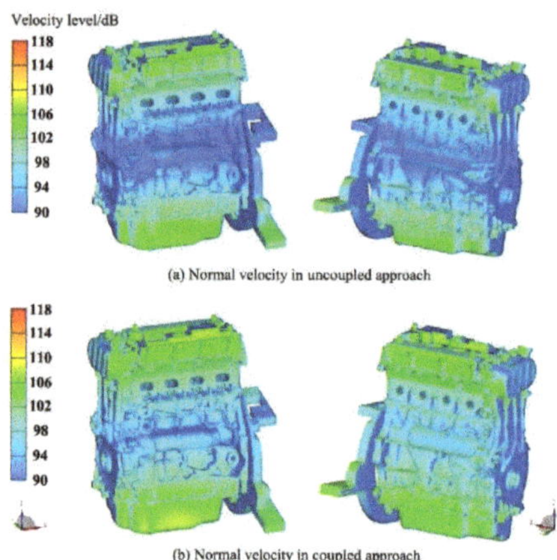

Figure 28. The colormaps of the vibration normal velocity level comparisons between the two approaches at the frequency of 166.7 Hz.

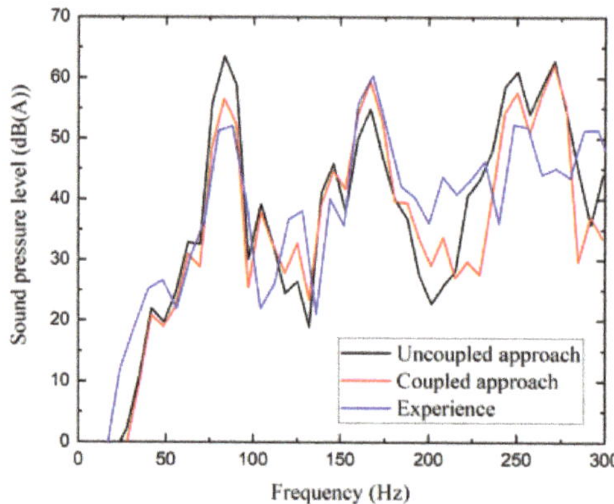

Figure 29. The sound pressure level at the top measuring point in the frequency domain comparison between the two approaches.

The sound pressure level comparisons of the five measuring points between the two simulation approaches and the experimental values are shown in Figure 30. Both two simulation results of all the measuring points are larger than the experimental values, except for the intake side measuring point. It can be inferred that the predicted noise values at the intake side measuring point are relatively underestimated because the engine models ignore high-pressure oil pumps and rails, the filter bases, and some front-end accessories (alternators and air-conditioning compressors). Compared with the uncoupled approach, the results in the coupled approach have a better accuracy at most measuring points, as shown in Figure 31.

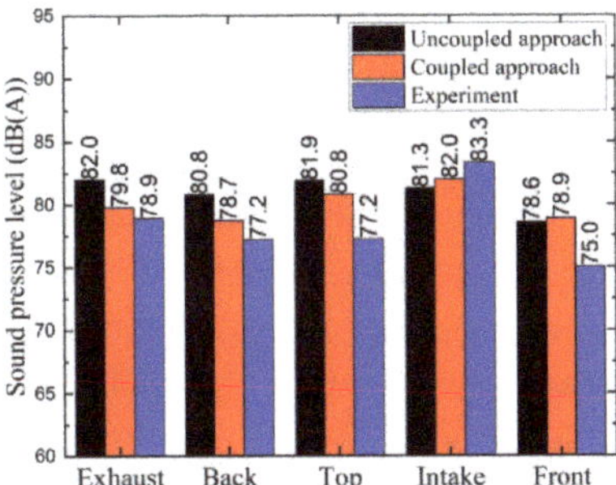

Figure 30. The sound pressure level comparisons of the five measuring points between the two simulation approaches and the measurements.

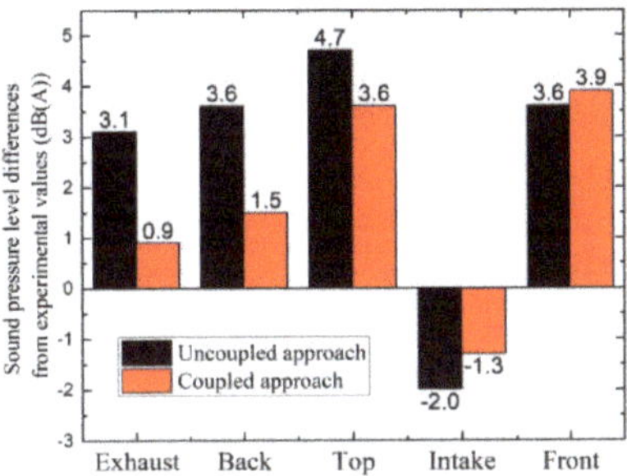

Figure 31. The sound pressure level differences of the two simulation approaches from the experimental values.

The overall sound power level comparison is shown in Figure 32. The simulation results in the uncoupled approach and the coupled approach are 95.4 dB(A) and 94.5 dB(A), respectively, while the experimental value is 93.6 dB(A). After using the coupled approach, the difference from the experiment is reduced from 1.8 dB(A) to 0.9 dB(A). The analyses show that the vibration and noise results obtained by the proposed coupled approach are closer to the experimental value.

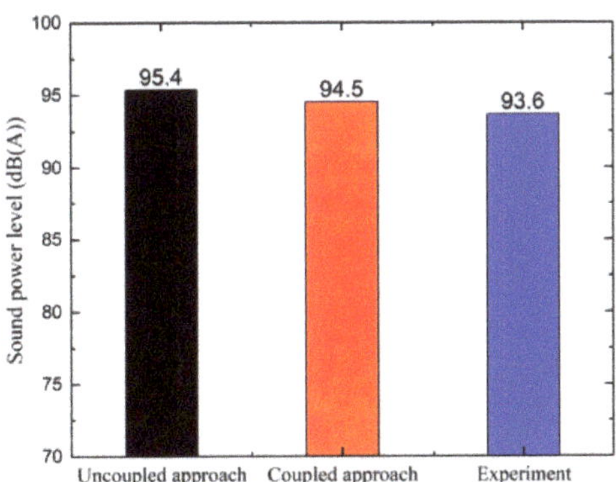

Figure 32. The overall sound power level comparison between the two simulation approaches and the experimental value.

6. Conclusions

A completed coupled engine dynamic model with the valve train system is proposed. The valve train system is integrated into the engine model, forming the full engine dynamic model. Based on the coupled multi-body dynamic model, the dynamic simulation is performed and the vibration and noise are predicted. The results of the coupled approach are compared with those of the uncoupled approach and the experimental values. The conclusions are summarized as follows:

1. The coupled approach considers interaction ignored in the uncoupled approach, which can improve the fidelity of the model. The impact of the cylinder head deformation to the valve train and the support and constraints of the valve train on the cylinder head are taken into consideration in the coupled approach. The camshaft speed is obtained from the crankshaft speed in the simulation, instead of a constant or preset speed. Therefore, the influence of speed fluctuation on the valve train is taken into consideration as well.
2. Applying the coupled approach, more detailed dynamic results of the valve train are presented. The influence of the interaction between the valve train system and the cylinder head can be observed in the simulation results of the parts displacements, velocities, accelerations, and forces. The coupled approach is more instructive for analyzing the fatigue problems and abnormal noise problems of the valve train system.
3. Compared with the uncoupled approach, the results of the coupled approach have a better accuracy in vibration simulations. The vibration of the cylinder head measuring point at the second-order frequency (83.3 Hz) is severely overestimated in the uncoupled approach, resulting in the overestimation of the overall vibration and noise. In the coupled approach, the results of vibration at the second-order frequency are closer to the experimental values.
4. The coupled approach has a better accuracy in noise prediction as well. The calculated overall sound power level values in the uncoupled and coupled approaches are 95.4 dB(A) and 94.5 dB(A), respectively, while the experimental value is 93.6 dB(A). After using the coupled approach, the difference from the experimental value is reduced from 1.8 dB(A) to 0.9 dB(A).

Author Contributions: For this research, X.Z. and X.L. put forward the approaches and established the simulation models; Y.Q. and Z.H. provided the experiment support; X.Z. and X.L. wrote the paper; while Y.Q. and Z.H. further improved the manuscript. All authors have read and agreed to the published version of the manuscript.

Funding: This research was funded by the National Natural Science Foundation of China grant number 51876188 and 51705454 and the National Key Research and Development Program of China grant number 2016YFB0101604.

Conflicts of Interest: The authors declare no conflict of interest.

References

1. Dudley, W.M. New method in valve cam design. *SAE Tech. Pap.* **1948**, 480170. [CrossRef]
2. Eiss, N.S., Jr. Vibration of cams having two degrees of freedom. *J. Eng. Ind.* **1964**, *86*, 343–349. [CrossRef]
3. Chan, C. Dynamic model of a fluctuating rocker-arm ratio cam system. *J. Mech. Des.* **1987**, *109*, 356–365. [CrossRef]
4. Seidlitz, S. Valve train dynamics-a computer study. *SAE Tech. Pap.* **1989**, 890620. [CrossRef]
5. Nagaya, K. Vibration analysis of high rigidity driven valve system of internal combustion engines. *J. Sound Vib.* **1993**, *165*, 31–43. [CrossRef]
6. Pisano, A.P. An experimental and analytical investigation of the dynamic response of a high-speed cam-follower system. *J. Mech. Transm. Autom. Des.* **1983**, *105*, 699–704. [CrossRef]
7. Lassaad, W. Nonlinear dynamic behaviour of a cam mechanism with oscillating roller follower in presence of profile error. *Front. Mech. Eng.* **2013**, *8*, 127–136. [CrossRef]
8. Guo, J. A new numerical method for developing the lumped dynamic model of valve train. *J. Eng. Gas Turbines Power* **2015**, *137*, 101507. [CrossRef]
9. Zhou, C. An enhanced flexible dynamic model and experimental verification for a valve train with clearance and multi-directional deformations. *J. Sound Vib.* **2017**, *410*, 249–268. [CrossRef]

10. Meuter, M. Multi-Body engine simulation including elastohydrodynamic lubrication for non-conformal conjunctions. *Proc. Inst. Mech. Eng. Part K J. Multi-Body Dyn.* **2017**, *231*, 457–468. [CrossRef]
11. Offner, G. Modelling of condensed flexible bodies considering non-linear inertia effects resulting from gross motions. *Proc. Inst. Mech. Eng. Part K J. Multi-Body Dyn.* **2011**, *225*, 204–219. [CrossRef]
12. Angeles, J.; Zakhariev, E. *Computational Methods in Mechanical Systems: Mechanism Analysis, Synthesis, and Optimization*; Springer: Berlin/Heidelberg, Germany, 1998.
13. Morrison, D. A framework for modal synthesis. *Comput. Music J.* **1998**, *17*, 45–56. [CrossRef]
14. Craig, R.R., Jr.; Bampton, M.C.C. Coupling of substructures for dynamic analyses. *AIAA J.* **1968**, *6*, 1313–1322. [CrossRef]
15. Liu, R. A study of the influence of cooling water on the structural modes and vibro-acoustic characteristics of a gasoline engine. *Appl. Acoust.* **2016**, *104*, 42–49. [CrossRef]
16. Teodorescu, M. Multi-Physics analysis of valve train systems: From system level to microscale interactions. *Proc. Inst. Mech. Eng. Part K J. Multi-Body Dyn.* **2007**, 221. [CrossRef]
17. Haslinger, J. Non-Smooth dynamics of coil contact in valve springs. *J. Appl. Math. Mech.* **2014**, *94*, 957–967. [CrossRef]
18. Beloiu, D.M. Modeling and analysis of valve train, part I-conventional systems. *SAE Int. J. Engines* **2010**, *3*, 850–877. [CrossRef]
19. ISO. *Acoustics—Determination of Sound Power Levels and Sound Energy Levels of Noise Sources Using Sound Pressure—Engineering Methods for an Essentially Free Field Over a Reflecting Plane*; International Organization Standardization: Geneva, Switzerland, 2010; Volume 3744.
20. Desmet, W. A Wave Based Prediction Technique for Coupled Vibro-Acoustic Analysis. Ph.D. Thesis, Katholieke Universiteit Leuven, Leuven, Belgium, 1998.
21. AVL Workspace. Excite-Power-Unit Theory: AVL User Manuals. 2017. Available online: http://www.avl-ast-china.com/upload/ueditor/file/20170516/1494905319788032969.pdf (accessed on 24 July 2020).
22. Andreatta, E.C. Valve train kinematics and dynamics simulation. *SAE Tech. Pap.* **2016**, *36*, 0213. [CrossRef]

Article

Identification of the Dynamic Parameters of a Parallel Kinematics Mechanism with Prismatic Joints by Considering Varying Friction

Abdur Rosyid [1] and Bashar El-Khasawneh [2,*]

1 Khalifa University Center for Autonomous Robotic Systems (KUCARS), Khalifa University of Science and Technology, P.O. Box 127788 Abu Dhabi, UAE; abdur.patrum@ku.ac.ae
2 Mechanical Engineering Department, Khalifa University of Science and Technology, P.O. Box 127788 Abu Dhabi, UAE
* Correspondence: bashar.khasawneh@ku.ac.ae

Received: 8 June 2020; Accepted: 9 July 2020; Published: 14 July 2020

Abstract: This study proposed a novel approach for the offline dynamic parameter identification of parallel kinematics mechanisms in which the friction is significant and varying. Since the friction is significant, it should be incorporated to provide an accurate dynamic model. Furthermore, the varying normal forces as a result of the changing posture of the mechanism lead to varying friction forces, specifically varying static and Coulomb friction forces. By considering this variation, the static and Coulomb friction parameters are identified as coefficients instead of forces. A bound-constrained optimization technique using an iterative global search tool was employed in this work to minimize the residual errors while maintaining the physical feasibility of the solutions. Moreover, the friction was modeled by using the nonlinear Stribeck friction model since a linear friction model was not sufficient, whereas the variation of the friction followed the variation of the normal forces, which were evaluated through the Lagrange multipliers in the constrained dynamic model of the mechanism. The solutions obtained were verified by using some trajectories that were different from those used in the identification.

Keywords: dynamic parameter identification; dynamic parameters; parallel kinematics; parallel robot; parallel manipulator; varying friction

1. Introduction

The dynamic parameters of a rigid body system typically consist of the inertial and friction parameters. In a system where its inertial parameters are very dominant and its friction parameters are relatively very small, the friction parameters can be neglected in the identification. However, when the friction is significant, one should incorporate it in the dynamics, and accordingly include the friction parameters in the identification. Otherwise, the corresponding dynamic model is not adequate to capture the real dynamics. An identification usually estimates the inertial parameters as the so-called barycentric parameters, which consist of the masses, the first moments of inertia, and the moments of inertia relative to the origin of the component frames. Upon obtaining the masses and the first moments of inertia, one can easily obtain the centers of masses (COMs). The moments of inertia relative to the COM can be obtained from the moments of inertia relative to the component frame by utilizing the Huygens–Steiner theorem, which is commonly called the parallel axis theorem.

The dynamics of a mechanism can typically be modeled as a linear system in terms of the dynamic parameters. As a result, linear least squares can be conveniently used to identify the dynamic parameters [1]. Based on measurements along a certain exciting trajectory, an overdetermined linear system can be composed, and subsequently, the dynamic parameters can be estimated by using the closed-form linear least squares solution, which involves the evaluation of the pseudo-inverse of the observation matrix. In this case, accurate estimates of the dynamic parameters can only be obtained if the observation matrix has a full rank. In fact, the observation matrix is often rank-deficient. In such a case, a numerical or symbolic mathematical treatment is commonly performed to transform the rank-deficient observation matrix into a full-rank one. This results in a new full-ranked linear system with a reduced size. As the rank-deficiency implies a linear dependency among some dynamic parameters, the reduced linear system can only present the dynamic parameters as linear combinations of the parameters, commonly known as the base parameters.

When friction is included, the linear least squares technique can only incorporate a linear friction model. The use of nonlinear optimization techniques can overcome this issue. There are at least three main advantages of this approach. First, one can directly utilize the nonlinear dynamic model, including a nonlinear friction model, to identify the dynamic parameters instead of transforming the nonlinear dynamic model into a linear model, as described earlier. In other words, one can directly use the nonlinear inverse dynamics for identification. The second advantage is the capability of this approach to provide the dynamic parameters as lonely parameters, i.e., not as linear combinations. The third advantage of this approach is the flexibility to identify the dynamic parameters as either the standard parameters or the barycentric parameters. One of the commonly used nonlinear optimization techniques is the nonlinear least squares, in which the Jacobian of the system with respect to the dynamic parameters should be evaluated. While in the linear identification method, the observation matrix should be full-rank to obtain a trustworthy solution from the linear least squares, in a similar fashion, the system Jacobian matrix should be full-rank to get a trustworthy solution from the nonlinear least squares.

Although the linear least squares methods applied to a full-rank reduced linear system or the nonlinear least squares applied to a nonlinear system with a full-rank system Jacobian matrix can theoretically give a trustworthy solution, in practice, it is very likely that it cannot give a physically feasible solution due to the presence of measurement noise and/or the inadequacy of the dynamic model. The authors noticed that both the linear and nonlinear least squares techniques can retrieve synthesized (simulated) measurements but fail to provide physically feasible estimates when real, noisy measurements are used. For that reason, constrained optimization is used in this work to obtain a physically feasible solution. A similar approach has been used by Farhat et al. [2] and Thanh et al. [3], who respectively used a constrained optimization algorithm and a direct search to identify the dynamic parameters of a parallel kinematics mechanism (PKM). Moreover, Poignet et al. [4] used interval analysis to identify the dynamic parameters of an H4 manipulator, which is a three-translational and one-rotational (3T1R) PKM.

Furthermore, this work considered a PKM with significant friction due to the presence of sliding (prismatic) joints subject to a significant load. This load was comprised of a constant gravity load and varying reaction forces. Different from a system having sliding joints with constant normal forces, such as in Pan et al. [5], in this work, the total load resulted in varying normal forces, and therefore, varying friction forces. In fact, the normal forces varied because of the changing posture of the PKM. This behavior applies to all types of PKMs, regardless of the magnitude of the variation. However, to the authors' knowledge, there has been no published work on the dynamic parameter identification of PKMs that considers the variation of the friction forces. Moreover, most of the works, including References [3,4,6–13], have also used a linear friction model to enable the use of linear least squares algorithms, including the unweighted linear least squares, weighted linear least squares, and total linear least squares algorithms, as the identification tool. Recently, Farhat et al. [2] used a nonlinear friction model to identify the dynamic parameters of a 3RPS (3 Revolute-Prismatic-Spherical joint topology) PKM. Furthermore, Briot and Gautier [13] identified not only the inertial and friction parameters of an Orthoglide PKM but also its drive gains. All of the mentioned works considered the friction in the PKMs constant. This study investigated the identification of the dynamic parameters of a PKM by considering nonlinear, varying friction. In short, this work proposed an approach for the identification of a PKM with noisy measurements, as well as nonlinear and varying friction. In this work, the dynamics were derived by using a Lagrange approach. A set of constraint equations corresponding to the required Lagrange multipliers were formulated. These Lagrange multipliers corresponded to the reaction forces at all joints, including the active and passive ones. Since the friction on the active, prismatic joints were assumed to be much more significant than that on the passive joints, only the former was considered. To describe the approach, a non-symmetric planar 3PRR (3 Prismatic-Revolute-Revolute joint topology) PKM [14,15], as shown in Figure 1, was considered as the study case. This paper is organized as follows. First, the inverse dynamics of the mechanism at hand with and without friction are derived and the friction model is presented. Second, the identification technique is discussed, where the discussion includes the optimization of the exciting trajectory and the physical feasibility of the estimation. Finally, the identification results are discussed.

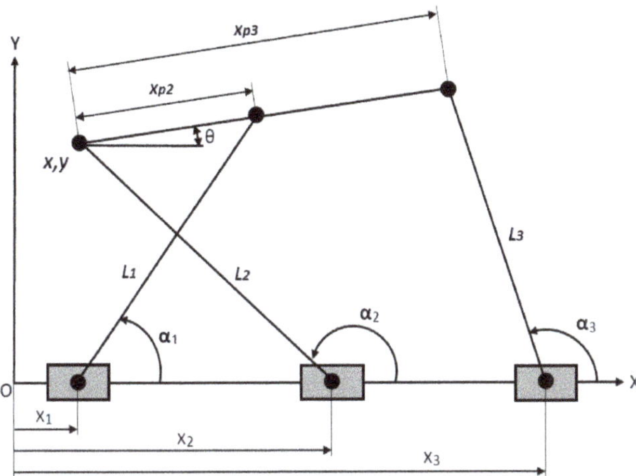

Figure 1. Planar 3PRR parallel kinematics mechanism (with the Z-axis perpendicular to the paper).

2. Dynamic Model

The dynamics of a mechanism can be modeled by using different sets of generalized coordinates. Since a PKM has constrained dynamics, the dynamic model is usually modeled by using a set of redundant generalized coordinates that can be divided into independent and dependent coordinates. In the mechanism at hand, twelve generalized coordinates, as shown in Rodríguez et al. [11] are used. For more convenience, the Cartesian coordinates of the moving platform x, y, and θ are used to represent the pose of the manipulator. Since the mechanism has three degrees of freedom without a kinematic redundancy, only three coordinates are independent, while the remaining coordinates are dependent. The closed-loop topology of the mechanism can be maintained by introducing several constraint equations in the dynamic model. The number of required constraint equations is $m = n–d$, where n is the number of generalized coordinates and d is the number of degrees of freedom, which also represents the number of independent coordinates.

The rigid body model of the mechanism with a fixed payload mounted on the moving platform is depicted in Figure 2. The mass of the payload is assumed to be constant, whereas its COM position is fixed relative to the origin of the local frame X′Y′. The component masses are lumped to their COMs. For all of the legs and the moving platform, the COMs are only defined in the XY plane since the mechanism is planar. This planar assumption is valid since the geometry of all the legs and the moving platform is symmetric relative to the XY plane. Moreover, the COMs of the legs are defined along the longitudinal axes of the legs since the geometry of the legs is symmetric relative to their longitudinal axes. However, the COMs of the moving platform should be defined relative to both of the X′ and Y′

axes since its COM is not located along the X′ axis. The distance between the COM of the moving platform to the origin of the local frame X′Y′ is given by:

$$X_{pG} = \sqrt{X_{pG(x)}^2 + X_{pG(y)}^2}. \tag{1}$$

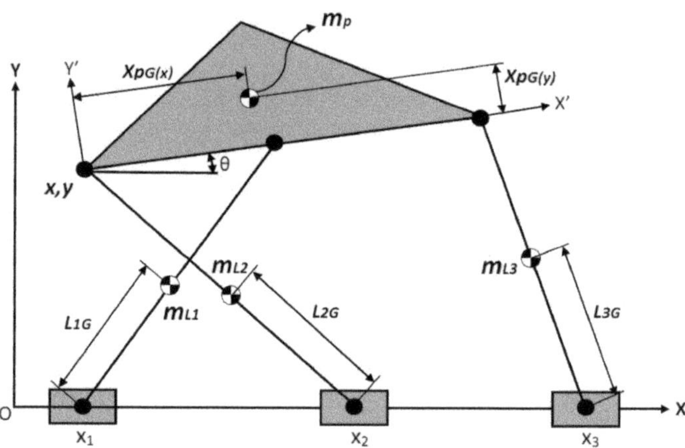

Figure 2. Centers of masses (COMs) of a 3PRR parallel kinematics mechanism.

A general rigid body typically has ten barycentric parameters, which include a mass, three first moments of inertia, and six elements of the inertia matrix. If the inertia matrix is evaluated with respect to the principal axes of the body, all the off-diagonal elements vanish to zero, i.e., the inertia matrix becomes a diagonal matrix. In such a case, only three elements of the inertia matrix should be defined. In a planar mechanism in which only rotation about the Z-axis is available, only the inertia elements related to the Z-axis are retained. Accordingly, when a planar mechanism is considered and the inertia matrix is evaluated with respect to the principal axis, which is the case for the mechanism at hand, the inertia matrix turns into a single scalar inertia. For this reason, the moment of inertia of each of the legs and the moving platform of the mechanism at hand is defined by a single scalar.

2.1. Dynamic Model without Friction

The total kinetic energy T of the mechanism is the sum of the kinetic energy corresponding to all the moving components of the mechanism, i.e., the actuators (sliders), the legs, and the moving platform:

$$T = \sum_{i=1}^{3} T_{si} + \sum_{i=1}^{3} T_{Li} + T_p, \tag{2}$$

$$T_{si} = \frac{1}{2} m_{si} \dot{x}_i^2, \tag{3}$$

$$T_{Li} = \frac{1}{2} m_{Li} V_{Gi} . V_{Gi} + \frac{1}{2} I_{Gi} \omega_i^2, \tag{4}$$

$$T_p = \frac{1}{2} m_p \left(\dot{x}_p^2 + \dot{y}_p^2 \right) + \frac{1}{2} I_{Gp} \omega_p^2, \tag{5}$$

where m_{si}, m_{Li}, and m_p denote the masses of slider i, leg i, and the moving platform, respectively; $\dot{x}_i$ is the translational velocity of the slider i; V_{Gi} is the translational velocity of the COM of leg i relative to the inertial frame; ω_p and ω_i are the angular velocities of the platform and leg i, respectively; I_{Gp} and I_{Gi} are the moments of inertia of the platform and leg i relative to their COMs, respectively; and $\dot{x}_p$ and

$\dot{y}_p$ are the translational velocities of the COM of the moving platform in x and y directions, respectively. The COM of the moving platform is given by:

$$x_p = x + x_{pG(x)} \cos\theta - x_{pG(y)} \sin\theta;\ y_p = y + x_{pG(x)} \sin\theta + x_{pG(y)} \cos\theta \tag{6a,6b}$$

It can be observed in both Equations (4) and (5) that the kinetic energy of the legs and the moving platform both have translational and rotational components due to the translation and rotation relative to the inertial frame. On the other hand, it can be observed in Equation (3) that the sliders only have kinetic energy from translations as they do not undergo rotation relative to the inertial frame.

Similarly, the total potential energy V is the sum of the potential energy of the legs and the moving platform. The potential energy considered here comes from the altitude of the mechanism components in the direction of gravity. Let us first assume that the mechanism as shown in Figures 1 and 2 is standing vertically, i.e., the Y-axis is vertical and therefore parallel with the direction of gravity. By considering the sliding path of the sliders as the datum in the potential energy evaluation, it can be observed that the sliding path of the sliders is coincident with the datum. As a result, the potential energy of the sliders vanishes and accordingly does not make any contribution to the total potential energy. The potential energy of the legs and the moving platform are respectively:

$$V_{Li} = m_{Li} g L_{Gi} \sin\alpha_i, \tag{7}$$

$$V_p = m_p g y_p, \tag{8}$$

where g denotes the gravitational acceleration, L_{Gi} denotes the altitude of the COM of leg i relative to the datum, and y_p denotes the altitude of the COM of the moving platform relative to the datum. The potential energy expressions in Equations (7) and (8) hold when the mechanism is standing vertically, as described earlier. When the mechanism is oriented on a horizontal plane such that all of the COMs have an identical altitude in the direction of gravity, the equations are no longer valid. In such a case, when the datum of the potential energy evaluation has an identical altitude to all of the COMs, then the potential energy due to gravity vanishes. In this case, the rigid body motion of the mechanism is not affected by gravity. A practical way to suppress the effect of gravity in a general model is by imposing a zero value to the gravitational acceleration constant g.

Using the Euler–Lagrange approach, the equations of motion are given by:

$$\frac{d}{dt}\left(\frac{\partial L}{\partial \dot{q}_i}\right) - \frac{\partial L}{\partial q_i} = Q_i + \sum_{k=1}^{m} \lambda_k \frac{\partial \Gamma_k}{\partial q_i}, \tag{9}$$

$$q = \begin{bmatrix} x_1 & x_2 & x_3 & y_1 & y_2 & y_3 & \alpha_1 & \alpha_2 & \alpha_3 & x & y & \theta \end{bmatrix}^T, \tag{10}$$

$$Q = \begin{bmatrix} Q_{x1} & Q_{x2} & Q_{x3} & Q_{y1} & Q_{y2} & Q_{y3} & Q_{\alpha 1} & Q_{\alpha 2} & Q_{\alpha 3} & Q_x & Q_y & Q_\theta \end{bmatrix}^T, \tag{11}$$

where q is the vector of generalized coordinates; Q is the vector of external loads corresponding to the generalized coordinates; and y_1, y_2, and y_3 denote the displacement of the sliders in the y-direction. The values of Q_{y1}, Q_{y2}, and Q_{y3} are zero since there is no external load exerted to the sliders in the y-direction, whereas the values of $Q_{\alpha 1}$, $Q_{\alpha 2}$, and $Q_{\alpha 3}$ are zero since they correspond to passive joints. If the end effector is not subject to any external loading, Q_x, Q_y, and Q_θ are also zero. The last term on the right-hand side of Equation (9) represents the generalized constraint forces, which can be seen as joint reaction forces. These constraint forces are the products of the Lagrange multipliers λ and the Jacobian of position constraints Γ.

The following nine position constraint equations are required to maintain the closed chain of the mechanism:

$$\Gamma = \begin{bmatrix} \Gamma_1 & \Gamma_2 & \Gamma_3 & \Gamma_4 & \Gamma_5 & \Gamma_6 & \Gamma_7 & \Gamma_8 & \Gamma_9 \end{bmatrix}^T, \tag{12}$$

where:

$$\begin{aligned}
\Gamma_1 &= y_1 = 0, \\
\Gamma_2 &= y_2 = 0, \\
\Gamma_3 &= y_3 = 0, \\
\Gamma_4 &= x + x_{p2}\cos\theta - x_1 - l_1\cos\alpha_1 = 0, \\
\Gamma_5 &= y + x_{p2}\sin\theta - l_1\sin\alpha_1 = 0, \\
\Gamma_6 &= x - x_2 - l_2\cos\alpha_2 = 0, \\
\Gamma_7 &= y - l_2\sin\alpha_2 = 0, \\
\Gamma_8 &= x + x_{p3}\cos\theta - x_3 - l_3\cos\alpha_3 = 0, \\
\Gamma_9 &= y + x_{p3}\sin\theta - l_3\sin\alpha_3 = 0.
\end{aligned} \tag{13}$$

The following equations of motion can be obtained by substituting Equations (10)–(13) into Equation (9):

$$M(q)\ddot{q} + N(q,\dot{q}) = Q + B(q,t)^T\lambda, \tag{14}$$

where M is the inertia matrix; N is a force vector containing the Coriolis, centrifugal, and gravitational forces; Q is the external force vector; and the last term is the constraint force vector, which maintains the closed chain of the mechanism. The system inertia matrix M, vector N, and matrix B depend on the mechanism posture. Although the inertia matrix of each component of the mechanism relative to either its COM or its local frame is constant, the system inertia matrix M varies with the change of the mechanism posture.

The position constraint equations can be differentiated twice with respect to time, as follows:

$$\frac{d}{dt}(B\dot{q}) = \dot{B}\dot{q} + B\ddot{q} = 0. \tag{15}$$

Finally, Equation (14) can be combined with Equation (15) to obtain a descriptor form that is presented in the following differential-algebraic equations:

$$\begin{bmatrix} M & B^T \\ B & 0 \end{bmatrix}\begin{bmatrix} \ddot{q} \\ \lambda \end{bmatrix} = \begin{bmatrix} Q-N \\ \dot{B}\dot{q} \end{bmatrix}. \tag{16}$$

Using the coordinate partitioning, we can rewrite Equation (16) as follows:

$$\begin{bmatrix} M_{aa} & M_{ap} & B_a{}^T \\ M_{pa} & M_{pp} & B_p{}^T \\ B_a & B_p & 0 \end{bmatrix}\begin{bmatrix} \ddot{q}_a \\ \ddot{q}_p \\ \lambda \end{bmatrix} = \begin{bmatrix} Q_a - N_a \\ Q_p - N_p \\ \dot{B}\dot{q} \end{bmatrix}, \tag{17}$$

where the subscripts a and p correspond to the active and passive joints, respectively.

Accordingly, the Lagrange multipliers can be obtained as follows:

$$\lambda = [B_p{}^T]^{-1}(Q_p - N_p - M_{pa}\ddot{q}_a - M_{pp}\ddot{q}_p), \tag{18}$$

whereas the inverse dynamics can be written in the following compact expression:

$$\tau = Q_a = \overline{M}\ddot{q}_a + \overline{N}, \tag{19}$$

where:

$$\overline{M} = M_{aa} - M_{ap}B_p{}^{-1}B_a - B_a{}^T(B_p{}^T)^{-1}(M_{pa} - M_{pp}B_p{}^{-1}B_a), \tag{20}$$

$$\overline{N} = \left(N_a + B_a{}^T(B_p{}^T)^{-1}\right)(Q_p - N_p) + \left(M_{ap}B_p{}^{-1} - B_a{}^T(B_p{}^T)^{-1}M_{pp}B_p{}^{-1}\right)\dot{B}\dot{q}. \tag{21}$$

Equation (19) provides the input forces τ required by the mechanism to undergo the specified motion.

Among the nine Lagrange multipliers in Equation (18), six multipliers correspond to the internal reaction forces in x- and y-directions at the three upper joints of the mechanism, whereas the other three multipliers correspond to the reaction forces at the three lower joints. These reaction forces are shown in Figure 3. Since the lower joints are prismatic joints with only one degree of freedom, i.e., translation in the x-direction, the reaction forces are exerted only in the y-direction, i.e., normal to the guideway of the sliders.

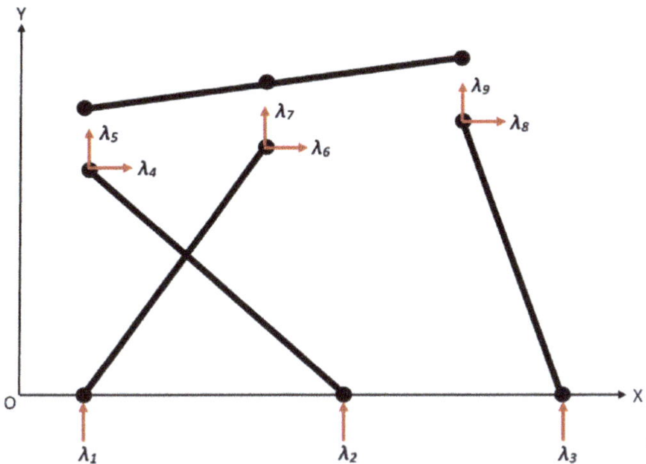

Figure 3. Joint reaction forces corresponding to the Lagrange multipliers.

Hence, there is one reaction force required to be evaluated at every lower joint, which means there are three reaction forces at all the lower joints. The generalized joint reaction forces R, which include two reaction forces in the x- and y-directions at every upper revolute joint and one reaction force in the y-direction at every prismatic joint, are given by:

$$R = B(q,t)^T \lambda. \tag{22}$$

From Equation (22), it is understood that the joint reaction forces are dependent on the Jacobian of the mechanism $B(q,t)^T$, which represents the change of the mechanism's posture.

2.2. Friction Modeling

Since the friction forces cannot be derived from the energy expressions, their contribution to the dynamics is presented separately here. Although the friction can be modeled as being either linear or nonlinear and either symmetric or asymmetric, it is commonly modeled as symmetric and linear. Furthermore, the friction at the prismatic joints is typically much higher than that at the revolute joints. This is particularly true when roller or ball bearings are used as the revolute joints since the friction in such bearings is very small. Therefore, when the friction should be considered, one can either consider the friction at all of the joints or the friction at only the prismatic joints, while neglecting the friction at the revolute joints. The mechanism at hand uses roller bearings as the revolute joints. As a result, the friction at the revolute joints is not significant and therefore neglected.

In this work, the Stribeck friction model, as shown in Figure 4b, is used. This friction model is modified from the static-viscous-Coulomb friction model shown in Figure 4a by adding the Stribeck friction effect to the static, Coulomb, and viscous friction components. This can be written as:

$$f\begin{cases} = f_s sign(\dot{q}) \text{ for } \dot{q} = 0^+ \text{ or } \dot{q} = 0^-, \\ = f_c sign(\dot{q}) + (f_s - f_c)e^{-\left(\frac{\dot{q}}{v_s}\right)^{\delta}} sign(\dot{q}) + f_v\dot{q} \text{ for } \dot{q} \neq 0, \end{cases} \tag{23}$$

where f_s, f_c, and f_v denote diagonal matrices containing the static, Coulomb, and viscous friction parameters, respectively. For the case of a nonzero velocity in Equation (23), the first and the third terms in the right-hand side are the Coulomb and viscous friction components, whereas the second term is the Stribeck-effect friction component. Furthermore, some values of δ have been recommended. In this work, the value of 1 was used. Using this friction model, there are four friction parameters to be identified: f_s, f_c, f_v, and v_s. In the case of varying normal forces, the static and Coulomb friction forces vary accordingly. As a result, the static and Coulomb friction cannot be identified as forces in the usual manner. In such a case, which is the case in this work, the static and Coulomb frictions should be identified as coefficients, i.e., μ_s and μ_c.

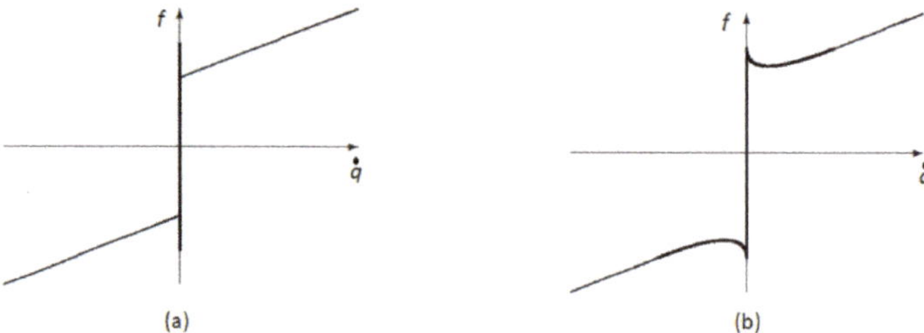

Figure 4. (**a**) Static-viscous-Coulomb friction and (**b**) Stribeck friction models.

The static and Coulomb friction parameters in the case of a translational joint, i.e., prismatic joint, are given by multiplying each of a dimensionless static friction coefficient μ_s and a dimensionless Coulomb friction coefficient μ_c by the normal force f_n exerted on the contact surface at the joint:

$$f_s = \mu_s f_n\,, \tag{24a}$$

$$f_c = \mu_c f_n \tag{24b}$$

The static and Coulomb friction coefficients typically depend on the types of materials in contact with each other and the lubrication condition.

In general, the normal force in a posture-changing mechanism, such as a PKM like the mechanism at hand, is not constant. In the prismatic joint of the mechanism at hand, as illustrated in Figure 5, three force components are acting, namely, the weight of the prismatic joint $m_{si}g$, which is given by the total weight of the slider and the moving part of the linear motor; the input force Q_{xi} exerted on the prismatic joint by the linear motor; and the force R_i exerted to the prismatic joint due to the weight of the other components mounted on the slider, as well as the external payload and the external force, if any. The exerted force R_i can be decomposed into horizontal and vertical components, namely, R_{xi} and R_{yi}. At any time, the total force in the horizontal direction is given by the sum of R_{xi} and Q_{xi}. In a static case, R_{xi} is overcome by Q_{xi} such that the total force in the horizontal direction is zero. However, when the prismatic joint is in motion, the magnitude of Q_{xi} is different from the magnitude of R_{xi}. If the magnitude of R_{xi} is larger, the slider is moving toward the equilibrium state of the mechanism, i.e., the mechanism is collapsing to the base in the case of mechanism working in XY plane. Conversely, if the magnitude of Q_{xi} is larger, this means that the slider is moving due to an actuation point.

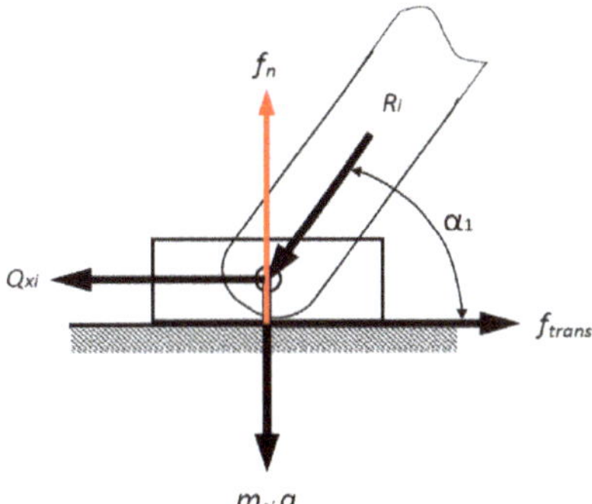

Figure 5. Free-body diagram at the slider i, including the translational friction f_{trans}.

The normal force f_n is equal but in the opposite direction to the sum of all the vertical force components. This can be written as follows:

$$f_n = m_{si}g + R_{yi}\ . \tag{25}$$

Although the weight of the prismatic joint, i.e., the first term on the right-hand side of Equation (25), is constant, the force component R_{yi} varies in general, depending on the mechanism posture. As a result, the normal force f_n, and accordingly the static and Coulomb friction parameters, f_s and f_c, are not constant.

Since the varying components of the normal forces are mainly due to the effect of gravity and the external force, the posture dependency of the static and Coulomb friction parameters as described above is more significant when the rigid body motion of the mechanism is against a gravitational load, i.e., the rigid body motion is in the XY plane and/or under an external force. However, when the mechanism is free from an external force and working in the XY plane while it is statically balanced by a gravitational compensation mechanism or the mechanism is working in XZ plane, i.e., horizontally, then the posture dependency is less significant. The external force can be excluded during the dynamic parameter identification by simply moving the mechanism without any working load. In such a case, only the gravitational effect remains, if any. When the mechanism working in the XY plane is statically balanced, the gravity effect is eliminated.

When the mechanism is working in the XZ plane, the gravitational effect is working in the off-plane direction and most of the weight of all the mechanism components is borne by the structure of the mechanism. This does not mean that the gravitational effect is eliminated from the joints. In fact, the prismatic and revolute joints are still subject to the gravity-induced load but it works in the off-plane direction. At the prismatic joint, this gravity-induced load can be seen as a vertical reaction force that is exerted on the base of a cantilever due to the weight of the cantilever. In addition to this vertical reaction force, there are also a horizontal reaction force and a reaction moment at the prismatic joint due to the weight of the cantilever. If the external force is excluded, the vertical reaction force will be the main force component, which results in an off-plane normal force working at the prismatic joint. In addition to this normal force component, there is also an in-plane normal force component that results from the motion of the mechanism. All of these normal forces lead to a friction force, which should be overcome during the actuation of the mechanism. Similarly, the off-plane normal forces that are equal but in the opposite direction to the bearing thrust forces also arise in the revolute joints. In addition to this, there are also in-plane normal forces at the revolute joints resulting from the motion of the mechanism. Furthermore, if an in-plane external force exists, it will be additionally exerted to the prismatic and revolute joints. If the external force does not have a varying vertical (off-plane) component, the vertical (off-plane) normal forces working at the three prismatic and revolute joints may be assumed to be uniform. Accordingly, the static and Coulomb friction parameters may be assumed to be constant. A more detailed friction formulation for the mechanism working in the XZ plane can be seen in Section 5.

2.3. Dynamic Model with Friction

By incorporating the friction forces, the equations of motion in Equation (14) can be expanded to the following:

$$M(q)\ddot{q} + N(q,\dot{q}) + f(\dot{q}) = Q + B(q,t)^T\lambda. \tag{26}$$

The differential-algebraic equations in Equation (16) then become:

$$\begin{bmatrix} M & B^T \\ B & 0 \end{bmatrix}\begin{bmatrix} \ddot{q} \\ \lambda \end{bmatrix} = \begin{bmatrix} Q-N-f \\ \dot{B}\dot{q} \end{bmatrix}. \tag{27}$$

The partitioned differential-algebraic equations are accordingly given by:

$$\begin{bmatrix} M_{aa} & M_{ap} & {B_a}^T \\ M_{pa} & M_{pp} & {B_p}^T \\ B_a & B_p & 0 \end{bmatrix}\begin{bmatrix} \ddot{q}_a \\ \ddot{q}_p \\ \lambda \end{bmatrix} = \begin{bmatrix} Q_a - N_a - f_a \\ Q_p - N_p - f_p \\ \dot{B}\dot{q} \end{bmatrix}. \tag{28}$$

The inverse dynamics can still be presented in the compact expression given in Equation (19), but the vector $\overline{N}$ should be modified to the following:

$$\overline{N} = \left(N_a + f_a + {B_a}^T({B_p}^T)^{-1}\right)(Q_p - N_p - f_p) + \left(M_{ap}{B_p}^{-1} - {B_a}^T({B_p}^T)^{-1}M_{pp}{B_p}^{-1}\right)\dot{B}\dot{q}. \tag{29}$$

It is worth mentioning that the friction forces can only be evaluated after the corresponding normal forces are evaluated, whereas the normal forces can only be computed after the corresponding Lagrange multipliers are computed. In fact, the Lagrange multipliers are not available at the initial time. For this reason, a statics formulation of the mechanism is used at the initial time to compute the normal forces. This is valid since the mechanism is at rest at the initial time. The detailed statics formulation used to obtain the static normal forces is not presented here to reduce the paper's length. In short, the static equations can be written based on the static equilibrium principle, and subsequently, can be presented as a linear system that can be solved for the normal forces, i.e., the reaction forces, by inverting the linear system.

3. Identification Algorithm

The mathematical model function Y, which serves as the base of the nonlinear identification, such as in the case of this work, is given directly by the nonlinear inverse dynamics model of the system, i.e.:

$$Y = Q_a = f(q,\dot{q},\ddot{q},\Phi) = \overline{M}\ddot{q}_a + \overline{V}. \tag{30}$$

The residual error of the estimation is given by the difference between the estimated active joint forces $\hat{Y}$ and the measured active joint forces $\widetilde{Y}$:

$$e = \hat{Y} - \widetilde{Y}, \tag{31}$$

where the estimated active joint forces $\hat{Y}$ are defined similarly to Equation (30) but using the estimated parameters $\hat{\Phi}$ instead:

$$\hat{Y} = \hat{\tau} = \hat{Q}_a = f(q,\dot{q},\ddot{q},\hat{\Phi}). \tag{32}$$

An identification algorithm is defined to minimize the residual error in Equation (31). This is most commonly achieved by minimizing a cost function J given by the square of the residual error, i.e.:

$$J = e^T e. \tag{33}$$

The dynamic parameter identification can be stated as the following optimization problem: find the estimates of the dynamic parameters Φ that minimize the cost function J defined in Equation (33).

A bound-constrained optimization method was applied in this work to minimize the residual error. An iterative bound-constrained global optimization solver in MATLAB® R2020a, namely, a constrained *pattern search* [16,17] was applied. The global search tool was used instead of a local search tool to avoid getting trapped in local minima.

To ensure that the estimates are physically feasible, some physical feasibility constraints should be incorporated into the estimation algorithm. This changes the least squares problem into a constrained optimization problem as follows: minimize the residual error, which is commonly represented by the square of the residual error, subject to the physical feasibility constraints. In this work, the physical feasibility is imposed by predefining bounds of the estimates in the estimation algorithm. Since the bound constraints are used as the physical feasibility constraints, the estimation problem can be written as:

$$\text{Minimize } J = e^T e \text{ subject to } LB \leq \Phi \leq UB\,, \tag{34}$$

where LB and UB denote the lower and upper bounds, respectively. The bounds are introduced to narrow the search space and avoid getting trapped at unexpected local minima.

This approach requires a priori knowledge of the numerical values of the estimated parameters. This can typically be obtained by using CAD data, manufacturer's data, or direct measurements of the main components. In general, this a priori data is not accurate, and therefore some lower and upper bounds are determined to capture the expected true values of the parameters. In this case, one should provide acceptable bounds, which will ensure the physical feasibility but one should not provide bounds that are too tight as they may fall beyond the true values of the parameters. Moreover, the span from the a priori values to the lower bounds does not need to be equal to that from the a priori values to the upper bounds; the spans depend on how much the tendency of the true values to be shifted down or up from the a priori values. The true value of a mass, for example, tends to be shifted up from the a priori value rather than shifted down due to additional components or accessories mounted on the main components, such as fasteners, cables, etc.

This approach is used since the design or nominal values of the parameters in this work can be obtained easily for some reasons: the masses of the components are known and the CAD models of the components are available. The masses of the components can be obtained by weighing them or calculating them from the volume and the material density of the components. The earlier is more practical when it is possible to disassemble the components and weigh them, whereas the latter can be done if weighing is not possible. The determination of the COM positions of the components in this work was simplified to one dimension as the geometry of the components was symmetric in the other two dimensions. These COM positions could be easily obtained using CAD software, which is PTC Creo Parametric 4.0 in this work, since the CAD models are available. Once the masses and the COM positions of the components are known, the first moments and the moments of inertia of the components can be easily calculated.

4. Optimal Exciting Trajectory

In the dynamic parameter identification, the exciting trajectory should be optimized to persistently excite the dynamics of the mechanism. For convenience, either a polynomial or sinusoidal exciting trajectory is typically used. Since the latter is periodic and can in general generate more complicated trajectories, it was used in this work. This trajectory is parameterized by using the following Fourier series:

$$q_i(t) = q_{i0} + \sum_{j=1}^{n_H} \left(\frac{a_{ij}}{2\pi f j} \sin(2\pi f j t) - \frac{b_{ij}}{2\pi f j} \cos(2\pi f j t) \right), \tag{35}$$

where n_H and f are the harmonic number and the fundamental frequency in Hz, respectively, whereas the subscript i corresponds to the ith generalized coordinate q. The parameters n_H and f should be predefined. For the mechanism at hand, $i = 1, 2, 3$, where q_1, q_2, and q_3 are given by the coordinates of the end-effector, i.e., x, y, and θ.

It can be observed in Equation (35) that the parameters that should be determined for an optimized exciting trajectory are q_{i0}, a_{ij}, and b_{ij}. The total number of these parameters is:

$$s = l(2n_H + 1). \tag{36}$$

The optimization of the exciting trajectory is performed to optimize an objective function, which indicates the rank of the system subject to some constraints of the mechanism. The constraints used are typically the reachable workspace; collision avoidance; and the limits of the position, velocity, acceleration, and forces of the active joints, zero initial and final velocity, and zero initial and final acceleration. Since the parameterized exciting trajectory is expressed in terms of the new parameters q_{i0}, a_{ij}, and b_{ij}, the objective function G should also be expressed in terms of these new parameters. In this work, the exciting trajectory optimization could be stated as the following constrained optimization problem: find $q_i, \dot{q}_i$, and $\ddot{q}_i$ that optimize the objective function G subject to:

$$\begin{gathered} q_{imin} \leq q_i \leq q_{imax} \\ |\dot{q}_i| \leq \dot{q}_{imax} \\ |\ddot{q}_i| \leq \ddot{q}_{imax} \\ \dot{q}_i(t=0) = \dot{q}_i(t=t_f) = 0 \\ \ddot{q}_i(t=0) = \ddot{q}_i(t=t_f) = 0 \\ |\tau_i| \leq \tau_{imax} \end{gathered} \tag{37}$$

The objective function G used in this work was a weighted composite of the condition number and inverse of the minimum singular value of the observation matrix W in the linear form of the inverse dynamics equations with respect to the dynamic parameters [13], i.e.:

$$G(q_i, \dot{q}_i, \ddot{q}_i) = G(q_{i0}, a_{ij}, b_{ij}) = w_1 cond(W) + w_2 \frac{1}{\sigma_{\min}}, \tag{38}$$

where w_1 and w_2 denote the corresponding weights.

5. Experimental Results and Discussions

The upper mechanism of a hybrid-kinematics machine shown in Figure 6 was used in the experiment. This 3PRR mechanism has legs with lengths of $L_1 = 600$ mm, $L_2 = 475$ mm, and $L_3 = 604$ mm. The position of the upper revolute joints on the moving platform was given by $x_{p2} = x_{p3} = 280$ mm. The COMs of the legs were $L_{1G} = 240.1$ mm, $L_{2G} = 207.1$ mm, and $L_{3G} = 272.8$ mm, whereas the COM of the moving platform was $x_{PG} = 239.5$ mm. In this experiment, the upper mechanism was oriented horizontally, i.e., in the XZ plane, and therefore its rigid body motion was not affected much by gravity, as discussed earlier.

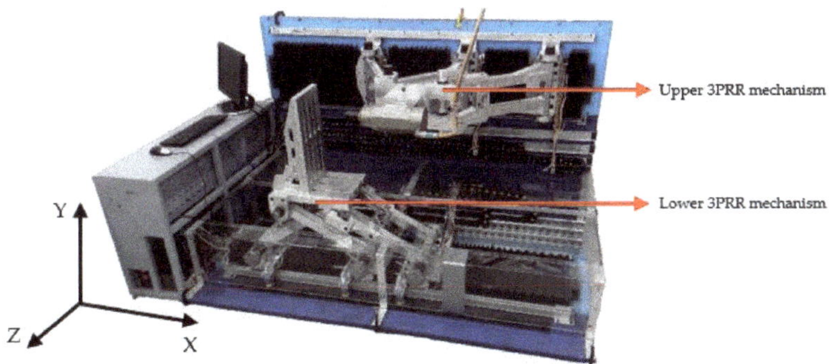

Figure 6. Hybrid-kinematics machine composed of lower and upper 3PRR mechanisms.

5.1. Optimization Setup

A bound-constrained optimization using MATLAB *pattern search* with a *GPS Positive basis 2N* poll method and default stopping criteria was implemented to estimate the dynamic parameters of the mechanism based on the nonlinear form of the inverse dynamics model with a varying nonlinear Stribeck friction model. For simplicity, the masses of the three sliders were assumed to be identical and the friction on the three prismatic joints was assumed to be uniform. Therefore, there was only a single value of each friction parameter instead of three values corresponding to the three sliders. Aside from simplicity concerns, this assumption was justified because all of the sliders moved on the same guideway.

The iterative optimization algorithm was executed by using the design values as the initial values. Table 1 shows the available design values of the inertial parameters, along with the determined bounds. The bounds of the inertial parameters were determined to be wider toward the upper limits since it was expected that the actual values of the parameters were larger than the design values due to additional parts being attached to the main components. On the other hand, wide non-negativity bounds from zero to infinity were applied to the friction parameters since there was no a priori information on their expected values.

Table 1. Estimates of the inertial and friction parameters using a nonlinear model with varying friction.

Inertial Parameter		Lower Bound	Upper Bound	Design Value	Estimate
m_{si}	kg	6.2550	9.3825	6.2550	7.2614
m_{L1}	kg	11.1280	16.6920	11.1280	12.1299
m_{L2}	kg	9.0800	13.6200	9.0800	10.0817
m_{L3}	kg	13.3720	20.0580	13.3720	14.4665
m_p	kg	30.3510	45.5265	30.3510	31.3359
$m_{L1}L_{1G}$	kg·m	2.0039	4.0077	2.6718	3.3970
$m_{L2}L_{2G}$	kg·m	1.4104	2.8208	1.8805	2.4049
$m_{L3}L_{3G}$	kg·m	2.7355	5.4711	3.6474	3.6484
$m_p x_{pG}$	kg·m	5.4517	10.9035	7.2690	8.2238
I_{L1}	kg·m^2	0.7323	1.4648	0.9765	1.2206
I_{L2}	kg·m^2	0.4453	0.8906	0.5937	0.7421
I_{L3}	kg·m^2	1.0424	2.0847	1.3898	1.5088
I_p	kg·m^2	1.7989	3.5979	2.3986	2.9509
Friction Parameter					
u_c	-	0	∞		0.1449
f_v	Ns/m	0	∞		83.1797
u_s	-	0	∞		0.3016
v_s	m/s	0	∞		0.3281

5.2. Friction and Input Forces

Since the mechanism was oriented horizontally, the friction at the prismatic joints, i.e., between the sliders and the guideway, could be decomposed into two main components: the friction component due to the normal forces in the y-direction and the friction component due to the normal force in the z-direction. The former normal forces $f_{n,yi}$ were exerted vertically on the prismatic joints due to the weight of the mechanism:

$$f_{n,yi} = (m_{si} + m_{Li} + \alpha_i m_p)g;\ i = 1,2,3. \tag{39}$$

The parameter α_i represents the portion of the moving platform weight that was supported by the slider i. The sum of all values of α_i should be 1. The best constant value of α_i was obtained by attempting various values. After some attempts, the obtained best values for i = 1, 2, 3 were 0.7, 0.2, and 0.1, respectively. For simplicity, the friction forces corresponding to these normal forces, i.e., the normal forces in the y-direction, were assumed to be constant and modeled using the static friction at the beginning of the motion and using Coulomb and viscous friction for the subsequent motion:

$$f_{const}\begin{cases} = diag(\mu_{s1}f_{n,y1}, \mu_{s2}f_{n,y2}, \mu_{s3}f_{n,y3})\begin{bmatrix} sign(\dot{x}_1) & sign(\dot{x}_2) & sign(\dot{x}_3) \end{bmatrix}^T \text{ for } \dot{q} = 0 \\ = diag(\mu_{c1}f_{n,y1}, \mu_{c2}f_{n,y2}, \mu_{c3}f_{n,y3})\begin{bmatrix} sign(\dot{x}_1) & sign(\dot{x}_2) & sign(\dot{x}_3) \end{bmatrix}^T \\ \quad + diag(f_{v1}, f_{v2}, f_{v3})\begin{bmatrix} \dot{x}_1 & \dot{x}_2 & \dot{x}_3 \end{bmatrix}^T \text{ for } \dot{q} \neq 0. \end{cases} \tag{40}$$

The latter normal forces, i.e., the normal force in the z-direction, act horizontally and vary with changes in the mechanism's posture. These varying normal forces result in varying static and Coulomb friction forces. Since the static and Coulomb components of the friction were varying, the overall friction corresponding to the horizontal normal forces was varying.

Furthermore, the friction forces due to the normal forces in the y-direction needed to be added to Equation (19) when evaluating the input forces. Thus the input forces could be written as follows:

$$\tau = Q_a = \overline{M}\ddot{q}_a + \overline{N} + f_{const}. \tag{41}$$

5.3. Data Acquisition and Filtering

The sinusoidal trajectory given by Equation (35) and shown in Figure 7 was used as the optimal exciting trajectory of the end effector. Its parameters were obtained by optimizing Equation (38) subject to Equation (37). The definition of the optimized exciting trajectory included the position, velocity, and acceleration of the end effector within a defined timespan (4 s). Using the inverse kinematics, the motion variables of the trajectories defined in the task space were transformed into the active joint space. While the active joints were moving to make the trajectories, their motion variables were measured using linear encoders. The active joint positions were directly measured by using the encoders. The uncertainty of the encoders was ±0.0005 mm. The active joint velocities and accelerations were derived from the measured positions. The measured positions, velocities, and accelerations of the active joints are shown in Figures 8–10, respectively. It is shown that the measured active joint positions were so smooth that they did not need any filtering. However, some noise was found in the measured velocities and accelerations. Therefore, they were filtered, as shown in Figures 9 and 10. It should be noticed that the measured trajectory positions, velocities, and accelerations were used in the identification instead of the planned (commanded) ones.

At the same time, the active joint forces were also measured. The active joint forces were not directly measured by using dedicated force sensors attached to the sliders. Instead, the measurement of the active joint forces was based on the voltage output from the controller to the servo drives. The active joint forces were calculated based on a proportional relationship between the voltage output from the controller to the servo drives and the resulting current in the servo drives, as well as a proportional relationship between the current drawn from the servo drives and the resulting actuation forces. This indirect measurement of the active joint forces was much cheaper than a direct measurement as no dedicated force sensors were required. The uncertainty of the active joint force measurement was ±0.0381 N. The measured active joint forces are shown in Figure 11. As the measured active joint forces were noisy, they were filtered, as shown in Figure 11. The filtering of the velocities, accelerations, and forces were all performed by using a Savitzky–Golay smoothing filter in MATLAB. Thus, all the raw data were acquired from the hardware and subsequently filtered in MATLAB using the aforementioned filter.

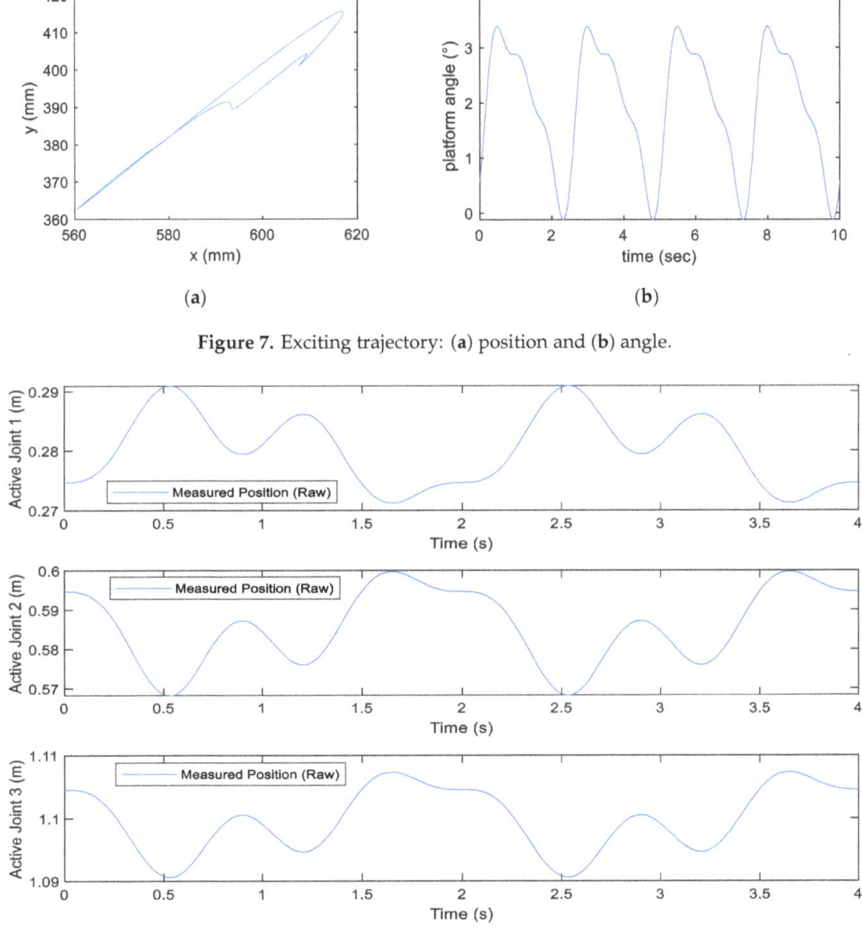

Figure 7. Exciting trajectory: (**a**) position and (**b**) angle.

Figure 8. Raw measured active joint positions.

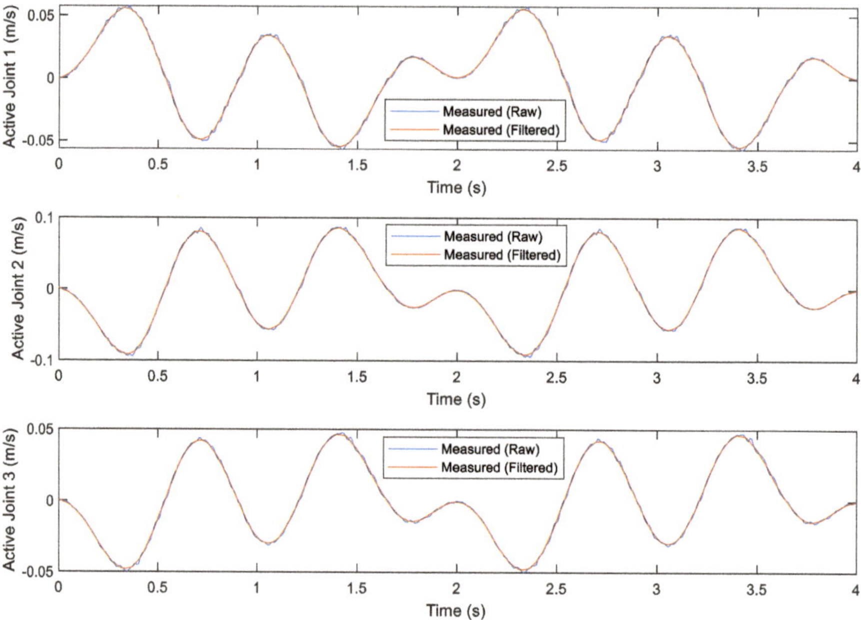

Figure 9. Raw and filtered measured active joint velocities.

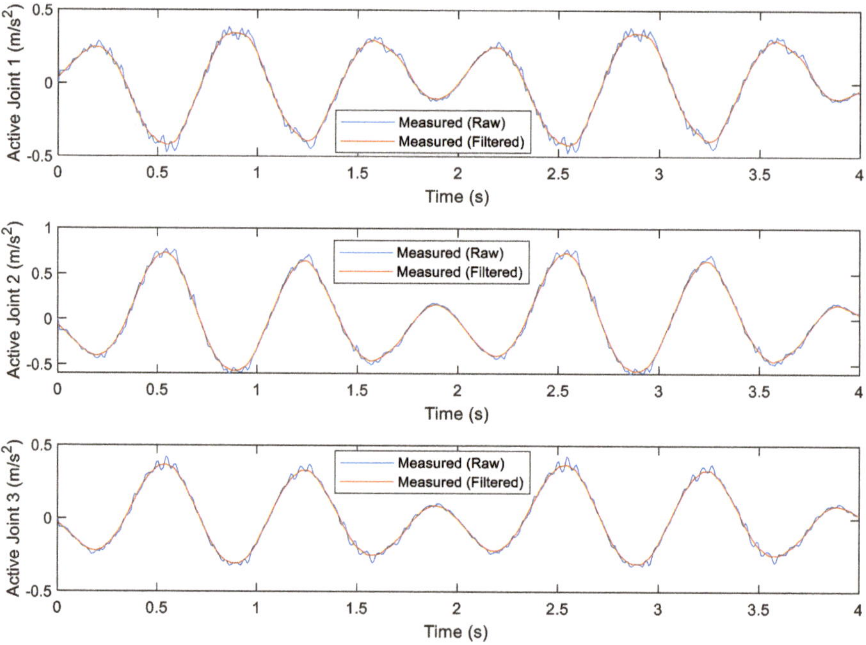

Figure 10. Raw and filtered measured active joint accelerations.

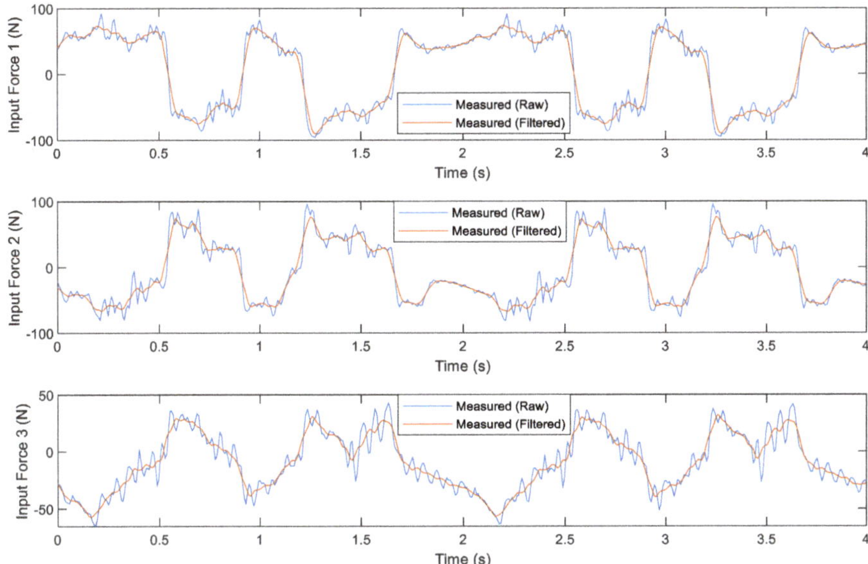

Figure 11. Raw and filtered measured input forces.

5.4. Results

In the following presentation of results, the estimated input forces are plotted together with the raw, measured input forces, not the filtered ones, for a more genuine comparison, although the identification used the filtered, measured input forces. Table 1 shows the estimates of the dynamic parameters, which consisted of the barycentric inertial parameters and the friction parameters. It can be seen that the estimated inertial parameters were larger than their design values. This was expected since the actual masses should be larger than their design values due to the attached additional parts, such as the nuts, cables, etc. The plots of the estimated and measured input forces created using the Coulomb-viscous friction and Stribeck friction models are shown in Figure 12a,b, respectively. It is shown that the latter friction model provided a better fit, whereas the former friction model was not adequate to represent the real dynamics. This justified the use of the Stribeck friction model instead of the Coulomb-viscous friction model.

The variation of the absolute values of the vertical reaction forces R_y on the three prismatic joints is shown in Figure 13. The absolute values are plotted here instead of the signed values to show the variation of the values more clearly. The variation of these normal forces resulted in the variation of the friction forces. The solution obtained by considering the varying friction, as shown in Table 1, was validated by applying some other exciting trajectories that were different from the one used in the identification. Figure 14 shows a full-circle trajectory executed for the validation, while Figure 15 shows the corresponding estimated and measured input forces. It is shown that the input forces estimated by considering the varying friction forces had a better fit than the input forces estimated by considering constant friction forces. Finally, Figure 16 shows the plots of the estimated input forces along with the estimated friction forces. The figure clearly shows the contribution of the significant friction to the overall dynamics.

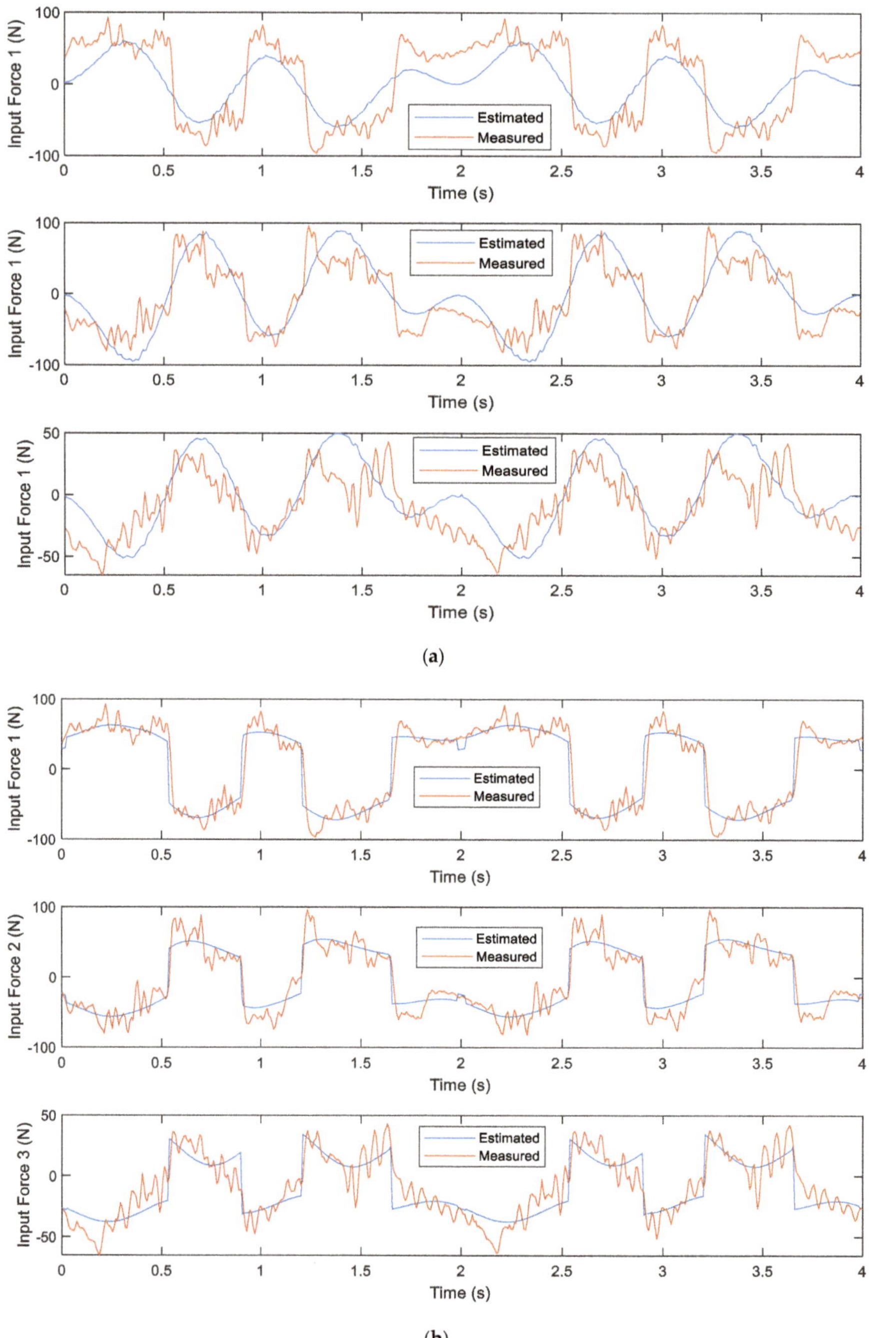

Figure 12. Comparison between the estimated and measured input forces: (**a**) using the Coulomb-viscous friction model and (**b**) using the Stribeck friction model.

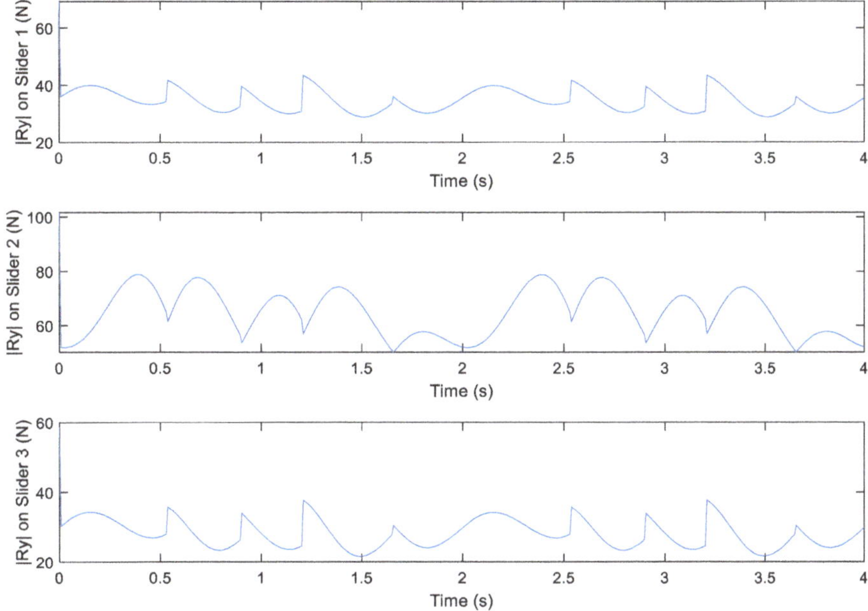

Figure 13. Variation of the absolute values of the vertical reaction forces on the three prismatic joints during the identification motion.

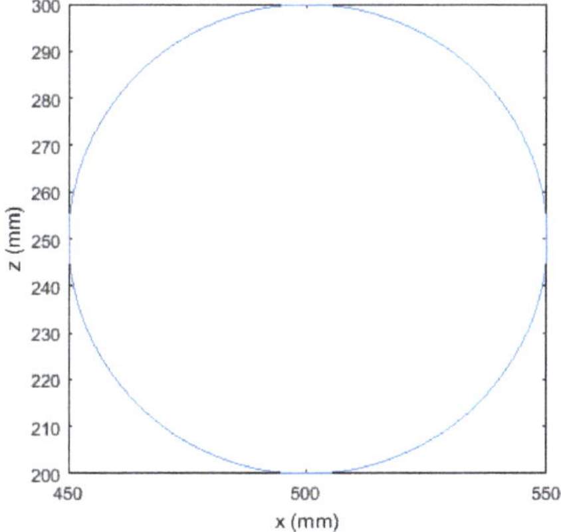

Figure 14. Full-circle validation trajectory.

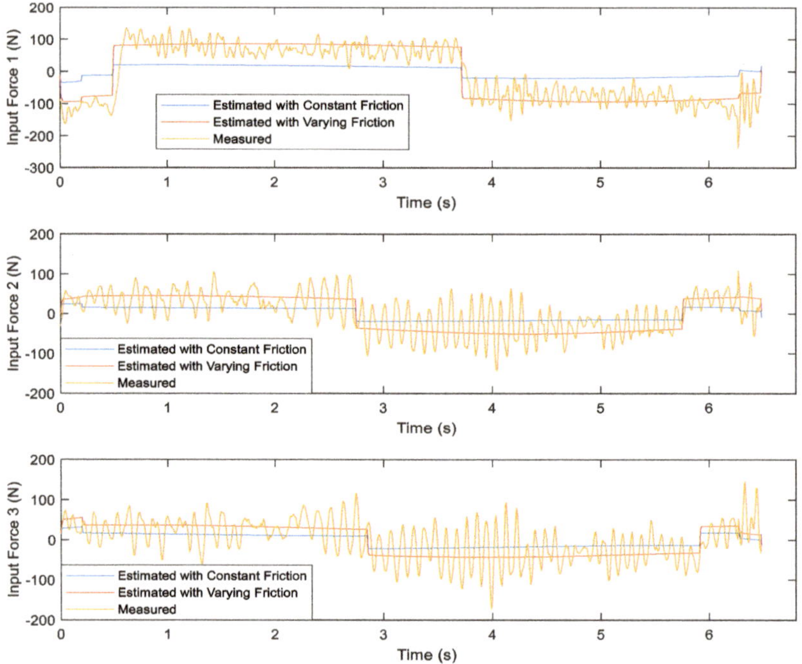

Figure 15. Comparison between the estimated and measured input forces in the validation motion.

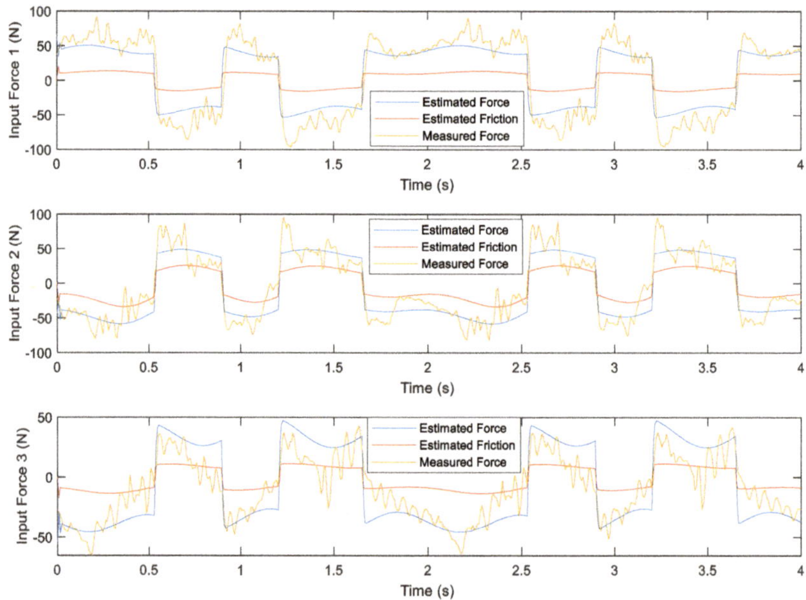

Figure 16. Comparison between the estimated friction forces and the estimated input forces.

6. Conclusions

It was shown that the iterative bound-constrained optimization method provided physically feasible estimates in the dynamic parameter identification with noisy measurements. The predefined bounds of the parameters were used as the constraints to ensure the physical feasibility of the solutions. These bounds can be obtained based on any a priori knowledge of the system, such as the design values obtained from a CAD model or the measurement data of the individual main components before the assembly. In a PKM with sliding joints, it was shown that the friction on the slider guideway was significant and therefore could not be neglected. Furthermore, the Stribeck friction model was shown to more accurately represent the friction and therefore provide a better dynamic model. In this work, the variation of the normal forces, which resulted in the variation of the static and Coulomb friction forces, were taken into consideration in the dynamic modeling and dynamic parameter identification of the PKM. This variation, unless it is negligible, should always be taken into consideration when dealing with PKMs since this variation is an inherent characteristic of the PKMs due to the changing posture of the mechanism. As a consequence of considering the variation of the static and Coulomb friction forces, which was the main contribution of this work, the static and Coulomb friction parameters f_s and f_c could not be identified as constant parameters since they were varying. Instead, the static and Coulomb friction coefficients μ_s and μ_c were identified since they were constant parameters. The product between these coefficients and the varying normal forces gives the varying static and Coulomb friction parameters f_s and f_c.

Author Contributions: Both authors discussed the idea of this work and made scientific contributions. A.R. performed the work and wrote the paper. B.E.-K. guided the work, gave advice, and reviewed the paper. Both authors have read and agreed to the published version of the manuscript.

Funding: This publication was based upon work supported by the Khalifa University of Science and Technology under Award No. RC1-2018-KUCARS.

Conflicts of Interest: The authors declare no conflict of interest.

References

1. Hoang, N.-B.; Kang, H.-J. Observer-based Dynamic Parameter Identification for Wheeled Mobile Robots. *Int. J. Precis. Eng. Manuf.* **2015**, *16*, 1085–1093. [CrossRef]
2. Farhat, N.; Mata, V.; Page, A.; Valero, F. Identification of dynamic parameters of a 3-DOF RPS parallel manipulator. *Mech. Mach. Theory* **2008**, *43*, 1–17. [CrossRef]
3. Thanh, T.D.; Kotlarski, J.; Heimann, B.; Ortmaier, T. Dynamics identification of kinematically redundant parallel robots using the direct search method. *Mech. Mach. Theory* **2012**, *52*, 277–295. [CrossRef]
4. Poignet, P.; Ramdani, N.; Vivas, O.A. Robust estimation of parallel robot dynamic parameters with interval analysis. In Proceedings of the 42nd IEEE International Conference on Decision and Control, Maui, HI, USA, 9–12 December 2003. IEEE Cat. No.03CH37475.
5. Pan, Y.-R.; Shih, Y.-T.; Horng, R.-H.; Lee, A.-C. Advanced Parameter Identification for a Linear-Motor-Driven Motion System Using Disturbance Observer. *Int. J. Precis. Eng. Manuf.* **2009**, *10*, 35–47. [CrossRef]
6. Gautier, M.; Khalil, W.; Restrepo, P.P. Identification of the Dynamic Parameters of a Closed Loop Robot. In Proceedings of the IEEE International Conference on Robotics and Automation, Nagoya, Japan, 21–27 May 1995.
7. Guegan, S.; Khalil, W.; Lemoine, P. Identification of the Dynamic Parameters of the Orthoglide. In Proceedings of the IEEE International Conference on Robotics and Automation—ICRA 2003, Taipei, Taiwan, 14–19 September 2003; pp. 3272–3277.
8. Vivas, A.; Poignet, P.; Marquet, F.; Pierrot, F.; Gautier, M. Experimental dynamic identification of a fully parallel robot. In Proceedings of the IEEE International Conference on Robotics & Automation, Taipei, Taiwan, 14–19 September 2003.
9. Renauda, P.; Vivas, A.; Andreff, N.; Poignet, P.; Martinet, P.; Pierrot, F.; Company, O. Kinematic and dynamic identification of parallel mechanisms. *Control Eng. Pract.* **2006**, *14*, 1099–1109. [CrossRef]

10. Rodríguez, M.D.; Iriarte, X.; Mata, V.; Ros, J. On the Experiment Design for Direct Dynamic Parameter Identification of Parallel Robots. *Adv. Robot.* **2009**, *23*, 329–348. [CrossRef]
11. Rodríguez, M.D.; Mata, V.; Valera, A.; Page, A. A methodology for dynamic parameters identification of 3-DOF parallel robots in terms of relevant parameters. *Mech. Mach. Theory* **2010**, *45*, 1337–1356. [CrossRef]
12. Shang, W.; Cong, S.; Kong, F. Identification of dynamic and friction parameters of a parallel manipulator with actuation redundancy. *Mechatronics* **2010**, *20*, 192–200. [CrossRef]
13. Briot, S.; Gautier, M. Global identification of joint drive gains and dynamic parameters of parallel robots. *Multibody Syst. Dyn.* **2015**, *33*, 3–26. [CrossRef]
14. Rosyid, A.; El-Khasawneh, B.; Alazzam, A. Optimized Planar 3PRR Mechanism for 5 Degrees-Of-Freedom Hybrid Kinematics Manipulator. In Proceedings of the ASME 2015 International Mechanical Engineering Congress and Exposition IMECE2015, Houston, YX, USA, 13–19 November 2015.
15. Rosyid, A.; El-Khasawneh, B.; Alazzam, A. Identification of lonely barycentric parameters of parallel kinematics mechanism with rank-deficient observation matrix. In Proceedings of the 11th International Symposium on Mechatronics and its Applications (ISMA), Sharjah, UAE, 4–6 March 2018.
16. Kelley, C.T. *Iterative Methods for Optimization;* Society for Industrial and Applied Mathematics: Philadelphia, PA, USA, 1999; pp. 87–108.
17. Audet, C.; Dennis, J.E., Jr. Analysis of Generalized Pattern Searches. *SIAM J. Optim.* **2003**, *13*, 889–903. [CrossRef]

Article

Modeling and Identification of an Industrial Robot with a Selective Modal Approach

Matteo Bottin [1], Silvio Cocuzza [1,*], Nicola Comand [2] and Alberto Doria [1]

1 Department of Industrial Engineering, University of Padova, 35131 Padova, Italy; matteo.bottin@unipd.it (M.B.); alberto.doria@unipd.it (A.D.)
2 Department of Management and Engineering, University of Padova, 36100 Vicenza, Italy; nicola.comand@phd.unipd.it
* Correspondence: silvio.cocuzza@unipd.it; Tel.: +39-049-827-6793

Received: 29 May 2020; Accepted: 28 June 2020; Published: 3 July 2020

Abstract: The stiffness properties of industrial robots are very important for many industrial applications, such as automatic robotic assembly and material removal processes (e.g., machining and deburring). On the one hand, in robotic assembly, joint compliance can be useful for compensating dimensional errors in the parts to be assembled; on the other hand, in material removal processes, a high Cartesian stiffness of the end-effector is required. Moreover, low frequency chatter vibrations can be induced when low-stiffness robots are used, with an impairment in the quality of the machined surface. In this paper, a compliant joint dynamic model of an industrial robot has been developed, in which joint stiffness has been experimentally identified using a modal approach. First, a novel method to select the test configurations has been developed, so that in each configuration the mode of vibration that chiefly involves only one joint is excited. Then, experimental tests are carried out in the selected configurations in order to identify joint stiffness. Finally, the developed dynamic model of the robot is used to predict the variation of the natural frequencies in the workspace.

Keywords: robot; compliance; machining; modes of vibration; impulsive testing

1. Introduction

In robotic processes, the compliance of the robot arm plays a very important role. In some conditions, for example, in robotic assembly, robot arm compliance can compensate for small position and orientation errors of the end-effector. In other processes, like machining, robot compliance may generate chatter vibrations, with an impairment in the quality of the machined surface. Indeed, still very few robots have been applied in this economically important application area [1], mainly due to their low stiffness [2]. In industrial robots, the compliance of the end-effector is mainly due to joint compliances [3–5], even if there are some examples of robots [6] having structural modes (dominated by link or bearing compliance) in the band of frequency that contains the modes dominated by joint compliance. The compliance properties of robots are very important in emerging fields as well, such as flexible assembly systems [7,8] and collaborative robotics [9]. Once robot joint stiffness has been identified, the kinematic redundancy of the arm with respect to the robotic task can be exploited to minimize (or maximize) the Cartesian compliance of the end-effector through optimization methods [10–13], as already proposed for the minimization of base reactions and kinetic energy.

In material removal processes, low stiffness causes imprecise products, due to the robot deflections during the robotic task. From a dynamic point of view, low frequency chatter vibrations [3,14,15] can be induced when low-stiffness robots are used, with an impairment in the quality of the machined surface. Moreover, vibrations cause a reduction of tool life and can lead to breakages in robot joint transmission. In particular, a low joint stiffness of the robot is one of the main issues in using robotic machining instead of computer numerical control (CNC) machining.

In robotic processes, the directions along which the robot arm shows large compliance are important. The stiffness matrix in the joint space does not directly give this piece of information; moreover, for a serial robot, the stiffness matrix in the Cartesian space is not diagonal and it is configuration-dependent. This means that the force and deformation in the Cartesian space are coupled and this can generate some counterintuitive results. A geometric and intuitive interpretation of the end-effector displacements in different robot operative conditions has been presented in [16].

Static tests are widely used to obtain the joint stiffness of industrial robots [17]. In the static tests, a set of forces is applied to the robot end-effector in different robot configurations, while the displacement sensors (laser sensors, vision systems, coordinate measuring machines) measure the static deformation of the end-effector. Therefore, the Cartesian stiffness of the robot can be calculated. Then, the stiffness of the joints is obtained through the analytical relation between the joint and Cartesian stiffness based on the kinematic model of robot. Various identification methods (least squares or genetic algorithms) are employed to numerically solve the problem.

Most of the researches (e.g., see [4,18–21]) neglect link flexibility and identify the joints' rotational stiffness (modeled with linear rotational springs) using the above-mentioned method. In [22], the robot arm is modeled considering rigid links and three lumped rotational springs for each joint to take into account joint compliance, bearings compliance, and link deflections. In [23], an analysis method in which both joint and link stiffness are considered is presented. In [24,25], two types of robot dynamic models are presented by considering either only joint compliance or its combination with link compliance, in order to predict the robot dynamic behavior in different postures. In [6] a hybrid approach is adopted, in which a critical structural compliance in a direction perpendicular to a joint axis is taken into account, together with joint compliance.

In [26], experimental modal analysis is used to identify the joint and base stiffness of an industrial robot represented with a four-degrees-of-freedom (four-DOF) planar model. Experimental modal analysis was also used in [6,16], to identify the stiffness of the first three joints of an industrial robot.

The static tests for the identification of joint stiffness have some drawbacks. The equipment for the measurement of end-effector displacements along three directions may be complex and expensive, especially if high accuracy is required and the workspace of the robot is large. The application of vertical forces on the end-effector is rather simple, since weights can be used, but the application of horizontal forces may require specific actuators that have to be connected with the end-effector without introducing additional constraints. Since Cartesian stiffness is configuration dependent, many test configurations are needed to map the robot workspace. The identification of joint stiffness from the measured Cartesian stiffness requires, in most cases, an optimization method. Finally, the measurements are affected by the structural compliance of the links, and it is difficult to separate the contribution of end-effector displacement caused by joint compliance from the one due to link compliance.

For the above-mentioned reasons, this research aims to use the methods of impulsive modal analysis to identify the joint stiffness of industrial robot arms. In future, the same methodology could be extended to other kinds of robots e.g., walking robots [27]. Some factors foster the development of testing methods based on impulsive excitation. The basic equipment (which includes an instrumented hammer, an accelerometer, a data acquisition board, and a computer) is nowadays affordable for most laboratories. Hammer excitation can be exerted in three directions, even if the robot workspace is wide. Modal analysis identifies all the modes of vibration, both the ones dominated by joint compliance, and the ones dominated by link compliance (if present), hence, the modal method allows a better understanding of the importance of link compliance than the static method.

The modes of vibration of a robot typically involve the rotations of some joints and the natural frequencies are functions of the stiffness of these joints. Therefore, the main issue of the modal methods for the identification of joint stiffness is the possibility of finding robot configurations with modes of vibration dominated by the stiffness of only one joint. In this specific case, modal stiffness coincides with joint stiffness, and the latter can be identified from the measured natural frequency and the calculated value of moment of inertia. Other important issues are the choice of proper excitation directions for the selected configurations and the minimization of test configurations.

This paper is organized as follows. In Section 2, a serial six-DOF robot is considered (Omron Adept Viper s650), since it is representative of industrial robots used for a wide range of operations, and a dynamic model of robot vibrations is developed. Section 3 deals with selective modal testing, and some criteria for finding robot configurations with modes of vibration dominated by only one joint, or by a couple of joints, are presented and discussed. Experimental tests carried out in the selected configurations are presented in Section 4, which significantly extends a previous work of the authors [16], in which only the first three joints of the robot were considered. Section 5 deals with the identification of joint stiffness, and an optimization technique is presented to cope with the cases in which the identified mode is dominated by a couple of joints. The use of the robot dynamic model with the identified stiffness values is presented in Section 6, and the variation of the natural frequencies in the robot workspace is analyzed. Finally, in Section 7, conclusions are drawn.

2. Dynamic Model

In the framework of this research an Adept Viper s650 robot manufactured by Omron Adept Inc. (Pleasanton, CA, USA) is considered. It is a medium size six-DOF industrial robot, its workable space is roughly a hemisphere with a radius of about 600 mm. Figure 1 shows that this robot has a typical anthropomorphic structure. There are three main joints that define the position of the wrist and three minor joints that define the orientation of the end-effector. All joints are driven by alternate current (AC) servomotors.

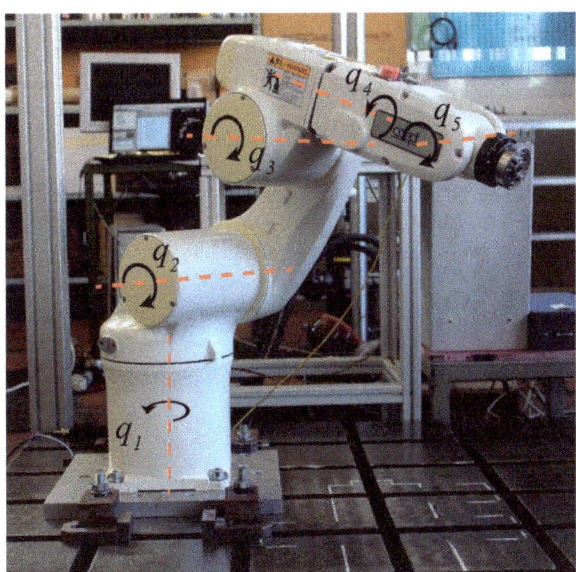

Figure 1. The robot Adept Viper s650 with the tested axes.

The vertical plane passing through the axis of joint 1, perpendicular to the axes of joint 2 and 3, is named the meridian plane of the robot arm.

This paper aims at developing a dynamic model for the study of vibrations around a working configuration of the robot defined by vector $\boldsymbol{q}_0$ of joint variables, therefore joint variables, velocities, and accelerations are defined by the following equations:

$$\begin{aligned} \boldsymbol{q} &= \boldsymbol{q}_0 + \Delta\boldsymbol{q} \\ \dot{\boldsymbol{q}} &= \Delta\dot{\boldsymbol{q}} \\ \ddot{\boldsymbol{q}} &= \Delta\ddot{\boldsymbol{q}} \end{aligned} \tag{1}$$

where vector $\Delta\boldsymbol{q}$ contains the small variations in joint variables around the selected configuration.

Joint compliance around the rotation axis is considered, whereas link and bearing compliance is neglected.

The axis of joint 6 coincides with the approach axis of the end-effector. Since, in most cases, the external force exerted on the end-effector is aligned with or intersects the approach axis, compliance about joint 6 is neglected.

If the non-linear Coriolis and centrifugal terms [28] are neglected, according to the assumption of small oscillations, the equations of free vibrations take the form [6]:

$$\boldsymbol{M}(\boldsymbol{q}_0)\Delta\ddot{\boldsymbol{q}} + \boldsymbol{C}_q\Delta\dot{\boldsymbol{q}} + \boldsymbol{K}\Delta\boldsymbol{q} = 0 \tag{2}$$

where $\boldsymbol{M}(\boldsymbol{q}_0)$ is the mass matrix that depends on $\boldsymbol{q}_0$. The elements on the diagonal of $\boldsymbol{M}(\boldsymbol{q}_0)$ are the direct inertia terms, the off-diagonal terms are the inertial cross-coupling terms.

$\boldsymbol{C}_q$ is a diagonal matrix that accounts for joint damping, $\boldsymbol{K}$ is a stiffness matrix that accounts for joint compliance and configuration-depending gravity torques. Gravity can generate large restoring or unbalancing torques about joint axes only when the robot arm is close to a vertical configuration (upwards or downwards).

If configuration-depending gravity torques are negligible the equations of free vibrations become:

$$\boldsymbol{M}(\boldsymbol{q}_0)\Delta\ddot{\boldsymbol{q}} + \boldsymbol{C}_q\Delta\dot{\boldsymbol{q}} + \boldsymbol{K}_q\Delta\boldsymbol{q} = 0 \tag{3}$$

where $\boldsymbol{K}_q$ is a diagonal matrix that accounts for joint compliance. Specific calculations have been be carried out to estimate the effect of gravity torques in the selected configurations.

The identification of matrices $\boldsymbol{K}_q$ and $\boldsymbol{C}_q$ is the main task of this research.

The mass matrix $\boldsymbol{M}(\boldsymbol{q}_0)$ can be calculated using the inertial data provided by the computer aided design (CAD) models of the robot. These CAD files have been retrieved from the website of the manufacturer and the links are assumed to have uniform density and no free spaces within the external shell. This is due to the uncertainty of the mass distribution within the links of the robot that could not be eliminated without a complete disassembly of the robot.

Starting from the CAD files, the following inertial parameters have been retrieved:

- Mass of link i (m_i) has been calculated by distributing the total mass m_{tot} based on the volume of each link;
- Center of mass of link i ($\boldsymbol{G}_i$) has been calculated with respect to the corresponding Denavit-Hartenberg reference frame (see Figure 2);
- Inertia tensor of link i ($\boldsymbol{I}_i$) has been calculated at the center of mass aligned with the correspondent Denavit-Hartenberg reference frame [28].

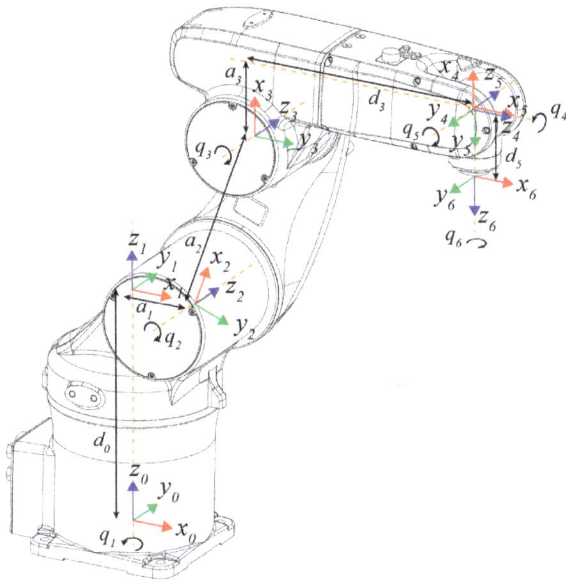

Figure 2. Robot reference frames.

On the basis of the above-mentioned parameters mass matrix $M(q_0)$ can be calculated as follows:

$$M(q_0) = \sum_{i=1}^{5} \left(m_i J_{p,i}^T J_{p,i} + J_{O,i}^T R_i I_i R_i^T J_{O,i} \right) \tag{4}$$

where R_i is the rotation matrix from link i to the base frame and $J_{p,i}$ and $J_{O,i}$ are the Jacobians:

$$\begin{aligned} J_{p,i} &= [z_1 \times (G_i - O_1) \cdots z_i \times (G_i - O_i)\, 0 \cdots 0] \\ & J_{O,i}[z_1 \cdots z_j\, 0 \cdots 0] \end{aligned} \tag{5}$$

3. Selective Modal Testing

Vibration signals contain information about mechanical systems [29], and this paper aims to use impulsive modal analysis to identify the stiffness properties of an industrial robot, and to develop a compliant robot model.

Modal analysis is a mature technology for the study of vibrations [30], which has been successfully used in many fields of aerospace [31], vehicle, and automation engineering [16,32,33].

In the field of industrial robotics, it is well recognized that joint stiffness has a larger effect on robot performance than link stiffness, therefore the main task of impulsive modal analysis is the identification of the stiffness of the joints.

A robot with assigned joint values (q) is considered. When an impulse force is exerted on a point of the robot by means of a hammer for modal testing and N acceleration components are measured by means of accelerometers, a set of N modes of vibrations can be identified. These modes of vibration transform the equations of motions (3) into a set of N independent equations in the modal coordinates $\eta_k\ k = 1, \ldots, N$:

$$\begin{bmatrix} M*_{11}(\boldsymbol{q}) & 0 & 0 \\ 0 & M*_{kk}(\boldsymbol{q}) & 0 \\ 0 & 0 & M*_{NN}(\boldsymbol{q}) \end{bmatrix} \begin{Bmatrix} \ddot{\eta}_1 \\ \ddot{\eta}_k \\ \ddot{\eta}_N \end{Bmatrix} + \begin{bmatrix} C*_{11}(\boldsymbol{q}) & 0 & 0 \\ 0 & C*_{kk}(\boldsymbol{q}) & 0 \\ 0 & 0 & C*_{NN}(\boldsymbol{q}) \end{bmatrix} \begin{Bmatrix} \dot{\eta}_1 \\ \dot{\eta}_k \\ \dot{\eta}_N \end{Bmatrix} + \begin{bmatrix} K*_{11}(\boldsymbol{q}) & 0 & 0 \\ 0 & K*_{kk}(\boldsymbol{q}) & 0 \\ 0 & 0 & K*_{NN}(\boldsymbol{q}) \end{bmatrix} \begin{Bmatrix} \eta_1 \\ \eta_k \\ \eta_N \end{Bmatrix} = \begin{Bmatrix} 0 \\ 0 \\ 0 \end{Bmatrix} \tag{6}$$

where the mass, damping and stiffness matrices are diagonal, and elements $M*_{kk}(\boldsymbol{q})$, $C*_{kk}(\boldsymbol{q})$ and $K*_{kk}(\boldsymbol{q})$ represent modal mass, damping and stiffness associated to the kth mode.

When the robot is tested in a generic configuration the modes of vibration typically involve rotations about many joints of the robot. Therefore, modal coordinate η_k is a combination of joint rotations, which is defined by the corresponding mode of vibration. Modal mass, damping and stiffness do not have a direct physical meaning, and are not very useful to identify the stiffness and damping characteristics of the robot joints.

Conversely, when the same robot is modally tested in a specific configuration in which the mass matrix shows that a joint variable i has negligible inertial cross coupling terms, one of the modes of vibration is dominated by the rotation about joint i, and the corresponding modal equation becomes:

$$I_{i_{zz}}(\boldsymbol{q})\ddot{q}_i + c_i\dot{q}_i + k_i q_i = F_i(t) \tag{7}$$

The modal coordinate η_i coincides with joint variable q_i, modal mass coincides with inertia about joint i ($I_{i_{zz}}(\boldsymbol{q})$), modal stiffness and damping coincide with joint stiffness (k_i) and damping (c_i), respectively. Mode i is very useful to identifying the stiffness of joint i.

Inertia about joint i ($I_{i_{zz}}(\boldsymbol{q})$) depends on the inertias of the following links and can be calculated by using the Huygens–Steiner theorem:

$$I_{i_{zz}}(\boldsymbol{q}) = \sum_{j=i}^{N} \left(\boldsymbol{R}_{ji} \boldsymbol{I}_i \boldsymbol{R}_{ji}^T + m_j \boldsymbol{P}_{ji} \right) \tag{8}$$

$$\boldsymbol{P}_{ji} = \begin{bmatrix} y_{ji}^2 + z_{ji}^2 & -x_{ji}y_{ji} & -x_{ji}z_{ji} \\ -y_{ji}x_{ji} & x_{ji}^2 + z_{ji}^2 & -y_{ji}z_{ji} \\ -z_{ji}x_{ji} & -z_{ji}y_{ji} & x_{ji}^2 + y_{ji}^2 \end{bmatrix} \tag{9}$$

where $\boldsymbol{R}_{ji}$ is the rotation matrix that transforms the reference frame of link j to the reference frame of link i, and $x_{l_{ji}}$, $y_{l_{ji}}$ and $z_{l_{ji}}$ are the coordinates of the center of mass of link j defined in the reference frame of link i.

It is worth noticing that usually mode i (dominated by joint i) includes small contributions of the other joints, and sometimes of structural compliance of the links, therefore assuming $M*_{ii}(\boldsymbol{q}) = I_{i_{zz}}(\boldsymbol{q})$, $K*_{ii}(\boldsymbol{q}) = k_i$ and $C*_{ii}(\boldsymbol{q}) = c_i$ is only an approximation.

For the above-mentioned reasons, a specific modal testing method is proposed, which is based on robot configurations and excitation directions that make possible the excitation of modes of vibrations dominated by only one joint.

In order to have an insight into the properties of the mass matrix of the tested robot, the elements of the mass matrix were evaluated in random configurations. In particular 1,000,000 random combinations of $q_1 \ldots q_5$ with values within the actual joint ranges were considered.

Since the inertial cross-coupling terms may be positive or negative, depending on the configuration, the mean values and standard deviations of the moduli of the elements of the mass matrix were calculated to avoid cancellations.

The results of this analysis are summarized in Figure 3, and show the following properties.

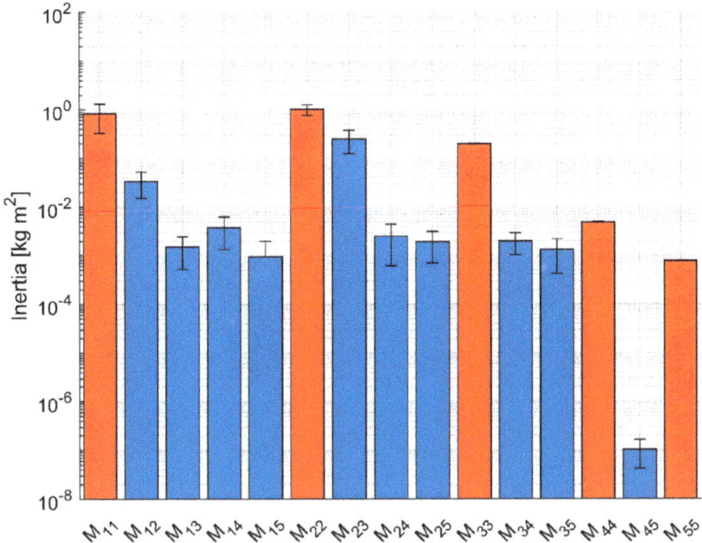

Figure 3. Mean values and standard deviations of mass matrix terms (10^6 random configurations), direct inertia terms (red bars) and inertial cross-coupling terms (blue bars).

(a) The direct inertia terms of the first three joints are by far larger than the ones of the wrist joints—notice that a logarithmic scale is adopted in Figure 3. Joint 2 (J2) has the largest mean value followed by Joint 1 (J1).

(b) The direct inertia term of joint i (M_{ii}) depends on the joint variables of the following links, hence M_{11} depends on $q_2 \ldots q_N$, whereas M_{NN} is constant. Therefore, the standard deviation decreases from J1 to J5.

(c) The inertial cross-coupling terms between the first three joints are much smaller than the direct inertia terms with the exception of M_{23}, which is comparable with M_{33}.

(d) M_{23} has a large standard deviation, and this means that there are robot configurations with a minimum coupling between J2 and J3.

(e) The mean values of the inertial cross-coupling terms of the first three joints with the wrist joints are very small in comparison with the direct inertia terms of the first three joints. They are more important if compared with the direct inertia terms of the wrist joints

(f) There is a negligible coupling between wrist joints (M_{45}).

In a specific configuration, the inertial cross-coupling terms of joint i (M_{ij}, $j \neq i$) are small or negligible if the velocity of joint i generates a distribution of robot arm velocities that cannot be generated by the velocities of the other joints. Even if the previous condition is satisfied, cross-coupling terms may be present, if the axes of the joints are not principal axes.

According to the results of Figure 3, the configurations for selective modal testing summarized in Figure 4 and Table 1 were chosen.

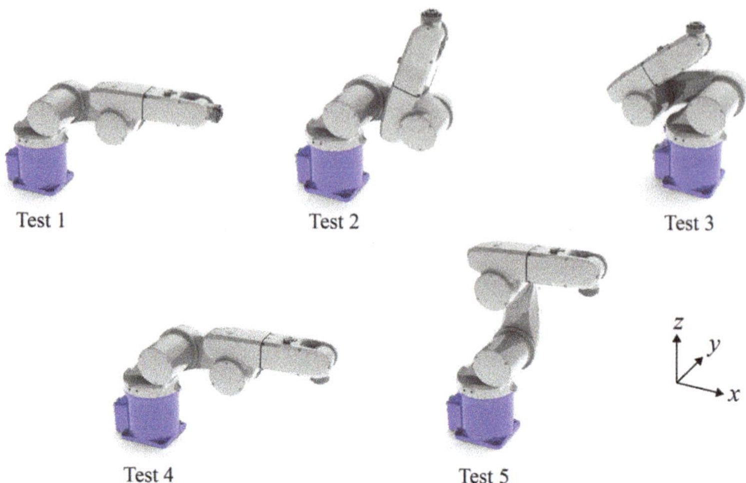

Figure 4. Robot configurations.

Table 1. Robot testing configurations.

Configuration	Joint 1 [°]	Joint 2 [°]	Joint 3 [°]	Joint 4 [°]	Joint 5 [°]	Joint 6 [°]
Test 1	0	0	90	0	0	0
Test 2	0	0	17	0	−17	0
Test 3	0	−189	243	0	−54	0
Test 4	0	0	90	0	90	0
Test 5	0	−90	180	0	90	0

Joint 1 (J1) is always perpendicular to joints 2 and 3 (J2 and J3). If joint variables q_2, q_3 are set to 0° and 90°, respectively, the robot arm is horizontal, and if q_4 q_5 are set to 0°, the approach axis is horizontal as well. In this configuration (Test 1) only J1 is able to generate relevant velocities in the horizontal plane. This configuration minimizes the inertial coupling between J1 and the other joints (which is usually small) and enhances the excitation of J1, when a lateral force (in y direction) is exerted on the end-effector.

Joints J2 and J3 are always parallel, and generate velocity components in the same plane, Figure 3 shows that the inertial cross coupling may be large. The mass matrix of a robot with two parallel joints [28] shows an inertial cross-coupling term depending on joint variable q_3 (the angle between the links). A numerical calculation of the ratios between the inertial cross-coupling term M_{23} and the direct inertias of J2 and J3 (M_{22} and M_{33} respectively) is depicted in Figure 5. They show that the inertial cross-coupling term tends to decrease when the two links tend to overlap, notice that the two links are aligned when $q_3 = 90°$. Minimum coupling configurations can be found for $q_3 = -38°$ and $q_3 = 293°$. These theoretical values have to be compared with the actual range of the joint ($-29° \leq q_3 \leq 256°$), and, for this reason, a configuration with $q_3 = 243°$ (Test 3) is suited to identifying the stiffness both of J2 and J3. It is worth noticing that, in this configuration, a vertical force generates a large moment about J3, thus, it is able to excite the mode dominated by this joint. Conversely, in this configuration a vertical force produces a small moment about J2, thus, the excitation of the mode dominated by this joint is poor. Figures 3 and 5 show that the inertial cross-coupling term is less important for J2 than for J3, since the direct inertia M_{22} term is much larger than M_{23}. For this reason, configuration Test 2 of Table 1, in which links 2 and 3 are almost perpendicular, gives a good compromise between the reduction of inertial cross coupling and the generation of a large moment about J2.

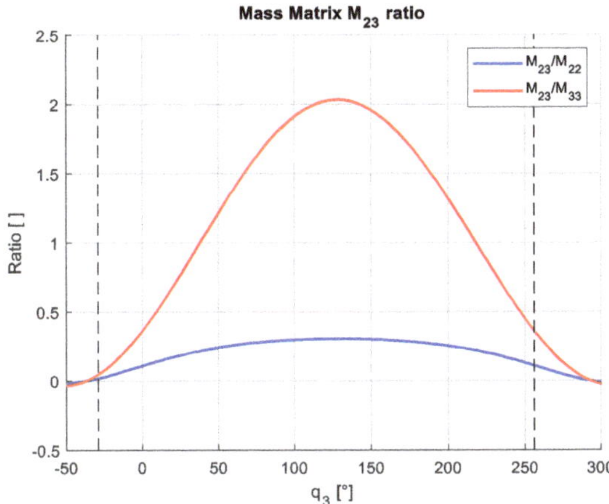

Figure 5. Effect of joint variable q_3 on inertial cross coupling between J2 and J3.

Joint J4 is able to generate velocity components perpendicular to the vertical plane that contains the robot arm (meridian plane). J2, J3 generate velocity components in meridian plane, and the inertial cross coupling with J4 is small. Conversely, J1 is able to generate velocity components in the same direction as J4, and there is a significant cross coupling with J4 (see Figure 3). The inertial cross coupling between J4 and J5 is always negligible (see Figure 3). In order to enhance the moment that a lateral hit can generate about J4, q_5 was set to ±90° and configuration Test 4 was defined.

Finally, when q_4 is set to zero, Joint 5 (J5) is able to generate velocity components of the last links in the meridian plane of the robot arm, but these components can be generated by J2 and J3 as well. When q_4 is set to 90°, J5 is able to generate velocity components out of the meridian plane of the robot arm, but these components can be generated by J1 as well. Therefore, Figure 3 shows that there is always a significant coupling between J5 and the arm joints. The test configuration can be chosen only with the aim of increasing the moment about J5 generated by the hammer hit. Test configuration Test 5 satisfies this condition for a hammer hit in the longitudinal direction.

The relevance of gravity restoring (or unbalancing) torques was evaluated in the selected configurations. The calculated values resulted much smaller than reference values of joint torques caused by joint stiffness [16].

4. Experimental Tests

The equipment needed to carry out the modal analysis of a robot with the impulsive method is rather simple and cheap. It includes a hammer for modal testing, an accelerometer, and a data acquisition board. In the framework of this research, a PCB 086C01 hammer (with load cell sensitivity 0.2549 mV/N), a PCB 356A17 tri-axial accelerometer (sensitivity 50.5 mV/(m/s^2)), and a NI9234 data acquisition board were used. The hammer and the accelerometer were made by PCB Piezotronics Inc., Depew, NY, USA, whereas the board was made by National Instruments Inc., Austin, TX, USA. Measured signals were analyzed in the frequency domain by means of ModalVIEW R2 (2017), which is a specific software for modal analysis developed by ABSignal-National Instruments Inc., Austin, TX, USA.

In order to characterize the vibrations of the robot 10 testing points were defined. Figure 6 shows that testing point 1 is located on the flange of the end-effector, testing point 10 is located on the base, before joint 1, and the other testing points are evenly distributed on the links. The number of testing points is a trade-off between the need for a large number of measurement locations, to accurately describe robot vibrations, and the need for a quick testing procedure. Modal analysis was carried out with the rowing response approach [34] Excitation was always exerted on the end-effector, whereas the tri-axial accelerometer was sequentially moved to the 10 testing points. The three axes of the accelerometer were always parallel or perpendicular to the robot surface.

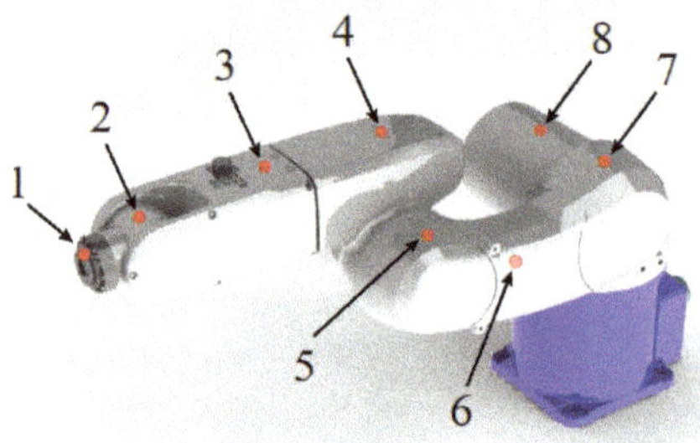

Figure 6. Measurement points for modal testing.

Since motions of the base of the robot may have a negative effect on the quality of measurements, the robot was rigidly fastened to a large steel base.

After some preliminary tests, data acquisition was carried out with a sampling frequency of 1024 Hz and 2048 samples.

Figure 7 deals with the experimental results obtained in configuration Test 1, with hammer excitation in the lateral direction (y). Plots (a) and (b) show the modulus and phase of the direct point frequency response function (FRF), which is the FRF measured in the excitation point between the acceleration in the direction of the hammer force and the hammer force. Plot (c) shows the complex mode indicator function (CMIF) calculated over the 30 FRFs [30]. Both the modulus of the FRF and the CMIF show a high and isolated peak at about 13 Hz. The phase of the direct point FRF shows a large phase change at the same frequency. These properties are hints of the presence of a well-defined mode of vibration at about 13 Hz. ModalVIEW made it possible to identify a mode of vibration at 12.8 Hz with a damping ratio of 2.3% [30]. This mode, which is represented at the bottom of Figure 7, is dominated by the rotation of the whole robot arm about J1.

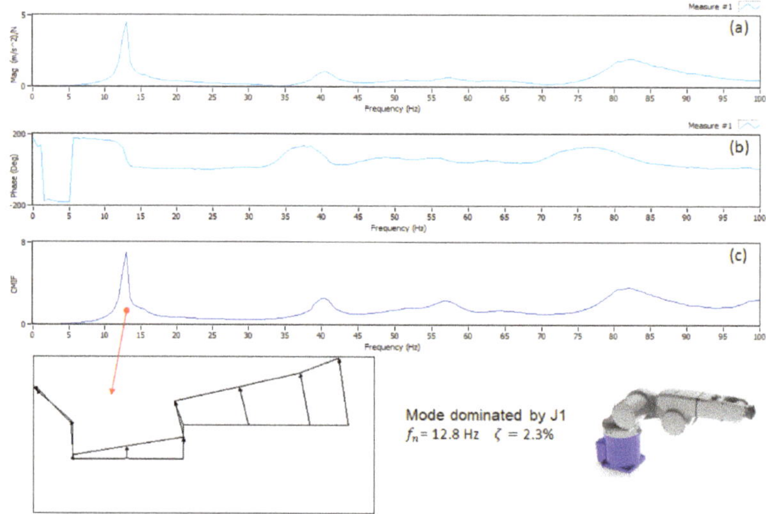

Figure 7. Test 1 configuration, (**a**) direct point frequency response function (FRF) modulus, (**b**) phase, (**c**) complex mode indicator function (CMIF).

Figure 8 shows the experimental results obtained in configuration Test 2 with hammer excitation in the vertical direction (z). In this case, the direct point FRF and the CMIF show a high and isolated peak at about 17 Hz. ModalVIEW made it possible to identify at 17.5 Hz a mode of vibration dominated by the rotation of the whole robot arm about J2, which is represented in Figure 8. The corresponding damping ratio is 1.9%, this value is congruent with the damping ratio of J1.

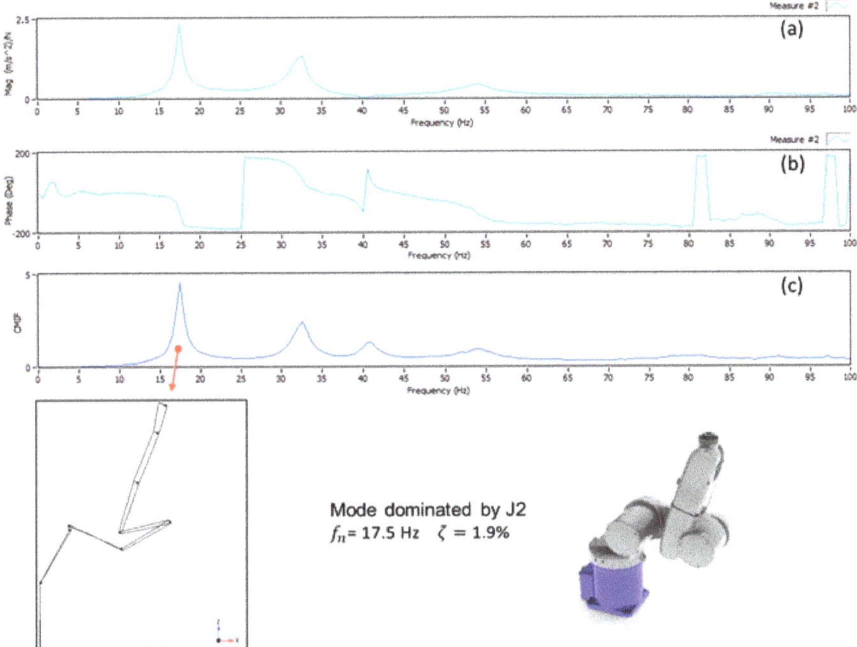

Figure 8. Test 2, (**a**) direct point frequency response function (FRF) modulus, (**b**) phase, (**c**) complex mode indicator function (CMIF).

The experimental results obtained in configuration Test 3 are summarized in Figure 9. The direct point FRF shows a high peak at about 30 Hz with a large phase change, but in this case the main peak is not alone, since there is another peak at about 35 Hz. The CMIF plot confirms this scenario. ModalVIEW made it possible to identify two modes. The first mode is at 30.8 Hz with a damping ratio of 2.1%. This mode, which is represented in Figure 9, is dominated by the compliance of J3, and shows some contribution of J2, owing to the inertial-cross coupling between the two joints. The second mode at 34.7 Hz is influenced by the structural compliance. In particular, structural compliance in the direction perpendicular to the axes J2 and J3 leads to an apparent torsion of the robot arm. The large damping ratio (4.4%) is in agreement with the contribution of structural deformation. The presence of structural modes in the range of frequencies of the modes dominated by joint compliance has been detected by other researchers [6].

Figure 10 deals with experimental results obtained in configuration Test 4 with lateral excitation (y direction). Both the direct point FRF and the CMIF highlight the presence of an important resonance peak beyond the 10 ÷ 40 Hz frequency band, which contains the previously identified modes. Modal analysis made it possible to identify a mode of vibration at 56.8 Hz with a 3.0% damping ratio, which is similar to the ones of the other joints. This mode, which is represented in Figure 10, shows a large contribution of the rotation about J4, which leads to lateral displacements of the points of the following links.

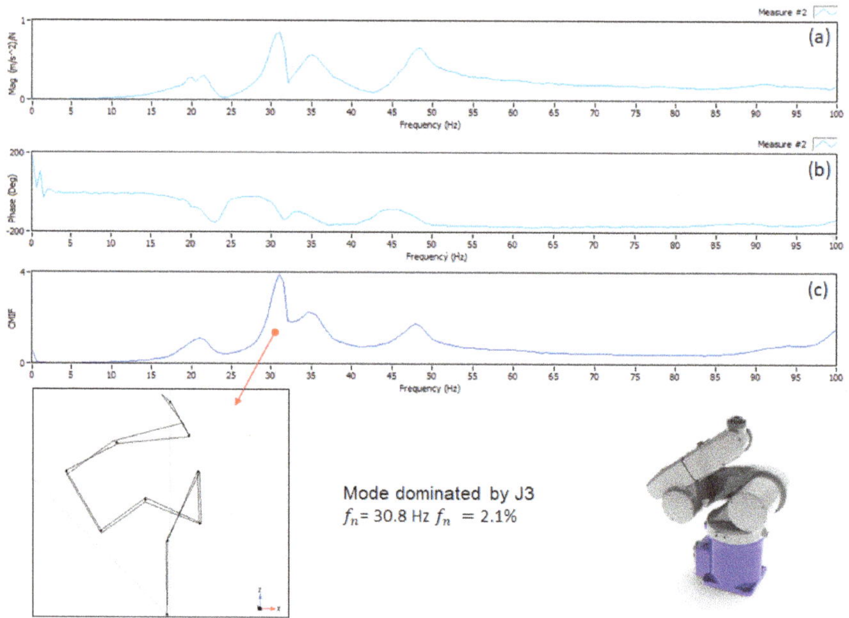

Figure 9. Test 3, (**a**) direct point frequency response function (FRF) modulus, (**b**) phase, (**c**) complex mode indicator function (CMIF).

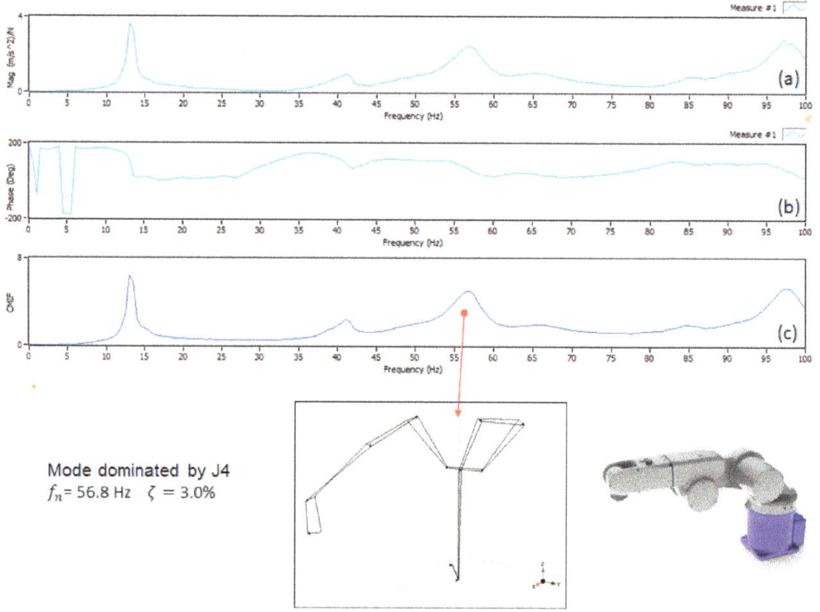

Figure 10. Test 4, (**a**) direct point frequency response function (FRF) modulus, (**b**) phase, (**c**) complex mode indicator function (CMIF).

Finally, Figure 11 shows experimental results obtained in configuration Test 5 with longitudinal excitation (x direction). At high frequency, there is a very important peak at about 170 Hz. Modal analysis showed that at 171.4 Hz there is a mode of vibration with large displacements localized in the last link and caused by the compliance of J5 (see Figure 11). The damping ratio of this mode is 2.8%.

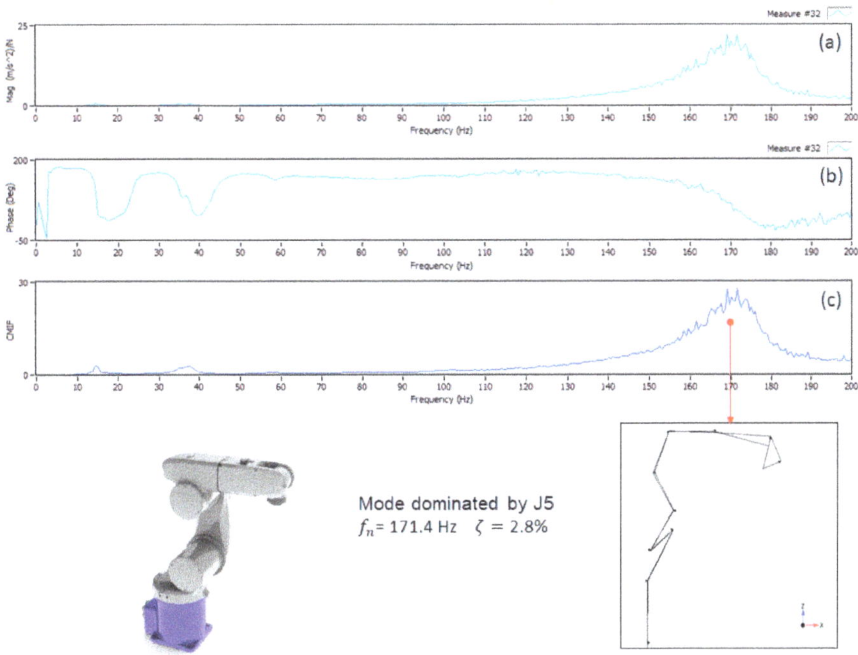

Figure 11. Test 5, (**a**) direct point frequency response function (FRF) modulus, (**b**) phase, (**c**) complex mode indicator function (CMIF).

5. Identification

Testing configurations Test 1 and Test 2 made it possible to identify modes of vibration clearly dominated by the stiffness properties of J1 and J2 respectively, and these stiffness properties are the most important from the point of view of robot operations (e.g., machining or assembly). Testing configurations Test 4 and Test 5 made it possible to identify modes of vibration with a large contribution of the stiffness of J4 and J5. Owing to the small moments of end-effector forces about J4 and J5, the stiffness of wrist joints is less critical than the stiffness of arm joints. Testing configuration Test 3 minimized the coupling between J2 and J3, nevertheless the identified modal shape a at 30.8 Hz showed some influence of the rotation about J2 and of structural deformation, since in this configuration the robot has a structural mode ad 34.7 Hz. Therefore, the measured frequency is chiefly influenced by the stiffness of J3, but it includes the contribution of other stiffness properties.

For the aforementioned reasons, the stiffness about J1, J2, J4 and J5 was directly calculated from Equation (7), assuming modal coordinate η_i coincident with q_i and using a 1 DOF model:

$$f_{ni} = \frac{1}{2\pi}\sqrt{\frac{k_i}{I_{i_{zz}}(\boldsymbol{q}_i)}} \tag{10}$$

$$k_i = (2\pi f_{ni})^2 I_{i_{zz}}(\boldsymbol{q}_i) \tag{11}$$

where f_{ni} is the natural frequency measured in testing configuration $\boldsymbol{q}_i$.

The damping coefficient of joint i (c_i) was calculated with the following equation:

$$c_i = \frac{2\zeta_i k_i}{2\pi f_{ni}} \tag{12}$$

where ζ_i is the identified damping ratio of mode i.

The results of these calculations are summarized in Table 2.

Table 2. Robot stiffness and damping properties obtained with the 1-degree-of-freedom (DOF) model.

	Joint 1	Joint 2	Joint 3	Joint 4	Joint 5
Stiffness [Nm/rad]	12,602	9257	7490	643	898
Damping [Nms/rad]	7.21	3.20	1.63	0.11	0.05

For joint J3, the value given by Equation (11) was considered only a first estimate of the actual joint stiffness (k_3). The identification of this stiffness was then improved carrying out an optimization. The following penalty function was defined:

$$F_p(k_3) = \sum_{j=1}^{5} \sum_{k=1}^{5} \left(f_{njk} - f_{pjk}(k_3)\right)^2 \tag{13}$$

This is the sum of the squared differences between the measured natural frequencies and the natural frequencies f_{pjk} predicted by the model with assigned k_3. Five modes of vibration ($j = 1, \ldots, 5$) in the five testing configurations are considered ($k = 1, \ldots, 5$). Minimization was carried out with the function fmincon of MATLAB. The optimized stiffness k_3 = 6485 Nm/rad was rather different from the first attempt value. Finally, damping coefficient was calculated according to Equation (12) and resulted c_3 = 1.41 Nms/rad.

The identified stiffness and damping properties were the implemented in the mathematical model of the robot (Equation (3)), and the natural frequencies and damping ratios were calculated in some validation configurations that were experimentally tested as well. Joint variables of validation configurations are summarized in Table 3.

Table 3. Robot validation configurations.

Configuration	Joint 1 [°]	Joint 2 [°]	Joint 3 [°]	Joint 4 [°]	Joint 5 [°]	Joint 6 [°]
Validation 1	0	30	30	0	90	0
Validation 2	30	−50	160	0	90	0

Table 4 shows the comparison between numerical values and experimental values in the validation configurations.

Table 4. Validation results.

Configuration		Mode 1	Mode 2	Mode 3	Mode 4	Mode 5
Validation 1	Calc. frequency [Hz]	15.4	16.1	33.0	57.2	171.9
	Exp. frequency [Hz]	16.3	17.8	36.2	58.4	170.5
	Calc. damping %	2.6	1.8	2.4	3.0	2.8
	Exp. damping %	3.9	2.1	3.1	3.4	3.2
Validation 2	Calc. frequency [Hz]	12.9	15.7	46.1	57.2	172.3
	Exp. frequency [Hz]	13.3	14.5	40.2	59.1	171.8
	Calc. damping %	1.33	2.8	3.4	3.0	2.8
	Exp. damping %	2.7	2.4	2.5	3.7	2.7

In validation configuration 1, the predicted natural frequencies of the first four modes are a bit lower than the experimental natural frequencies. The largest error is less than −10%, and it takes place in the second mode, which is a mode on the meridian plane dominated by the rotation about J2. The predicted natural frequency of Mode 5 is a bit larger than the experimental value.

In validation configuration 2, the predicted natural frequencies of Mode 1 and Mode 4 are a bit lower than the experimental values (about −3%), whereas the other natural frequencies are higher than the experimental values. The largest error (+15%) takes place in the frequency of Mode 3. This happens because modal analysis (Figure 9) showed that in this range of frequencies structural deformability affects the modes of vibration, and this effect is not taken into account by the numerical model, which is based on joint stiffness.

Looking at modal damping, there is a general agreement between numerical and experimental values in both configurations. The modes that show a large experimental damping typically show a large numerical damping. Only in the first mode the calculated value is significantly lower than the experimental value. The presence of configuration dependent friction forces, due to large gravity loads, may be the cause of this phenomenon.

6. Numerical Simulations

Recent studies [2,6,35,36] have highlighted that, in some robot operations (e.g., milling tasks), the dynamic properties of the robot arm are important, and that a simple optimization of robot configurations based on static stiffness does not lead to the best performance [2]. The modes of vibration of a robot are configuration-dependent, and the variation of natural frequency and modal damping in the workspace is an important feature of robot dynamics. It gives some hints useful to improve robot performance selecting robot configurations with highly damped modes having natural frequencies far from the excitation frequencies.

Experimental modal analysis is time consuming, thus, the mathematical model based on modal analysis in a small number of selected configurations is a very useful tool to predict the variation in the dynamic properties of the robot over the whole workspace.

The first numerical analysis aimed to understand the effect of robot configuration on the natural frequency and damping of the modes of vibration. A total of 10,000 random configurations of the robot corresponding to 10,000 random combinations of joint variables J2, J3, J4, and J5 (within joint ranges) were defined, and the modal properties were calculated. Since the robot model is axial-symmetric about J1, the variation of J1 was not taken into account, and this joint variable was set to zero. The results of these calculations are collected in Figures 12 and 13, which depict the mean values and standard deviations of natural frequencies and modal dampings, respectively.

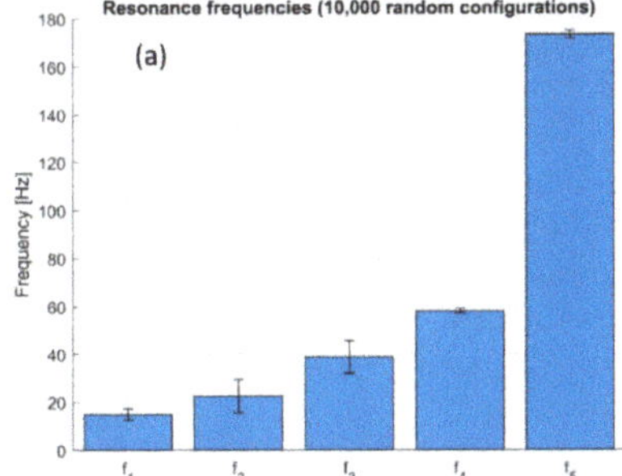

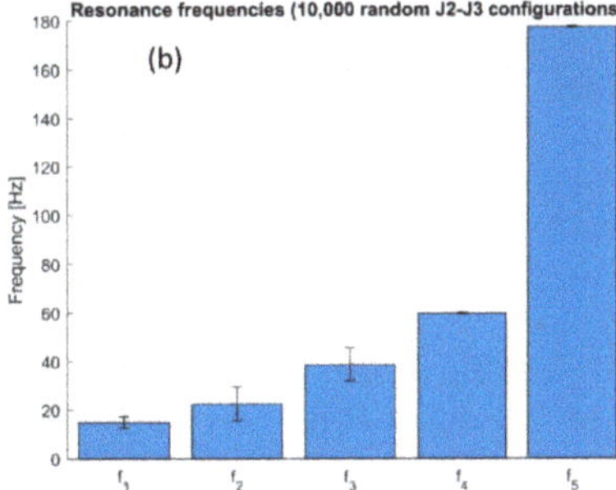

Figure 12. Mean values and standard deviations of natural frequencies in the workspace, (**a**) random variations of J2, J3, J4, J5, (**b**) random variations of J2 and J3.

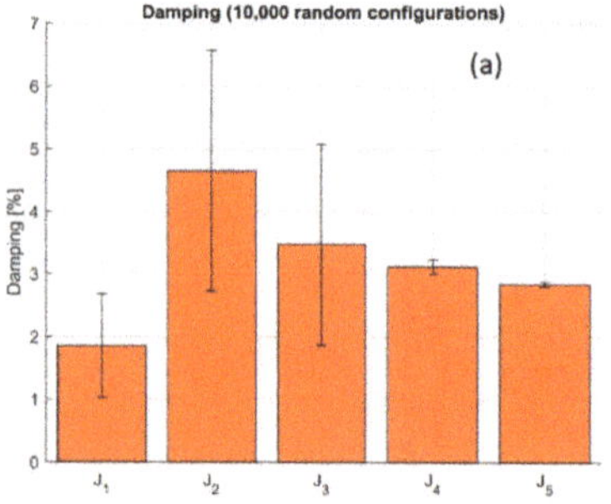

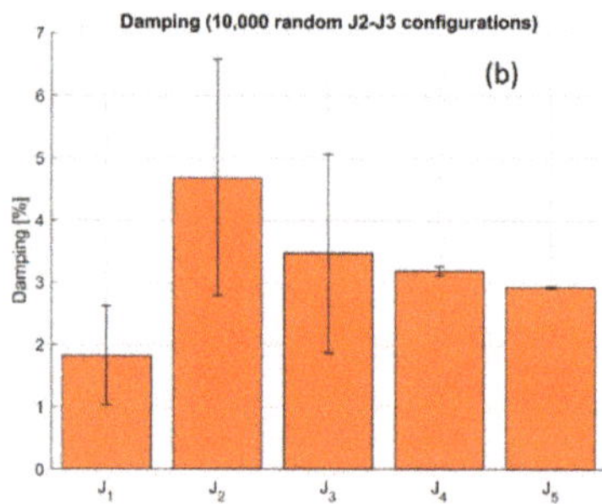

Figure 13. Mean values and standard deviations of modal damping in the workspace, (**a**) random variations in J2, J3, J4, J5, (**b**) random variations in J2 and J3.

The natural frequencies of the second and third mode, which are usually characterized by large displacements in the meridian plane, show the largest standard deviations. Therefore, these modes are influenced by the robot configuration. The first natural frequency, which usually corresponds to a mode of vibration with large displacements in the horizontal plane, shows a mild standard deviation, since the robot configuration affects the moment of inertia about J1. The natural frequencies of the last two modes, which are chiefly dominated by J4 and J5, show very small standard deviations. This happens because the robot configuration has a small effect on the direct inertia term of J4 and no effect on the direct inertia term of J5, but it only changes the inertial cross-coupling terms (see Figure 3).

The statistical analysis of modal damping variation with robot configurations is depicted in Figure 13. This figure shows that robot configuration affects the modal damping of the first three modes, since standard deviations are comparable with the mean values. Conversely, robot configuration has a very small effect on the damping of the last modes, which chiefly involve wrist joints, because standard deviations are very small.

Figures 12b and 13b show the result of a similar calculation, in which only random variations in J2 and J3 were performed. The standard deviations of the natural frequencies and modal dampings are very similar to the ones of Figures 12a and 13a, therefore, this result points out that the variations in J2 and J3 have the largest effect on the modal properties of the robot.

Figures 12 and 13 highlight another important feature of the robot. Variations in joint variables are able to modify the natural frequencies, but they do not change the typical frequency bands of the modes. In other words, the frequencies of the modes associated to the wrist joints are always higher than the frequencies associated to the arm joints. The first two modes belong to the same frequency band (15 ÷ 30 Hz), and this band has a negligible overlap with the bands containing the other modes.

The statistical analysis showed mild variations in the properties of the first modes of vibrations of the robot, due to variations in J2 and J3. These joint variables determine the configuration of the robot in the meridian plane and in particular the location of the wrist (the origin of coordinate system 4 in Figure 2). Therefore, a further numerical analysis was carried out to study the dependence of the first three natural frequencies on the position of the wrist in the meridian plane. The results are presented in terms of contour plots, in which the darker colors represent the lower frequencies and the lighter colors represent the higher frequencies. Since some locations can be reached with two robot configurations (elbow-up and elbow-down), both configurations are considered in separate plots.

Figure 14 shows that the first natural frequency decreases in a regular way when the distance between the wrist and the robot base increases, and this effect takes place both in the elbow-up and in the elbow-down configuration.

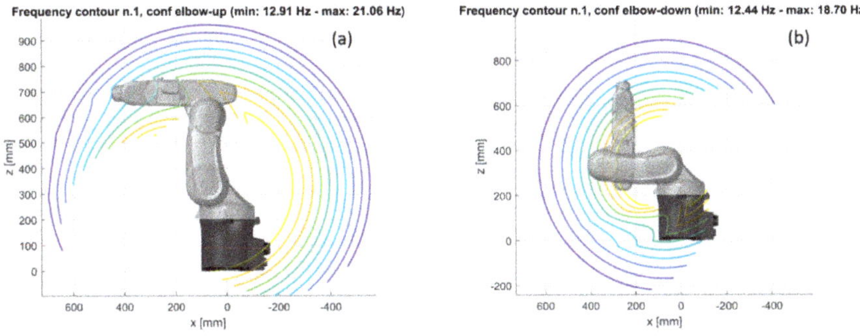

Figure 14. Contour plot of the natural frequency of mode 1 in the meridian plane, (**a**) elbow-up, (**b**) elbow-down.

The contour plot of the second natural frequency is more complex (Figure 15). The second natural frequency reaches the lowest values when the arm is extended in order to reach locations in front of the base or on the back of the base. In these configurations, links 2 and 3 tend to align. The maximum value of the second natural frequency takes place when the arm has to reach locations above the base. This effect is present in both the considered configurations.

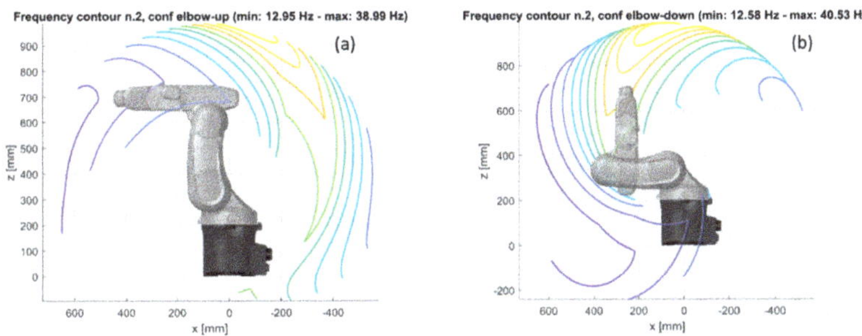

Figure 15. Contour plot of the natural frequency of mode 2 in the meridian plane, (**a**) elbow-up, (**b**) elbow-down.

Figure 16 shows a rather regular trend of the third natural frequency. In both configurations, this natural frequency tends to increase when the wrist location is far from the base and links 2 and 3 tend to align (joint variable J3 is close to 90°).

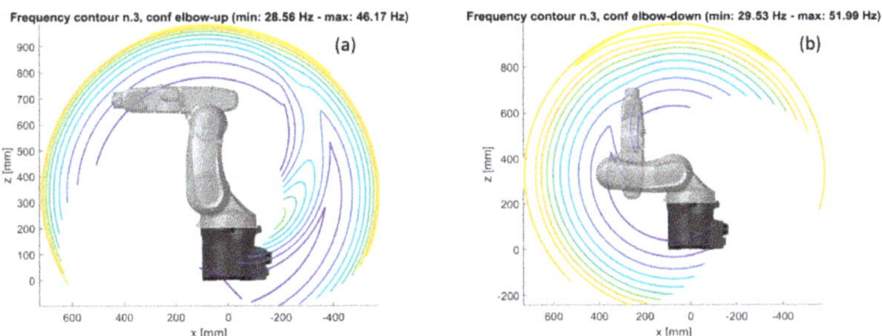

Figure 16. Contour plot of the natural frequency of mode 3 in the meridian plane, (**a**) elbow-up, (**b**) elbow-down.

Therefore, the angle (q_3) between link 2 and 3 has opposite effects on the second and third natural frequencies that usually correspond to modes that take place in the meridian plane. The second natural frequency decreases when the two links tend to align, because the direct inertia term M_{22} increases. The third natural frequency increases when the two links tend to align, because the inertial cross coupling term (M_{23}) between J2 and J3 increases (see Figure 5). This effect takes place even if a simple 2 DOF model of the robot in the meridian plane is adopted [28].

Finally, the analysis of Figures 15 and 16 shows that it is rather difficult to maximize both the second and the third natural frequencies. Only in a limited region of the workspace above the robot it is possible to achieve large values of both frequencies.

7. Conclusions

The proposed testing method based on selective modal testing requires simple equipment, does not take a long testing time, can be implemented in every robot configuration, and makes it possible to detect the presence of modes of vibration affected by structural and bearing compliance.

The combination of the results of selective modal testing with a mathematical model of mass distribution of the robot makes it possible a straightforward identification of joint stiffness and damping. A simple optimization procedure on a small set of parameters is carried out, to identify joint stiffness in the presence of a strong inertial cross-coupling between the joints.

The mathematical model implemented with the identified stiffness and damping properties gives information about the variation of the dynamic characteristics of the robot in the workspace. This information is useful to optimize the performance of the robot in machining tasks.

In future, the method will be improved, in order to cope with the presence of the structural mode in the band of frequency of interest.

Author Contributions: Conceptualization, M.B., S.C., N.C., A.D.; methodology, M.B., S.C., A.D.; software, M.B., N.C.; investigation, S.C., N.C., A.D.; writing—original draft preparation, M.B., S.C.; writing—review and editing, N.C., A.D.; supervision, A.D. All authors have read and agreed to the published version of the manuscript.

Funding: This research was funded by Fondazione Aldo Gini (Call 2018), Ermenegildo Zegna Founder's Scholarship 2019/2020, and by University of Padova DII project BIRD187930.

Conflicts of Interest: The authors declare no conflict of interest.

References

1. Chen, H.; Dong, F. Robot machining: Recent development and future research issues. *Int. J. Adv. Manuf. Technol.* **2013**, *66*, 1489–1497. [CrossRef]
2. Leonesio, M.; Villagrossi, E.; Beschi, M.; Marini, A.; Bianchi, G.; Pedrocchi, N.; Tosatti, L.M.; Grechishnikov, V.; Ilyukhin, Y.; Isaev, A. Vibration Analysis of Robotic Milling Tasks. *Procedia CIRP* **2018**, *67*, 262–267. [CrossRef]
3. Pan, Z.; Zhang, H.; Zhu, Z.; Wang, J. Analysis of robotic machining process. *J. Mater. Process. Technol.* **2006**, *173*, 301–309. [CrossRef]
4. Zhang, H.; Wang, J.; Zhang, G.; Gan, Z.; Pan, Z.; Cui, H.; Zhu, Z. Machining with flexible manipulator: Toward improving robotic machining performance. In Proceedings of the IEEE/ASME International Conference on Advanced Intelligent Mechatronics, Monterey, CA, USA, 24–28 July 2005.
5. Carbone, G. Stiffness analysis and experimental validation of robotic systems. *Front. Mech. Eng. Chin.* **2011**, *6*, 182–196. [CrossRef]
6. Huynh, H.N.; Assadi, H.; Rivière-Lorphèvre, E.; Verlinden, O.; Ahmadi, K. Modelling the dynamics of industrial robots for milling operations. *Robot. Comput. Integr. Manuf.* **2020**, *61*, 101852. [CrossRef]
7. Rosati, G.; Faccio, M.; Carli, A.; Rossi, A. Fully flexible assembly systems (F-FAS): A new concept in flexible automation. *Assembly Autom.* **2013**, *33*, 8–21. [CrossRef]
8. Rosati, G.; Faccio, M.; Finetto, C.; Carli, A. Modelling and optimization of fully flexible assembly systems (F-FAS). *Assembly Autom.* **2013**, *33*, 165–174. [CrossRef]
9. Faccio, M.; Bottin, M.; Rosati, G. Collaborative and traditional robotic assembly: A comparison model. *Int. J. Adv. Manuf. Technol.* **2019**, *102*, 1355–1372. [CrossRef]

10. Cocuzza, S.; Pretto, I.; Debei, S. Least-squares-based reaction control of space manipulators. *J. Guid. Control Dynam.* **2012**, *35*, 976–986. [CrossRef]
11. Tringali, A.; Cocuzza, S. Predictive control of a space manipulator through error and kinetic energy expectation. In Proceedings of the 69th International Astronautical Congress (IAC 2018), Bremen, Germany, 1–5 October 2018.
12. Cocuzza, S.; Pretto, I.; Debei, S. Novel reaction control techniques for redundant space manipulators: Theory and simulated microgravity tests. *Acta Astronaut.* **2011**, *68*, 1712–1721. [CrossRef]
13. Cocuzza, S.; Tringali, A.; Yan, X. Energy-efficient motion of a space manipulator. In Proceedings of the 67th International Astronautical Congress (IAC 2016), Guadalajara, Mexico, 26–30 September 2016; pp. 8788–8804.
14. Gasparetto, A. Eigenvalue analysis of mode-coupling chatter for machine-tool stabilization. *J. Vib. Control* **2001**, *7*, 181–197. [CrossRef]
15. Gasparetto, A. A system theory approach to mode coupling chatter in machining. *J. Dyn. Syst. Meas. Control* **1998**, *120*, 545–547. [CrossRef]
16. Doria, A.; Cocuzza, S.; Comand, N.; Bottin, M.; Rossi, A. Analysis of the compliance properties of an industrial robot with the Mozzi axis approach. *Robotics* **2019**, *8*, 80. [CrossRef]
17. Ni, H.; Zhang, C.; Hu, T.; Wang, T.; Chen, Q.; Chen, C. A dynamic parameter identification method of industrial robots considering joint elasticity. *Int. J. Adv. Rob. Syst.* **2019**, *16*, 1729881418825217. [CrossRef]
18. Dumas, C.; Caro, S.; Garnier, S.; Furet, B. Joint stiffness identification of six-revolute industrial serial robots. *Rob. Comput. Integr. Manuf.* **2011**, *27*, 881–888. [CrossRef]
19. Abele, E.; Weigold, M.; Rothenbücher, S. Modeling and identification of an industrial robot for machining applications. *CIRP Ann.* **2007**, *56*, 387–390. [CrossRef]
20. Rafieian, F.; Liu, Z.; Hazel, B. Dynamic model and modal testing for vibration analysis of robotic grinding process with a 6DOF flexible-joint manipulator. In Proceedings of the International Conference on Mechatronics and Automation, Changchun, China, 9–12 August 2009.
21. Alici, G.; Shirinzadeh, B. Enhanced stiffness modeling, identification and characterization for robot manipulators. *IEEE Trans. Rob.* **2005**, *21*, 554–564. [CrossRef]
22. Abele, E.; Rothenbücher, S.; Weigold, M. Cartesian compliance model for industrial robots using virtual joints. *Prod. Eng.* **2008**, *2*, 339. [CrossRef]
23. Carbone, G.; Ceccarelli, M. A stiffness analysis for a hybrid parallel-serial manipulator. *Robotica* **2004**, *22*, 567–576. [CrossRef]
24. Mousavi, S.; Gagnol, V.; Bouzgarou, B.C.; Ray, P. Dynamic behaviour model of a machining robot. In Proceedings of the ECCOMAS Multibody Dynamics, Zagreb, Croatia, 1–4 July 2013; pp. 771–779.
25. Mousavi, S.; Gagnol, V.; Bouzgarrou, B.C.; Ray, P. Stability optimization in robotic milling through the control of functional redundancies. *Robot. Comput. Integr. Manuf.* **2018**, *50*, 181–192. [CrossRef]
26. Behi, F.; Tesar, D. Parametric identification for industrial manipulators using experimental modal analysis. *IEEE Trans. Rob. Autom.* **1991**, *7*, 642–652. [CrossRef]
27. Corral, E.; García, M.J.; Castejon, C.; Meneses, J.; Gismeros, R. Dynamic Modeling of the Dissipative Contact and Friction Forces of a Passive Biped-Walking Robot. *Appl. Sci.* **2020**, *10*, 2342. [CrossRef]
28. Craig, J.J. *Introduction to Robotics, Mechanics & Control*; Addison-Wesley Publishing Company: Reading, MA, USA, 1986.
29. Gómez, M.J.; Corral, E.; Castejon, C.; García-Prada, J.C. Effective crack detection in railway axles using vibration signals and WPT energy. *Sensors* **2018**, *18*, 1603. [CrossRef] [PubMed]
30. Ewins, D.J. *Modal testing: Theory and Practice*; Research Studies Press: Hertfordshir, UK, 1984.
31. Verbeke, J.; Debruyne, S. Vibration analysis of a UAV multirotor frame. In Proceedings of the ISMA 2016 International Conference on Noise and Vibration Engineering, Leuven, Belgium, 19–21 September 2016.
32. Cossalter, V.; Doria, A.; Basso, R.; Fabris, D. Experimental analysis of out-of-plane structural vibrations of two-wheeled vehicles. *Shock Vib.* **2004**, *11*, 433–443. [CrossRef]
33. Belotti, R.; Caneva, G.; Palomba, I.; Richiedei, D.; Trevisani, A. Model updating in flexible-link multibody systems. *J. Phys. Conf. Ser.* **2016**, *744*, 012073. [CrossRef]
34. Cossalter, V.; Doria, A.; Mitolo, L. Inertial and modal properties of racing motorcycles. *SAE Tech. Pap.* **2002**. [CrossRef]

35. Soriano, E.; Rubio, H.; Castejón, C.; García-Prada, J.C. Design of a low-cost manipulator arm for industrial fields. In *New Trends in Mechanism and Machine Science*; Springer: Cham, Switzerland, 2015; pp. 839–847.
36. Rahman, N.; Carbonari, L.; Caldwell, D.; Cannella, F. Kinematic Analysis, Prototypation and Control of a Novel Gripper for Dexterous Applications. *J. Intell. Robot. Syst.* **2018**, *91*, 193–206. [CrossRef]

Article

A Three-Dimensional Parametric Biomechanical Rider Model for Multibody Applications

Matteo Bova, Matteo Massaro * and Nicola Petrone

Department of Industrial Engineering, University of Padova, Via Venezia 1, 35131 Padova, Italy; matteo.bova@phd.unipd.it (M.B.); nicola.petrone@unipd.it (N.P.)

* Correspondence: matteo.massaro@unipd.it

Received: 29 May 2020; Accepted: 27 June 2020; Published: 29 June 2020

Abstract: Bicycles and motorcycles are characterized by large rider-to-vehicle mass ratios, thus making estimation of the rider's inertia especially relevant. The total inertia can be derived from the body segment inertial properties (BSIP) which, in turn, can be obtained from the prediction/regression formulas available in the literature. Therefore, a parametric multibody three-dimensional rider model is devised, where the four most-used BSIP formulas (herein named Dempster, Reynolds-NASA, Zatsiorsky–DeLeva, and McConville–Young–Dumas, after their authors) are implemented. After an experimental comparison, the effects of the main posture parameters (i.e., torso inclination, knee distance, elbow distance, and rider height) are analyzed in three riding conditions (sport, touring, and scooter). It is found that the elbow distance has a minor effect on the location of the center of mass and moments of inertia, while the effect of the knee distance is on the same order magnitude as changing the BSIP data set. Torso inclination and rider height are the most relevant parameters. Tables with the coefficients necessary to populate the three-dimensional rider model with the four data sets considered are given. Typical inertial parameters of the whole rider are also given, as a reference for those not willing to implement the full multibody model.

Keywords: rider; body segment inertial parameters; biomechanics; human body; motorcycle; bicycle; multibody

1. Introduction

It is well-known that motorcycles and bicycles are characterized by a large rider-to-vehicle mass ratio. On motorcycles, this ratio is usually in the range of 0.25–0.35, the lower figure being related to heavy bikes and the upper to sport bikes; while, on bicycles, the ratio is always greater than one. The inertial properties of the rider are, thus, especially relevant when it comes to dynamic simulations, such as those carried out in a multibody environment for handling and stability analyses of the vehicle–rider system [1–6].

Historically, motorcycle–rider and bicycle–rider systems have been simulated by considering the rider to be rigidly fixed to the chassis, while it has now become standard to consider the rider 'suspended' on the chassis (with springs tuned to give the typical frequencies and damping ratios of riders) with their hands interacting with the handlebar [7–20]. Regardless of the approach (i.e., fixed or suspended rider), estimation of the rider's inertial properties is necessary in both cases. Moreover, in the most advanced motorcycle driving simulators [21–27], the contribution of the rider's motion on the saddle to the vehicle dynamics are accounted for. Most of the time, such motion is monitored by cameras. The related inertial effects are introduced into the vehicle model after the inertial properties of the rider driving the simulator are assumed. Even in this case, estimation of the inertial properties is relevant.

There are different methods to estimate the body segment inertial parameters (BSIP) of a given subject [28–34], including direct measurements on cadavers [35–37] or photogrammetry and medical imaging on living humans [38–48]; however, they are more generally estimated by regression and scaling equations (which are derived from such measurements). The inertial properties of the whole body can, then, be computed after defining the subject's posture. A practical approach is to build a three-dimensional multibody model of the whole human body, which consists of the body segments, properly connected to one another. The next step is defining the formulas to be used to estimate the properties of such segments.

A number of studies have been reported in the literature focused on the calculation of the inertial properties (e.g., mass, center of mass, moments of inertia, or gyration radii) and lengths of body segments of human bodies using regression/prediction formulas, which are given as functions of the mass and height (see, e.g., Reference [43,49–56]). Most of the time, the center of mass is assumed to lie on the longitudinal axis of each body segment, and the longitudinal and transverse axes are assumed principal of inertia.

The regression/prediction formulas of body segments in Reference [49] were derived based on the combination of the data sets in Reference [28,57]: the first considered a sample of American males in the age range of 20–40 years, while the second considered a random selection of adults, young males and females 20–30 years of age, some females in the 40–50 age bracket, and a number of subjects with disabilities in all age ranges. The center of mass was assumed along the longitudinal axis and a single gyration radius was given for each body segment.

Nine old male cadavers (average age 69) were used in Reference [50] to estimate the inertial properties of body segments from the human mass and height; a total of 14 body segments were considered (head, trunk, arms, forearms, hands, thighs, shanks, feet). This model is very popular for the two-dimensional estimation of inertial properties of human bodies. The center of mass was assumed along the longitudinal axis, but no moments of inertia were given (as they were deferred to future studies). This paper updates the results of Reference [35], which are still in use (see, e.g., Reference [33]), combined with those of Reference [58].

The lengths of body segments were obtained by combining the data of Reference [35,59–61] in Reference [51]. The location of the center of mass of each segment was derived from Reference [35–37,62,63], while the inertial properties were derived from Reference [37] (6 male cadavers, average age 54). The number of body segments is 14.

The data reported in Reference [36] were adjusted, in Reference [52], to reference the body segment proportions to joint centers, instead of the originally employed bony landmarks. The original data were taken by measuring thirteen cadavers; the body segments affected by these adjustments were the trunk, upper arms, forearms, thigh, and calves.

In Reference [53], the mean relative center of mass positions and radii of gyration reported by Reference [43] were adjusted, in order to reference them to the joint centers or other commonly used landmarks; 14 body segments were considered. The population consisted of 100 young males (average age 24) and 15 young females (average age 19), whose biomechanical properties were identified in-vivo by gamma-ray scanning techniques. Three gyration radii for each body segment were given and the center of mass was assumed to be aligned with the longitudinal axis of the segment.

In Reference [54,55], similarly to Reference [43,53], the data reported in Reference [39] (31 adult males, mean age 27.5) and [40] (46 women, mean age 31) were adjusted in order to reference them to the joint centers or other commonly used landmarks, according to the recommendations in Reference [64,65]. A relevant characteristic of these latter data sets is that they did not assume that the center of mass lay on the longitudinal axis of the segment, nor that the segment axes were principal of inertia. Therefore, the related regression/prediction formulas did not include such assumptions. Overall, 15 body segments were considered, as the trunk was divided in pelvis and torso. The body model was further updated in Reference [56], where an adjustment procedure to split the torso into thorax and abdomen segments was included.

When it comes to the literature specifically related to motorcycles, the works in Reference [18,19] are considered two classics. In Reference [19], data collected in vivo by the Japan Automobile Research

Institute and data from cadavers of Reference [37] were combined to estimate the distribution of mass in 15 body segments. The resulting positions of the centers of mass and moments of inertia of riders were obtained by means of photographic measurements of rider positions. In Reference [18], an experimental characterization of 35 riders was carried out, with the locations of footrest, saddle, and handgrip corresponding to four motorcycles. The centers of mass and moments of inertia were given; however, the footpegs, saddle, and handgrip positions were not. Finally, in Reference [66], a multibody rider model was briefly described, although no reference to the data set used was given.

Regarding the literature specifically related to bicycles, in Reference [67], a rider model based on the mass data of Reference [35] and 31 grid points (mapping the skeleton of the rider and the bicycle) was presented. The inertial properties of each body segment, aligned along lines connected to appropriate grid points, was computed by assuming its shape (e.g., spheres and cuboids).

A review of the literature shows that at least three works have been alternatively widely employed for the estimation of the inertial properties of a human body in a generic configuration: Reference [50,51,53]. These are hereafter referred to as 'Dempster-Model', 'Reynolds-NASA-Model', and 'Zatsiorsky–DeLeva-Model', respectively, after their authors. In addition, the more recent [54,55] was also included in the multibody models considered, and is referred to as the 'McConville–Young–Dumas' model.

The aim of this work was to compare the four most widespread regression/prediction formulas available in the literature, in order to estimate the BSIP of human bodies in the case of the typical rider postures (sport, touring, and scooter). Unfortunately, such formulas are provided using different reference frames, landmarks, and even slightly different assumptions, thus making a direct comparison based only on their coefficients impossible. Therefore, a three-dimensional parametric multibody model is built, and the four selected regression/prediction formulas are adjusted to reference them to the same set of frames and landmarks. Tables are given to make it easier for other researchers to build the four models discussed in this work. As the target application is the estimation of the inertial properties of motorcycle and bicycle riders, the subject's posture is defined by giving the location of the footpegs, saddle, and handgrips together with the torso inclination, the knee distance, and the elbow distance. However, the model can potentially also be used for car/truck drivers. In this latter scenario, the footpegs become the positions of feet near the gas and brake pedals, the saddle becomes the seat, the handgrips become the hand positions on the steering wheel, and the torso inclination becomes the inclination of the seat. The numerical findings of the numerical model are compared against experimental estimations of rider inertial properties, for validation purposes. Finally, a sensitivity analysis on the most important model parameters is carried out, in order to identify the most relevant factors and compare the trends provided by the four different models considered, as well as to compare the trends of the different data sets.

The work is organized as follows: In Section 2, the multibody model is built, while the model is validated against experiments in Section 3. Section 4 discusses the sensitivity analysis on the main model parameters, and Section 5 provides a set of baseline rider parameters for multibody simulations.

2. Multibody Model

The multibody model of the whole human body is built as an assembly of body segments, using the BSIP computed from the rider mass and height through four different regression/prediction formulas. In the following sections, the key aspects of the model are given, namely the frames and bodies used, the constraints and degrees of freedom considered, the data sets considered, and the implementation details.

2.1. Frames and Bodies

Whole-body inertial properties are usually given with respect to a reference frame centerd on the pelvis. Three planes are defined through such point for a standing body: the sagittal plane (body symmetry plane), the transverse plane (parallel to the floor), and the frontal or coronal plane (normal to the other two planes); see Figure 1a. The x-axis (sagittal axis) is the intersection between the sagittal and transverse plane and points forward, the z-axis (longitudinal axis) is the intersection between the sagittal and frontal plane and points upward, and the y-axis (transverse axis) is defined such that $x \times y = z$.

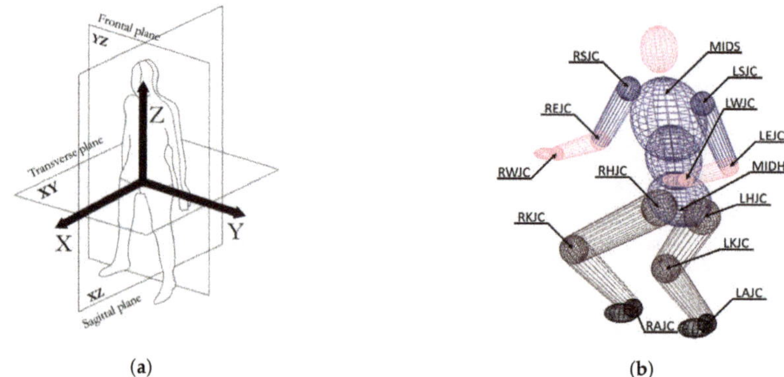

(a) (b)

Figure 1. Rider multibody model: reference frame for the whole body [51], with the sagittal (x), transverse (y), and longitudinal (z) axes and planes (**a**); and main points of the multibody model (**b**).

The multibody model of a human body consists of 14 body segments, namely trunk, forearms (2×), arms (2×), thighs (2×), shanks (2×), hands (2×), feet (2×), and head (including neck). Some authors have employed 16 segments by dividing the trunk into upper and lower trunk and not including the neck as part of the head. The location and orientation of each segment is defined using a reference frame centerd on the proximal point of the segment, as reported in Table 1. The reference frame of each body segment is parallel to that centerd in the pelvis, as shown in Figure 1a, when the human body is in the standard anatomical position (i.e., standing with the arms and legs parallel to the trunk and the palms pointing forward).

Table 1. Proximal and distal points for each body segment. * indicates the origin of the reference frames attached to each body segment, while / indicates that the distal point is not defined, as it is not necessary for the model construction.

Segment	Proximal Point	Distal Point
Head	MIDS *	/
Trunk	MIDH *	MIDS
Arm	RSJC/LSJC *	REJC/LEJC
Forearm	REJC/LEJC *	RWJC/LWJC
Hand	RWJC/LWJC *	/
Thigh	RHJC/LHJC *	RKJC/LKJC
Shank	RKJC/LKJC *	RAJC/LAJC
Foot	RAJC/LAJC *	/

When the hands are placed on the handgrips, the x-axis (palm direction) points downward (vertically), the z-axis (fingers direction) points backward (horizontally), and the y-axis completes the frame. When the feet are on the footpegs, the x-axis points forward (horizontally), the z-axis upward (vertically), and the y-axis completes the frame.

2.2. Constraints

The multibody model consists of 84 unconstrained degrees of freedom (DOF), as there are 14 three-dimensional rigid bodies. The following points are introduced to define the configuration of the human body on the bike: the saddle point, the handgrip points (2×), and the footpeg points (2×).

The hands are fixed to the handlebar in the handgrip points, defined on the hands, and the feet are fixed to the chassis in the footpeg points, defined on the feet and, so, 24 DOF are constrained. The head center is fixed and aligned with the trunk's longitudinal axis, at a given distance and, so, 6 DOF are constrained. All these fixed constraints are enforced by the following standard equations [68–72]:

$$(\boldsymbol{P}_i - \boldsymbol{P}_j) \cdot \boldsymbol{x}_i = 0, \quad (1)$$

$$(\boldsymbol{P}_i - \boldsymbol{P}_j) \cdot \boldsymbol{y}_i = 0, \quad (2)$$

$$(\boldsymbol{P}_i - \boldsymbol{P}_j) \cdot \boldsymbol{z}_i = 0, \quad (3)$$

$$\boldsymbol{x}_i \cdot \boldsymbol{y}_j = 0, \quad (4)$$

$$\boldsymbol{z}_i \cdot \boldsymbol{x}_j = 0, \quad (5)$$

$$\boldsymbol{z}_i \cdot \boldsymbol{y}_j = 0, \quad (6)$$

where $\boldsymbol{P}_i$ is the connection point on body i to which a frame with axes $\boldsymbol{x}_i, \boldsymbol{y}_i, \boldsymbol{z}_i$ is attached, while $\boldsymbol{P}_j$ is the connection point on body j to which a frame with axes $\boldsymbol{x}_j, \boldsymbol{y}_j, \boldsymbol{z}_j$ is attached.

Spherical joints constrain the trunk to the arms on RSJC/LSJC and the forearms to the hands on RWJC/LWJC, constraining 12 DOF; see Figure 1b. Similarly, spherical joints constrain the thighs to the trunk on RHJC/LHJC and the shank to the feet on RAJC/LAJC, also constraining 12 DOF; see Figure 1b. The trunk is attached to the saddle point through a spherical joint, constraining 3 DOF. The equations related to all these 9 spherical joints are given as (1)–(3).

The arms are connected to the forearms through revolute joints centerd in REJC/LEJC, with the rotation axis perpendicular to the plane defined by SJC, EJC, and WJC, constraining 10 DOF. Similarly, the thighs are connected to the shanks through revolute joints centerd in RKJC/LKJC with the rotation axis perpendicular to the plane defined by HJC, KJC, and AJC, constraining 10 DOF. The equations related to these four revolute joints are given as (1)–(3), (5), and (6).

The remaining 7 DOF are the inputs for the numerical analyses (together with human height and mass) and can be related to the rotation of the trunk along the horizontal (roll), vertical (yaw), and transverse (pitch) axes through the saddle point; the distance between elbows and the distance between knees; and the distance between the right elbow and the sagittal plane and the distance between the right knee and the sagittal plane. In more detail, the trunk axes are constrained to have a fixed orientation with respect to the saddle point (and related frame), as follows:

$$\boldsymbol{x}_i \cdot \boldsymbol{z}_j = \cos\left(\frac{\pi}{2} + \alpha\right), \quad (7)$$

$$\boldsymbol{x}_i \cdot \boldsymbol{y}_j = \cos\left(\frac{\pi}{2} - \beta\right), \quad (8)$$

$$\boldsymbol{y}_i \cdot \boldsymbol{z}_j = \cos\left(\frac{\pi}{2} + \gamma\right), \quad (9)$$

where the indices i and j refer to the trunk and saddle frames, respectively, while α, β, and γ are the rider's longitudinal lean (or pitch) angle, yaw angle, and roll angle, respectively, with respect to the saddle. The distance between elbows and knees and their position with respect to the sagittal plane are defined by the following equations:

$$(\boldsymbol{P}_{el} - \boldsymbol{P}_{er}) \cdot (\boldsymbol{P}_{el} - \boldsymbol{P}_{er}) = d_e^2, \quad (10)$$

$$(\boldsymbol{P}_{kr} - \boldsymbol{P}_{MIDH}) \cdot \boldsymbol{y}_{MIDH} = d_{kr}, \quad (11)$$

$$(\boldsymbol{P}_{kl} - \boldsymbol{P}_{kr}) \cdot (\boldsymbol{P}_{kl} - \boldsymbol{P}_{kr}) = d_k^2, \quad (12)$$

$$(\boldsymbol{P}_{er} - \boldsymbol{P}_{MIDH}) \cdot \boldsymbol{y}_{MIDH} = d_{er}, \quad (13)$$

where $\boldsymbol{P}_{el}$ and $\boldsymbol{P}_{er}$ are the left and right elbow points, d_e is the distance between elbows, and d_{er} is the distance between the right elbow and the sagittal plane, while $\boldsymbol{P}_{kl}$ and $\boldsymbol{P}_{kr}$ are the left and right knee points, d_k is the distance between knees, and d_{kr} is the distance between the right knee and the sagittal plane. At this stage, the model is kinematically determinate.

2.3. Data Sets

The four most popular data sets reported in the literature were employed to compute the length of body segments, the location of the center of mass, and the moments of inertia. Details with the specific tables employed are reported in Appendix A, in order to make it easier to practically implement the model independently. Indeed, such data are spread throughout multiple references, which partially explains the confusion sometimes related to the definition of the exact sources employed for the computation of the inertial properties.

2.4. Implementation

The configuration of the multibody model is obtained by solution of the 84 non-linear algebraic equations associated to the constraints described in Section 2.2. The problem can either be solved using a root-finding solver or by implementation on one of the several available commercial multibody software packages. The authors decided to implement the model in the well-known Adams through a macro. Most constraints are standard and, thus, already available (e.g., fixed, spherical, and revolute joints); however, the related equations are given in Section 2.2, while others were written by hand using the function GCON (generic constraints). Again, such constraints are given in Section 2.2. The configuration was obtained without much trouble, with a solution error of 10^{-10} in a number of steps which did not exceeds 50.

3. Validation

In this section, the numerical results of the multibody model described in Section 2 are compared against the experimental findings. The center of mass of nine riders (age in the range 24–27, mean body mass index (BMI) 23.87 kg/m^2 with standard deviation 2.82 kg/m^2, mean height 1.76 m with standard deviation 0.05 m, mean mass 73.93 kg with standard deviation 6.87 kg, mean elbow distance 0.53 m with standard deviation 0.05 m, and knee distance 0.48 m, equal to the fuel tank width, for all riders) was measured in riding position using the testing rig developed at the University of Padova shown in Figure 2.

3.1. Test Procedure

The machine consisted of a rigid frame, four load cells (two fixed to the frame and two adjustable to fit the vehicle wheelbase), an inclinometer, and an hydraulic lifting system. The motorcycle was fixed to the load cells (two for each wheel) at the wheel axles. Both front and rear suspensions were locked, in order to avoid dynamic effects during measurements.

For the estimation of the center of mass components (horizontal x_G and vertical z_G, with respect to the rear wheel axle), the vehicle was pitched at different angles by lifting the rigid frame. For each position, a static measurement of the loads normal to the rigid frame (e.g., vertical when the pitch was zero) was taken and the results were interpolated as follows:

$$N_f(w - x_G) - N_r x_G + (N_f + N_r)\tan(\mu) z_G = 0, \tag{14}$$

where w is the distance between the wheel axles, N_f and N_r are the front and rear loads (measured with the load cells), and μ is the vehicle pitch angle (measured with an inclinometer). The center of mass of the rider was obtained by subtraction of the measurement of the bike from the measurement of the bike and rider. Foam cushions between the trunk and the tank (visible in Figure 2) were employed to prevent involuntary rider motions when pitching the bike during the test, as well as to guarantee a comfortable riding position while standing in a static posture for a long time (the test duration was about 15 min). A sporty posture (mean lean angle 62 degrees with standard deviation 4.5 degrees) was chosen, allowing the rider to get a more sturdy leaning point on the bike, thus reducing vibration and, ultimately, yielding better measurements. For the same reasons, the maximum pitch angle during tests was limited to 16 degrees. The procedure consisted of 17 measures for each rider, from 0 degrees to 16 degrees and return, with steps of 2 degrees.

The elbow distance was measured during the test, in order to make sure that the rider kept a fixed posture with respect to the bike. Trunk lean angle and trochanter center position with respect to the saddle point were derived from photographic data. The photos taken (one at the beginning of each test) were elaborated to correct distortions and aberration phenomena. This was done by means of a reference photo depicting a chessboard, through which a scaling matrix (pixel to mm) was built using a standard open source software for image processing. As a further check, for each rider's photo, three measures are verified: front and rear wheel radii and wheels axle distance. Finally, it was assumed that the rider's sagittal plane was on the bike's vertical symmetry plane. The model MIDH is the mean point between the right and left throcanter anatomical landmarks; see point S in Figure 2. Therefore, different riders had slightly different postures, although sitting on the same bike.

Figure 2. A rider on the test rig, with the main kinematic points used in the multibody model indicated.

3.2. Data Analysis

The comparison between the numerical model results and the experimental data for the four data sets considered is reported in Figure 3. The test subjects were selected among the students of the Department of Industrial Engineering of University of Padova, who had experience in cycling and possibly in motorcycle riding. Estimation of the measurement errors was done by repeating the entire procedure with the same rider eight times and evaluating the standard deviations of the measured center of mass coordinates. The measurement error calculated for both the x and z components was equal to ± 2.7 mm with a degree of confidence of 99.7%. These errors were assumed to be independent of the rider measured.

The RMS of the difference (numerical vs. experimental) in the estimation of the center of mass x co-ordinate for the four data sets were: 4.72% for Reynolds-NASA, 2.78% for Dempster, 2.74% for Zatsiorsky–DeLeva, and 2.72% for McConville–Young–Dumas (with respect to the nominal experimental values). The RMS of the difference (numerical vs. experimental) in the estimation of the center of mass z co-ordinate for the four data sets were: 2.64% for Reynolds-NASA, 1.75% for Dempster, 1.66% for McConville–Young–Dumas, and 1.56% for Zatsiorsky–DeLeva (with respect to the nominal experimental values).

The simulated x co-ordinate of G was almost identical in all data sets, with the exception of Reynolds-NASA, which gave larger estimates (i.e., G shifted towards the front wheel). More than half of the experiments lay in between Reynolds-NASA and the other models, while the remaining experimental estimates were smaller (i.e., shifted towards the rear wheel) than all of the numerical estimates. The predictions of the four models were clearly distinguished from one another in the z co-ordinate. The tallest estimation was with Reynolds-NASA and the shortest with McConville–Young–Dumas, while the other two models lay in between. Two-thirds of the experiments lay in between the Reynolds-NASA and McConville–Young–Dumas models, while the remaining experiments were shorter than all the numerical estimates. The maximum difference between the simulations and experiments was 0.05 m (both horizontally and vertically), obtained with the Reynolds-NASA model (test 9).

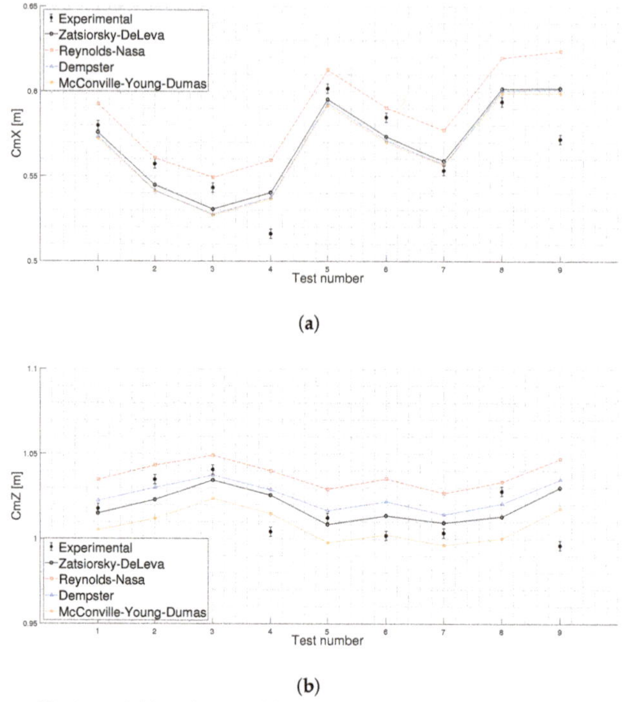

(**a**)

(**b**)

Figure 3. Horizontal (**a**) and vertical (**b**) position of *G*: numerical vs. experimental.

4. Sensitivity Analysis

The multibody model was employed to investigate the effect of the main model parameters—namely knee distance, elbow distance, lean angle, and rider height—on the rider's inertial properties. The rider's yaw and roll (β and γ in (8) and (9)) were assumed to be zero, while the elbows and knees were assumed to be symmetric with respect to the sagittal plane ($d_{er} = d_e/2$ and $d_{kr} = d_k/2$). The analysis was performed over three different types of motorcycle: sport (tucked position), touring, and a scooter (upright position). A 74 kg rider was assumed, whose posture was defined with the parameters in Table 2. Four data sets were considered.

Sensitivity with respect to the rider mass was not considered, ase the center of mass does not change with it, while the gyration radii are linearly proportional to it. Indeed, the position of the whole center of mass **G** can be computed from the positions of the center of mass of its *i* components $\boldsymbol{P}_i$:

$$\boldsymbol{G} = \frac{\sum m_i \boldsymbol{P}_i}{\sum m_i} = \frac{\sum \alpha_i m \boldsymbol{P}_i}{m} = \sum \alpha_i \mathcal{T}_i \begin{bmatrix} \beta_i^x l_i \\ \beta_i^y l_i \\ \beta_i^z l_i \\ 1 \end{bmatrix} = \sum \alpha_i \mathcal{T}_i \begin{bmatrix} \beta_i^x \gamma_i h \\ \beta_i^y \gamma_i h \\ \beta_i^z \gamma_i h \\ 1 \end{bmatrix}. \quad (15)$$

The mass m_i of each body segment is proportional to the whole-body mass m through the coefficient α_i (Tables A1–A3 and A5), while the position of the center of mass of each body segment is proportional to the body segment length l_i through the coefficients β_i^z, β_i^x, and β_i^y (Tables A1–A3 and A5), when expressed in the local reference frame. The projection onto the global reference frame is obtained through a 4 × 4 roto-translation matrix $\mathcal{T}_i$, which has the 3 × 3 rotation matrix $\mathcal{R}_i$ of the local reference frame in its upper-left corner (whose columns consist of the unit vectors of the local reference frame, expressed in the global reference frame) and the location of the origin of the reference frame $\mathcal{O}_i$ in its fourth column (again expressed in the global frame); the (4,4) value is always one [1,70]. Finally, the length of the body segment l_i is proportional to the body height h through the coefficient γ_i (Tables A1–A3 and A5). Equation (15) indicates that the position of the center of mass of the whole body only depends on the height (and, of course, on the data set selected for the coefficients α_i, β_i, and γ_i). Similarly, the whole-body inertia matrix, J_G, can be obtained from the parallel axis theorem as:

$$
\begin{aligned}
J_G &= \sum \left(\mathcal{R}_i \begin{bmatrix} m_i\rho_x^2 & m_i\rho_{xy}^2 & m_i\rho_{xz}^2 \\ m_i\rho_{xy}^2 & m_i\rho_y^2 & m_i\rho_{yz}^2 \\ m_i\rho_{xz}^2 & m_i\rho_{yz}^2 & m_i\rho_z^2 \end{bmatrix} \mathcal{R}_i^T + m_i(\boldsymbol{r}_i^T\boldsymbol{r}_i\boldsymbol{I} - \boldsymbol{r}_i\boldsymbol{r}_i^T) \right) \\
&= m\sum \left(\mathcal{R}_i \begin{bmatrix} \alpha_i\eta_x^2 l_i^2 & \alpha_i\eta_{xy}^2 l_i^2 & \alpha_i\eta_{xz}^2 l_i^2 \\ \alpha_i\eta_{xy}^2 l_i^2 & \alpha_i\eta_y^2 l_i^2 & \alpha_i\eta_{xz}^2 l_i^2 \\ \alpha_i\eta_{xz}^2 l_i^2 & \alpha_i\eta_{yz}^2 l_i^2 & \alpha_i\eta_z^2 l_i^2 \end{bmatrix} \mathcal{R}_i^T + \alpha_i(\boldsymbol{r}_i^T\boldsymbol{r}_i\boldsymbol{I} - \boldsymbol{r}_i\boldsymbol{r}_i^T) \right) \\
&= m\sum \left(h^2\alpha_i\mathcal{R}_i \begin{bmatrix} \eta_x^2\gamma_i^2 & \eta_{xy}^2\gamma_i^2 & \eta_{xz}^2\gamma_i^2 \\ \eta_{xy}^2\gamma_i^2 & \eta_y^2\gamma_i^2 & \eta_{yz}^2\gamma_i^2 \\ \eta_{xz}^2\gamma_i^2 & \eta_{yz}^2\gamma_i^2 & \eta_z^2\gamma_i^2 \end{bmatrix} \mathcal{R}_i^T + \alpha_i(\boldsymbol{r}_i^T\boldsymbol{r}_i\boldsymbol{I} - \boldsymbol{r}_i\boldsymbol{r}_i^T) \right),
\end{aligned} \tag{16}
$$

where $\rho_x, \rho_y, \rho_z, \rho_{xy}, \rho_{xz}$, and ρ_{yz} are the gyration radii in the segment reference frame, which are proportional to the segment length l_i through the coefficients $\eta_x, \eta_y, \eta_z, \eta_{xy}, \eta_{xz}$, and η_{yz} (Tables A1 and A2, A4–A6); $\boldsymbol{I}$ is the 3 × 3 identity matrix; and $\boldsymbol{r}_i$ is the distance between the center of mass of the body segment and the whole center of mass

$$\boldsymbol{r}_i = \boldsymbol{P}_i - \boldsymbol{G}, \tag{17}$$

which has been shown to be dependent on h only; see (15). Therefore, given the data set of regression/prediction coefficients, the moments of inertia of the whole body are proportional to m (given the posture and height) and change with h (indeed, the posture changes with height).

Table 2. Baseline parameters for the rider on the three bikes. The coordinates are given with respect to the saddle point.

Parameter	Sport	Touring	Scooter
mass	74 kg	74 kg	74 kg
height	1.76 m	1.76 m	1.76 m
lean	62 deg	20 deg	0 deg
d_k	0.48 m	0.48 m	0.48 m
d_e	0.53 m	0.69 m	0.58 m
MIDH	x: 0 m y: 0 m z: 0.075 m	x: 0 m y: 0 m z: 0.075 m	x: 0 m y: 0 m z: 0.075 m
footpegs	x: 0.085 m y: ±0.205 m z: −0.450 m	x: 0.085 m y: ±0.210 m z: −0.535 m	x: 0.430 m y: ±0.210 m z: −0.470 m
handgrips	x: 0.745 m y: ±0.230 m z: 0.030 m	x: 0.540 m y: ±0.305 m z: 0.300 m	x: 0.460 m y: ±0.260 m z: 0.285 m

4.1. Effect of Knees Distance

The knee distance was varied throughout the range 0.20–0.70 m in 0.01 m increments, keeping all other parameters constant. Figure 4 shows the variation of G for the four data sets for each rider posture. When the knee distance increased, G moved rearward in all bikes, while the vertical displacement depended on the bike. The mean variations (end-to-end) in the x direction were: −0.96 cm for sport posture, −1.08 cm for touring posture, and −1.31 cm for scooter posture. The mean variations in the z direction were: 0.70 cm for sport posture, 0.53 cm for touring posture, and −0.33 cm for scooter posture. Overall, the variation in the location of G, as related to the variation of the knee distance, was of the same order of magnitude of the variation in the location of G related to changes between McConville–Young–Dumas and Reynolds-NASA data sets (up to 4 cm).

Figure 5 shows the variation of the gyration radii for the four data sets for each rider posture. An increase of the knee distance corresponded to an increase of ρ_x (roll) and ρ_z (yaw)—as intuition suggests—while ρ_y had a slightly decreasing trend. The mean variations (end-to-end) in ρ_x were: 2.16 cm for sport posture, 1.46 cm for touring posture, and 2.36 cm for scooter posture. The mean variations in ρ_y were: −0.32 cm for sport posture, −0.73 cm for touring posture, and −0.40 cm for scooter posture. The mean variations in ρ_z were: 3.30 cm for sport posture, 3.44 cm for touring posture, and 2.06 cm for scooter posture. The variations of the $\boldsymbol{\rho}$ components between models (up to 6 cm) were, again, on the same order of magnitude of the variations related to the knee distance. The model at the upper end depended on the component and posture selected, while the lower-end model was always the Dempster; probably due to the lack of local inertia tensors for body segments. As expected, on the sport bike, ρ_x (roll) and ρ_y (pitch) were minimum, due to the tucked position. The touring position had the maximum ρ_x, due to the erect position (similarly to scooter), combined with the larger distance between saddle and footpegs.

4.2. Effect of Elbow Distance

The elbow distance was varied, in 0.01 m steps, throughout the range 0.30–0.70 m, keeping all other parameters constant. Figure 6 shows the variation of G for the four data sets for each rider posture. With an increase in elbow distance, G moved slightly upward, while the horizontal displacement depended on the posture (moved slightly rearward in touring and slightly forward in sport/scooter). The mean variations (end-to-end) in the x direction were: 0.08 cm for sport posture, −0.21 cm for touring posture, and 0.04 cm for scooter posture. The mean variations in the z direction were: 0.18 cm for sport posture, 0.15 cm for touring posture, and 0.25 cm for scooter posture. The variations in the location of G related to the elbow distance were an order of magnitude smaller than the variations related to changes in the data set.

Figure 7 shows the variation of the gyration radii for the four data sets for each rider posture. An increase of the elbow distance value corresponded to a slight increase of ρ_x and ρ_z, while ρ_y was almost independent from this parameter. The mean variations in ρ_x were: 0.82 cm for sport posture, 0.69 cm for touring posture, and 0.75 cm for scooter posture. The mean variations in ρ_y were: 0.06 cm for sport posture, 0.06 cm for touring posture, and 0.12 cm for scooter posture. The mean variations in ρ_z were: 0.96 cm for sport posture, 1.08 cm for touring posture, and 0.96 cm for scooter posture.

4.3. Effect of (Longitudinal) Lean Angle

The lean angle was varied, in 1 degree steps, throughout the range 45–85 degrees for sport posture, 5–45 degrees for touring posture, and from −5 (slightly leaned rearward) to +35 degrees for scooter posture, keeping all other parameters constant. Figure 8 shows the variation of G for the four data sets for each rider posture. With an increase of the lean angle, G moved significantly forward (except for the sport posture, due to the high lean angle values) and significantly downward (except for the scooter posture, due to the low lean angle values). The mean variations (end-to-end) in the x direction were: 6.09 cm for sport posture, 12.75 cm for touring posture, and 13.58 cm for scooter posture.

The mean variations in the z direction were: −14.19 cm for sport posture, −6.79 cm for touring posture, and −4.10 cm for scooter posture. The variations of $\boldsymbol{G}$ between models (up to 4 cm) were almost a third of the variations related to the lean angle.

Figure 9 shows the variation of the gyration radii for the four data sets for each rider posture. An increase in the lean angle value corresponded to a decrease of ρ_x (roll) and ρ_y (pitch)—as intuition suggests—while ρ_z exhibited a decreasing behavior for small lean-angle values (<20 degrees) and then increased for high lean-angle values (>40 degrees). The mean variations (end-to-end) in ρ_x were: −8.52 cm for sport posture, −4.95 cm for touring posture, and −2.98 cm for scooter posture. The mean variations in ρ_y were: −4.74 cm for sport posture, −4.44 cm for touring posture, and −5.05 cm for scooter posture. The mean variations in ρ_z were: 3.55 cm for sport posture, 1.18 cm for touring posture, and −3.07 cm for scooter posture.

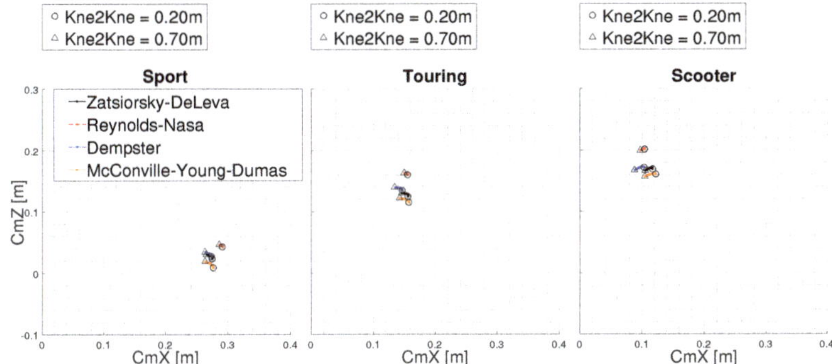

Figure 4. Position of the center of mass as a function of knee distance.

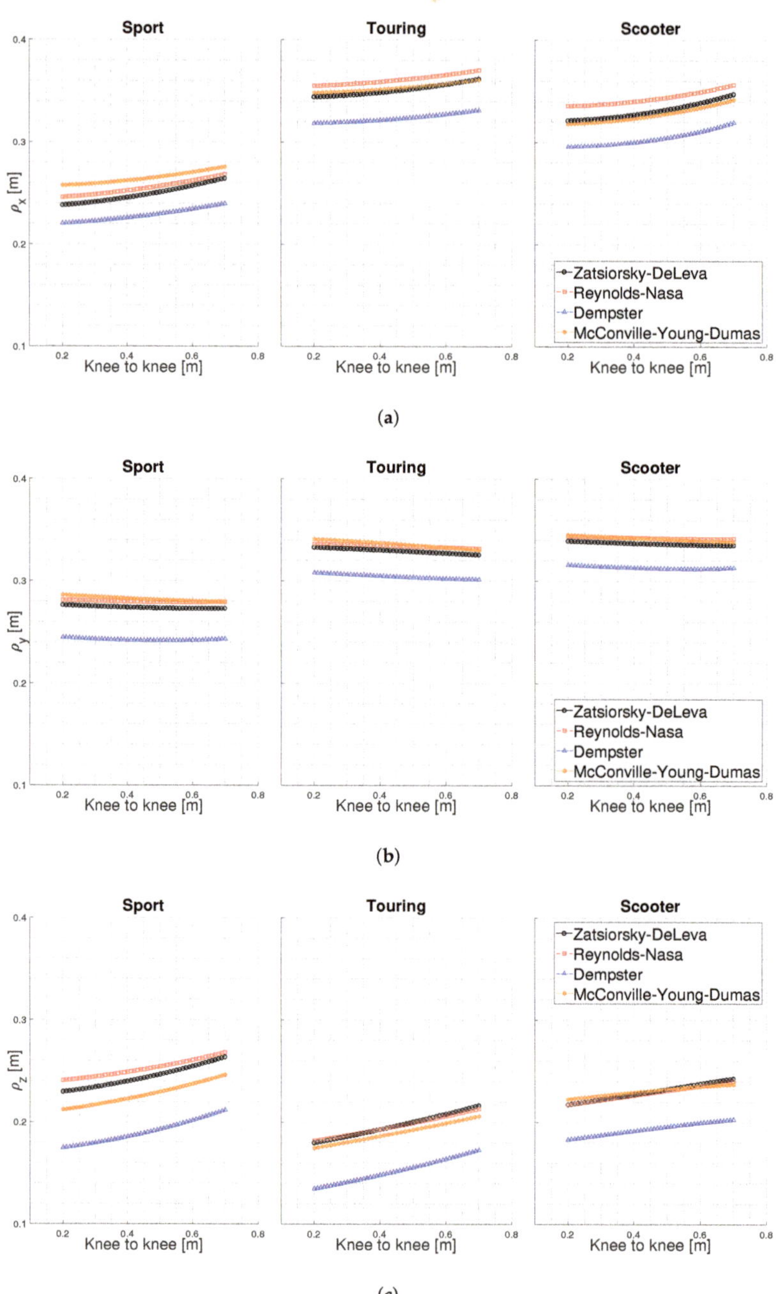

Figure 5. Roll ρ_x (**a**), pitch ρ_y (**b**), and yaw ρ_z (**c**) gyration radii as a function of knee distance.

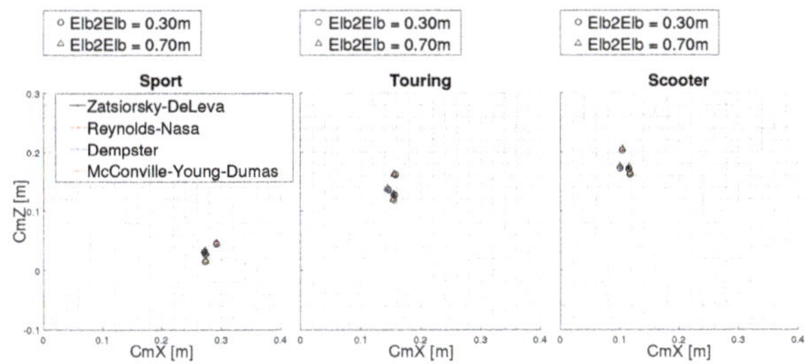

Figure 6. Position of the center of mass as a function of elbow distance.

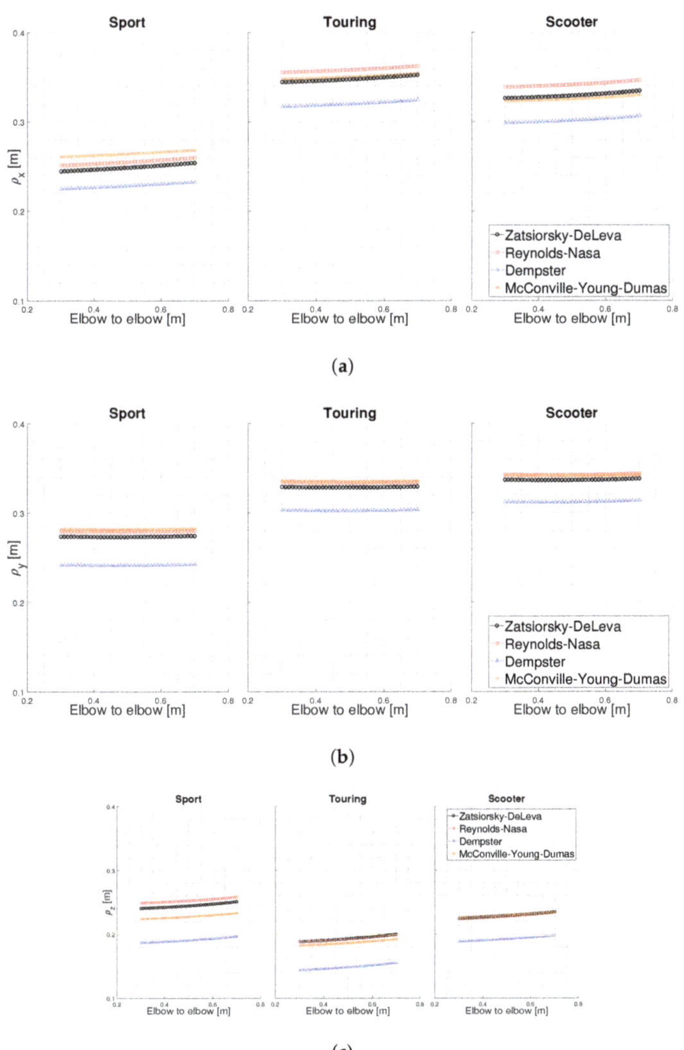

Figure 7. Roll ρ_x (**a**), pitch ρ_y (**b**), and yaw ρ_z (**c**) gyration radii as a function of elbow distance.

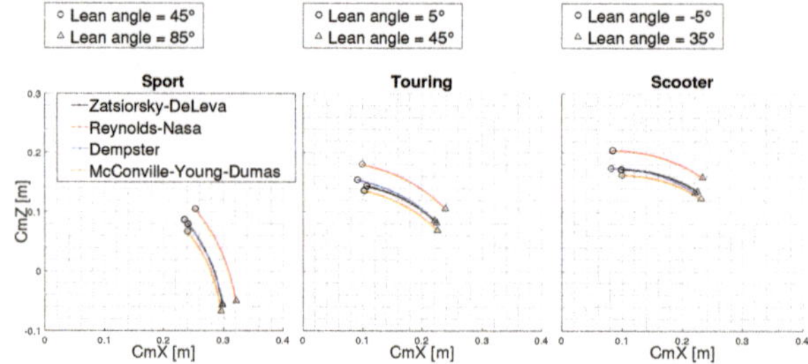

Figure 8. Position of the center of mass as a function of lean angle.

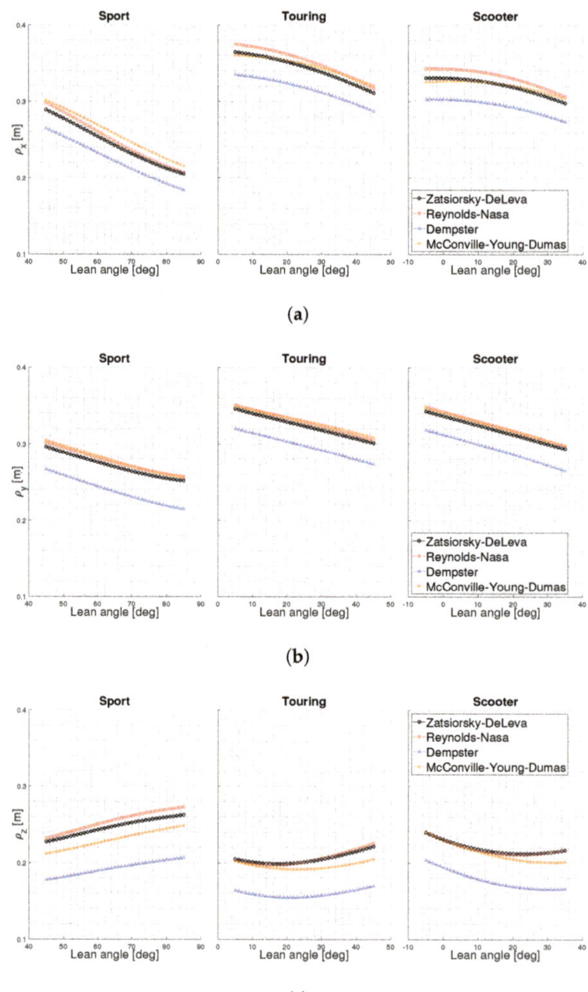

Figure 9. Roll ρ_x (**a**), pitch ρ_y (**b**), and yaw ρ_z (**c**) gyration radii as a function of lean angle.

4.4. Effect of Rider Height

The rider height was varied throughout the range 1.55–1.95 m in 0.01 m steps, keeping all other parameters constant. Figure 10 shows the variation of **G** for the four data sets for each rider posture. When increasing the rider height, **G** moved forward and upward, with a slope that depended on the rider posture (maximum for scooter and minimum with sport). The mean variations (end-to-end) in the x direction were: 5.88 cm for sport posture, 3.66 cm for touring posture, and 1.88 cm for scooter posture. The mean variations in the z direction were: 2.34 cm for sport posture, 4.67 cm for touring posture, and 6.40 cm for scooter posture. The variations of **G** between models (up to 4 cm) were comparable with the variations related to the rider height.

Figure 11 shows the variation of the gyration radii for the four data sets for each rider posture. An increase of rider height value corresponded to an increase of all ρ components. The mean variations (end-to-end) in ρ_x were: 2.27 cm for sport posture, 3.88 cm for touring posture, and 2.83 cm for scooter posture. The mean variations in ρ_y were: 2.85 cm for sport posture, 4.06 cm for touring posture, and 3.59 cm for scooter posture. The mean variations in ρ_z were: 2.29 cm for sport posture, 1.69 cm for touring posture, and 2.79 cm for scooter posture.

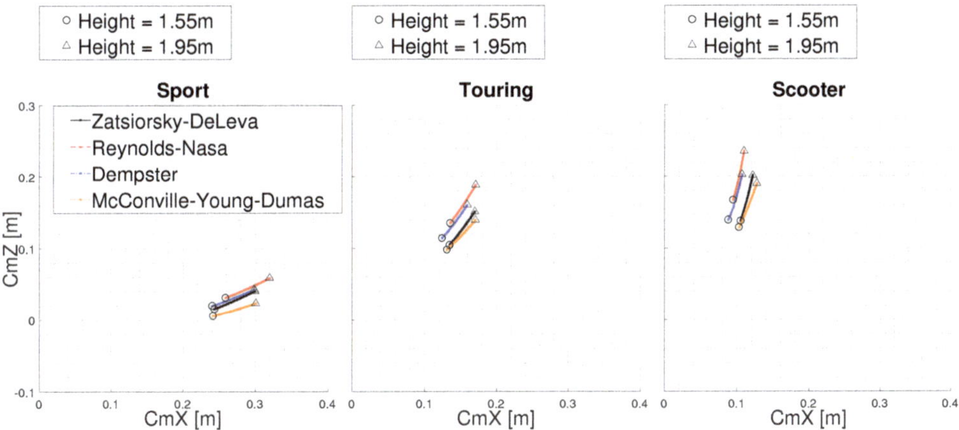

Figure 10. Position of the center of mass as a function of rider height.

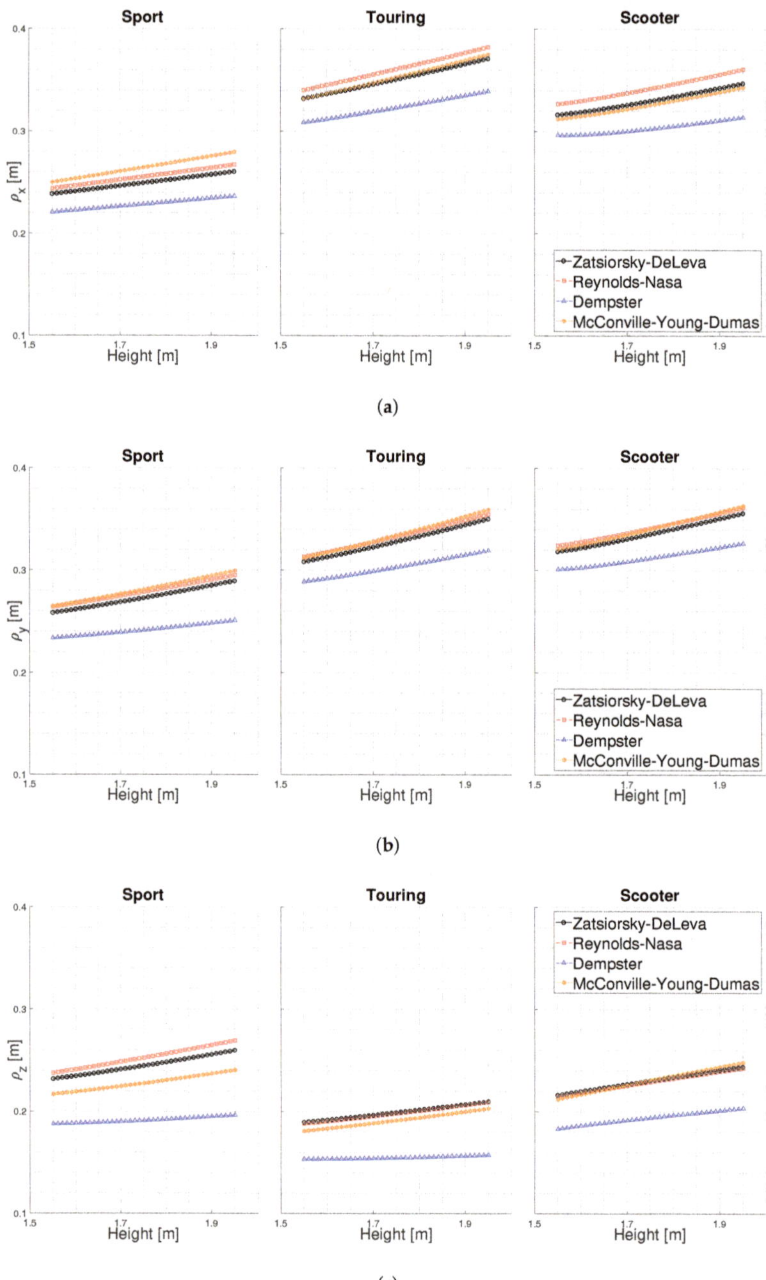

Figure 11. Roll ρ_x (**a**), pitch ρ_y (**b**), and yaw ρ_z (**c**) gyration radii as a function of rider height.

5. Baseline Parameters

Typical inertial parameters for sport, touring, and scooter postures are given in Table 3 for all four data sets considered, as a reference for multibody simulations. The posture parameters, the mass and height used, as well as the location of the saddle points, footpeg points, and handgrip points, are given in Table 2. For the four data sets considered, **G** had the highest position in the case of the scooter and the lowest in the case of the sport posture; when it came to the horizontal position, the sequence of locations was scooter–touring–sport, when moving from the rear- to the front-end. The sport posture gave the minimum values of both roll and pitch gyration radii for all data sets, while the maximum values are obtained with the touring (maximum roll) and scooter (maximum pitch) postures, respectively. The touring posture also gave the minimum values for the pitch gyration radii.

In Reference [3], the typical sport rider gyration radii are given. The roll ρ_x and pitch ρ_y radii were in the range 0.23–0.28 m, which the values computed in this work fell within; see the sport column of Table 3. The yaw radius ρ_z in Reference [3] was in the range 0.15–0.19 m, a bit smaller than the values computed in this work (all larger than 0.19 m); see again the sport column of Table 3.

The centers of mass, together with the roll and yaw moments of inertia, of 35 riders were measured in Reference [18], both in upright and leaned posture. The locations of footpegs, handgrips, and saddles were not given. For the upright posture, the mean **G** coordinates were $G_x = 0.09$ m and $G_z = 0.26$ m (with respect to the saddle), while the mean gyration radii were $\rho_x = 0.338$ m and $\rho_z = 0.201$ m. These values were close to those reported in the scooter column of Table 3. For the leaned posture, the mean **G** coordinates were $G_x = 0.23$ m and $G_z = 0.20$ m, while the computed mean gyration radii were $\rho_x = 0.277$ m and $\rho_z = 0.205$ m. These values were between those reported in the sport and touring columns of Table 3.

The rider posture of a typical bicycle rider was analyzed in Reference [67]. Comparison of the locations of the saddle, footpegs, and handgrips suggests that the posture was close to that of the touring vehicle considered in this work. Indeed, in Reference [67], the computed radii were $\rho_x = 0.333$ m, $\rho_y = 0.335$ m, and $\rho_z = 0.181$ m, which were close to the values reported in the touring column of Table 3.

Table 3. Typical rider parameters for sport, touring, and scooter postures. **G** coordinates are given with respect to the saddle reference frame.

	Sport	**Touring**	**Scooter**
Zatsiorsky—De Leva	G_x: 0.273 m G_z: 0.103 m ρ_x: 0.250 m ρ_y: 0.273 m ρ_z: 0.245 m	G_x: 0.155 m G_z: 0.203 m ρ_x: 0.352 m ρ_y: 0.329 m ρ_z: 0.199 m	G_x: 0.115 m G_z: 0.246 m ρ_x: 0.331 m ρ_y: 0.337 m ρ_z: 0.231 m
Reynolds—NASA	G_x: 0.291 m G_z: 0.120 m ρ_x: 0.256 m ρ_y: 0.279 m ρ_z: 0.253 m	G_x: 0.156 m G_z: 0.237 m ρ_x: 0.362 m ρ_y: 0.334 m ρ_z: 0.199 m	G_x: 0.104 m G_z: 0.278 m ρ_x: 0.343 m ρ_y: 0.343 m ρ_z: 0.230 m
Dempster	G_x: 0.271 m G_z: 0.107 m ρ_x: 0.229 m ρ_y: 0.242 m ρ_z: 0.192 m	G_x: 0.144 m G_z: 0.213 m ρ_x: 0.324 m ρ_y: 0.304 m ρ_z: 0.155 m	G_x: 0.100 m G_z: 0.248 m ρ_x: 0.303 m ρ_y: 0.313 m ρ_z: 0.195 m
McConville—Young—Dumas	G_x: 0.271 m G_z: 0.089 m ρ_x: 0.266 m ρ_y: 0.281 m ρ_z: 0.226 m	G_x: 0.151 m G_z: 0.192 m ρ_x: 0.353 m ρ_y: 0.335 m ρ_z: 0.191 m	G_x: 0.115 m G_z: 0.234 m ρ_x: 0.325 m ρ_y: 0.341 m ρ_z: 0.233 m

6. Discussion

The investigation carried out showed that there are four main BSIP formulas available and that such formulas need to be adjusted to a common set of frames and assumptions before being employed. After such an adjustment (see Appendix A), the BSIP coefficients are not far from one another and provide similar trends (although with slightly different figures) when varying the main posture parameters of motorcycle and bicycle riders. The numerical estimates provided in this work for baseline riders were consistent with those in similar studies reported in the literature. The minor effect related to elbow distance and the small effect related to knee distance suggest that the inertial estimation of riders based only on the height and lean is sufficient, in most cases. When it comes to the BSIP models selected, the 'McConville–Young–Dumas' model offers clearer implementation, together with the 'Dempster' model. The 'Reynolds-NASA' model requires the inspection of different sources, where there are different conventions even within the main source. The 'Zatsiorsky–DeLeva' model has some missing points and requires some assumptions during implementation. Finally, the 'McConville–Young–Dumas' model was the only model featuring a center of mass away from the longitudinal axis of the body segment and cross moments of inertia. However, such effects are secondary when compared with the transportation moments related to the parallel axis theorem.

7. Conclusions

The main BSIP prediction/regression formulas available in the literature were reviewed, with special emphasis on implementations for two-wheelers. A parametric rider model suitable for multibody applications has been developed. The BSIP prediction/regression formulas of the four most-used models were employed to investigate the estimation of the location of the center of mass and gyration radii. The location of the center of mass was also compared against experimental data on a sample of riders. It was found that all four BSIP data sets considered gave similar trends, although with slightly different figures. The sport posture had the minimum roll and pitch moments of inertia, the touring posture had the maximum roll, and the scooter posture had the maximum pitch moment of inertia. The elbow distance had a minor effect in the estimation of the location of the center of mass and the gyration radii, while the effect of the knee distance was on the same order of magnitude as changing the data set. The (longitudinal) lean angle and the rider height were the most relevant factors. Leaning forward lowers and moves forward the center of mass—this lowering is the main effect in the case of the sport motorcycle, while forward motion is the main effect in the case of touring and scooter postures. The lean angle also lowered the roll and pitch gyration radii, while the yaw gyration radius increased only for high values of lean angle (above 40 degrees). Increasing the height of the rider moved the center of mass forward and upward—the forward motion is the main effect in the case of the sport bike, while rising is the main effect in the case of the touring and scooter. Height also increased all gyration radii. Finally, typical parameters for a sport posture, touring posture, and scooter posture were provided.

This work, in addition to suggesting the main parameters to be considered when estimating the variation of the inertial properties of bicycle and motorcycle riders, provides tables for the implementation of the most popular BSIP formulas in a consistent set of frames and landmarks, in order to serve as a practical reference for future analyses by other researchers.

Author Contributions: Conceptualization, M.M. and N.P.; methodology, M.B., M.M. and N.P.; software, M.B.; validation, M.B. and M.M.; formal analysis, M.B. and M.M.; investigation, M.B. and M.M.; resources, M.M.; data curation, M.B.; writing—original draft preparation, M.B. and M.M.; writing—review and editing, M.B. and M.M.; visualization, M.B. and M.M.; supervision, M.M.; project administration, M.M. All authors have read and agreed to the published version of the manuscript.

Funding: This research received no external funding.

Conflicts of Interest: The authors declare no conflict of interest.

Appendix A

The coefficients of the data sets are reported in Tables A1–A6. Given a human body with height h and mass m, the length of segment i is $\gamma_i h$, which has mass $\alpha_i m$, center of mass in co-ordinate $[\beta_x^i \gamma_i h, \beta_y^i \gamma_i h, \beta_z^i \gamma_i h]^T$ (expressed in the local reference frame), gyration radii of the main moments of inertia $[\eta_x^i \gamma_i h, \eta_y^i \gamma_i h, \eta_z^i \gamma_i h]^T$ (again in the local reference), and the radii of the cross moments of inertia are $[\eta_{xy}^i \gamma_i h, \eta_{yz}^i \gamma_i h, \eta_{xz}^i \gamma_i h]^T$ (Dumas model only). Details on the specific models are given in the sections below.

Appendix A.1. Zatsiorsky–DeLeva Model Implementation

Most of the data necessary to build the Zatsiorsky-De Leva model are taken from Table 4 in Reference [53]. For arms, forearms, hands, thighs, shanks, and feet, the upper part of the table is used (male columns) without any adjustment, while alternative endpoints are selected for the head and trunk: the head is measured from the vertex to cervicale (CERV) and the trunk from CERV to MIDH. Implementation in the multibody model requires replacing CERV with MIDS: these two points are vertically aligned (along the trunk longitudinal axis) and their distance can be computed using Table 4 in Reference [53], which gives the trunk in terms of MIDS and MIDH.The HEEL point is not defined with respect to the shank endpoint (LMAL). However, this point is given in Reference [73], to which [53] also refers. In addition, the HEEL point is assumed to be vertically aligned with LMAL and shifted downward by half of the LMAL height, which is measured in Table 75 (male mean value has been used) of Reference [73]. The ratio, with respect to the foot length, is evaluated using Table 51, again from Reference [73]. Tables from Reference [73] are also used to estimate the lateral displacements of SJC and HJC with respect to MIDS and MIDH, respectively. The lateral displacement of SJC is assumed to be equal to half of the biacromial breadth (Table 10), while the lateral displacement of HJC is assumed to be equal to half of the bispinous breadth (Table 14). The ratios are evaluated using Table 99 for the stature. Handgrip points are assumed coincident with the hand endpoints, while footpeg points are assumed vertically aligned with LMAL and shifted downward of LMAL height (Table 75).

Table A1. Zatsiorsky–DeLeva model parameters for each body segment.

	α_i	γ_i	β_x^i	β_y^i	β_z^i	η_x^i	η_y^i	η_z^i
Head and neck	6.94	18.99	0	0	63.26	22.26	23.14	19.17
Trunk	43.46	29.61	0	0	56.90	38.39	35.81	19.78
Arm	2.71	16.18	0	0	−57.72	28.50	26.90	15.80
Forearm	1.62	15.45	0	0	−45.74	27.60	26.50	12.10
Hand	0.61	4.95	0	0	−79.00	62.80	51.30	40.10
Thigh	14.16	24.25	0	0	−40.95	32.90	32.90	14.90
Shank	4.33	24.93	0	0	−44.59	25.50	24.90	10.30
Foot	1.37	14.82	44.15	0	−12.44	12.40	24.50	25.70

Appendix A.2. Dempster Model Implementation

The data necessary to build the Dempster model are taken from Reference [50]. Mass percentages of body segments are taken from Table 4 (Dempster column), with the exception of the trunk coefficient, which is obtained by subtracting the other segments from the whole body; otherwise, the total mass is more than 100% (a well-known issue). The segments lengths and G co-ordinate coefficients are taken from Tables 2 and 3 and Figure 3 in [50]. The trunk center of mass distance from MIDH is evaluated as the weighted average of thorax, shoulder, and lumbocoxal segment CoM positions. The gyration radii are not reported and, thus, are assumed to be zero. Handgrip points are assumed to be coincident with the hand centers of mass, while footpeg points are assumed coincident with AJC points.

Table A2. Dempster model parameters for each body segment.

	α_i	γ_i	β_x^i	β_y^i	β_z^i	η_x^i	η_y^i	η_z^i
Head and neck	7.92	20.33	0	0	63.85	0	0	0
Trunk	50.412	27.98	0	0	55.63	0	0	0
Arm	2.64	17.35	0	0	−43.74	0	0	0
Forearm	1.531	15.72	0	0	−42.93	0	0	0
Hand	0.612	3.422	0	0	−100	0	0	0
Thigh	10.008	23.99	0	0	−42.78	0	0	0
Shank	4.612	25.05	0	0	−42.64	0	0	0
Foot	1.431	3.694	100	0	0	0	0	0

Appendix A.3. Reynolds-NASA Model Implementation

The data necessary to build the Reynolds-NASA model are taken from Reference [51,74]. Mass percentages of body segments are taken from Table 15 in Reference [51], while $\boldsymbol{G}$ position and ρ components are taken from Tables 12 and 28, respectively. For some body segments, the lengths given in these tables have different definitions (e.g., bone lengths instead of link lengths). Therefore, data are taken from Tables 1 and 3 in Reference [51] for the location of the center of mass and from tables in Reference [74] for the moments of inertia (the mean of the data contained in each table are used, taking into account only western males). As in the Zatsiorsky–DeLeva model, the heel is not defined with respect to the shank endpoint: similar assumptions are thus made. The heel point is assumed to be vertically aligned with AJC and shifted downward by half of the LMAL height, which is measured in Reference [74], Table 543. The ratio, with respect to the foot length, is evaluated using Table 362. The same reference is used to estimate the lateral displacements of SJC and HJC with respect to MIDS and MIDH, respectively. The lateral displacement of SJC is assumed to be equal to half of the biacromial breadth (Table 103), while the lateral displacement of HJC is assumed to be equal to half of the biliocristale breadth (Table 130). Handgrip points are assumed coincident with the hand centers of mass, while footpeg points are assumed to be vertically aligned with AJC and shifted downward of the LMAL height.

Table A3. Reynolds-NASA model parameters for each body segment (mass, length, and center of mass).

	α_i	γ_i	β_x^i	β_y^i	β_z^i
Head and neck	8.40	18.07	0	0	65.41
Trunk	50.00	29.31	0	0	58.00
Arm	2.80	$89.94(0.185\,h + 0.01338)$	0	0	−48.00
Forearm	1.698	$98.70(0.140\,h + 0.02688)$	0	0	−41.00
Hand	0.602	6.19	0	0	−51.00
Thigh	10.00	$90.34(0.281\,h - 0.01902)$	0	0	−41.00
Shank	4.30	$107.76(0.268\,h - 0.08369)$	0	0	−44.00
Foot	1.40	15.18	44.00	0	−12.80

Table A4. Reynolds-NASA model parameters for each body segment (length and gyration radii).

	γ_i	η_x^i	η_y^i	η_z^i
Head and neck	11.14	31.60	30.90	34.20
Trunk	29.31	43.00	35.20	20.80
Arm	18.64	26.10	25.40	10.40
Forearm	15.17	29.60	29.20	10.80
Hand	5.00	50.40	45.60	26.60
Thigh	27.03	27.90	28.40	12.20
Shank	25.12	28.20	28.20	7.60
Foot	15.18	12.20	24.90	26.10

Appendix A.4. McConville–Young–Dumas Model Implementation

The data necessary to build the McConville–Young–Dumas model are taken from Tables 1 and 2 in Reference [54], along with the corrections in Reference [55]. Torso and pelvis body segments are combined into a single trunk element. All the frames in Reference [54] have the y-axis pointing upward and the x-axis pointing forward in the standard anatomical position: a 90-degree rotation along the x-axis is applied to make it consistent with the convention of the multibody model. The trunk and head bodies are defined with respect to a frame located in the cervical joint center (CJC), while the MIDS point is required. MIDS is assumed to lie along the trunk longitudinal axis and translated downward, with respect to CJC, as reported in Table 2 of Reference [54]. Handgrip points are assumed coincident with the projection of the hand centers of mass along the hand longitudinal axes, while footpeg points are assumed to be vertically aligned with AJC and horizontally aligned with the foot centers of mass.

Table A5. McConville–Young–Dumas model parameters for each body segment.

	α_i	γ_i	β_x^i	β_y^i	β_z^i	η_x^i	η_y^i	η_z^i
Head and neck	6.70	19.77	1.58	-0.08	63.28	22.16	23.74	16.62
Trunk	47.50	28.14	−2.26	0.17	56.24	36.79	37.14	22.91
Arm	2.40	15.31	1.70	2.60	−45.20	33.00	32.00	14.00
Forearm	1.70	15.99	1.00	−1.40	−41.70	28.00	27.00	11.00
Hand	0.60	4.52	8.20	−7.40	−83.90	61.00	56.00	38.00
Thigh	12.30	24.41	−4.10	−3.30	−42.90	29.00	30.00	15.00
Shank	4.80	24.46	−4.80	−0.70	−41.00	28.00	28.00	10.00
Foot	1.20	10.34	38.20	−2.60	−15.10	17.00	36.00	37.00

Table A6. McConville–Young–Dumas model parameters for each body segment (cross gyration radii).

	η_{xy}	η_{yz}	η_{xz}
Head and neck	1.58	−2.37	−5.54
Trunk	−0.99	3.16	15.89
Arm	−5.00	−2.00	6.00
Forearm	−2.00	8.00	3.00
Hand	−15.00	20.00	22.00
Thigh	2.00	7.00	7.00
Shank	2.00	−5.00	−4.00
Foot	8.00	0	13.00

References

1. Limebeer, D.J.N.; Massaro, M. *Dynamics and Optimal Control of Road Vehicles*; Oxford University Press: Oxford, UK, 2018.
2. Kooijman, J.D.G.; Schwab, A.L. A review on bicycle and motorcycle rider control with a perspective on handling qualities. *Veh. Syst. Dyn.* **2013**, *51*, 1722–1764. [CrossRef]
3. Cossalter, V. *Motorcycle Dynamics*, 2nd ed.; Lulu.com: Morrisville, NC, USA, 2006.
4. Sharp, R.S.; Evangelou, S.; Limebeer, D.J.N. Advances in the modelling of motorcycle dynamics. *Multibody Syst. Dyn.* **2004**, *12*, 251–283. [CrossRef]
5. Pacejka, H.; Besselink, I.J.M. *Tire and Vehicle Dynamics*, 3rd ed.; Butterworth-Heinemann: Oxford, UK, 2012.
6. Cossalter, V.; Lot, R.; Massaro, M. An advanced multibody code for handling and stability analysis of motorcycles. *Meccanica* **2011**, *46*, 943–958. [CrossRef]
7. Doria, A.; Marconi, E.; Massaro, M. Identification of rider's arms dynamic response and effects on bicycle stability. In Proceedings of the AMSE IDETC International Design Engineering Technical Conferences & Computers and Information in Engineering Conference, St. Louis, MO, USA, 16–19 August 2020.
8. Tomiati, N.; Colombo, A.; Magnani, G. A nonlinear model of bicycle shimmy. *Veh. Syst. Dyn.* **2019**, *57*, 315–335. [CrossRef]
9. Uchiyama, H.; Tanaka, K.; Nakagawa, Y.; Kinbara, E.; Kageyama, I. *Study on Weave Behavior Simulation of Motorcycles Considering Vibration Characteristics of Whole Body of Rider*; SAE Technical Paper; SAE International: Warrendale, PA, USA, 2018.
10. Klinger, F.; Nusime, J.; Edelmann, J.; Plöchl, M. Wobble of a racing bicycle with a rider hands on and hands off the handlebar. *Veh. Syst. Dyn.* **2014**, *52*, 51–68. [CrossRef]
11. Massaro, M.; Cole, D.J. Neuromuscular-Steering Dynamics: Motorcycle Riders vs. Car Drivers. In Proceedings of the ASME 2012 5th Annual Dynamic Systems and Control Conference, Fort Lauderdale, FL, USA, 17–19 October 2012; pp. 217–224.
12. Massaro, M.; Lot, R.; Cossalter, V.; Brendelson, J.; Sadauckas, J. Numerical and experimental investigation of passive rider effects on motorcycle weave. *Veh. Syst. Dyn.* **2012**, *50*, 215–227. [CrossRef]
13. Doria, A.; Formentini, M.; Tognazzo, M. Experimental and numerical analysis of rider motion in weave conditions. *Veh. Syst. Dyn.* **2012**, *50*, 1247–1260. [CrossRef]
14. Schwab, A.L.; Meijaard, J.P.; Kooijman, J.D.G. Lateral dynamics of a bicycle with a passive rider model: Stability and controllability. *Veh. Syst. Dyn.* **2012**, *50*, 1209–1224. [CrossRef]
15. Plöchl, M.; Edelmann, J.; Angrosch, B.; Ott, C. On the wobble mode of a bicycle. *Veh. Syst. Dyn.* **2012**, *50*, 415–429. [CrossRef]
16. Cossalter, V.; Doria, A.; Lot, R.; Massaro, M. The effect of rider's passive steering impedance on motorcycle stability: Identification and analysis. *Meccanica* **2011**, *46*, 279–292. [CrossRef]
17. Sharp, R.S.; Limebeer, D.J.N. On steering wobble oscillations of motorcycles. *Proc. Inst. Mech. Eng. Part J. Mech. Eng. Sci.* **2004**, *218*, 1449–1456. [CrossRef]
18. Katayama, T.; Aoki, A.; Nishimi, T.; Okayama, T. *Measurements of Structural Properties of Riders*; SAE Technical Paper; SAE International: Warrendale, PA, USA, 1987.
19. Nishimi, T.; Aoki, A.; Katayama, T. *Analysis of Straight Running Stability of Motorcycles*; SAE Technical Paper; SAE International: Warrendale, PA, USA, 1985; p. 856124.
20. Cossalter, V.; Lot, R.; Massaro, M. The influence of frame compliance and rider mobility on the scooter stability. *Veh. Syst. Dyn.* **2007**, *45*, 313–326. [CrossRef]
21. Cheng, J.; Su, H.; Chen, K. Driver Posture Detection Method in Motorcycle Simulator. In Proceedings of the 2019 International Conference on Artificial Intelligence and Advanced Manufacturing (AIAM), Dublin, Ireland, 17–19 October 2019; pp. 622–626.
22. Nagasaka, K.; Ichikawa, K.; Yamasaki, A.; Ishii, H. *Development of a Riding Simulator for Motorcycles*; SAE Technical Papers; SAE International: Warrendale, PA, USA, 2018. [CrossRef]
23. Will, S.; Schmidt, E.A. Powered two wheelers' workload assessment with various methods using a motorcycle simulator. *IET Intell. Transp. Syst.* **2015**, *9*, 702–709. [CrossRef]
24. Massaro, M.; Cossalter, V.; Lot, R.; Rota, S.; Ferrari, M.; Sartori, R.; Formentini, M. A portable driving simulator for single-track vehicles. In Proceedings of the 2013 IEEE International Conference on Mechatronics (ICM), Vicenza, Italy, 27 February–1 March 2013; pp. 364–369. [CrossRef]

25. Stedmon, A.W.; Brickell, E.; Hancox, M.; Noble, J.; Rice, D. MotorcycleSim: A user-centerd approach in developing a simulator for motorcycle ergonomics and rider human factors research. *Adv. Transp. Stud.* **2012**, 31–48. [CrossRef]
26. Cossalter, V.; Lot, R.; Massaro, M.; Sartori, R. Development and validation of an advanced motorcycle riding simulator. *Proc. Inst. Mech. Eng. Part J. Automob. Eng.* **2011**, *225*, 705–720. [CrossRef]
27. Nehaoua, L.; Arioui, H.; Mammar, S. Review on single track vehicle and motorcycle simulators. In Proceedings of the 2011 19th Mediterranean Conference on Control Automation (MED), Corfu Island, Greece, 20–23 June 2011; pp. 940–945.
28. Drillis, R.; Contini, R. *Body Segment Parameters*; Technical Report 116.03; New York Univ. School of Engineering and Science: New York, NY, USA, 1964.
29. Reid, J.G.; Jensen, R.K. Human body segment inertia parameters: A survey and status report. *Exerc. Sport Sci. Rev.* **1990**, *18*, 225–241. [CrossRef] [PubMed]
30. Jensen, R.K. Human Morphology: Its role in the mechanics of movement. *J. Biomech.* **1993**, *26*, 81–94. [CrossRef]
31. Pearsall, D.J.; Reid, G. The Study of Human Body Segment Parameters in Biomechanics: An Historical Review and Current Status Report. *Sport. Med. Eval. Res. Exerc. Sci. Sport. Med.* **1994**, *18*, 126–140. [CrossRef] [PubMed]
32. Erdmann, W.S. Geometry and inertia of the human body-review of research. *Acta Bioeng. Biomech.* **1999**, *1*, 23–35.
33. Winter, D.A. *Biomechanics and Motor Control of Human Movement*, 4th ed.; John Wiley & Sons: Hoboken, NJ, USA, 2009.
34. Dumas, R.; Wojtusch, J. *Handbook of Human Motion*; chapter Estimation of the Body Segment Inertial Parameters for the Rigid Body Biomechanical Models Used in Motion Analysis; Springer International Publishing: Cham, Switzerland, 2017; pp. 1–31. [CrossRef]
35. Dempster, W.T. *Space Requirements of the Seated Operator*; Technical Report WADC-TR-55-159; Wright Air Development Center, Wright-Patterson Air Force Base: Dayton, OH, USA, 1955.
36. Clauser, C.E.; McConville, J.T.; Young, J.W. *Weight, Volume and Center of Mass of Segments of the Human Body*; Technical Report 69-70; Aerospace Medical Research Laboratories, Wright-Patterson Air Force Base: Dayton, OH, USA, 1969.
37. Chandler, R.F.; Clauser, C.E.; McConville, J.T.; Reynolds, H.M.; Young, J.W. *Investigation of Inertial Properties of the Human Body*; Technical Report ADA016485; Air Force Aerospace Medical Research Lab Wright-Patterson AFB OH, Wright-Patterson Air Force Base: Dayton, OH, USA, 1975.
38. Jensen, R.K. Estimation of the biomechanical properties of three body types using a photogrammetric method. *J. Biomech.* **1978**, *11*, 349–358. [CrossRef]
39. McConville, J.T.; Churchill, T.D.; Kaleps, I.; Clauser, C.E.; Cuzzi, J. *Anthropometric Relationships of Body and Body Segment Moments of Inertia*; Technical Report AFAMRL-TR-80-119; Aerospace Medical Research Laboratory, Wright–Patterson Air Force Base: Dayton, OH, USA, 1980.
40. Young, J.W.; Chandler, R.F.; Snow, C.C.; Robinette, K.M.; Zehner, G.F.; Lofberg, M.S. *Anthropometric and Mass Distribution Characteristics of the Adults Female*; Technical Report FA-AM-83-16; FAA Civil Aeromedical Institute: Oklaoma City, OK, USA, 1983.
41. Ackland, T.R.; Blanksby, B.A.; Bloomfield, J. Inertial characteristics of adolescent male body segments. *J. Biomech.* **1988**, *21*, 319–327. [CrossRef]
42. Mungiole, M.; Martin, P.E. Estimating segment inertial properties: Comparison of magnetic resonance imaging with existing methods. *J. Biomech.* **1990**, *23*, 1039–1046. [CrossRef]
43. Zatsiorsky, V.M.; Seluyanov, V.N.; Chugunova, L.G. *Contemporary Problems of Biomechanics*; chapter Methods of determining mass-inertial characteristics of human body segments; CRC Press: Boca Raton, FL, USA, 1990; pp. 272–291.
44. Pearsall, D.J.; Reid, J.G.; Livingston, L.A. Segmental inertial parameters of the human trunk as determined from computed tomography. *Ann. Biomed. Eng.* **1996**, *24*, 198–210. [CrossRef]
45. Cheng, C.K.; Chen, H.H.; Chen, C.S.; Lee, C.L.; Chen, C.Y. Segment inertial properties of Chinese adults determined from magnetic resonance imaging. *Clin. Biomech.* **2000**, *15*, 559–566. [CrossRef]
46. Durkin, J.L.; Dowling, J.J.; Andrews, D.M. The measurement of body segment inertial parameters using dual energy X-ray absorptiometry. *J. Biomech.* **2002**, *35*, 1575–1580. [CrossRef]
47. Dumas, R.; Aissaoui, R.; Mitton, D.; Skalli, W.; De Guise, J.A. Personalized body segment parameters from biplanar low-dose radiography. *IEEE Trans. Biomed. Eng.* **2005**, *52*, 1756–1763. [CrossRef]

48. Bauer, J.J.; Pavol, M.J.; Snow, C.M.; Hayes, W.C. MRI-derived body segment parameters of children differ from age-based estimates derived using photogrammetry. *J. Biomech.* **2007**, *40*, 2904–2910. [CrossRef]
49. Contini, R. Body segment parameters. II. *Artif. Limbs* **1972**, *16*, 1–19. [PubMed]
50. Dempster, W.T.; Gaughran, G.R.L. Properties of body segments based on size and weight. *Am. J. Anat.* **1967**, *120*, 33–54. [CrossRef]
51. Reynolds, H.M. Chapter the Inertial Properties of the Body and Its Segments. In *Anthropometric Source Book Volume I: Anthropometry for Designers*; National Aeronautics and Space Administration, Scientific and Technical Information Office: Yellow Springs, OH, USA, 1978; Volume 1, pp. 1–75.
52. Hinrichs, R.N. Adjustments to the segment center of mass proportions of Clauser et al. (1969). *J. Biomech.* **1990**, *23*, 949–951. [CrossRef]
53. De Leva, P. Adjustments to zatsiorsky-seluyanov's segment inertia parameters. *J. Biomech.* **1996**, *29*, 1223–1230. [CrossRef]
54. Dumas, R.; Chèze, L.; Verriest, J.P. Adjustments to McConville et al. and Young et al. body segment inertial parameters. *J. Biomech.* **2007**, *40*, 543–553. [CrossRef] [PubMed]
55. Dumas, R.; Chèze, L.; Verriest, J.P. Corrigendum to "Adjustments to McConville et al. and Young et al. body segment inertial parameters" [J. Biomech. (2006) in press] (doi:10.1016/j.jbiomech.2006.02.013). *J. Biomech.* **2007**, *40*, 1651–1652. [CrossRef]
56. Dumas, R.; Robert, T.; Cheze, L.; Verriest, J.P. Thorax and abdomen body segment inertial parameters adjusted from McConville et al. and Young et al. *Int. Biomech.* **2015**, *2*, 113–118. [CrossRef]
57. Contini, R. *Body Segment Parameters (Pathological)*; Technical Report 1584.03; New York Univ. School of Engineering and Science: New York, NY, USA, 1970.
58. Plagenhoef, S. *Patterns of Human Motion: A Cinematographic Analysis*; Prentice-Hall: Upper Saddle River, NJ, USA, 1971.
59. Trotter, M.; Gleser, G.C. A re-evaluation of estimation of stature based on measurements of stature taken during life and of long bones after death. *Am. J. Phys. Anthropol.* **1958**, *16*, 79–123. [CrossRef] [PubMed]
60. Dempster, W.T.; Sherr, L.A.; Priest, J.G. Conversion scales for estimating humeral and femoral lengths and the lengths of functional segments in the limbs of american caucasoid males. *Hum. Biol.* **1964**, *36*, 246–262.
61. Geoffrey, S.P. A 2-D Mannikin—The Inside Story. X-Rays Used to Determine a New Standard for a Basic Design Tool. In Proceedings of the SAE International Congress and Exposition of Automotive Engineering, Detroit, MI, USA, 9–12 January 1961.
62. Becker, E.B. Measurement of Mass Distribution Parameters of Anatomical Segments. *SAE Trans.* **1972**, *81*, 2818–2833.
63. Braune, W.; Fischer, O. The center of gravity of the human body as related to the German infantryman. *ATI* **1889**, *138*, 452.
64. Wu, G.; Siegler, S.; Allard, P.; Kirtley, C.; Leardini, A.; Rosenbaum, D.; Whittle, M.; D'Lima, D.D.; Cristofolini, L.; Witte, H.; et al. ISB recommendation on definitions of joint coordinate system of various joints for the reporting of human joint motion—Part I: Ankle, hip, and spine. *J. Biomech.* **2002**, *35*, 543–548. [CrossRef]
65. Wu, G.; Van Der Helm, F.C.T.; Veeger, H.E.J.; Makhsous, M.; Van Roy, P.; Anglin, C.; Nagels, J.; Karduna, A.R.; McQuade, K.; Wang, X.; et al. ISB recommendation on definitions of joint coordinate systems of various joints for the reporting of human joint motion—Part II: Shoulder, elbow, wrist and hand. *J. Biomech.* **2005**, *38*, 981–992. [CrossRef] [PubMed]
66. Imaizumi, H.; Fujioka, T.; Omae, M. Rider model by use of multibody dynamics analysis. *JSAE Rev.* **1996**, *17*, 75–77. [CrossRef]
67. Moore, J.K.; Hubbard, M.; Kooijman, J.D.G.; Schwab, A.L. A method for estimating physical properties of a combined bicycle and rider. In Proceedings of the ASME IDETC/CIE International Design Engineering Technical Conferences & Computers and Information in Engineering Conference, San Diego, CA, USA, 30 August–2 September 2009.
68. Wittenburg, J. *Dynamics of Multibody Systems*, 2nd ed.; Springer: Berlin/Heidelberg, Germany, 2007.
69. Nikravesh, P.E. *Computer-Aided Analysis of Mechanical Systems*; Prentice Hall: Upper Saddle River, NJ, USA, 1988.
70. Shabana, A.A. *Dynamics of Multibody Systems*, 4th ed.; Cambridge University Press: Cambridge, UK, 2013.
71. de Jalon, G.; Bayo, J.E. *Kinematic and Dynamic Simulation of Multibody Systems*; Springer: Berlin/Heidelberg, Germany, 1994.

72. Cheli, F.; Pennestrì, E. *Cinematica e Dinamica dei Sistemi Multibody*; CEA: Milan, IT, USA, 2006.
73. Gordon, C.C.; Churchill, T.; Clauser, C.E.; Bradtmiller, B.; McConville, J.T.; Tebbetts, I.; Walker, R.A. *Anthropometric Survey of US Army Personnel: Summary Statistics, Interim Report for 1988*; Technical Report; Anthropology Research Project Inc.: Yellow Springs, OH, USA, 1989.
74. Webb Associates, A.R.P. *Anthropometric Source Book, Volume II: A Handbook of Anthropometric Data*; NASA reference publication; National Aeronautics and Space Administration, Scientific and Technical Information Office: Yellow Springs, OH, USA, 1978.

Article

Investigation of Zero Moment Point in a Partially Filled Liquid Vessel Subjected to Roll Motion

Muhammad Usman [1], Muhammad Sajid [1,*], Emad Uddin [1] and Yasar Ayaz [1,2]

1 School of Mechanical and Manufacturing Engineering (SMME), National University of Sciences and Technology (NUST), Islamabad 44000, Pakistan; 13msmemusman@smme.edu.pk (M.U.); emaduddin@smme.nust.edu.pk (E.U.); yasar@smme.nust.edu.pk (Y.A.)

2 National Center for Artificial Intelligence (NCAI), National University of Sciences and Technology (NUST), Islamabad 44000, Pakistan

* Correspondence: m.sajid@smme.nust.edu.pk

Received: 4 May 2020; Accepted: 2 June 2020; Published: 9 June 2020

Abstract: Liquid-handling robots are designed to dispense sub-microliter quantities of fluids for applications including laboratory tests. When larger amounts of liquids are involved, sloshing must be considered as a parameter affecting stability, which is of significance for autonomous vehicles. The measurement and quantification of slosh in enclosed volumes poses a challenge to researchers who have traditionally resorted to tracking the air–liquid interface for two-phase flow analysis. There is a need for a simpler method to predict rollover in these applications. In this work, a novel solution to address this problem is proposed in the form of the Zero Moment Point (ZMP) of a dynamic liquid region. Computational experiments of a partially filled, two-dimensional liquid vessel were carried out using the Volume of Fluid (VOF) method in a finite volume based open-source computational fluid dynamics solver. The movement of the air–liquid interface was tracked, while the Center of Mass and the resulting Zero Moment Point were determined from the numerical simulations at each time step. The computational model was validated by comparing the wall pressure and movement of the liquid-free surface to experimentally obtained values. It was concluded that for a dynamic liquid domain, the Zero Moment Point can be instrumental in determining the stability of partially filled containers subjected to sloshing.

Keywords: computational fluid dynamics; fluid sloshing; table cart method; zero moment point

---

1. Introduction

Recent research suggests that in the future humanoid robots will be able to perform tasks that are physically intense as well as technically complex. While humanoids have been configured for clearing heavy solid blocks [1,2], very little attention has been paid to liquid cargo. The presence of partially filled liquid containers such as large gallons, barrels, or cylindrical vessels [3,4] introduces a new parameter of liquid slosh to the problem of humanoid stability. Previous work on liquid-handling robots has been focused on small quantities of fluids with applications in the automation of chemical or biochemical laboratories [5,6], where they are required to dispense a designated amount of reagent, samples, and/or other liquids into a selected container. Recent advancements in liquid-handling robots for small liquid quantities focus on modular and open-source design [7,8], while hydrodynamics of sloshing liquids have been applied to improve microtiter plate assays [9,10]. In small quantities, the presence of liquids does not pose a significant challenge for the stability of a robot. However, when the amount of fluid is comparable to or larger than the mass of the robot, e.g., the transportation of substantial liquid cargo [11,12], maintaining stability becomes a challenge at greater rates of motion, which introduce sloshing.

Ongoing research concerning the stability of load-carrying humanoid robots can be divided into two groups: feedback control systems, which largely ignore the physical parameters of robots and load such as mass, Center of Mass (CoM) and body geometry [13]; and Zero Moment Point (ZMP), which directly deals with all the aforementioned parameters. ZMP is a point where the influence of all forces acting on the mechanism can be replaced by one single force [14]. While the primary use of ZMP is to dynamically balance the robot, considerable advancement has been made in ZMP applications since its discovery. In 2003 Kajita et al. [15] introduced the table cart method to design a ZMP based controller in order to generate walking patterns of biped robots. A preview controller was utilized that in turn used the future ZMP reference. Errors introduced in ZMP determination due to the differences between the table cart and multi-body model were compensated for by the preview control. Guangbin Sun et al. [16] extended previous work on the table cart method to calculate ZMP's and the vertical displacement/motion of the CoM on uneven ground, providing dynamic balancing on all types of terrains. Using the novel approach of extended ZMP analysis, an Atlas robot was simulated that successfully demonstrated its ability to walk on uneven terrains. More recently, Scianca et al. [17] incorporated a constraint on stability in the formulation of model predictive control system for humanoid gait generation and obtained recursive feasibility conditions guaranteeing the stability of CoM and ZMP dynamics ensuring intrinsically stability.

An exhaustive search for liquid-handling robots involving relatively larger quantities of fluids did not reveal any previous research, providing the impetus for this study. Sloshing in partially filled vessels is an aggressive movement of the agitated quantity of fluid within a container. It produces dynamic forces on the outer walls, thus growing into a principal concern for the operation of the vessel. The standard approaches to resolving sloshing of inviscid and/or incompressible liquids in a partially filled vessels subjected to sway, surge, pitch, and roll motions using differential equations have been covered in the literature, notably by Faltinsen [18] and Ibrahim [19]. Computational fluid dynamics (CFD) is a powerful numerical tool that facilitates the analysis of liquids under dynamic forces and has been used to study sloshing in partially filled containers leading to more accurate results [20,21]. Within CFD, various methods have been used to study sloshing. For instance, smoothed particle hydrodynamics (SPH) was employed by L. Delorme et al. [11] in an experimental and computational study of water sloshing in a partially filled tank under a forced rolling motion. The Arbitrary Lagrangian Eulerian (ALE) approach with finite element methods (FEM) has been used [21] to simulate fluid structure interaction problems for baffled and non-baffled containers. More recent works on sloshing [22–24] provide insight into sloshing for computing force on the walls of the vessel; however, they do not comment on its effect on stability.

In the present study, the effect of sloshing is linked with stability by calculating the ZMP of the fluid domain at different fill levels. In order to simulate the sloshing phenomena and obtain the fluid properties required for calculating ZMP, the open source CFD tool OpenFOAM [25] was employed. The numerical description of sloshing in partially filled vessel was validated against experimental data available in literature.

2. Methodology

2.1. Problem Description

The present study aims to determine the ZMP of the fluid region in a partially filled rectangular container subjected to sloshing motion. The experimental work by Souto-Iglesias et al. [26] was utilized to validate the numerical simulation of sloshing motion in the fluid domain. The computational domain consisted of a 1:60 scaled model of a longitudinal section of a 138,000 mm^3 tanker hold resulting in a rectangular region of 900 mm breadth and 520 mm height, as shown in Figure 1. Due to the two-dimensional nature of the problem, the front and back faces were not resolved in the numerical simulation.

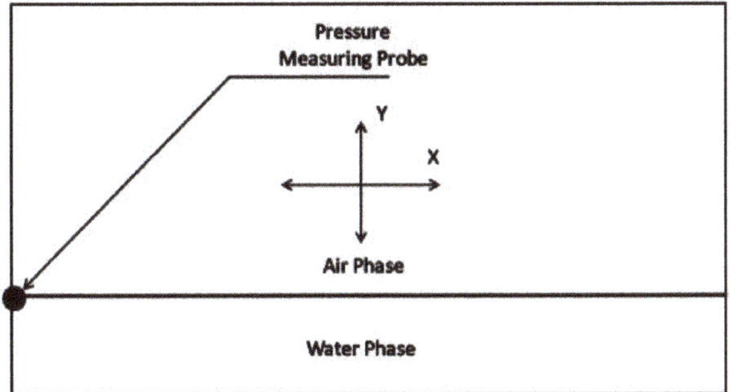

Figure 1. Computational domain employed in the study.

In the experimental work, the pressure measuring probe was placed on the left boundary at the liquid–air interface (Figure 1) for an 18% fill level, after which the setup was subjected to a ±4° rotation. This necessitates the simulation of not only the two-phase flow but also requires rotation of the mesh. For this purpose, the "Inter-DyMFOAM" solver of OpenFOAM was employed. A uniformly structured mesh with 2400 cells was prepared in the "blockMesh" utility, with an aspect ratio of 1.54074 and a non-orthogonality of 0. A grid independence study was carried out with twice as many cells, which did not indicate a significant effect on the output parameters of Center of Mass and pressure relative to our chosen configuration, as shown in Figure 2; therefore, the initial mesh density was retained for further work.

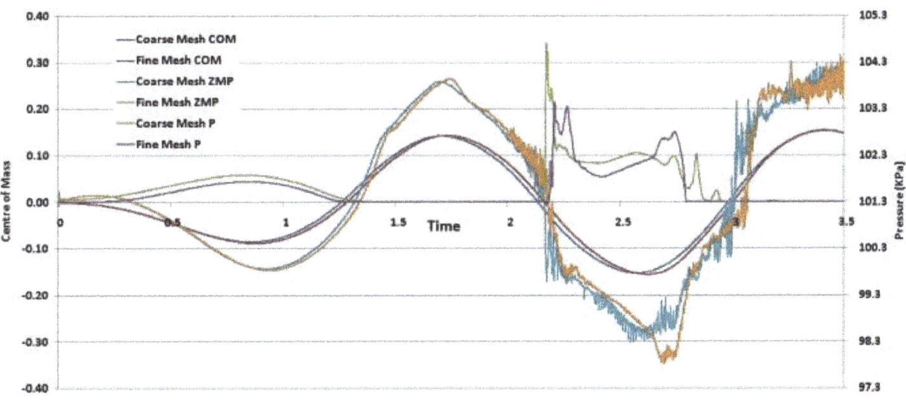

Figure 2. Grid independence study comparing center of mass (COM), zero moment point (ZMP) and pressure P, for coarse and fine meshes.

2.2. Mathematical Modelling

The mathematical modelling employed in this study had three aspects: the ZMP description of a dynamic liquid region, computationally capturing the dynamics of the air and liquid regions in a domain subjected to periodic motion, and extracting essential information from the computational model for the calculation of ZMP of the dynamic liquid region.

The ZMP [14] methodology describes the stability of bipeds by defining a point where the gravitational force acting on the biped balances the resultant horizontal inertia, which results in no moment along the horizontal plane at the point of contact of the biped with the ground. The ZMP can be calculate using the approach presented by Miomir Vukobratovic [14] along with the table cart method [15,16]. Between these two techniques, the table cart method can be seamlessly integrated to the liquid domain due to its dependence on the Center of Mass.

A basic outline of a liquid tank carrying robot is depicted in Figure 3A, highlighting the CoM and the free surface between the air and liquid regions that is subjected to extreme topological changes during the robot's motion. Figure 3B shows the cart table description of this robot where a massless table is arranged within a robot and a cart is shown moving on it. Here the cart containing mass "m" at position (x, y) on the table represents the liquid domain of the container, while the table is considered to have the same shape and dimensions of support polygon as that of a robot foot print [27].

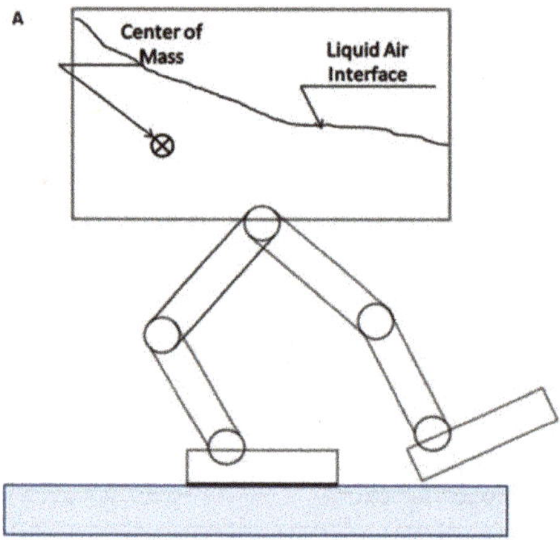

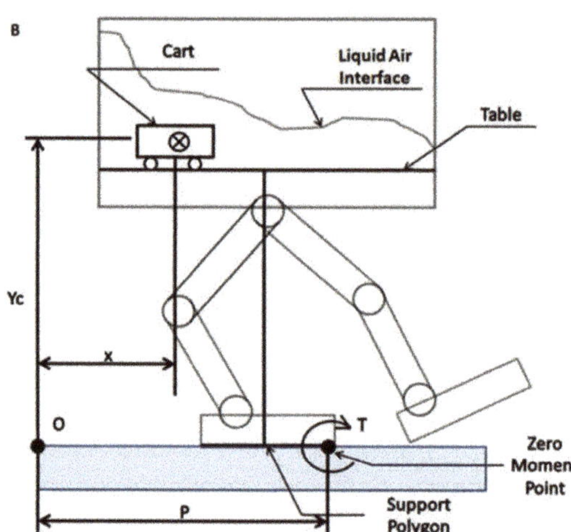

Figure 3. Liquid carrying robot: (**A**) simplified model, and (**B**) table cart description.

The principle of ZMP was adapted from Kajita [15] and Siciliano [27], who employ the following relationship to locate the ZMP with respect to the origin:

$$ZMP_x = x - \frac{y_c}{g}\ddot{x} \tag{1}$$

where g is the gravitational constant, the location of CoM of the cart along the horizontal axis measured from the origin is x, while $\ddot{x}$ is its acceleration and y_c is the height of CoM of the cart. For the computation of *ZMP* for a liquid-carrying robot, the value of y_c and x are extracted from CoM, while $\ddot{x}$ is obtained from a second order time differencing scheme.

The interDyMFoam ("DyM" stands for "Dynamic Mesh") solver of OpenFOAM (OF), generally employed to simulate problems involving multiphase flow coupled with moving domains, was modified for this research to calculate the ZMP while solving the three-dimensional Navier–Stokes equations over two fluid domains consisting of incompressible air and water by employing the Volume of Fluid (VOF) approach over a finite volume discretization. The VOF method describes each phase by the fraction α_i of the volume of the i'th cell that is occupied by the dispersed phase. The InterDyMFoam solver uses a PIMPLE approach, which combines both the Pressure Implicit with Splitting of Operators (PISO) and the Semi-Implicit Method for Pressure Linked Equations (SIMPLE) algorithms. Its main structure is inherited from the original PISO, but it allows equation under-relaxation to ensure the convergence of all the equations at each time step, as outlined by Jasak [25]. The fluids (air and water) are treated as incompressible, immiscible, and isothermal, while the interface between them is captured using a VOF technique coupled with a moving mesh function. To resolve fluid behavior under dynamic conditions, the conservation of mass and momentum equations are first solved over the liquid and air regions. These are as follows:

$$\nabla \cdot u = 0 \tag{2}$$

$$\frac{\partial \rho U}{\partial t} + \nabla \cdot (\rho U U) - \nabla \cdot \left(\mu \left(\nabla U + (\nabla U)^T \right) \right) = \rho g - \nabla p \tag{3}$$

where U denotes the velocity, μ denotes the dynamic viscosity, ρ is the density, g is the gravitational acceleration, and t is the time. In order to the capture the air–liquid interface, the VOF method introduces a function α with a value of zero if the mesh cell is completely filled with air and unity if it is completely filled with water [28]. The free surface is formed by the cells containing the value between unity and zero. A special technique can be used to find the derivative of α, keeping in view that it is a step function. These derivatives give the normal direction of quick change in the value of α. Eventually, after determining the normal direction and value of α, a line can be drawn in the boundary cell to show the free surface of the fluid and estimate the interface. The time-dependent governing equation of α is given below:

$$\frac{\partial \alpha}{\partial t} + \nabla \cdot (U\alpha) + \nabla \cdot (U_C \alpha (1 - \alpha)) = 0. \tag{4}$$

The Volume of Fluid method in the solver makes use of an artificial compression term $\nabla \cdot U_C\ \alpha$ $(1 - \alpha)$ which is added in phase Equation (4). U_C is the artificial compressive velocity, and equals to |UC| = min[Cα |U|,max(|U|)]. The fluid density and dynamic viscosity are calculated using following relationships:

$$\rho = \alpha \rho_{water} + (1 - \alpha)\mu_{air}) \tag{5}$$

$$\mu = \alpha \mu_{water} + (1 - \alpha)\mu_{air}) \tag{6}$$

where ρ_{water} is the density of water, ρ_{air} is the density of air, α is the fluid fraction of water, ρ is the overall density of the control volume, μ is the kinematic viscosity of the overall control volume, and μ_{water} and μ_{air} are the kinematic viscosities of water and air, respectively. The term $(1 - \alpha)$ shows that if a cell is α% filled with water then the remaining portion will be filled with air.

The physical properties used in the study include the density of water, ρ_{water} of 998 kg/m^3, and dynamic viscosity μ_{water} of 8.94×10^{-4} kg/(m.s). The density and dynamic viscosity of air are 1.225 kg/m^3 and 1.983×10^{-5} kg/(m.s) respectively, and the surface tension, σ between them is 0.0728 N/m. The frequency of rotation is kept at 2 Hz, while the roll motion amplitude is 4° degree. The upper portion of the container has an atmospheric pressure of 101,000 Pa. The roll center is taken in the middle of the tank (with coordinates of 0 0 −0.225), whereas this study excludes heave and sway motion.

The next phase in the mathematical modelling was the extraction of necessary data from the multiphase description of a dynamic fluid domain. This included the calculation of the CoM and the Cartesian components of its acceleration. The solver was modified to calculate the CoM at each time instant in order to solve Equation (1). For this purpose, let us suppose we have two masses M_1 and M_2 and their distance from origin X_1 and X_2 as shown in Figure 4.

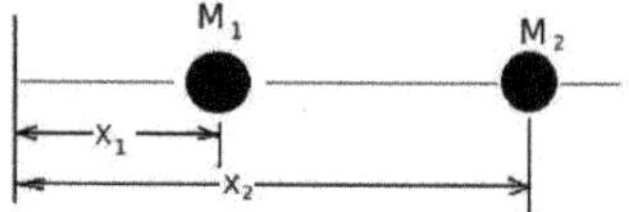

Figure 4. Center of mass two bodies.

Here, the CoM was between the center of both masses and was calculated by summing the products of the initial masses and distances and dividing this sum with the sum of the masses.

$$X_{COM} = \frac{M_1X_1 + M_2X_2}{M_1 + M_2} \tag{7}$$

This CoM was calculated for two bodies only for the horizontal distance. While dealing with fluids, CFD gives us several control volumes that are individual masses, and they are three dimensional and can be arranged in two dimensions considering the problem at hand. Equation (7) can be used to obtain the CoM in a three-dimensional space:

$$COM = \frac{\sum m_i \vec{r}_i}{\sum m_i} \tag{8}$$

In this equation, the *COM* is obtained in vector format, r_i is the position vector of the *i*'th mass from the origin, and m_i is mass of that *i*'th mass or control volume. In interFoam density is constant, so we can write the mass as ρV_i and implement the above equation as in Equation (9).

$$COM = \frac{\sum \rho V_i \vec{r}_i}{\sum \rho V_i} \tag{9}$$

To implement this condition in openFoam, the following code is added in interStabilityFoam.C:

```
CoM = sum ( rho * mesh.V() * mesh.C().dimensionedInternalField())/sum( rho * mesh.V());
```

The spatiotemporal acceleration of the CoM was computed from the following second order numerical difference scheme:

$$a_x = \Delta_k^2[f](x) = \sum_{i=0}^{2}(-1)^i \binom{2}{i} f(x - ih) \tag{10}$$

where the equation is using binomial coefficients after the summation sign, x is the location of CoM, h is the time step, and f finds the CoM. Equation (10) provided a numerical estimate of the acceleration, which was implemented in the solver after the start of time loop by first assigning the x, y, and z components of the CoM values calculated previously to scalar variables as follows:

```
scalar CoMx=CoM.value().x();
scalar CoMy=CoM.value().y();
scalar CoMz=CoM.value().z();
```

Initially, for the first two steps there was no acceleration based on the backward differencing approach. From the third time step onwards, the acceleration along the x-axis was calculated by the function:

$$Ax = (CoMx3 - 2.0 * CoMx2 + CoMx1)/pow((runTime.deltaTValue()),2);$$

where CoMx3 is the x-coordinate of CoM at the current time step "i", while CoMx2 and CoMx1 are the x-coordinates of the previous time steps "i−1" and "i−2", respectively. An if-else loop over a parallel time step was added to the code, in which three time step values of the CoM are stored to calculate acceleration using backward differencing. At every time step "i" the values from the previous time step "i−1" and "i−2" are updated into the parallel time domain. The values of the CoM component are redefined at every time step. The ensuing information was then employed in the estimation of the zero-moment point using Equation (1) as follows:

$$Xzmp = CoMx - ((Ax * CMY)/9.81);$$

Now, the schematic for the ZMP solver can be drawn, as in Figure 5.

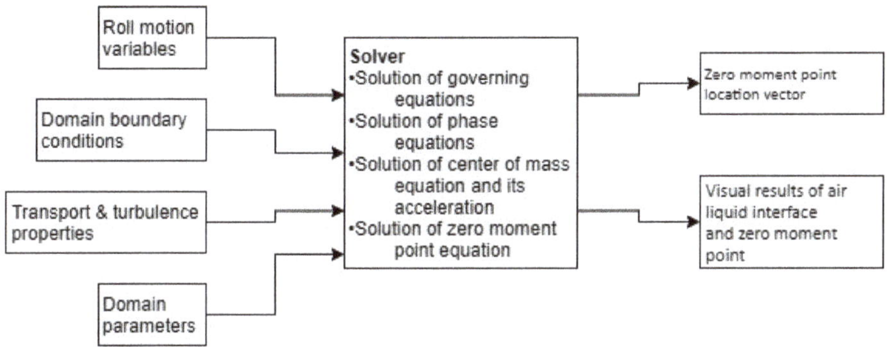

Figure 5. Outline of solver processes.

In this flow diagram, the roll motion parameters, domain boundary conditions, transport, and turbulence models, as well as domain parameters, are to be specified by the user. The solver solves fluid domain with the governing equations of fluids, i.e., the Navier–Stokes equation, the continuity equation, the phase equation for the air–liquid interface, and the CoM equation to obtain the ZMP. While the Courant number Co should be restricted ~0.25 for the stability of transient simulations in explicit interface capturing solvers, the semi-implicit solver used in the study was unconditionally stable and the value of Co was determined by the solver based on temporal accuracy requirements.

2.3. Test Cases

Seven experimental simulations were carried out in OpenFOAM. First, an experiment with an 18% filled tank was carried out to benchmark the computational method. Subsequent experiments were carried at different fill levels (20%, 30%, 40%, 50%, 60%, 70%, and 80%) in a rectangular tank of the same dimensions. Higher fill levels restrict the physical space available for liquid sloshing so were not significant to the present study.

3. Results and Discussion

The pressure data at the probe from the published results was compared with the pressure at the probe location obtained from the numerical study, as shown in Figure 6, in order to validate computation of air–water multiphase fluid dynamics, after which the ZMP was calculated for the fluid domain to assess stability against tipping over. The maximum deviation between the experimental study and the numerical study was observed to be less than 5.5% in terms of pressure, while the occurrence of the peak pressure pulse showed a maximum of 0.1 s temporal distortion. The numerical implementation of the roll angle had very little deviation from experimental values (<0.01%).

This comparison shows that both domains have almost the same angle at same time in the study, leading to very similar pressure perturbations. Different test cases were prepared in this study and pressures extracted for all of them. A comparison of the pressure on the probe at different fill levels is shown in Figure 7. This highlights the behavior of pressure with fill levels indicating that while the occurrence of the peak pressure was over similar roll angles for all fill levels, the magnitude of the pressure peak at the probe position was inversely proportional to the fill level. The only deviation from this inverse relationship was observed between the 18% and the 30% fill level, where the 18% filled tank showed a slightly shorter pressure peak than 30% fill level. This is because the availability of volumetric space allows the fluid to experience greater levels of slosh at lesser fill levels, while at very low fill levels the amount of fluid subjected to slosh is small and unable to generate higher pressure peaks. The maximum value of peak pressure was observed in the 30% filled tank, while tanks that were more than 50% full showed a negligible pressure peak.

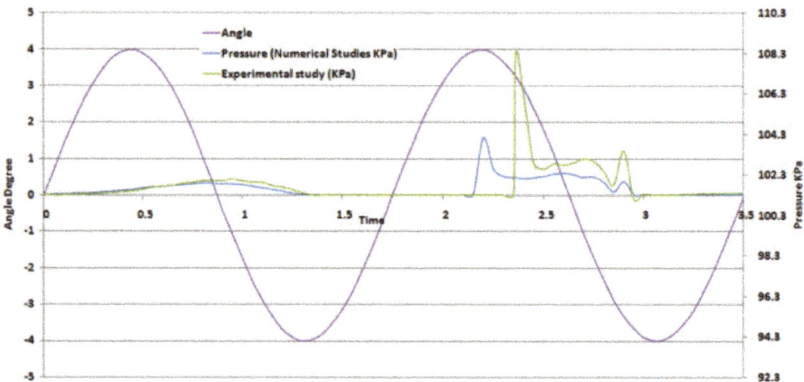

Figure 6. Pressure comparison (experimental vs. numerical).

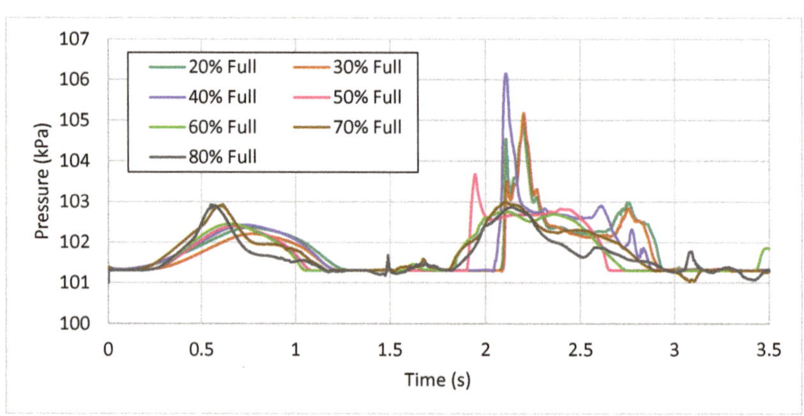

Figure 7. Pressure curves at different fill levels.

The CoM has a key role in the application of the ZMP. Therefore, after analyzing the pressure it was necessary to observe the behavior of the CoM in relation to the motion of the fluid domain. The change in position of the CoM from the origin with respect to time is shown in Figure 8, along with the pressure sensed by two probes placed on the left- and right-side walls at a distance from the bottom coinciding with the air water interface for an 18% fill level. This comparison shows that pressure is exerted on the walls due to the motion of fluid and, ultimately, the CoM.

It was observed that in the first slosh, corresponding to a half roll motion, the CoM did not deviate much, while for the second roll motion the CoM moved a greater value due to a larger amount of liquid slosh. The CoM thus indicates a sinusoidal behavior with increasing amplitude.

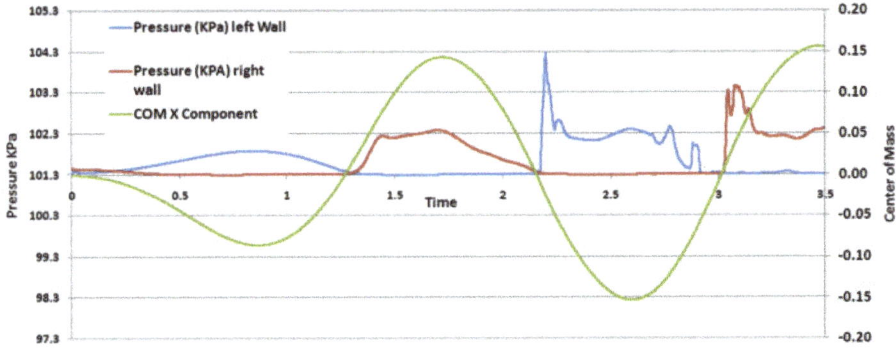

Figure 8. Center of Mass (CoM) and pressure at the left and right probes.

We also observed that, as the liquid bulk moved from one side of the container to the other, the pressure sensed by the probe on that side experienced a significant jump. When pressure on left wall increased in first slosh the CoM moved to the left side, giving negative value. After that, the fluid bulk moved towards right side of the tank, due to which the pressure on the left wall probe dropped to zero and right wall probe increased, as shown in Figure 9. Whereas the CoM showed a positive value, the displacement of the CoM was negative in the left half of container and positive in the right half. It was also observed that pressure was higher in the second slosh than that in the first slosh. Thus, the observed behavior of the CoM was in perfect harmony with the roll motion as indicated by the pressures sensed by the two probes.

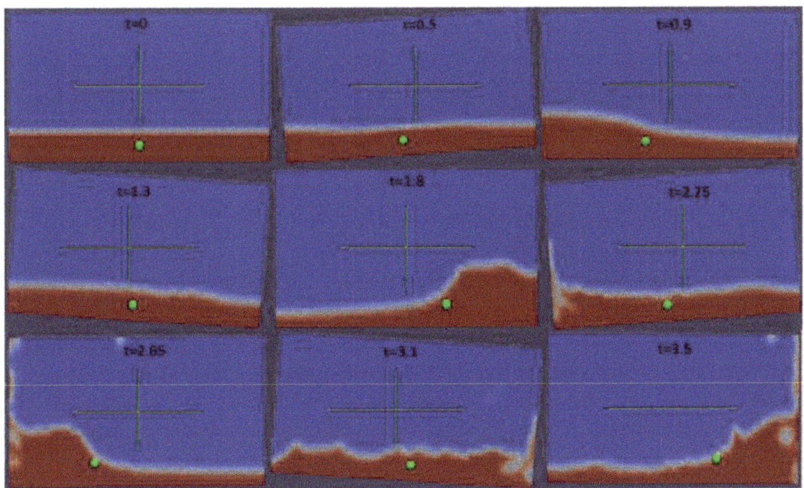

Figure 9. Center of mass and air–liquid interface.

Next, we summarize the movement of the CoM along with the fluid bulk and the air–liquid interface in Figure 10. The position of the CoM at different time steps can be visually observed in the figure at different angles corresponding to previously described range of roll angles.

The results of the present study highlight the close relation between liquid slosh and CoM, which has hitherto received little attention in the literature due to experimental limitations in determining the CoM. Figure 11 shows the change in behavior of the CoM across the different fill levels of the tanker. This indicates that with the decrease in fill level the CoM moves outwards from the center of domain. Similar to the pressure curves in Figure 8, fill levels greater than 50% exhibited minor changes in the displacement of the CoM, while, contrary to the pressure curve, the maximum displacement was shown by a 18% fill level (instead of 30%). Overall, the two characteristics exhibited similar trends and complemented each other.

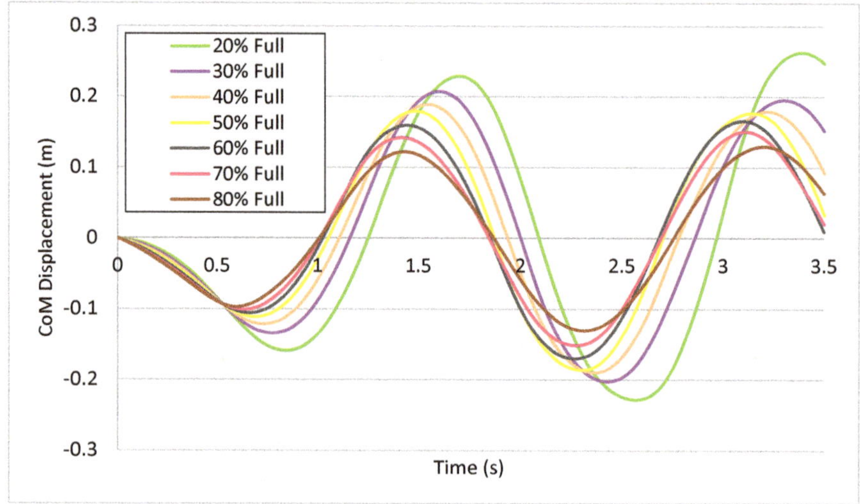

Figure 10. Center of Mass vs. fill level of container.

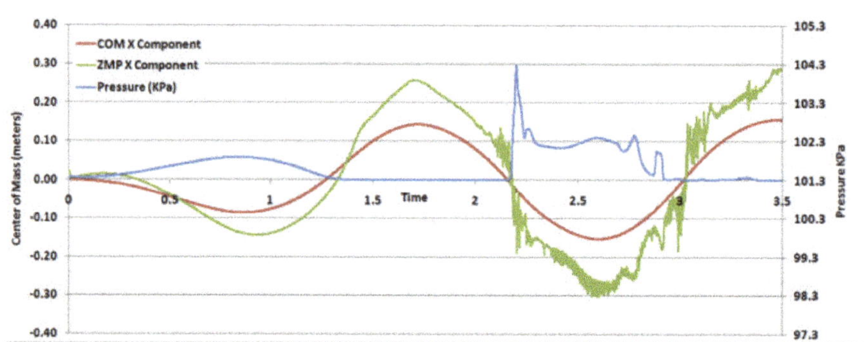

Figure 11. Zero Moment Point (ZMP), Center of Mass, and pressure comparison.

Using the CoM information and estimating the ZMP according to Equation (1), we plotted the ZMP, the CoM, and the left pressure probe with respect to time for 18% fill level (Figure 11). This showed alternation of the ZMP around the center of domain corresponding to the movement of the fluid bulk. In later roll motions, oscillations were observed in the ZMP location which correspond to sloshing frequencies, while, overall, the Zero Moment Point curve follows the CoM curve (Figure 12).

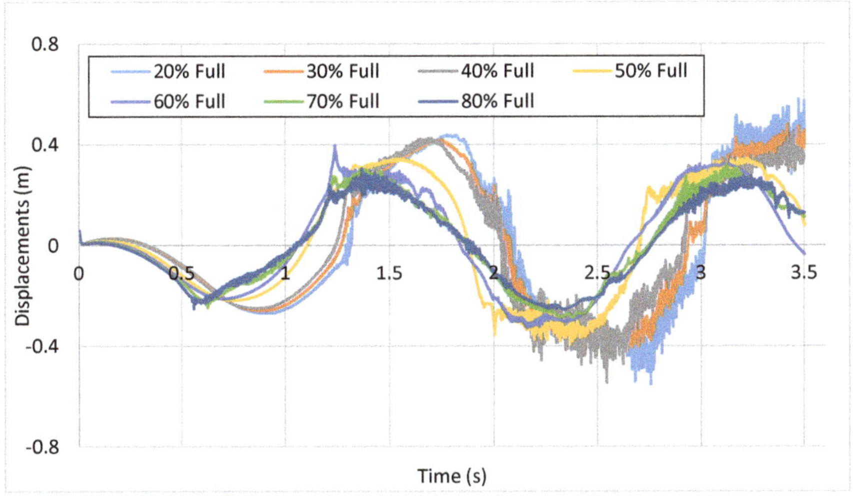

Figure 12. Displacement of horizontal component of ZMP vs. time at different fill levels.

The evolution of the ZMP across different fill levels is summarized in Figure 12. We observed oscillations in 20%, 30%, and 40% fill levels, while the onset of these oscillations was earliest in the smallest fill level. It can be seen in the case of a 30% fill level that after the onset of oscillations there was slight dampening before it peaked again, while at fill levels of 50% or greater there was no oscillation in the ZMP curve. Hence, we suggest that the ZMP of a dynamic liquid domain can be calculated to estimate its stability, determined by retaining the ZMP within the support polygon as in the present study cases.

4. Conclusions

The stability of dynamic fluid cargo for autonomous vehicles and liquid-handling robots poses a significant hurdle for the advancement of robotics. The indeterministic nature of sloshing liquids that obstruct development of rigorous motion planning has been addressed in this work by a novel approach that incorporates the Zero Moment Point. Computational fluid dynamics (CFD) were employed to capture the displacements of the Center of Mass of a rectangular liquid domain subjected to roll motion by solving the Navier–Stokes equations coupled with the Volume of Fluid (VOF). The computational results were first validated against experimental data prior to the estimation of ZMP. Analysis of the results shows a strong agreement between the fluid's sloshing motion, the CoM and its ZMP. This enables us to determine the stability of partially filled liquid containers. Further research can be carried out on determining support polygons for complex liquid transport problems involving humanoid robots or autonomous vehicles.

Author Contributions: Conceptualization, M.S. and Y.A.; Methodology, M.U., M.S., Y.A. and E.U.; Software, M.U. and M.S.; Validation, M.U. and M.S.; Investigation, M.U.; Resources, M.S. and E.U.; Data Curation, M.U. and M.S.; Writing—Original Draft Preparation, M.U.; Writing—Review & Editing, M.S.; Supervision, M.S. All authors have read and agreed to the published version of the manuscript.

Funding: This research received no external funding.

Conflicts of Interest: The authors declare no conflict of interest.

References

1. Harada, K.; Kajita, S.; Saito, H.; Morisawa, M.; Kanehiro, F.; Fujiwara, K.; Kaneko, K.; Hirukawa, H. A Humanoid Robot Carrying a Heavy Object. In Proceedings of the 2005 IEEE International Conference on Robotics and Automation, Barcelona, Spain, 18–22 April 2005; pp. 1712–1717.
2. Hirukawa, H. Walking biped humanoids that perform manual labour, Philosophical Transactions of the Royal Society A: Mathematical. *Phys. Eng. Sci.* **2007**, *365*, 65–77. [CrossRef] [PubMed]
3. Kam, L.W.; Gvildys, J. Fluid-structure interaction of tanks with an eccentric core barrel. *Comput. Methods Appl. Mech. Eng.* **1986**, *58*, 51–77.
4. Kargbo, O.; Xue, M.-A.; Zheng, J. Multiphase Sloshing and Interfacial Wave Interaction with a Baffle and a Submersed Block. *J. Fluids Eng.* **2019**, 141. [CrossRef]
5. Kong, F.; Yuan, L.; Zheng, Y.F.; Chen, W. Automatic Liquid Handling for Life Science:A Critical Review of the Current State of the Art. *J. Lab. Autom.* **2012**, *17*, 169–185. [CrossRef] [PubMed]
6. Lorenz, M.G.O. Liquid-Handling Robotic Workstations for Functional Genomics. *J. Assoc. Lab. Autom.* **2004**, *9*, 262–267. [CrossRef]
7. Faiña, A.; Nejati, B.; Stoy, K. EvoBot: An Open-Source, Modular, Liquid Handling Robot for Scientific Experiments. *Appl. Sci.* **2020**, *10*, 814. [CrossRef]
8. Barthels, F.; Barthels, U.; Schwickert, M.; Schirmeister, T. FINDUS: An Open-Source 3D Printable Liquid-Handling Workstation for Laboratory Automation in Life Sciences. *SLAS Technol. Transl. Life Sci. Innov.* **2020**, *25*, 190–199. [CrossRef]
9. Samad, A.; Khan, A.A.; Sajid, M.; Zahra, R. Assessment of biofilm formation by pseudomonas aeruginosa and hydrodynamic evaluation of microtiter plate assay. *J. Pak. Med Assoc.* **2019**, *69*, 666–671.
10. Zahra, R.; Khan, A.A.; Sajid, M. Hydrodynamic Evaluation of Microtiter Plate Assay Using Computational Fluid Dynamics for Biofilm Formation. In *Proceedings of the Asme-Jsme-Ksme 2019 8th Joint Fluids Engineering Conference*; American Society of Mechanical Engineers: New York, NY, USA, 2019.
11. Delorme, L.; Colagrossi, A.; Souto-Iglesias, A.; Zamora-Rodriguez, R.; Botia-Vera, E. A set of canonical problems in sloshing, Part I: Pressure field in forced roll—Comparison between experimental results and SPH. *Ocean Eng.* **2009**, *36*, 168–178. [CrossRef]
12. Liu, D.; Lin, P. Three-dimensional liquid sloshing in a tank with baffles. *Ocean Eng.* **2009**, *36*, 202–212. [CrossRef]
13. Sugihara, T.; Nakamura, Y.; Inoue, H. Real-time humanoid motion generation through ZMP manipulation based on inverted pendulum control. In Proceedings of the 2002 IEEE International Conference on Robotics and Automation (Cat. No.02CH37292), Washington, DC, USA, 11–15 May 2002; Volume 2, pp. 1404–1409.
14. Vukobratovic, M.; Borovac, B. Zero Moment Point—Thirty Five Years of its Life. *Int. J. Hum. Robot.* **2004**, *1*, 157–173. [CrossRef]
15. Kajita, S.; Kanehiro, F.; Kaneko, K.; Fujiwara, K.; Harada, K.; Yokoi, K.; Hirukawa, H. Biped walking pattern generation by using preview control of zero-moment point. In Proceedings of the 2003 IEEE International Conference on Robotics and Automation (Cat. No.03CH37422), Taipei, Taiwan, 14–19 September 2003; Volume 2, pp. 1620–1626.
16. Sun, G.; Wang, H.; Lu, Z. A Novel Biped Pattern Generator Based on Extended ZMP and Extended Cart-Table Model. *Int. J. Adv. Robot. Syst.* **2015**, *12*, 94. [CrossRef]
17. Scianca, N.; Simone, D.D.; Lanari, L.; Oriolo, G. MPC for Humanoid Gait Generation: Stability and Feasibility. *IEEE Trans. Robot.* **2020**, 1–18. [CrossRef]
18. Faltinsen, O.M.; Lagodzinskyi, O.E.; Timokha, A.N. Resonant three-dimensional nonlinear sloshing in a square base basin. Part 5. Three-dimensional non-parametric tank forcing. *J. Fluid Mech.* **2020**, *894*, A10. [CrossRef]
19. Ibrahim, R.A. Recent Advances in Physics of Fluid Parametric Sloshing and Related Problems. *J. Fluids Eng.* **2015**, *137*, 090801. [CrossRef]
20. Sanapala, V.S.; Rajkumar, M.; Velusamy, K.; Patnaik, B.S.V. Numerical simulation of parametric liquid sloshing in a horizontally baffled rectangular container. *J. Fluids Struct.* **2018**, *76*, 229–250. [CrossRef]
21. Ansari, M.; Firouz-Abadi, R.; Ghasemi, M. Two phase modal analysis of nonlinear sloshing in a rectangular container. *Ocean Eng.* **2011**, *38*, 1277–1282. [CrossRef]

22. Boroomand, B.; Bazazzadeh, S.; Zandi, S.M. On the use of Laplace's equation for pressure and a mesh-free method for 3D simulation of nonlinear sloshing in tanks. *Ocean Eng.* **2016**, *122*, 54–67. [CrossRef]
23. Jäger, M. *Fuel Tank Sloshing Simulation Using the Finite Volume Method;* Springer: Berlin/Heidelberg, Germany, 2019.
24. Tang, Y.; Yue, B.; Yan, Y. Improved method for implementing contact angle condition in simulation of liquid sloshing under microgravity. *Int. J. Numer. Methods Fluids* **2019**, *89*, 123–142. [CrossRef]
25. Jasak, H.; Jemcov, A.; Tukovic, Z. OpenFOAM: A C++ library for complex physics simulations, in International workshop on coupled methods in numerical dynamics. In Proceedings of the Ternational Workshop on Coupled Methods in Numerical Dynamics IUC, Dubrovnik, Croatia, 19–21 September 2007; pp. 1–20.
26. Souto-Iglesias, A.; Botia-Vera, E.; Martín, A.; Pérez-Arribas, F. A set of canonical problems in sloshing. Part 0: Experimental setup and data processing. *Ocean Eng.* **2011**, *38*, 1823–1830. [CrossRef]
27. Siciliano, B.; Khatib, O. *Springer Handbook of Robotics;* Springer: Berlin/Heidelberg, Germany, 2016.
28. Hu, Z.Z.; Greaves, D.; Raby, A. Numerical wave tank study of extreme waves and wave-structure interaction using OpenFoam®. *Ocean Eng.* **2016**, *126*, 329–342. [CrossRef]

Article

Hierarchical Synchronization Control Strategy of Active Rear Axle Independent Steering System

Bin Deng [1], Han Zhao [1], Ke Shao [2], Weihan Li [1,*] and Andong Yin [1]

[1] School of Automotive and Transportation Engineering, Hefei University of Technology, Hefei 230009, China; ddc2013@mail.hfut.edu.cn (B.D.); hanzhao@hfut.edu.cn (H.Z.); yinandong@hfut.edu.cn (A.Y.)

[2] Center for Artificial Intelligence and Robotics, Tsinghua Shenzhen International Graduate School, Tsinghua University, Shenzhen 518055, China; shao.ke@sz.tsinghua.edu.cn

* Correspondence: liweihan@hfut.edu.cn

Received: 28 April 2020; Accepted: 19 May 2020; Published: 20 May 2020

Abstract: The synchronization error of the left and right steering-wheel-angles and the disturbances rejection of the synchronization controller are of great significance for the active rear axle independent steering (ARIS) system under complex driving conditions and uncertain disturbances. In order to reduce synchronization error, a novel hierarchical synchronization control strategy based on virtual synchronization control and linear active disturbance rejection control (LADRC) is proposed. The upper controller adopts the virtual synchronization controller based on the dynamic model of the virtual rear axle steering mechanism to reduce the synchronization error between the rear wheel steering angles of the ARIS system; the lower controller is designed based on an LADRC algorithm to realize an accurate tracking control of the steering angle for each wheels. Experiments based on a prototype vehicle are conducted to prove that the proposed hierarchical synchronization control strategy for the ARIS system can improve the control accuracy significantly and has the properties of better disturbances rejection and stronger robustness.

Keywords: active rear axle independent steering (ARIS) system; hierarchical synchronization control; virtual coupling control; linear active disturbance rejection control (LADRC)

---

1. Introduction

With the development of electronic and control technologies, the active rear-wheel steering (ARS) system for automobiles has been actively studied as an effective vehicle maneuvering technology that can enhance the handling stability of the vehicle and provide better comfort and active safety for the driver [1–3]. The general ARS system adopts an additional rear wheel steering gear (e.g., rack and pinion steering gear), which is basically the same as the front wheel steering gear. The electric actuator of the ARS system is widely used in electric vehicles; compared with an hydraulic actuator, it can greatly reduce the mass and complexity of the ARS system. However, the kinematic relationship between the left and right rear wheels is still fixed in traditional Ackerman geometry.

To achieve a more flexible steering control, active rear-axle independent steering (ARIS) system has drawn many attentions. The ARIS technology can be potentially applied in wheeled vehicles to realize a more intelligent and complex motion. Compared with the traditional steering trapezoid mechanism, the mechanical connection between the wheels of the rear axle is cancelled, and the flexible communication connection is replaced where each wheel is directly steered by an individual actuator motor [4]. Since each wheel of the rear axle has an independent actuator, it gets rid of the shortcomings and limitations of the traditional rack and pinion steering mechanism and further improves the dynamic performance of the vehicle. Explicitly, the ARIS system can provide the vehicle with a smaller turning radius and a more flexible driving mode. When working together with an anti-lock braking system (ABS), it can further shorten the braking distance of the vehicle on a slippery surface [5,6]. The rear axle steering system was introduced in the Porsche 911 Turbo. Two independent motors are used to control the two rear wheels, respectively, achieving more precise rear wheel steering angle control. In 2019, the company of AEV Robotics introduced a lightweight modular vehicle system (MVS), and an electric four-wheel independent steering system was installed.

The ARIS system improves the overall performance of the vehicle, but also increases the difficulty of control. Lots of literatures focus on control strategies of handling stability. In Reference [7–12], H∞ optimal control, game-based hierarchical cooperative control and receding horizon control methods were developed to enhance handling and stabilities. But few have studied steering angle tracking control.

The vehicle is affected by many uncertainties in real driving conditions and each steering wheel is subject to a complex external disturbance. Explicitly, in the process of steering motion, if there exists a large tracking control error or a serious lag of the rear wheel steering angle in the ARIS system, the Ackerman error between the rear axle steering angles will increase. In this case, the sideslip and drag may occur on the tire, which will further aggravate tire wear and the vehicle driving safety will not be guaranteed in serious cases [13]. In order to ensure vehicle stability and reduce tire wear, the steering angle between rear axle vehicles should conform to Ackermann principle. However, the Ackerman principle cannot be guaranteed by the steering mechanism because each steering wheel of the ARIS system is independent. So, it is necessary to ensure the rear wheel tracking target angle and achieve the Ackerman principle by the developed control method. In consideration of the uncertainty of the system parameters and the nonlinear characteristics of the tire [14], the controller design for ARIS system becomes more difficult. In addition, due to the complexity of vehicle driving conditions, the safety and stability of the vehicle will be significantly affected if the system fails, in which case fault diagnosis and fault-tolerant control can been considered in the controller design [15].

Therefore, to obtain a desired steering angle control performance, a robust controller with higher response speed and control accuracy is required. This is very important to the driving performance of the vehicle. In addition, not only should the steering angle tracking control be realized, but also the steering kinematics characteristics of the vehicle should be considered to ensure steering angle synchronized control of the left and right of the rear axle wheels [16]. In this paper, we define the process of realizing the steering angle between left and right rear wheels to satisfy the Ackerman principle as synchronized control. So, we mainly focus on the high-precision tracking control and synchronized control of the steering angle of the ARIS system in this paper.

In the four-wheel steering system of vehicles, the traditional control method is often used. The single wheel tracking controller is designed by decoupling multiple steering wheels through motion planning. Proportional-integral-derivative (PID) control, adaptive control, sliding mode control, fuzzy control and other methods are used in steering angle controller. The adaptive control method was used to realize the steering wheel angle tracking control in [17,18], but its prediction and estimation method for the road disturbance is based on the tire characteristics in a better linear region. Neither the nonlinearity of tires nor the uncertainty of road conditions in the actual situation are taken into account. Aiming at the nonlinearity of the steering system, sliding mode control (SMC) was adopted to realize the tracking control of the steering angle in [19–21], which improves the robustness of the controller. SMC can solve the nonlinearity of a control system and has strong robustness, but the chattering phenomenon at the balance point is still a problem to be overcome in practical application.

Moreover, even if the expected tracking performance of each independent steering wheel tracking controller is achieved, it cannot guarantee that the synchronization error between wheels can be reduced. A master-slave control strategy based on disturbance sliding mode control was proposed for a two motor coupled steer by wire system in [22]. However, when the slave steering wheel is disturbed by external disturbance, the master steering wheel will not respond. The master-slave control method is suitable for the system with an obvious master-slave relationship, but not for the system with a mutual coupling relationship. In Reference [23], the contour error model for the four-wheel independent steering system (4WIS) mobile robot system was established and the synchronization control was designed based on the cross coupling algorithm, which can reduce the synchronization error between wheels and improve the accuracy of the tracking trajectory. But complex sensors are needed to acquire the vehicle's attitude and position information to calculate the trajectory error. It is not suitable for general vehicles without navigation sensors. In Reference [24], Kevin Payette proposed the concept of virtual shaft control applied to the synchronization control of a multi-axis machine. The virtual shaft control and improved electronic line-shafting are widely applied in multi-axis machine systems [25,26]. In [27], fault diagnosis and fault-tolerant control techniques were proposed for an over-actuated electric vehicle to deal with the controller reconfiguration in the presence of system faults and structured and unstructured uncertainties.

In this study, firstly, we will establish the model of the dynamics and kinematics of the ARIS system. Then, considering the coupling between wheels, a hierarchical synchronization control strategy will be designed to reduce the synchronization error between rear wheels. The upper controller adopts the virtual synchronization controller based on the dynamic model of the virtual rear wheel steering mechanism, while the lower controller is designed based on the linear active disturbance rejection control (LADRC) algorithm. The LADRC method has the advantages that it does not depend on a model and is not sensitive to parameter change and disturbance [28,29]. Finally, the control strategy proposed in this paper will be verified by a real vehicle test.

The remainder of this paper is organized as follows. In Section 2, the dynamics and kinematics of the ARIS system are constructed. In Section 3, the hierarchical synchronization control strategy for the ARIS system is investigated. The virtual synchronization controller is designed based on a dynamic model of the virtual rear wheel steering mechanism, and the steering angle tracking controller for each steering wheel is developed based on LADRC. In Section 4, the experiment results on a real vehicle with an ARIS system is presented. Finally, conclusions and future work are given in Section 5.

2. Dynamic Modeling

2.1. Dynamic Modeling of the Active Rear-Axle Independent Steering (ARIS) System

The steering mechanism of the ARIS system is realized by two electric actuators with adjustable length installed on the left and right rear wheels. A linear electric drive motor with reducer (e.g., gear reducer and screw-nut transfer) is selected as the electric actuator. In order to facilitate the establishment of the model, the steering mechanism and actuator are simplified, and the simplified dynamic model is shown in Figure 1.

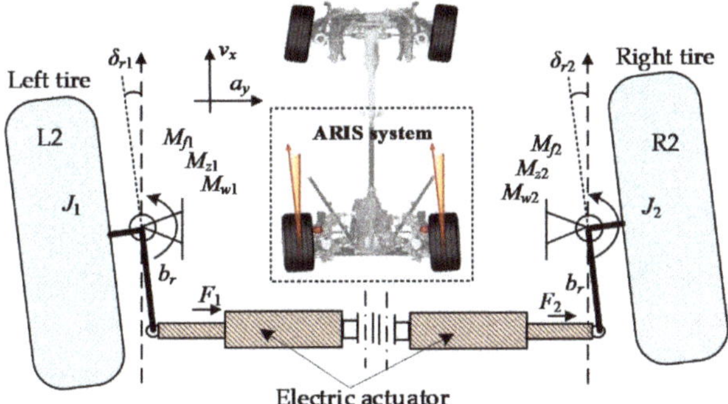

Figure 1. The dynamic model of active rear axle independent steering mechanism.

Through the mechanical analysis, the plant model of each wheel can be described by

$$\begin{cases} J_1\ddot{\delta}_{r1} + c_1\dot{\delta}_{r1} + k_1\delta_{r1} = F_1 \cdot b_r \cos\delta_{r1} + M_{f1} + M_{z1} + M_{w1} \\ J_2\ddot{\delta}_{r2} + c_2\dot{\delta}_{r2} + k_2\delta_{r2} = F_2 \cdot b_r \cos\delta_{r2} + M_{f2} + M_{z2} + M_{w2} \end{cases} \tag{1}$$

where J_i is the total moment of inertia of rotating parts; F_i is the driving force of electric actuator; M_{fi} is the friction moment; M_{zi} is the tire self-aligning torque around the kingpin; M_{wi} is the moment of uncertain force on the ground; c_i is the equivalent damping coefficient; k_i is the equivalent stiffness; b_r is the length from the connection point to the kingpin; δ_{ri} is the rear wheel angle. In the above parameter variables, subscript i= 1,2 indicates the left and right wheels of rear axle, respectively.

According to Coulomb's law, the friction moment produced by the rotation of the tire around the kingpin can be described as

$$M_{fi} = M_{sfi} + F_{zri}\mu_{ri} \tag{2}$$

where M_{sfi} is the initial value of friction moment; F_{zri} is the rear wheel vertical load; and μ_{ri} is the Coulomb friction coefficient. In order to ensure the straight-line driving stability of the vehicle, the rear wheel kingpin should have a certain inclination, which will generate self-aligning torque around the kingpin. According to [30], the self-aligning torque, M_{zi}, can be given by

$$M_{zi} = (F_{zi}s_i \sin 2\gamma_i \sin \delta_{ri})/2 \tag{3}$$

where γ_i is the kingpin inclination, and s_i is the horizontal distance from the tire center to the kingpin.

The moment of uncertain force on the ground, M_{wi}, includes the coupling force caused by the synchronization error between rear axle wheels. It is also related to vehicle speed v_x and lateral acceleration a_y. Explicit modeling of this moment is not possible. Instead, it can be described with implicit form as

$$M_{wi} = f_{wi}(\delta_{r1}, \delta_{r2}, v_x, a_y) \tag{4}$$

It can be seen from Equations (1) to (4) that the dynamic model of the ARIS system has strong nonlinearity and uncertainty. For this system, it must rely on a robust controller with strong anti-disturbance capacity to achieve high-precision tracking performance in the presence of system uncertainties.

2.2. Kinematic Model of Rear Wheel Steering

The front axle steering mechanism of the vehicle is a type of traditional rack and pinion trapezoid steering mechanism, while the rear axle adopts the ARIS system. There is no mechanical connection between the front and rear steering axles, nor between the left and right wheels of the rear axle. In order to ensure the steering performance of the vehicle and reduce the sideslip and the wear of tires, the steering angle of each rear wheels must conform to the kinematic constraint model of the whole vehicle. Firstly, the steering angle scale factor between the front axle and the rear axle is obtained according to the vehicle dynamics model. Then, the steering angles of the left and right wheels of the rear axle are obtained according to the Ackermann steering geometry.

In order to obtain the ratio of front and rear wheel steering angle at low speed steady driving, the dynamic model of four-wheel steering vehicle is simplified as a two degree of freedom vehicle model, given by [31].

$$\begin{cases} mv_x(\dot{\beta} + \omega) + (k_{\alpha f} + k_{\alpha r})\beta + \frac{bk_{\alpha r} - ak_{\alpha f}}{v_x}\omega = k_{\alpha f}\delta_f + k_{\alpha r}\delta_r \\ I_z\dot{\omega} + (ak_{\alpha f} - bk_{\alpha r})\beta + \frac{a^2k_{\alpha f} + b^2k_{\alpha r}}{v_x}\omega = ak_{\alpha f}\delta_f - bk_{\alpha r}\delta_r \end{cases} \tag{5}$$

where m is the mass of the vehicle body; I_z is the yaw moment of inertia of the vehicle; $k_{\alpha f}$ and $k_{\alpha r}$ are the cornering stiffness of the front and rear wheels; a and b are the distance from the front and rear axles to the vehicle mass center; δ_f is the front axle wheel angle; δ_r is the rear axle wheel angle; v_x is the vehicle longitudinal speed; ω is the vehicle yaw rate; and β is the vehicle sideslip angle.

Active rear wheel steering can produce an active sideslip angle, which can reduce the sideslip angle of the vehicle and improve the stability of the vehicle. In a steady state, $d\omega/dt$ = 0 and β = 0 are chosen as the objective functions, and the equivalent active steering target angle δ_r of rear wheel can be obtained by [32]

$$\delta_r = f_k(\delta_f, v_x) = \delta_f \frac{ak_{\alpha f}mv_x^2 - b(a+b)k_{\alpha f}k_{\alpha r}}{bk_{\alpha r}mv_x^2 + a(a+b)k_{\alpha f}k_{\alpha r}} \tag{6}$$

Figure 2 shows the kinematic model of four-wheel steering with two degrees of freedom. The dotted box in Figure 2 shows the steering angle kinematic relationship of the ARIS system. Generally, the front axle of the vehicle adopts the rack and pinion steering mechanism. In order to simplify the model, N is the reduction ratio from the driver's steering wheel angle to the steering angle of the front axle. f_k represents the ratio between front axle steering angle and rear axle steering angle. According to the Ackerman steering geometry principle [33], the ideal kinematic constraint equation of rear wheel steering is established as

$$\begin{cases} \delta_f = \theta_{sw}/N \\ \tan\delta_{r1} = \rho\tan\delta_r/(\rho - w/2) \\ \tan\delta_{r2} = \rho\tan\delta_r/(\rho + w/2) \\ \rho = l/(\tan\delta_f + \tan\delta_r) \end{cases} \tag{7}$$

where θ_{sw} is the driver's steering wheel angle; $l = a + b$ is the wheel base; w is the wheel track; ρ is the turning radius; δ_{r1} is the left rear wheel steering angle and δ_{r2} is the right rear wheel steering angle.

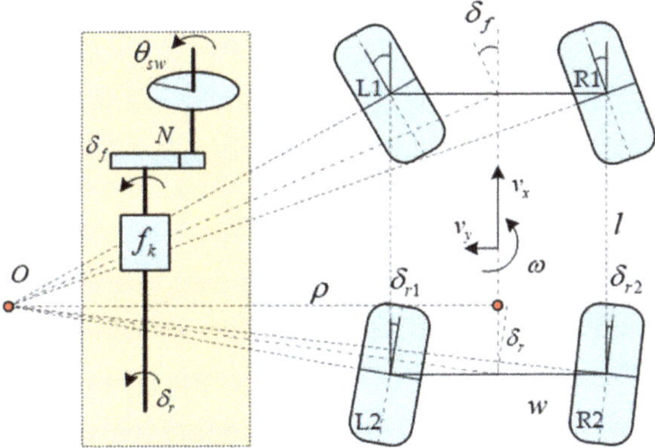

Figure 2. The kinematic model of four wheel steering with two degrees of freedom.

Substituting (6) into (7), the constraint equation can be obtained as

$$\begin{cases} \delta_{r1} = \arctan(\rho\tan\delta_r/(\rho - w/2)) \\ \delta_{r2} = \arctan(\rho\tan\delta_r/(\rho + w/2)) \end{cases} \tag{8}$$

where $\delta_r = \delta_f \frac{ak_{\alpha f}mv_x^2 - blk_{\alpha f}k_{\alpha r}}{bk_{\alpha r}mv_x^2 + alk_{\alpha f}k_{\alpha r}}$.

3. Hierarchical Synchronization Control Strategy Design

In order to solve the synchronization control problem of the independent rear wheel steering actuator, motivated by the virtual shaft control algorithm, a hierarchical synchronization control strategy based on the virtual rear axle steering dynamics model is proposed. The block diagram of the proposed control strategy of ARIS system is shown in Figure 3.

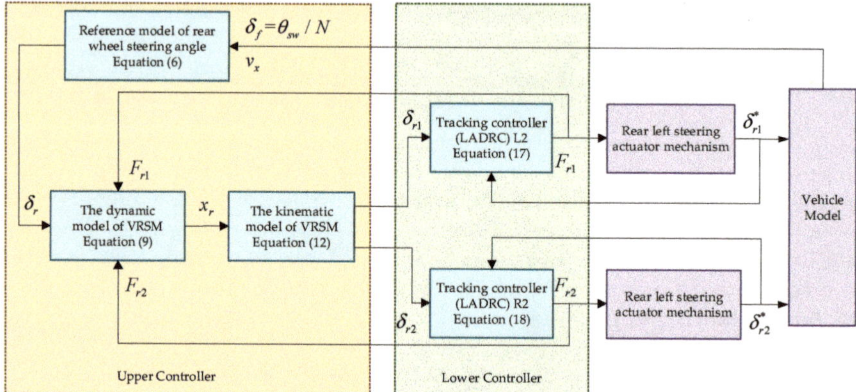

Figure 3. The hierarchical synchronization control strategy of active rear axle independent steering (ARIS) system.

The upper controller in the left dotted box of Figure 3 adopts the virtual synchronization controller to reduce the synchronization error between rear wheels. The upper controller consists of three parts. Firstly, according to the vehicle dynamics, the reference steering angle of the rear wheel is obtained, which is shown in Equation (6). Then, the dynamic model of the virtual rear axle steering mechanism (VRSM) including the synchronization control force and the external disturbance force is established, which is shown in Equation (9). Finally, the target tracking steering angles of the left and right rear axle wheels, δ_{r1} and δ_{r2}, are redistributed according to the Ackermann kinematic model, which is shown in Equation (12).

Considering the nonlinearity, uncertainty and external disturbance of the ARIS system, the lower controller adopts the LADRC controller to realize each steering wheel angle tracking control. The disturbance inside and outside the ARIS system is observed by the linear extended state observer (LESO) in the LADRC controller and can be compensated to the designed lower controller to reduce the influence of external disturbances on the control system.

3.1. Virtual Synchronization Controller

The dotted box in Figure 4 represents the established VRSM, which adopts the concept of a traditional rack and pinion steering gear. Considering system stiffness and damping, the dynamic model of VRSM can be described as Equation (9) according to Newton's law.

$$m_r\ddot{x}_r + c_r\dot{x}_r + k_r x_r = F_r - F_{r1} - F_{r2} \tag{9}$$

where m_r is the equivalent mass of the rack; x_r is the translational displacement of the rack; k_r and c_r are the equivalent stiffness coefficient and damping coefficient respectively; F_r is the virtual synchronization driving force of the steering rack; and F_{r1} and F_{r2} are the forces fed back to the steering gear by the rear wheels on both sides.

In order to realize steering angle synchronization control, Proportional-differential (PD) feedback control law is introduced by

$$\begin{aligned} F_r &= k_{rp}e_s + k_{rd}\dot{e}_s \\ &= k_{rp}(\delta_r - \delta_r^*) + k_{rd}(\dot{\delta}_r - \dot{\delta}_r^*) \end{aligned} \tag{10}$$

where e_s is defined as the synchronization error; k_{rp} and k_{rd} are the proportional coefficient and differential coefficient respectively; and δ_r^* is the equivalent rear axle steering angle derived from the measured rear left and right wheel steering angle. When only the relationship between the steering angles of the rear wheels is considered, according to the Ackerman principle, δ_r^* can be inversely solved by

$$\delta_r^* = \arctan\frac{2\tan\delta_{r1}^*\tan\delta_{r2}^*}{\tan\delta_{r1}^* + \tan\delta_{r2}^*} \tag{11}$$

where δ_{r1}^* and δ_{r2}^* represent the measured wheel steering angle, respectively.

x_r / g_r can be taken as virtual rear axle steering angle after synchronization controller. g_r is the radius of the virtual steering pinion gear. Referring to Equation (8), the output reference angle of VRSM can be obtained as

$$\begin{cases} \delta_{r1} = \arctan\left(\frac{\rho}{\rho - w/2}\tan\frac{x_r}{g_r}\right) \\ \delta_{r2} = \arctan\left(\frac{\rho}{\rho + w/2}\tan\frac{x_r}{g_r}\right) \end{cases} \tag{12}$$

In the proposed strategy, δ_r is not directly calculated to get the reference angle signal of each rear wheel, but after the synchronization controller. In the presence of external interference, the forces F_{r1} and F_{r2} are fed back to the steering rack, then the VRSM senses the external disturbance and immediately adjusts the reference steering angle of the left and right rear wheels to realize instantaneous synchronization control.

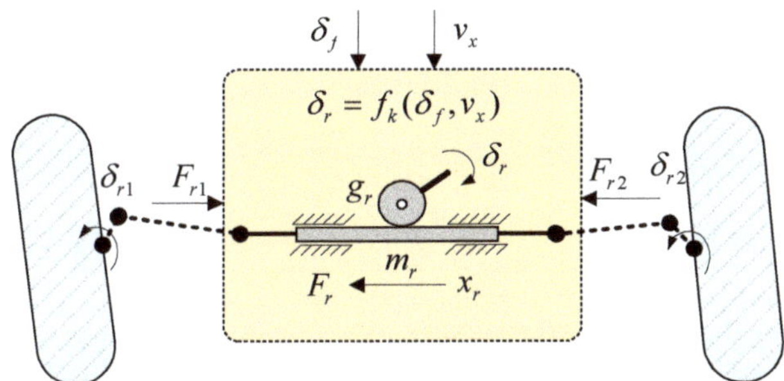

Figure 4. The virtual rack and pinion steering gear model of the rear axle.

3.2. Steering Angle Tracking Controller

LADRC is a control method for estimating and compensating uncertainties. The LADRC method takes the internal disturbance caused by the model parameter perturbation of the system together with the external disturbance of the system as the total disturbance of the system. It is observed by LESO and compensated to the controller. It is a kind of control method which is not completely dependent on the system model. It is suitable for angle tracking control of an ARIS system with uncertainty and nonlinearity.

It can be seen from Equation (1) that the plant is a nonlinear coupled dynamic model. Firstly, the terms in the model are written in the form of equivalent total disturbance except for the second-order term and the control input term in the model. Then, the plant is decoupled into the second-order single input single output plant. Furthermore, the total disturbance is considered as an extended new state, which is estimated by LESO in real time. Finally, the observation disturbance is compensated to the linear feedback control law and the final control output is obtained. Because the left and right wheels are decoupled by the LADRC control method, the steering angle tracking controller can be designed separately. In the framework of LADRC, Equation (1) can be rewritten as

$$\ddot{\delta}_{ri} = b_{0i}u_i + f_i \tag{13}$$

where $b_{0i} = b_r/J_i$,$f_i = (M_{fi} + M_{zi} + M_{wi} - c_i\dot{\delta}_{ri} - k_i\delta_{ri})/J_i$,$u_i = F_i \cdot \cos\delta_{ri}$, and i = 1,2.

The total disturbance term f_i in the model can be defined as an extended state, and the system state vector can be defined as

$$x = \begin{bmatrix} x_1 & x_2 & x_3 & x_4 & x_5 & x_6 \end{bmatrix}^T = \begin{bmatrix} \delta_{r1} & \dot{\delta}_{r1} & f_1 & \delta_{r2} & \dot{\delta}_{r2} & f_2 \end{bmatrix}^T \tag{14}$$

Equation (13) can be written in the form of state space equation, given by

$$\begin{cases} \dot{x} = \boldsymbol{A}x + \boldsymbol{B}u + \boldsymbol{E}\dot{f} \\ y = \boldsymbol{C}x \end{cases} \tag{15}$$

where y is the system output, $u = \begin{bmatrix} u_1 \\ u_2 \end{bmatrix}$ is the control input, $f = \begin{bmatrix} f_1 \\ f_2 \end{bmatrix}$, $\boldsymbol{A} = \begin{bmatrix} A_1 & 0 \\ 0 & A_2 \end{bmatrix}$, $\boldsymbol{B} = \begin{bmatrix} B_1 & 0 \\ 0 & B_2 \end{bmatrix}$, $\boldsymbol{C} = \begin{bmatrix} C_1 & 0 \\ 0 & C_2 \end{bmatrix}$, $\boldsymbol{E} = \begin{bmatrix} E_1 & 0 \\ 0 & E_2 \end{bmatrix}$ are the coefficient matrix with $A_i = \begin{bmatrix} 0 & 1 & 0 \\ 0 & 0 & 1 \\ 0 & 0 & 0 \end{bmatrix}$, $B_i = \begin{bmatrix} 0 \\ b_{0i} \\ 0 \end{bmatrix}$, $C_i = \begin{bmatrix} 1 \\ 0 \\ 0 \end{bmatrix}^T$, $E_i = \begin{bmatrix} 0 \\ 0 \\ 1 \end{bmatrix}$, $i = 1,2$.

Define $z = \begin{bmatrix} z_1 & z_2 & z_3 & z_4 & z_5 & z_6 \end{bmatrix}^T$ as the observed state of system. According to the theory of state observer, LESO can be designed as follows

$$\begin{aligned} \dot{z} &= \boldsymbol{A}z + \boldsymbol{B}u + \boldsymbol{L}(y - \boldsymbol{C}z) \\ &= (\boldsymbol{A} - \boldsymbol{LC})z + \boldsymbol{B}u + \boldsymbol{L}y \end{aligned} \tag{16}$$

where $\boldsymbol{L}_{6\times2} = [L_1\ L_2]^T$ is the gain matrix of the observer, $L_1 = L_2 = [\beta_1\ \beta_2\ \beta_3]^T$. Assuming that the total disturbance f is bounded, and an appropriate observer gain $\boldsymbol{L}$ can be selected to make $\boldsymbol{A}$-$\boldsymbol{LC}$ asymptotically stable, then state $\boldsymbol{x}$ can be estimated. To simplify parameter tuning process, L_1 and L_2 are set to $[3\omega_o\ 3\omega_o{}^2\ \omega_o{}^3]^T$ by the method of pole configuration. ω_o is called the observer bandwidth.

By introducing the estimated disturbance state z_3, the controlled system is compensated as a linear integral series system. The proportional-differential (PD) form is selected for the linear state error feedback control law, given by

$$\begin{cases} u_{01} = k_p(\delta_{r1} - z_1) - k_d z_2 \\ u_1 = (u_{01} - z_3)/b_{01} \\ F_{r1} = u_1/\cos z_1 \end{cases} \tag{17}$$

$$\begin{cases} u_{02} = k_p(\delta_{r2} - z_4) - k_d z_5 \\ u_2 = (u_{02} - z_6)/b_{02} \\ F_{r2} = u_2/\cos z_4 \end{cases} \tag{18}$$

where δ_{r1} and δ_{r2} are the given reference signal; u_{01} and u_{02} are the ideal control output; and k_p and k_d are the gain coefficients of the feedback control law. According to [28], the parameters of k_p and k_d can be selected by

$$k_p = \omega_c^2,\ k_d = 2\omega_c \tag{19}$$

where ω_c is the controller bandwidth.

The control block diagram of linear active disturbance rejection control (LADRC) tracking controller for ARIS system is shown in Figure 5. The stability of LADRC was analyzed and proved in detail in [34]. The upper controller adopts the PD feedback control law in order to realize steering angle synchronization control. It can be seen from Equation (10) that when the synchronization error e_s converges to 0, the equivalent rear axle steering angle δ_r^* will converge to the target angle of the rear axle wheel δ_r. It can be seen from Equations (17) and (18) that under the control of F_{r1} and F_{r2} of the lower controller, δ^*_{r1} and δ^*_{r2} converge to δ_{r1} and δ_{r2}, respectively. Therefore, ARIS system can achieve stability. The proposed control strategy above has the following advantages. (1) The lower controller output signals, including the external disturbance of both side wheels, are fed back to the upper controller and interact on the dynamic model of VRSM together. It reduces the synchronization error caused by the unilateral disturbance. (2) The LESO in the lower controller can not only observe the external disturbance, but also reconstruct the steering wheel angle and angular velocity. (3) The feedback control law uses the estimated states as the input to avoid the output oscillation of the controller in case of directly using the measurement value of the sensor with noise, especially in a differential term.

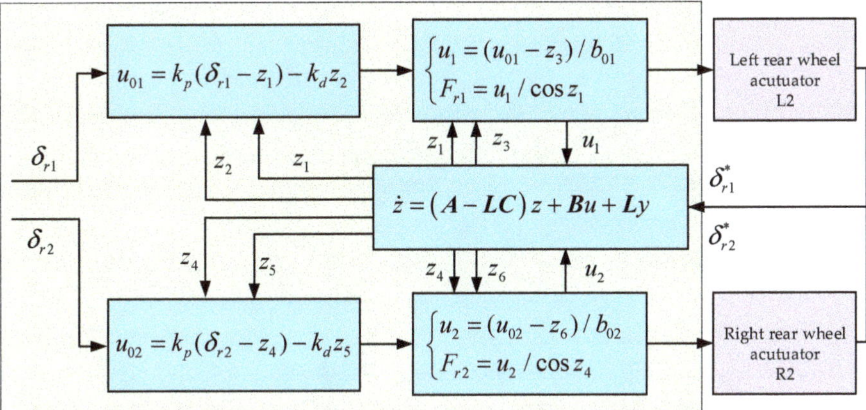

Figure 5. The block diagram of linear active disturbance rejection control (LADRC) tracking controller for ARIS system.

4. Experiments and Analysis

To demonstrate the proposed control strategy for the ARIS system, experiments were carried out on a real prototype vehicle shown in Figure 6. The suspension of vehicle was the double wishbone independent suspension. The power system of vehicle adopted four-in-wheel motor independent drive. The ARIS system was implemented in the prototype vehicle rear axle steering system, and the steering angle of each rear axle wheel was obtained and controlled by the real-time working stroke of the steering actuator. Each wheel steering angle could be indirectly measured by a magnetic encoder installed at the end of the actuator. The magnetic encoder was AS5047P with 12-bit resolution. Considering the measurement noise and the steering mechanical chain, the final actual resolution of steering angle was about 0.02deg. The main parameters of the prototype vehicle and AIRS system are shown in Table 1.

The model-based design (MBD) method was adopted to implement the proposed control strategy. The control strategy was modeled in a MATLAB/Simulink software (R2016b, Mathworks, Natick, Mass, USA) environment. The executable C code was generated by the automatic code generation technology, and then compiled and flashed into the embedded microprocessor. The microprocessor adopts stm32f407 (STMicroelectronics, Geneva, CH) based on Cortex-M4 Kernel. The PWM modulation frequency for the Metal Oxide Semiconductor Field Effect Transistor (MOSFET) of actuator motor was set to 20 kHz and the sampling frequency was 200 Hz. Figure 7 shows the control structure in the experiments.

(**a**) (**b**)

Figure 6. (**a**) The prototype vehicle with the ARIS system; (**b**) the structure of the ARIS system.

Table 1. Parameters of the prototype vehicle and ARIS system.

Parameters	Value	Parameters	Value
m	95 kg	b_r	0.16 m
a	0.65 m	g_r	0.03 m
b	0.7 m	c_r	120 N/deg/s
w	1.2 m	k_r	460 N/deg
N	10	m_r	12 kg
J_1, J_2	1.37 kg.m^2	ω_o	80 Hz
k_{r1}, k_{r2}	2.2 Nm/deg	ω_c	20 Hz
c_{r1}, c_{r2}	4.2 Nm/deg/s		

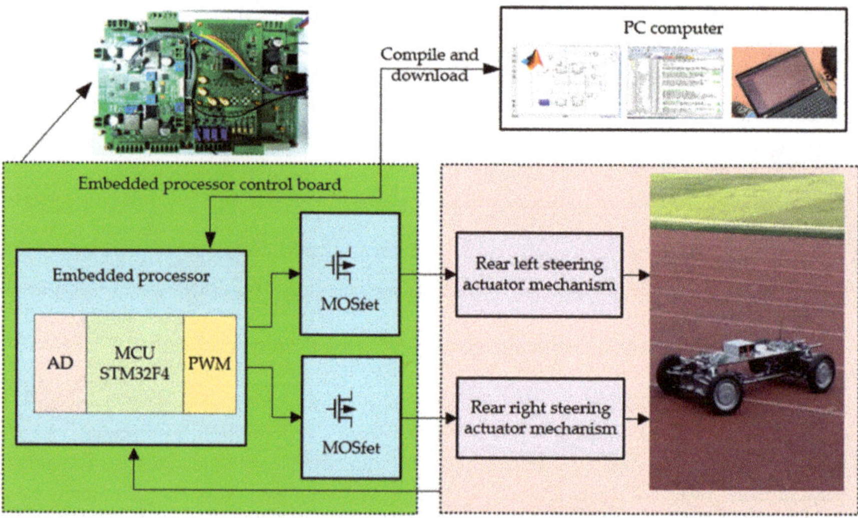

Figure 7. The control structure of the experiment.

4.1. Static Steering Test

In order to test the steering angle tracking performance under the proposed control strategy, the static experiment was studied for the ARIS system in three cases. The prototype vehicle mass was 95 kg. The reference steering angle is a sinusoidal sweep signal with a time-varying amplitude and frequency.

Case 1: The master-slave method (MS) [16] was adopted for the synchronization controller, while the PID method was adopted for the tracking controller. Case 2: The MS method was adopted for the synchronization controller, while the LADRC method was adopted for the tracking controller. Case 3: The VRSM method was adopted for the synchronization controller, while the LADRC method was adopted for the tracking controller.

In order to evaluate the effectiveness of the proposed hierarchical synchronization control strategy, the root mean square error (RSME) was defined to examine control performance as follows.

$$RSME_t = \sqrt{\frac{1}{2n}\sum_{k=1}^{n}\left(\left|e_{t1}(k)\right|^2 + \left|e_{t2}(k)\right|^2\right)} \tag{20}$$

$$RSME_s = \sqrt{\frac{1}{n}\sum_{k=1}^{n}\left|e_s(k)\right|^2} \tag{21}$$

where $RSME_t$ is the root mean square error of left and right steering wheel angle tracking error; $e_{ti}(k) = \delta_{ri}(k) - \delta^*_{ri}(k), i = 1,2$; $RSME_s$ is the root mean square error of the synchronization error; $RSME_c$ is defined to evaluate comprehensive performance; λ is the weight coefficient. Synchronization error is more important than tracking error for tire wear, so we assign $\lambda = 0.6$.

$$RSME_c = (1-\lambda)RSME_t + \lambda RSME_s \tag{22}$$

First of all, case 1 (Figures 8 and 9) and case 2 (Figures 10 and 11) are compared. Figures 8 and 10 show the rear wheel steering angle tracking profiles and errors under case 1 and case 2, respectively. Although the tracking errors of both controllers increase as the frequency of the steering angle command grows. Specifically, comparing Figure 8b,d with Figure 10b,d, it can be seen that the maximum tracking error of PID increases to 0.665 deg, which is much larger compared to that of LADRC (0.485 deg) and the peak error occurs when the steering direction changes.

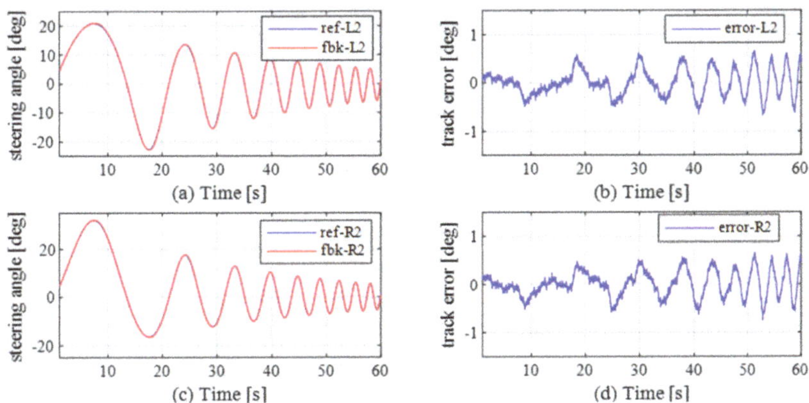

Figure 8. The steering angle of rear axle wheel profiles and tracking errors in case 1. L2 and R2 are marked as the left rear wheel and right rear wheel, respectively. (**a**) The steering angle of left rear wheel profiles; (**b**) the tracking error of left rear wheel profiles; (**c**) the steering angle of right rear wheel profiles; (**d**) the tracking error of right rear wheel profiles.

Case 1 and case 2 do not adopt the synchronization control strategy, and the synchronization errors are shown in Figures 9 and 11, respectively. The synchronization error is reduced in case 2 compared with case 1 under static and basically undisturbed conditions. This is because the tracking angle accuracy of LADRC method is improved.

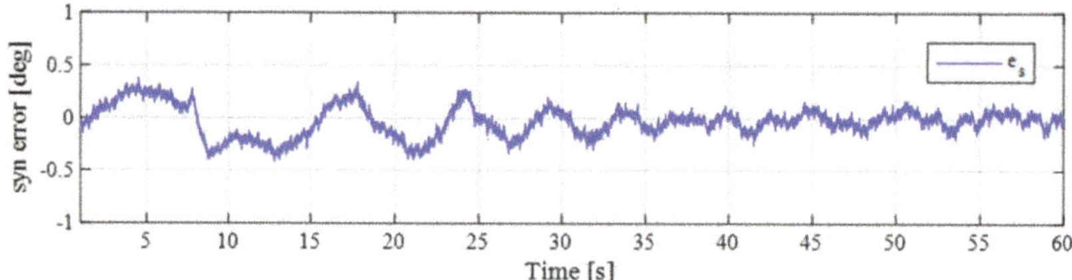

Figure 9. The synchronization errors between left and right wheels in case 1.

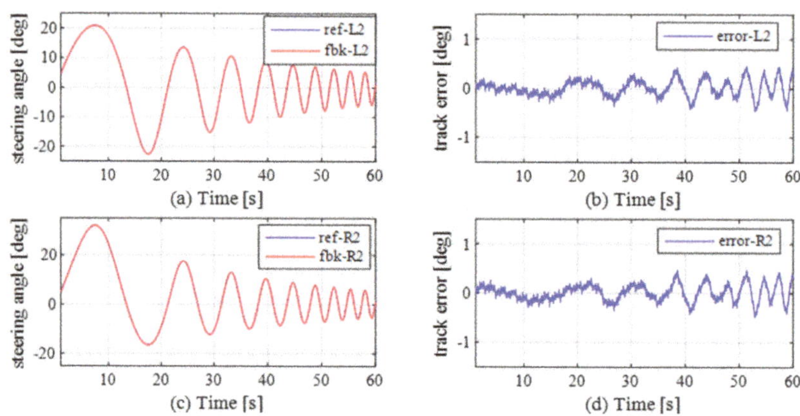

Figure 10. The steering angle of rear axle wheel profiles and tracking errors in case 2. (**a**) The steering angle of left rear wheel profiles; (**b**) the tracking error of left rear wheel profiles; (**c**) the steering angle of right rear wheel profiles; (**d**) the tracking error of right rear wheel profiles.

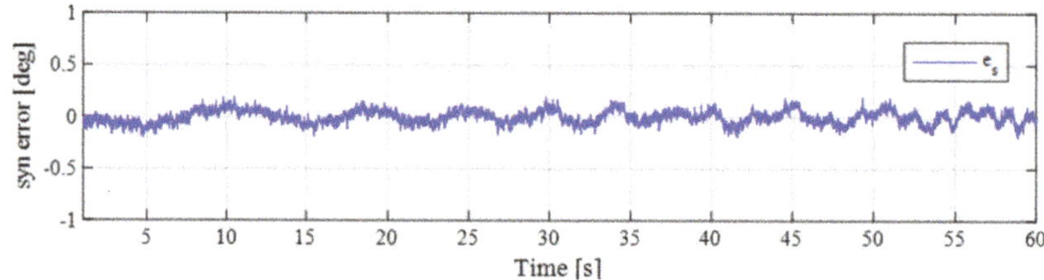

Figure 11. The synchronization errors between left and right wheels in case 2.

Compared with case 1 and case 2, we can conclude that at either large amplitudes or high frequencies, the control accuracy of LADRC is better than that of PID, with lower tracking error and synchronization error. It is because that LESO in LADRC can observe the external interference and compensate it to the control law.

An obvious problem can be observed in both Figures 9 and 11, that the synchronization error between the left and right wheel steering angle is not controlled and reduced well in both case 1 and case 2. This is because the adopted MS control strategy does not consider the interaction between rear wheels and external uncertainty disturbance. In particular, the error can reach 0.665 deg at the moment of the steering direction changes in case 1.

Then, by comparing the results shown in Figures 10 and 12, it can be seen that the tracking performance in case 3 is basically the same as in case 2, since the same lower control method (LADRC) is used. By comparing the results in Figures 11 and 13, it can be seen that case 3, which adopts the hierarchical synchronization control strategy proposed in this paper, has better synchronization control accuracy. The maximum synchronization error angle is less than 0.186 deg.

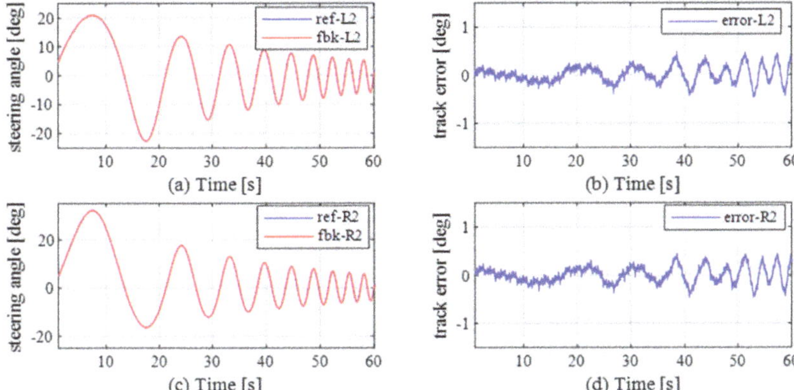

Figure 12. The steering angle of rear axle wheel profiles and tracking errors in case 3. (**a**) The steering angle of left rear wheel profiles; (**b**) the tracking error of left rear wheel profiles; (**c**) the steering angle of right rear wheel profiles; (**d**) the tracking error of right rear wheel profiles.

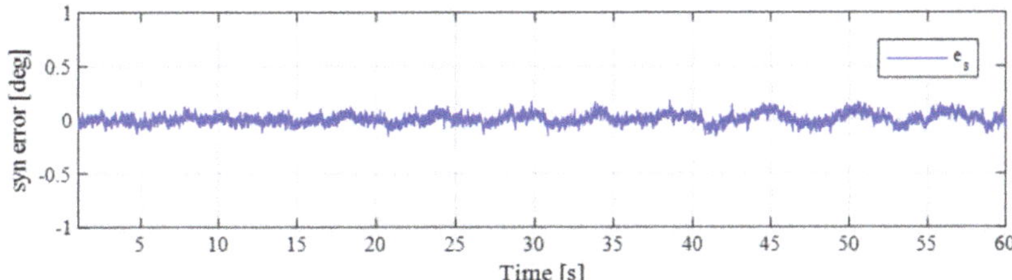

Figure 13. The synchronization errors between left and right wheels in case 3.

Table 2 is the root mean square comparison table under three cases according to Equations (20)–(22). In order to show the differences more intuitively, Table 2 is drawn into a histogram in Figure 14. From Figure 14, we can get the following conclusion: (1) comparing case 1 and case 2, it can be seen that in the case of without hierarchical synchronization control strategy, the synchronization error can be reduced to a certain extent by improving the tracking control accuracy. However, it is still unable to achieve a better suppression effect on the external uncertain disturbance; (2) comparing case 2 and case 3, on the basis of the LADRC tracking angle controller, the hierarchical synchronization control strategy based on the dynamic model of the virtual rear axle steering mechanism is introduced to further reduce the synchronization error of ARIS system, and slightly reduce the tracking error at the same time.

Table 2. Control performance in three cases.

Evaluation Index	Case 1	Case 2	Case 3
max e_t *[deg]*	0.665	0.485	0.483
mean absolute e_t *[deg]*	0.202	0.137	0.136
max e_s *[deg]*	0.514	0.210	0.186
mean absolute e_t *[deg]*	0.149	0.054	0.039
$RSME_t$	0.249	0.187	0.166
$RSME_s$	0.194	0.095	0.049
$RSME_c$	0.216	0.106	0.096

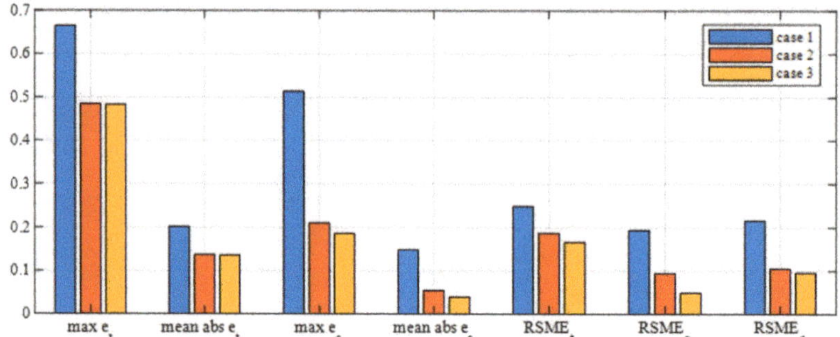

Figure 14. The control performance comparison of three cases.

4.2. *Dynamic Steering Test*

The driving condition of double lane change (DLC) is used to test the proposed strategy performance of the ARIS system under a dynamic uncertain environment. The mass of vehicle is 95 kg. The vehicle speed is 12 km/h. Figure 15 shows the steering reference angle of the rear axle wheel under the DLC condition. L2 represents the left rear wheel and R2 represents the right rear wheel. Figure 16 shows the driving status of the vehicle. In order to compare and analyze the control performance, the vehicle is tested under the condition of with and without synchronization control.

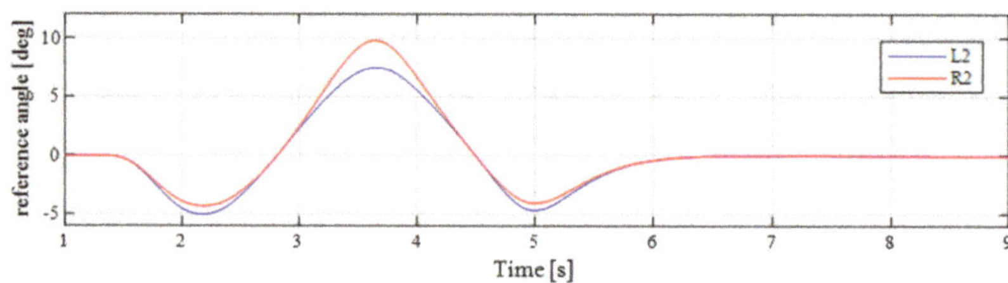

Figure 15. The rear axle wheels steering reference angle of the ARIS system under the condition of double lane change (DLC).

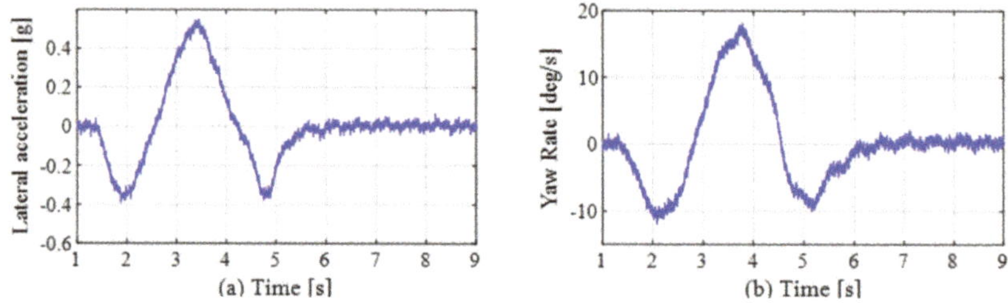

Figure 16. The vehicle state under the condition of DLC. (**a**) The lateral acceleration of the vehicle; (**b**) the yaw rate of the vehicle.

Figure 16 shows that in 3.4s, the lateral acceleration and yaw rate reach the maximum value. In the case of no synchronization control, the external uncertain disturbance caused by the lateral acceleration will further increase the steering tracking error and synchronization error of the left and right wheels. As shown in Figure 17, the maximum synchronization error reaches 0.45deg. It can be seen from Figure 18 that the synchronization error is significantly reduced and the maximum synchronization error is less than 0.15deg with the hierarchical synchronization control strategy. This is because the disturbance on either side of the left and right wheels will be fed back to the dynamic model of VRSM.

As a result, the left and right rear wheels are subject to the same level of total disturbance in the virtual model, which further makes the tracking controller show the same tracking error characteristics. Therefore, the synchronization error is reduced.

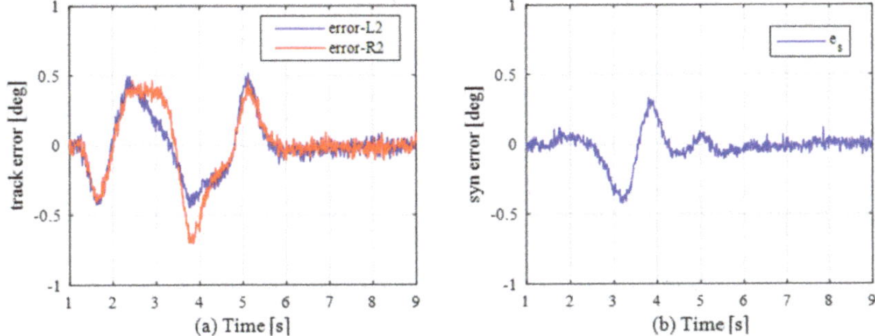

Figure 17. The tracking error and the synchronization errors without the proposed control strategy. (**a**) The tracking error of the rear wheels; (**b**) the synchronization error of the rear wheels.

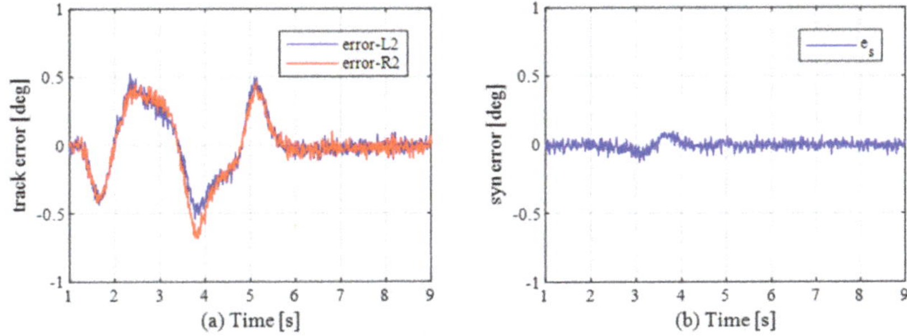

Figure 18. The tracking error and the synchronization errors with the proposed control strategy. (**a**) The tracking error of the rear wheels; (**b**) the synchronization error of the rear wheels.

The above experiments prove that the control strategy proposed in this paper can significantly reduce synchronization error under uncertain disturbance condition and is superior with the properties of disturbance rejection and robustness.

5. Conclusions

In this paper, a hierarchical synchronization control strategy is proposed to solve the problem of steering angle control for the ARIS system. The proposed control strategy contains two levels. In the upper controller, VRSM is constructed to reduce the synchronization error caused by the inconsistent disturbance of the left and right wheels. The external disturbance of the left and right wheels is fed back to the dynamic model of VRSM, which makes the left and right wheels suffer the same disturbance in the virtual model. In the VRSM, a PD control element is also introduced to further reduce the synchronization error. The lower controller is designed to track the steering angle of each wheels by using an LADRC, where the uncertainties and external disturbance of each single wheel steering system are observed and compensated by the proposed LESO. The experimental results on the prototype vehicle show that the hierarchical synchronization control strategy for ARIS systems realizes a significant reduction in the steering angle tracking error and synchronization error compared with the non-synchronization control strategies under external uncertain disturbance.

Author Contributions: Conceptualization, B.D.; Data curation, B.D. and K.S.; Formal analysis, W.L.; Funding acquisition, H.Z.; Investigation, B.D.; Project administration, H.Z. and A.Y.; Resources, A.Y.; Supervision, H.Z.; Validation, B.D. and K.S.; Writing – original draft, B.D. and W.L.; Writing – review & editing, K.S. and W.L. All authors have read and agreed to the published version of the manuscript.

Funding: This research was supported by the Fundamental Research Funds for the Central University of China (Grant No. PA2019GDZC0101).

Conflicts of Interest: The authors declare no conflict of interest.

Nomenclature

J_i	Total moment of inertia of rotating parts.
F_i	Driving force of electric actuator.
M_{fi}	Friction moment.
M_{zi}	Tire self-aligning torque around the kingpin.
M_{wi}	Moment of uncertain force on the ground.
c_i	Equivalent damping coefficient.
k_i	Equivalent stiffness.
b_r	Length from the connection point to the kingpin.
δ_{ri}	Rear wheel steering angle.
M_{sfi}	Initial value of friction moment.
F_{zri}	Rear wheel vertical load.
μ_{ri}	Coulomb friction coefficient.
γ_i	Kingpin inclination.
s_i	Horizontal distance from the tire center to the kingpin.
m	Mass of the vehicle body.
I_z	Yaw moment of inertia of the vehicle.
$k_{\alpha f}$	Cornering stiffness of front wheels.
$k_{\alpha r}$	Cornering stiffness of rear wheels.
a	Distance from the front axles to the vehicle mass center.
b	Distance from the rear axles to the vehicle mass center.
δ_f	Front axle wheel angle.
δ_r	Rear axle wheel angle.
v_x	Vehicle longitudinal speed.
a_y.	Vehicle lateral acceleration.
ω	Vehicle yaw rate.
β	Vehicle sideslip angle.
N	Reduction ratio.
f_k	Ratio between front axle steering angle and rear axle steering angle.
θ_{sw}	Driver's steering wheel angle.
l	Wheel base.
w	Wheel track.
ρ	Turning radius.
δ_{r1}	Left rear wheel steering angle.
δ_{r2}	Right rear wheel steering angle.
m_r	Equivalent mass of the rack.
x_r	Translational displacement of the rack.
g_r	Radius of the virtual steering pinion gear.
k_r	Equivalent stiffness coefficient.
c_r	Equivalent damping coefficient.
F_{ri}	Forces fed back to the steering gear by the rear wheels on both sides.
F_r	Virtual synchronization driving force of the steering rack.
e_s	Synchronization error.
e_t	Steering wheel tracking error.
k_{rp}	Proportional coefficient.

k_{rd}	Differential coefficient.
δ_r^*	Equivalent rear axle steering angle.
$\delta_{r1}^*,\delta_{r2}^*$	Measured rear wheel steering angles.
δ_{r1}, δ_{r2}	Given reference signal.
u_{01}, u_{02}	Ideal control output.
k_p, k_d	Gain coefficients of the feedback control law.
ω_c	Controller bandwidth.
ω_o	Observer bandwidth.
λ	Weight coefficient.

References

1. Bredthauer, L.; Lynch, D. Use of Active Rear Steering to Achieve Desired Vehicle Transient Lateral Dynamics. In *SAE Technical Paper Series*; SAE: Warrendale, PA, USA, 2018; Available online: https://www.sae.org/publications/technical-papers/content/2018-01-0565/ (accessed on 3 April 2018).
2. Volker, M.; Stadie, W. Electrohydraulic rear axle steering systems for agricultural vehicles. *ATZ Worldw.* **2017**, *10*, 22–27. [CrossRef]
3. Li, B.; Rakheja, S.; Feng, Y. Enhancement of vehicle stability through integration of direct yaw moment and active rear steering. *Proc. Inst. Mech. Eng. Part D: J. Automob. Eng.* **2016**, *230*, 830–840. [CrossRef]
4. Jo, H.Y.; Lee, U.K.; Kam, M.S. Development of the independent-type steer-by-wire system using HILS. *Int. J. Automot. Technol.* **2006**, *7*, 321–327.
5. Nagai, M.; Yamanaka, S.; Hirano, Y. Integrated control of active rear wheel steering and yaw moment control using braking forces. *Trans. Jpn. Soc. Mech. Eng.* **1999**, *42*, 301–308. [CrossRef]
6. Salman, M.; Zhang, Z.; Boustany, N. Coordinated Control of Four Wheel Braking and Rear Steering. In Proceedings of the 1992 American Control Conference, Chicago, IL, USA, 24–26 June 1992; IEEE: Piscataway, NJ, USA, 1992; pp. 6–10.
7. Guo, H.; Hao, N.; Chen, H. Lateral stability controller design for electrical vehicle based on active rear wheel steering. In Proceedings of the World Congress on Intelligent Control and Automation WCICA 2016, Changsha, China, 4–8 July 2018; pp. 1285–1290.
8. Shen, H.; Huang, M.; Tan, Y. Active Rear Wheel Steering Control Strategy Research Based on H∞ Optimal Control. *J. Comput. Theor. Nanosci.* **2016**, *13*, 2043–2048. [CrossRef]
9. Harada, H. Control strategy of active rear wheel steering in consideration of system delay and dead times. *JSAE Rev.* **2004**, *16*, 171–177. [CrossRef]
10. Zhao, L.; Lu, S.; Zhang, B. Game-Based Hierarchical Cooperative Control for Electric Vehicle Lateral Stability via Active Four-Wheel Steering and Direct Yaw-Moment Control. *Energies* **2019**, *12*, 3339. [CrossRef]
11. Zardin, B.; Borghi, M.; Gherardini, F.; Zanasi, N. Modelling and Simulation of a Hydrostatic Steering System for Agricultural Tractors. *Energies* **2018**, *11*, 230. [CrossRef]
12. Feng, J.; Bao, C.; Wu, J. Research on methods of active steering control based on receding horizon control. *Energies* **2018**, *11*, 2243. [CrossRef]
13. Chen, X.; Xu, N.; Guo, K. Tire wear estimation based on nonlinear lateral dynamic of multi-axle steering vehicle. *Int. J. Automot. Technol.* **2018**, *19*, 63–75. [CrossRef]
14. Wang, Y.; Gao, X.; Zhang, X. Static steering resisting moment of tire for heavy multi-axle steering vehicle. *Trans. Chin. Soc. Agric. Eng.* **2010**, *26*, 146–150.
15. Kamel, M.A.; Yu, X.; Zhang, Y.M. Formation control and coordination of multiple unmanned ground vehicles in normal and faulty situations: A review. *Annu. Rev. Control* 2020. [CrossRef]
16. Zong, C.F.; Sun, H.; Chen, G.Y. Steering angle allocation method for distributed independent steering vehicles. *J. South China Univ. Technol.* **2017**, *45*, 16–22.
17. Yamaguchi, Y.; Murakami, T. Adaptive Control for Virtual Steering Characteristics on Electric Vehicle Using Steer-by-Wire System. *IEEE Trans. Ind. Electron.* **2009**, *56*, 1585–1594. [CrossRef]
18. Zhang, R.Z.; Xiong, L.; Yu, Z.P. Active control of steering angle of smart car steering wheel. *J. Mech. Eng.* **2017**, *53*, 106–113. [CrossRef]

19. Wang, H.; Man, Z.; Kong, H. Terminal sliding mode control for steer-by-wire system in electric vehicles. In Proceedings of the 2012 7th IEEE Conference on Industrial Electronics and Applications (ICIEA), Singapore, 18–20 July 2012; IEEE: Piscataway, NJ, USA, 2012; pp. 919–924.
20. Wang, H.; Kong, H.; Man, Z. Sliding Mode Control for Steer-by-Wire Systems with AC Motors in Road Vehicles. *IEEE Trans. Ind. Electron.* **2013**, *61*, 1596–1611. [CrossRef]
21. Do, M.T.; Man, Z.; Zhang, C. Robust Sliding Mode-Based Learning Control for Steer-by-Wire Systems in Modern Vehicles. *IEEE Trans. Veh. Technol.* **2014**, *63*, 580–590. [CrossRef]
22. Li, Y.; Yang, L.; Yang, G. Network-Based Coordinated Motion Control of Large-Scale Transportation Vehicles. *IEEE/ASME Trans. Mechatron.* **2007**, *12*, 208–215. [CrossRef]
23. Ye, Y.; He, L.; Zhang, Q. Steering control strategies for a four-wheel-independent-steering bin managing robot. *IFAC PapersOnLine* **2016**, *49*, 39–44. [CrossRef]
24. Payette, K. The virtual shaft control algorithm for synchronized motion control. In *Proceedings of the 1998 American Control Conference. ACC*; IEEE: Piscataway, NJ, USA, 1998; Volume 5, pp. 3008–3012.
25. Zhang, C.F.; Xiao, Y.Y.; Wen, L.; He, J. Study on Control Strategy of Sliding Mode Variable Structure-based Electronic Virtual Line Shafting. In *Proceedings of the Intelligent System Design and Engineering Applications (ISDEA), 2013 Third International Conference*; IEEE Computer Society: Piscataway, NJ, USA, 2013; pp. 1326–1329.
26. Wolf, C.M.; Lorenz, R.D.; Valenzuela, M.A. Digital implementation issues of electronic line shafting. *IEEE Trans. Ind. Appl.* **2010**, *46*, 750–760. [CrossRef]
27. Djeziri, M.A.; Merzouki, R.; Bouamama, B.O. Fault Diagnosis and Fault-Tolerant Control of an Electric Vehicle Overactuated. *IEEE Trans. Veh. Technol.* **2013**, *62*, 986–994. [CrossRef]
28. Gao, Z. Active Disturbance Rejection Control: A paradigm shift in feedback control system design. *Am. Control Conf.* **2006**, *7*.
29. Miklosovic, R.; Radke, A.; Gao, Z. Discrete implementation and generalization of the extended state observer. In *Proceedings of the American Control Conference*; IEEE: Piscataway, NJ, USA, 2006; p. 6.
30. Yunchao, W.; Feng, P.; Pang, W.; Zhou, M. Pivot steering resistance torque based on tire torsion deformation. *J. Terramechanics* **2014**, *52*, 47–55. [CrossRef]
31. He, H.; Peng, J.; Xiong, R.; Fan, H. An acceleration slip regulation strategy for four-wheel drive electric vehicles based on sliding mode control. *Energies* **2014**, *7*, 3748–3763. [CrossRef]
32. Feng, K.T.; Tan, H.S.; Tomizuka, M. Automatic steering control of vehicle lateral motion with the effect of roll dynamics. In Proceedings of the American Control Conference, Philadelphia, PA, USA, 26–26 June 1998; Volume 4, pp. 2248–2252.
33. Choi, M.W.; Park, J.S.; Lee, B.S.; Lee, M.H. The performance of independent wheels steering vehicle (4WS) applied Ackerman geometry. In *Proceedings of the 2008 International Conference on Control, Automation and Systems*; IEEE: Piscataway, NJ, USA, 2008; pp. 197–202.
34. Carlos, A.I.; Hebertt, S.R.; José, A.A. Stability of active disturbance rejection control for uncertain systems: A Lyapunov perspective. *Int. J. Robust Nonlinear Control* **2017**, *27*, 4541–4553.

Article

Characterization and Modelling of Various Sized Mountain Bike Tires and the Effects of Tire Tread Knobs and Inflation Pressure

Andrew Dressel [1,*] and James Sadauckas [2]

1 Departments of Mechanical & Civil Engineering, University of Wisconsin-Milwaukee, Milwaukee, WI 53211, USA
2 Vehicle Dynamics & Simulation Group, Harley-Davidson Motor Company, Wauwatosa, WI 53222, USA; james.sadauckas@harley-davidson.com
* Correspondence: adressel@uwm.edu

Received: 11 April 2020; Accepted: 29 April 2020; Published: 1 May 2020

Abstract: Mountain bikes continue to be the largest segment of U.S. bicycle sales, totaling some USD 577.5 million in 2017 alone. One of the distinguishing features of the mountain bike is relatively wide tires with thick, knobby treads. Although some work has been done on characterizing street and commuter bicycle tires, little or no data have been published on off-road bicycle tires. This work presents laboratory measurements of inflated tire profiles, tire contact patch footprints, and force and moment data, as well as static lateral and radial stiffness for various modern mountain bike tire sizes including plus size and fat bike tires. Pacejka's Motorcycle Magic Formula tire model was applied and used to compare results. A basic model of tire lateral stiffness incorporating individual tread knobs as springs in parallel with the overall tread and the inflated carcass as springs in series was derived. Finally, the influence of inflation pressure was also examined. Results demonstrated appreciable differences in tire performance between 29 × 2.3″, 27.5 × 2.8″, 29 × 3″, and 26 × 4″ knobby tires. The proposed simple model to combine tread knob and carcass stiffness offered a good approximation, whereas inflation pressure had a strong effect on mountain bike tire behavior.

Keywords: bicycle; mountain bike; tire tread pattern; force and moment; e-bike; tyre; dynamics

1. Introduction

Mountain biking is a popular recreation and fitness activity that uses a bicycle and components designed to be rugged; to withstand off-road riding; and capable of handling unpaved surfaces, loose dirt, gravel, mud, and other terrains. In 2018, some 8.69 million people participated in mountain biking in the U.S. alone [1], which saw some USD 577.5 million in mountain bike sales the year prior [2].

Tire behavior is a critical factor in bicycle performance and safety. Like road or city bicycle tires, weight must be kept low because, except for e-bikes, the rider must propel the vehicle under their own power. Tire durability is important because a flat tire can ruin a ride. Ride comfort, gleaned from the tire deflection, is a consideration even for bicycles with front and rear suspension, whereas performance and grip become even more important when navigating up or down steep grades, dodging trees, and other obstacles. In addition to tire size options, a wide variety of tread patterns, made of up individual "knobs", that is, tread elements, of various shapes, sizes, and depths are available, depending on intended usage.

As with any sport, mountain biking has a large cadre of enthusiasts. There is much debate among racers, riders, and industry marketing lobbies over optimal tire size, tread pattern, inflation pressure, and, more recently, rim width. Although all these things are likely to affect a tire's performance, little or no scientific study of mountain bike tire properties exists in the literature.

In this work, four modern mountain bike tire sizes were characterized through force and moment measurement via the tire test device at the University of Wisconsin–Milwaukee [3]. Pacejka's Motorcycle Magic Formula tire model [4], which emphasizes the high camber achieved by single-track vehicles, was fitted to the data while a non-linear polynomial from Dynamotion's FastBike motorcycle tire model [5] was fitted to the twisting torque due to camber. The four sizes of knobby tires were tested at realistic inflation pressures for their respective size and intended use on modern (wider) rim widths. Tire cross-sectional profiles were captured and compared using a simple and effective method. Tire footprint analysis comparing contact patch area as well as shape via the major to minor axes of the fitted ellipse was carried out. Static lateral and radial stiffness were also measured. Further comparison was made between the full knobby tire and a less-treaded tire of similar carcass dimensions. These same tires then had their respective knobby and file-tread patterns removed to further quantify the tread influence. Trends in key tire properties across a range of inflation pressures were presented, along with results for two of the tires of interest on a slightly narrower rim.

Mountain bike tires are commonly marketed in Imperial units, with the first number indicating the approximate outer diameter in inches and the second number indicating the approximate overall tire width when mounted. Although the metric-based, ISO 5775 international sizing designation developed by the European Tyre and Rim Technical Organisation (ETRTO) is generally listed somewhere on the tire's sidewall, its prominence seems to vary depending on brand. Because all tires in this study prominently displayed sizing in Imperial units and maximum recommended inflation pressure in pounds per square inch (psi) on their sidewalls, this work refers to the tires as such. Table 1 shows both the Imperial-based, marketing size, and the metric-based ETRTO size for each tire considered herein, where double-apostrophes (″) signify inches.

Tire inflation pressures can be converted as follows:

$$1 \text{ psi} = 0.06894757 \text{ bar}. \quad (1)$$

Table 1. Imperial-based marketing size versus European Tyre and Rim Technical Organisation (ETRTO) designation.

Tire Size (in)	ETRTO (mm)
29 × 2.3″	58–622
29 × 2.5″	63–622
29 × 3.0″	76–622
27.5 × 2.8″	71–584
26 × 4.0″	102–559

2. Materials and Methods

A brief description of the types of measurement conducted as well as the tools used for these measurements is followed by a discussion of data analysis methods prior to examining the results.

2.1. Measurement

2.1.1. Tire Inflated Radius

Tire inflated radius was simply measured from ground to center of axle with the wheel in a vertical upright position using a rule or measuring tape. This value defines the tire radius from axle to the crown, that is, the outermost point on the circumference of the undeformed, inflated tire profile.

2.1.2. Tire Profile

Although various means exist in the automotive and motorcycle industry to measure tire profiles including laser scanning, coordinate measuring machine (CMM) arms, or other optical methods, this study chose a lower-tech, portable solution—a common machinist's or carpenter's contour gauge.

As shown in Figure 1, this device consisted of a series of pins captured between two plates. When the pins were pressed axially against an object, they slid between the plates with a small amount of frictional resistance, their tips conforming to the contour of the object. The user simply selects a contour gauge of adequate width and depth, extends and levels the pins toward the object intended for measurement, and then slowly presses the device onto the object until the desired section shape is captured. In the case of tire profiles, this process was conducted radially toward the wheel center to capture the tire crown to shoulder (outboard edge of tread) shape across its width, and was then repeated orthogonally to the wheel plane to capture the side-wall contour, tire height, and rim interface. In the case of staggered knob arrangements, the process can be repeated for each set of knobs.

Figure 1. Capturing mountain bike tire tread profile with carpenter's contour gauge.

The more complicated portion of the process was then digitizing and arranging the gauge results to accurately represent the tire's geometry. Following each gauge measurement, the gauge was aligned to, and overlaid on, a piece of graph paper and scanned. The images were then re-oriented, aligned, and scaled (if necessary) in a photo processing software package. Finally, a MATLAB script was used to identify the edge of the profile, trace it, and scale the result with respect to the grid on the paper.

2.1.3. Footprints

Footprints were collected by coating the tire surface containing the expected contact patch with ink using an office ink pad. The tire, which had been set to a specific inflation pressure, was then allowed to rest vertically on a piece of white cardstock paper with prescribed normal load applied through added weights. In this case, the force and moment fixture was used to accomplish the task; however, a bicycle with rear wheel mounted in a stationary trainer and over whose front wheel appropriate weights were applied could also be used.

2.1.4. Force and Moment

Force and moment measurements were performed with the tire test device at the University of Wisconsin–Milwaukee. As shown in Figure 2 it consisted of a welded steel frame and an aluminum fork to hold a bicycle wheel in a desired orientation on top of a small treadmill of flat-top chain. It had a two-degrees of freedom pivot, implemented with an automobile universal joint, far (1.3 m) forward of the bicycle tire so that slight variations in vertical or horizontal position produced negligible variations in orientation angle. The forward pivot was implemented with needle-bearings so that any friction in the bearings or seals generated a negligible lateral force at the contact patch.

Figure 2. Tire force and moment measuring device at the University of Wisconsin–Milwaukee.

This device allowed for sweeping slip and camber angles while measuring the lateral force, F_y, and vertical moment, M_z, generated in the contact patch. It used a force sensor to maintain the lateral location of the contact patch and a second force sensor acting on a lever arm of known length to prevent rotation of the fork that held the bicycle wheel about its steering axis.

In order to allow for the inevitable flexibility of the test frame and the bicycle wheel, the slip orientation of the bicycle rim was measured with a pair of laser position sensors mounted rigidly to the support platen for the flat-top chain near each end of the contact patch. Similarly, the camber orientation of the rim was measured with an accelerometer on the fork.

Slip angle was altered by pivoting the treadmill about a vertical axis under the center of the contact patch. Camber angle was altered by tilting the fixture about the longitudinal axis of the universal joint, which passed through the contact patch.

The vertical load borne by the tire was generated simply by the weight of the device's frame. Additional mass could be added above the fork as desired. For this study, the applied normal load was set to 418 N, which equated to front tire normal load of a 95 kg rider sitting on an 11 kg bicycle with 40% front and 60% rear weight distribution.

The lateral force generated in the contact patch was transmitted through the bicycle wheel to the fork, and thus to the test device frame. From the frame, the force was transmitted to both the lateral force sensor and the universal joint. A simple static summing of the moments about a vertical axis through the universal joint provided a relationship between the lateral force generated in the contact patch and the lateral force measured by the sensor. The mounting point on the frame for the lateral force sensor was positioned on the same axis through the center of the contact patch as the universal joint so that changes in camber angle had no effect on the lateral force sensor geometry.

2.1.5. Static Lateral and Radial Stiffness

Radial stiffness was measured by applying incremental weight to the top of the fork and measuring the resulting vertical deflection of the tire with a dial indicator.

Static lateral stiffness was measured by pulling on the rim immediately above the contact patch with a force and simultaneously recording the force magnitude and the resulting lateral deflection of the rim.

As illustrated in Section 3.2.4, the lateral shear stiffness of a specific number of knobs was measured in nearly the same way, except that the tire was not mounted on a wheel and the knobs were isolated from the rest of the tread by pressing them between two appropriately sized rectangular plates. The normal load was applied directly by the wheel and the test frame resting on the top rectangular plate. A lateral force was applied to the rim, as before, and the resulting lateral deflection was recorded.

Non-skid tape, by 3M, was applied to the top of the treadmill to maximize friction for both the static lateral stiffness and the force and moment measurements.

2.2. Fitting the Data

Pacejka's Magic Formula was fit to the experimental force and moment data [6] in a two-step process. First the data was smoothed using a 1D weighted, Blaise filter. Then, coefficients for Pacejka's Magic Formula, whose general form is depicted in Equation (2), were found to best fit the smoothed datasets for lateral force due to slip and camber, as well as to the self-aligning moment due to slip. Specifically, constraints from Pacejka's Motorcycle Magic Formula Tire model, which more equally emphasizes the camber and slip angles, were applied. The utility of the Pacejka's Motorcycle Magic Formula for this comparison were two-fold in that, given the specified constraints, each of the coefficients, and particular combinations thereof, had some physical significance, as explained below. Secondly, the fit data could be compared graphically and even extrapolated slightly to supplement the sometimes limited range of angles achieved during the physical test.

$$y = D\sin\{C\arctan[Bx - E(Bx - \arctan Bx)]\}, \tag{2}$$

$$Y(X) = y(x) + S_V, \tag{3}$$

$$x = X + S_H, \tag{4}$$

where

- Y: output variable. In this case, either lateral force, F_y, or self-aligning moment, $M_{Z_{SA}}$;
- X: input variable; here, either slip or camber angle in radians; and
- B: stiffness factor, which determined the slope at the origin;
- C: shape factor, which controlled the limits of the range of the sine function;
- D: peak value (when C was constrained as specified);
- E: curvature factor for the peak, controlling its horizontal position;
- S_H: horizontal shift;
- S_V: vertical shift.

The product BCD, obtained by multiplying the respective Pacejka coefficients, corresponded to the slope at the origin, that is, linear stiffness of the data, which was normalized by applied vertical load, and represented the cornering stiffness coefficient for slip, camber stiffness coefficient, and self-aligning moment coefficient, respectively.

Twisting torque due to camber, influenced predominantly by the difference in peripheral velocities across the bicycle tire's toroidal shape [7], was modelled using the nonlinear formula from the FastBike multibody simulation software depicted in Equation (5).

$$M_{Z_{TW}}(\varphi) = m_r \varphi N\left(1 + t_w \varphi^2\right), \tag{5}$$

where

- $M_{Z_{TW}}$: twisting torque due to camber, which was then normalized by
- N: the tire normal load in Newtons; and
- φ: camber angle in radians;
- m_r: linear twisting torque coefficient;
- t_w: non-linear twisting torque coefficient.

Figure 3 shows the results of the curve fitting for the 29 × 2.3″ knobby tire. The yellow dots represent the cloud of data collected during the fixture's separate sweeps of slip and camber. The thin blue line depicts the smoothed curve from the Blaise filter. Finally, the red dashed line represents the respective fit curve for that data, whereas the red plus symbol represents the origin of the fit curve given any offsets. In this case, the Magic Formula curves for lateral force for both slip and camber incorporated a small vertical offset, whereas that for self-aligning moment incorporated both a small vertical and horizontal offset. The twisting torque formulation also contained a vertical offset term.

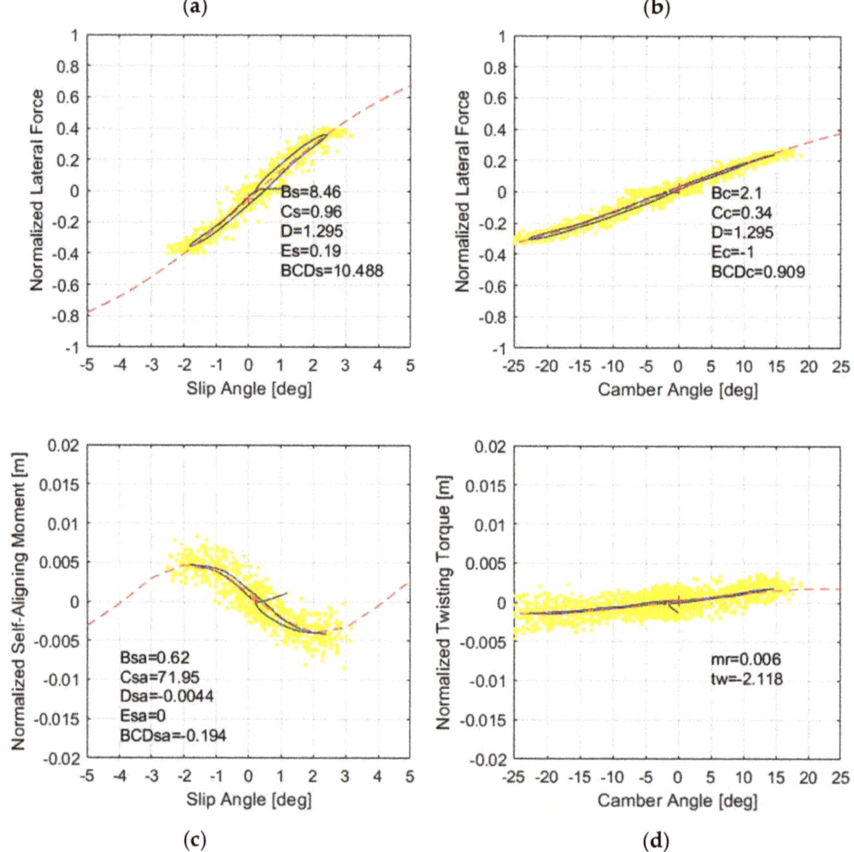

Figure 3. Force and moment raw data, smoothed lines, and fit curves for 29 × 2.3″ knobby on 25 mm rim at 25 psi (1.7 bar) with (**a**) normalized lateral force vs. slip angle, (**b**) normalized lateral force vs. camber angle, (**c**) normalized self-aligning moment vs. slip angle and (**d**) normalized twisting torque vs. camber angle.

As can be seen, the lateral force versus slip (Figure 3a) and, to an even greater extent, the self-aligning moment versus slip (Figure 3c) exhibited the characteristic shape of the Magic Formula, whereas the lateral force versus camber (Figure 3b) showed less curvature, as did the twisting torque (Figure 3d), with some negative (downward) curvature in this example. It should be noted that the plot scales were fixed to allow comparison across all datasets reported herein.

In this case, although the fixture was capable of 30 degrees of camber, the 29 × 2.3″ knobby was only measured to about +15 and −25 degrees, whereas slip angles achieved by the fixture for this particular tire were on the order of +/−2 degrees. In contrast large, heavy-duty automotive fixtures test slip angles up to, or in excess of, five or six degrees. Although such slip angles may be seen in aggressive transient or racing maneuvers, the added allure of measuring to these extremes is to better capture the peak and subsequent asymptote of the curves. In the case of the bicycle tire measurements and particularly the wider, knobby mountain bike tire measurements, the width of the treadmill, which was originally designed and built for testing relatively narrow road tires, was the limiting factor. As the wide tires were rotated in slip or in camber, care needed to be taken so that the tire did not encounter the platen that supported the treadmill. As such, the ability to accurately identify the curvature, peak, and asymptote terms for lateral force versus slip was limited and became even more limited with wider tires, which in the case of the 26x4″ may only achieve slip angles up to +/−1 degree with the studied treadmill arrangement. Regardless, the slope of the respective curves near the origin, that is, the stiffness or product of BCD Pacejka coefficients, was well captured and proved useful for the subsequent comparisons herein.

3. Results and Discussion

3.1. Various Sizes of Mountain Bike Tires

For decades, mountain bikes were equipped with so-called 26-inch wheels. This was in reference to the approximate outer tire diameter, consisting of a rim with approximately a 22-inch (559 mm) outer diameter and a tire that is about 2 inches tall at top and bottom. The so-called "29er" wheel and tire size, with a 24.5-inch (622 mm) diameter rim (equal to that of "700c" road bike wheels) gained popularity in the early 2000s on the basis of its purported ability to roll over obstacles with greater ease. However, some riders still yearned for the quick handling of smaller diameter wheels. After significant tooling investments from the bicycle industry, the so-called 27.5-inch or "650b" wheel size, with about a 23-inch (584 mm) rim diameter, offered what some thought was the optimal compromise.

As wheel diameters ebbed and flowed, pioneers within the industry have also explored various tire widths. Fat bike tires, which are generally 3.8 inches (96.5 mm) or wider, grew out of a desire to ride on soft snow and sand. The "plus tire", 2.6 inches (66 mm) to roughly 3 inches (76 mm) wide, split that difference, allegedly trading some of the original 29er's outright speed for more tire volume and confidence-inspiring traction. Now "29 plus" tires take an extreme to the extreme in terms of both diameter and width, and are a growing segment in both enduro mountain bike and bike packing applications.

As tire sizes have evolved, new rim widths have been adapted to follow suit. In the past, 18 or 19 mm inner width rims, similar to those used on past road racing bikes, were the norm. Over the past decade, rims gradually grew in width, chasing various trade-offs and trends in wheel stiffness, tire stiffness, tire profile, weight, and strength. For this study, "modern" rim widths appropriate to each tire size were selected on the basis of realistic use case and/or market benchmarking within that segment. Inner rim width will be referred to in millimeters where appropriate.

Similarly, each of the above tire and rim combinations have their own performance trade-offs in terms of "nominal" inflation pressure for a given rider weight, bike fitment, and terrain. Nominal inflation pressures used for each tire size in the following study are based on the author's experience with these tire setups for summer off-road trail riding.

For the remainder of the paper, the 29 × 2.3″ knobby tire on 25 mm inner rim width at 25 psi inflation pressure is considered the baseline to which data of the other tire configurations are compared.

3.1.1. Cross-Sectional Profiles

Figure 4 shows side-by-side comparisons of measured, inflated tire profiles for four different knobby mountain bike tire sizes. All the tires were from the same manufacturer, and the three tires on the left are all the same tire model only in different sizes. The tire on the right is a different model. As described previously, as the tread pattern often contains alternating sets of knobs at regular intervals along the tire's circumference, these profiles overlay each of the individual knob arrangements onto one cross section. This was done to better understand and compare the cross-sectional shapes, that is, the effective toroid radius of each tire, although it does potentially limit the ability to visually decipher spacing between individual knobs, which can be read from the footprints instead.

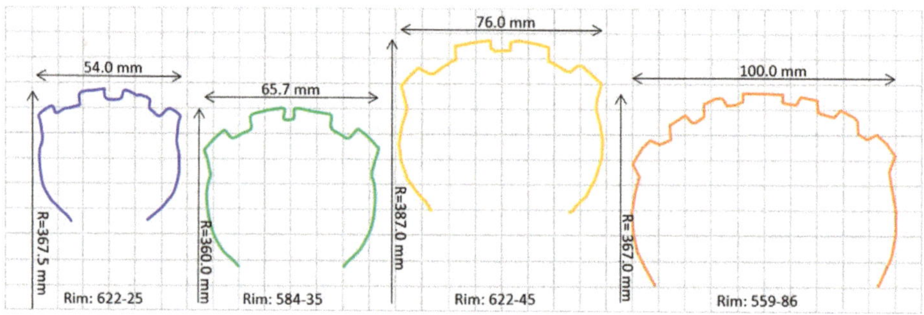

Figure 4. Cross sectional profiles for 29 × 2.3″, 27.5 × 2.8″, 29 × 3″, and 26 × 4″ knobby tires, respectively.

The tires shown increase in width from left to right. In terms of inflated outer radii (or diameters), the 29 × 3″ is by far the largest, followed by the 29 × 2.3″, which is very slightly larger than the 26 × 4″ fat bike tire, and finally the 27.5 × 2.8″ plus tire.

The shape of these mountain bike tires' inflated profiles is obviously influenced by rim width, as can be seen in the curvature of the tires' sidewalls and the angle that the lower sidewalls assume toward their respective bead interfaces. Here, rim width and rim-width-relative-to-tire-width increase from left to right.

Examining the treaded portion of each profile, intended to interact with the ground as the tire rolls and deforms and as the bicycle cambers and steers, several observations can be made. The 29 × 2.3″ tire profile was rather flat, that is, having a large toroidal radius, with limited drop from the two center rows of knobs to the shoulder knobs. It is also worth noting that the knob widths were relatively small, and the shoulder knobs were almost equal in height and width but with considerable draft (or bracing) down toward the tire sidewall to presumably support the knob. The middle two profiles represent the "plus" tires with the 27.5 × 2.8″ on the left and 29 × 3″ on the right. These tires had a similar tread pattern to the 29 × 2.3″, but both knob width and spacing increased as the tire size increased. It is uncertain if this tread scaling is for aesthetic or performance reasons. Notice the large gap between the center rows of knobs and shoulder knobs on the 29 × 3″ tire. Notice also that the angle of the shoulder knob was steeper and that (neglecting deformation) the tire would need to roll farther for the tangent (ground line) to engage the shoulder knobs. The 26 × 4″ fat bike tire on the right had a different tread pattern with seven rows of knobs as opposed to the four of the other tires, and included a center row.

3.1.2. Footprints

As described in the various literature on cars [8] and motorcycles [9], tire footprints can tell us a lot about the interaction between the tire and ground. Normal load is applied to the tire, which deforms to create a contact patch, or footprint, which at zero camber angle is quite elliptical on smooth, flat ground. The distribution of the normal load over the contact patch area yields the contact patch pressure (not measured here). In the absence of camber, the contact patch pressure may take on a fairly regular distribution. The presence of tread knobs obviously localizes contact pressure. Similarly, the knobs localize the shear stress used to generate lateral and longitudinal friction forces as the tire is slipped, cambered, driven or braked. Although not the focus of this work, it has been suggested that even for car tires, depending on (more closely spaced) tread pattern, the localized pressure can be 10× the average [10].

Figure 5 presents the knobby mountain bike tire footprints in the same order and on the same 10 mm grid as the cross sections in Figure 4. Here, the fitted ellipse was superimposed on the footprint and its axes were used to calculate length and width of the contact patch. Examining overall dimensions, it can be seen that the 26 × 4″ fat bike tire had both the longest and widest contact patch followed in length by the 29 × 3″ plus tire, the 29 × 2.3″ knobby, and 27.5 × 2.8″. Hence, the length of the contact patch depended not only on tire outer diameter, but presumably also on deflection, as influenced by inflation pressure, vertical load, and the tire's carcass. In terms of contact patch width, the 26 × 4″ was widest, whereas 27.5 × 2.8″ and 29 × 3″ were essentially equal, despite their size designation, and the 29 × 2.3″ at its nominal inflation pressure was narrower.

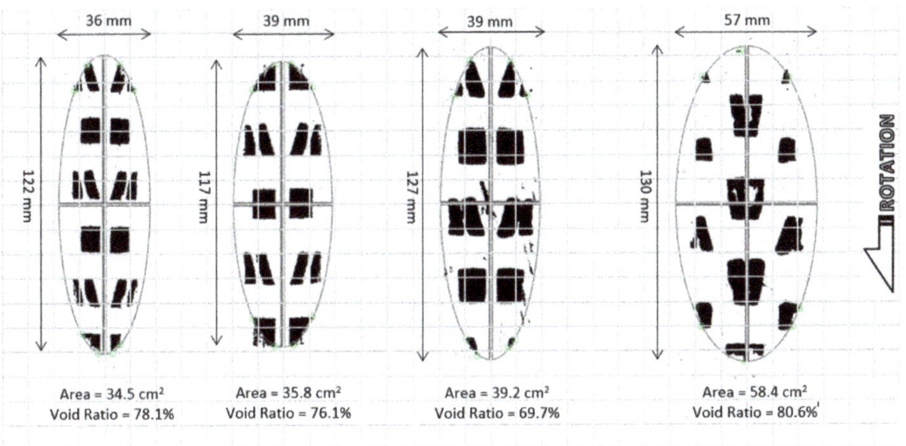

Figure 5. Footprints for 29 × 2.3″, 27.5 × 2.8″, 29 × 3″, and 26 × 4″ knobby tires, respectively.

These images bring very clear meaning to the tire industry term "void ratio" [11]. For these knobby tires, the amount of white, empty space within the contact patch ellipse considerably exceeded the amount of black tread interacting with ground. The same MATLAB script used for fitting the ellipse also computed the void ratio by comparing white space to contact patch area after converting the image to binary. The resulting void ratios in terms of percentage for each knobby tire are listed on the figure and ranged from 69.7% to 80.6%.

As noted in the cross-sectional profiles, the tread pattern of the 27.5 × 2.8″ and 29 × 3″ tires essentially scaled the size and spacing of the 29 × 2.3″ tread pattern upward in proportion to tire size. Again, the tread-pattern of these three tires offered no center ridge of knobs, whereas the 26 × 4″ tire did.

Note that the three tread patterns on the left employed essentially two different knob typologies, one that was rectangular and one that was slightly wedge-shaped with an hourglass-shaped longitudinal groove whose depth was roughly half of the 4.5 mm tread depth. The 26 × 4″ fat bike tire, on the right, showed four knob typologies within its footprint, two alternating central knob chevrons with varying degrees of central and trailing-edge relief, and two types of intermediate knobs similar to those of the other tires. Note that at zero camber angle and the inflation pressures shown, the footprints did not engage the shoulder knobs, nor the more outboard band of intermediate knobs on the 26x4″ fat tire shown in the profiles of Figure 4. It should be noted that purposefully positioning the wheel's rotation during footprint tests to engage leading and trailing knobs to better define borders of the contact patch greatly assisted in the identification of the ellipse.

Finally, these images of the footprint were useful in identifying how many knobs, pairs of knobs, or partial knobs are in contact at a given inflation pressure. This becomes relevant in a subsequent section.

3.1.3. Forces and Moments

Force and moment plots based on the Pacejka Magic coefficients and FastBike fit parameters are shown in Figure 6. Although each dataset retained a specific marker type, its line style was divided into two sections. The solid line portion, emanating from the origin, indicates values within the measurement range, whereas the dashed portion of the line represents any extrapolated values on the basis of the fit. As described previously, the limited treadmill width of the test fixture constrained slip and camber ranges, particularly for the wider knobby tires. Although the linear portion of the curve near the origin, whose slope (i.e., stiffness) was represented by the product of BCD Pacejka coefficients, was well identified, the curvature toward the peak and the eventual asymptote often had lower resolution, as was evident when comparing the 29 × 2.3″ solid versus dashed line to that of the 26x4″. As such, caution should be exercised if attempting to examine peak or curvature behavior of the wider tire variants, especially in terms of slip angle.

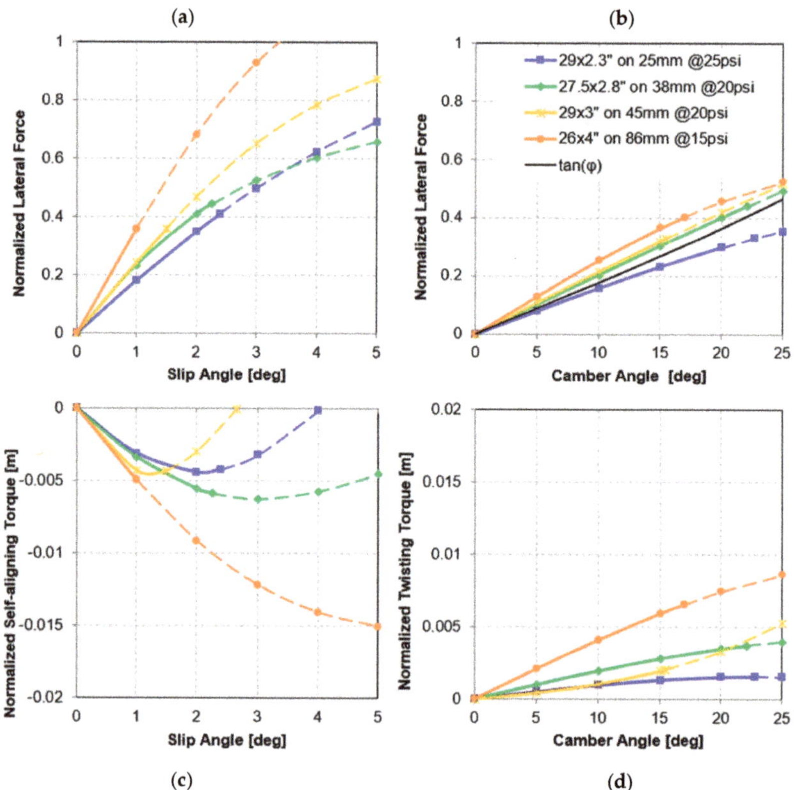

Figure 6. Force and moment fitted curves for 29 × 2.3″, 27.5 × 2.8″, 29 × 3″, and 26 × 4″ knobby tires with (**a**) normalized lateral force vs. slip angle, (**b**) normalized lateral force vs. camber angle, (**c**) normalized self-aligning moment vs. slip angle and (**d**) normalized twisting torque vs. camber angle.

Upon examining each of the plots in Figure 6, several interesting trends were evident. In Figure 6a, the upper left plot of normalized lateral force versus sideslip, the 29 × 2.3″ had the lowest cornering stiffness. The 27.5 × 2.8″and 29 × 3″ plus tires shared similar, steeper slopes but differed in curvature, whereas the 26 × 4″ fat bike tire had nearly double the cornering stiffness of the baseline tire. The vertical axis was limited at a normalized force of 1.0 to highlight another interesting trend. Although the peak of the curve, whose magnitude was captured by the Pacejka D coefficient, should represent the peak frictional value (approaching 1.3 here), the tires were tested on a non-skid tape (a particular formulation of sandpaper)-coated treadmill, not on actual asphalt or dirt, which would have a commensurately lower coefficient of friction. Hence, as with most test bench characterization data, it is useful for relative comparison, while various scaling methods can be employed to adjust to an appropriate friction level if desired.

In addition to the four knobby tire curves, whose camber stiffnesses also increased with width (Figure 6b), the upper right plot of normalized lateral force due to camber also included a reference line representing the so-called tangent rule, at which point the net ground reaction force, without slip, is in the plane of the wheel [7], as shown by

$$\frac{F_y}{N} = \tan(\varphi), \tag{6}$$

where F_y is the lateral force, N the normal load on the tire, and φ the camber angle.

Any tire below this line, such as the 29 × 2.3″ knobby, must make up the difference via a positive sideslip angle (into the turn). The other three tires would need negative sideslip angles (slipping to the outside of the turn) for equilibrium. Depending on the tire pairing, front versus rear, this may influence the vehicle's understeer/oversteer ratio.

The lower left plot (Figure 6c) shows the normalized self-aligning moment, which acts to align the wheel with its velocity vector, that is, diminish the tire's slip angle. In various physical tire models, this torque can be thought of as the product of the lateral force and pneumatic trail, where the pneumatic trail is the result of an offset in the shear stress distribution toward the rear of the contact patch. An approximation of the tire pneumatic trail is shown in Figure 7, obtained by dividing the aligning moment by the lateral force due to slip, and ignoring the point at zero slip for which no lateral force exists. The shape of the pneumatic trail curve is sometimes fit with a cosine function in some versions of Pacejka's Magic Formula. As can been seen, the tire with the longest contact patch, in this case the 26 × 4″, does not necessarily have the largest value of pneumatic trail near zero-slip, however, the other tires did follow that trend. It is possible that impending curvature of the 26 × 4″ self-aligning moment was not captured due the limited slip angle measurement range for that tire. In this case, the tires with a wider contact patch with respect to their length had a pneumatic trail that decayed more slowly.

The lower right plot in Figure 6d shows normalized tire twisting torque versus camber. This torque was due to fore aft shear stress distribution across the contact patch width driven by the toroidal tire shape and acted to steer the wheel into the corner in the direction of lean. Here, the 26 × 4″ fat bike tire still dominated, followed by the 27.5 × 2.8″ plus tire, whereas at low camber angles the 29 × 2.3″ tire trumped the 29 × 3″. It is interesting to note that at higher camber angles (or lower inflation pressures, shown later), the 29 × 3″ plus tire's twisting torque increased to a level just beyond that of the 27.5 × 2.8″. It was hypothesized that as the wider-spaced shoulder knobs of the 29 × 3″ tire encounter ground, they effectively increase the width of the patch thus augmenting the twisting torque. The high twisting torque of the 26 × 4″ fat bike tires seemed to corroborate the anecdotal depictions of heavy feeling, "autosteer" on many fat bikes as they were leaned into a corner, which requires the rider to counteract an increase in steering angle with significant steering effort.

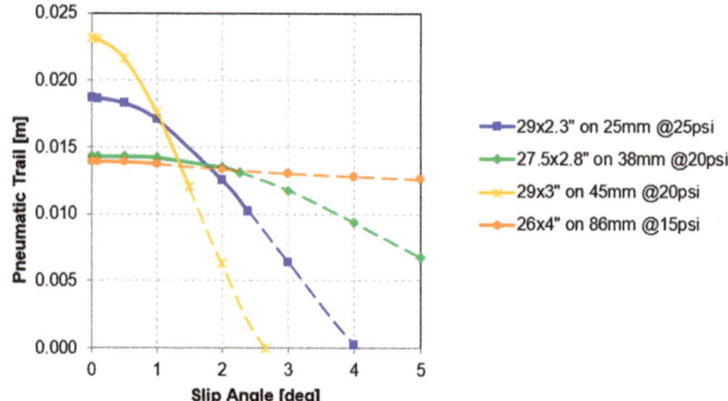

Figure 7. Calculated pneumatic trail for 29 × 2.3″, 27.5 × 2.8″, 29 × 3″, and 26 × 4″ knobby tires.

3.1.4. Static Lateral and Radial Stiffness

Although the static lateral and radial stiffness test was fairly straightforward and did not take into account any stiffening of the carcass due to centrifugal effects of the tire mass being accelerated radially as the wheel spins, these are still very useful parameters for understanding and modelling bicycle out of plane stability, in the case of lateral stiffness, and in-plane compliance, in the case of radial stiffness. Here, it seemed that for the given rim width and "nominal" inflation pressure selected for each size, the lateral stiffness, shown in Figure 8a, increased with tire (and/or rim) width and the radial stiffness, shown in Figure 8b for all four configurations, happened to converge around 60,000 N/m. The maximum value of the vertical axis on these plots was set to roughly the lower bound of stiffnesses for motorcycle tires [7]. Hence, bicycle tires at the inflation pressures shown were significantly less stiff than motorcycle tires.

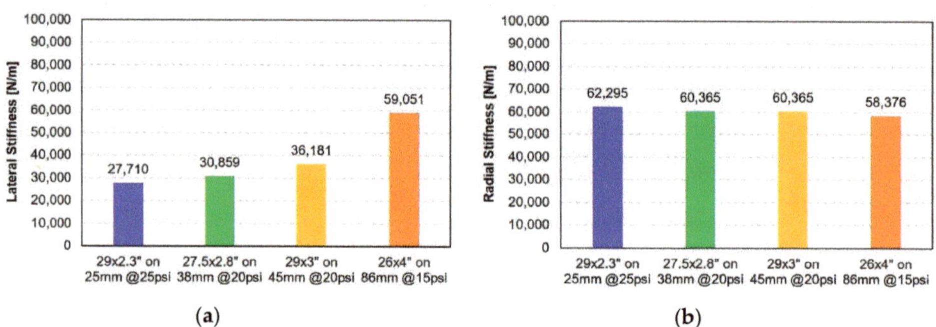

Figure 8. (**a**) Static lateral and (**b**) radial stiffness of 29 × 2.3″, 27.5 × 2.8″, 29 × 3″, and 26 × 4″ knobby bicycle tires.

3.2. *Effect of Tread Knobs*

In order to examine the effect of tread knobs, this study investigated two levels of resolution. First, a less treaded tire with similar cross-sectional profile and undeformed, inflated outer radius was measured for comparison to the 29 × 2.3″ knobby tire. Although this tire appeared much smoother than the knobby, it did indeed possess a "file-tread" pattern consisting of many small, pyramidal, and closely spaced knobs (each roughly 1.5 mm wide at their base and equally tall at their peak).

The lesser knob height required a slightly larger tire size of 29 × 2.5″ to achieve a similar outer diameter, as evident in the tire cross-sectional profiles.

After characterization of both the knobby baseline and file-tread tires, the second level of resolution involved sanding off the tread of both tires and re-characterizing what are referred to herein as the "bald" variants, as shown in Figure 9. In this case, the knobby tire became something like a true "slick" per Figure 9a, whereas the negative grooves of the file-tread in the Figure 9b tire remained. The various colors that appeared in the bald 29 × 2.3″ where the knobs previously existed suggested different compounds used to form the carcass and the knobs. The results are presented in the same format as the previous section.

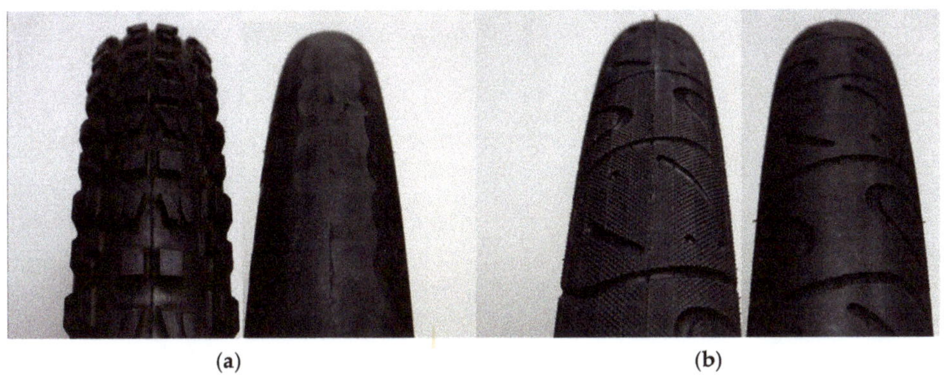

Figure 9. Images of (**a**) original treaded 29 × 2.3″ knobby and its bald tire counterpart and (**b**) the 29 × 2.5″ file-tread tire with its bald tire counterpart.

3.2.1. Cross-Sectional Profiles

Figure 10 again shows the cross-sectional profile of the baseline 29 × 2.3″ knobby tire, this time with the 29 × 2.3″ "bald" variant overlaid. As captured by the measurement, all tread knobs were completely removed, and this was done around the tire's entire circumference. Similarly, the 29 × 2.5″ file-tread and 29 × 2.5″ bald tire are shown in the overlay. Here, the removal of the more tightly spaced and shorter height knobs resulted in a more subtle, annular reduction in the treaded portion of the tire cross section.

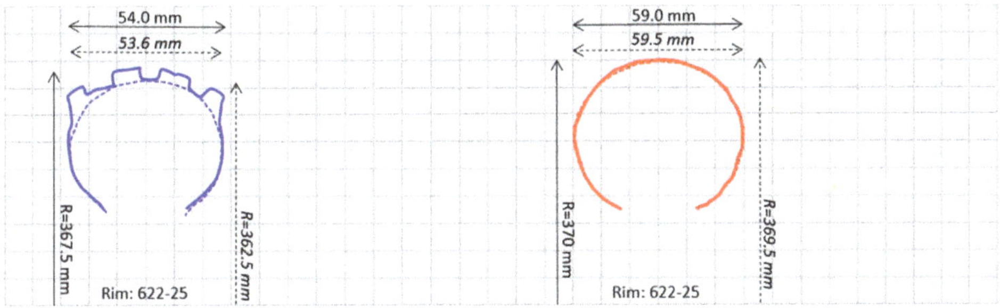

Figure 10. Cross sectional profiles of 29 × 2.3″ knobby tire (solid) overlaid with its bald variant (dashed), on the left, and 29 × 2.5″ file-tread tire (solid) overlaid with its bald variant (dashed), on the right.

3.2.2. Footprints

Figure 11 shows tire footprints for the 29 × 2.3″ baseline knobby tire and bald variant, as well as the 29 × 2.5″ file-tread and bald variant, all on 25 mm width rims at 25 psi (1.7 bar). The small, closely spaced file-tread knobs, as well as the negative tread grooves on the 29 × 2.5″, can be seen. The file-tread was effectively removed from the 29 × 2.5″ bald variant, but the negative grooves remained. The 29 × 2.3″ bald tire was essentially a slick.

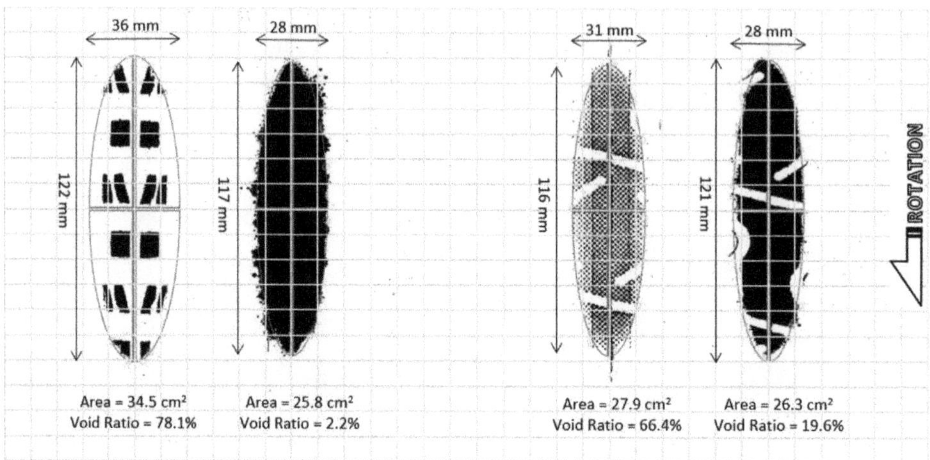

Figure 11. Footprints for 29 × 2.3″ knobby, 29 × 2.3″ bald, 29 × 2.5″ file-tread, and 29 × 2.5″ bald tires at 25 psi (1.7 bar) on 25 mm inner width rim.

The 29 × 2.3″ bald contact patch was both shorter and narrower than that of the 29 × 2.3″ knobby tire, whereas the 29 × 2.5″ bald contact patch was longer but narrower than that of the 29 × 2.5″ file-tread tire. It is interesting to note that, despite having a smaller outer radius than the 29 × 2.5″ tires, the 29 × 2.3″ knobby tire yielded the widest and longest contact patch of these four variants.

The relatively high (66.4%) void ratio of the 29 × 2.5″ file-tread might seem surprising, as this tire was chosen as a less-treaded comparison to the 29 × 2.3″ knobby (78.1%). Even after removing the file-tread pattern, the remaining negative tread grooves yielded an appreciable (19.6%) void ratio for the 29 × 2.5″ bald tire, as opposed to the nearly slick (2.2%) void ratio of its 29 × 2.3″ bald counterpart.

3.2.3. Forces and Moments

Various trends are evident in the fitted force and moment plots depicted in Figure 12. The upper left plot of normalized lateral force versus sideslip (Figure 12a) shows an increase in cornering stiffness as the tires became less treaded. There was a large increase in slope from the 29 × 2.3″ knobby to the 29 × 2.3″ bald tire, whereas the 29 × 2.5″ file-tread to bald modification showed an increase but of less magnitude.

The same pattern was repeated in the other three plots. Removing the knobs had a big effect and removing the file-tread had a small effect, both trending in similar directions. Overall, the slick tire was the stiffest, and the knobby tire was the least stiff. As noted previously, the shape of the curve beyond the recorded data, indicated by the dashed lines, was an extrapolation based on the fit coefficients. Thus, the curvature of some of the more limited datasets may have been exaggerated.

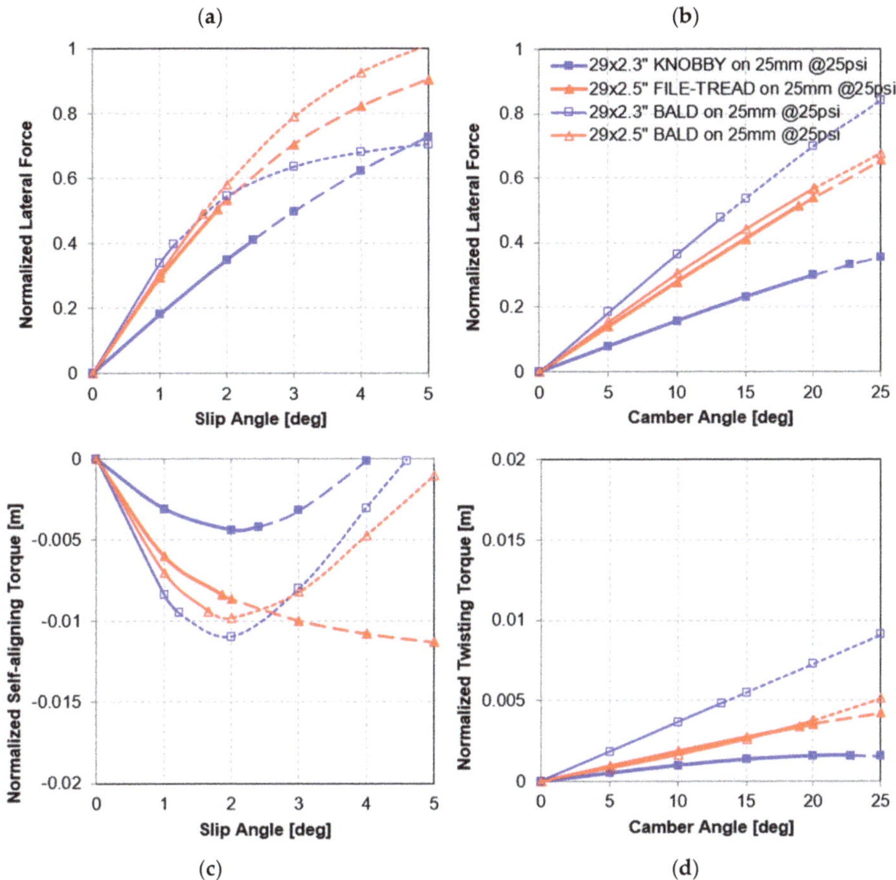

Figure 12. Force and moment fitted curves for 29 × 2.3″ knobby, 29 × 2.3″ bald, 29 × 2.5″ file-tread, and 29 × 2.5″ bald tires at 25 psi (1.7 bar) on 25 mm inner width rim with (**a**) normalized lateral force vs. slip angle, (**b**) normalized lateral force vs. camber angle, (**c**) normalized self-aligning moment vs. slip angle and (**d**) normalized twisting torque vs. camber angle.

3.2.4. Static Lateral Stiffness

To understand the contribution of tire tread knobs to total (treaded) tire lateral stiffness, knob shear stiffness, and bald carcass lateral stiffness were measured separately and then combined, in the manner of springs in series, as in Equation (7), to compare against the measured lateral stiffness of the knobby tire.

$$\frac{1}{K_{total\ tire}} = \frac{1}{K_{bald\ carcass}} + \frac{1}{K_{tread}}, \tag{7}$$

As illustrated in Figure 13, the shear stiffness of the tread knobs was measured by isolating an integral number of knobs (six shown) of an unmounted tire between matching aluminum plates. Vertical load was applied directly from the bare rim to the top aluminum plate. Then, the rim was pulled laterally while two cap screw heads prevented it from sliding relative to the upper plate. The applied lateral force and resulting lateral deflection of the rim were recorded simultaneously. The measured tread lateral stiffness was then divided by the number of knobs present between the plates to yield an approximate individual knob stiffness.

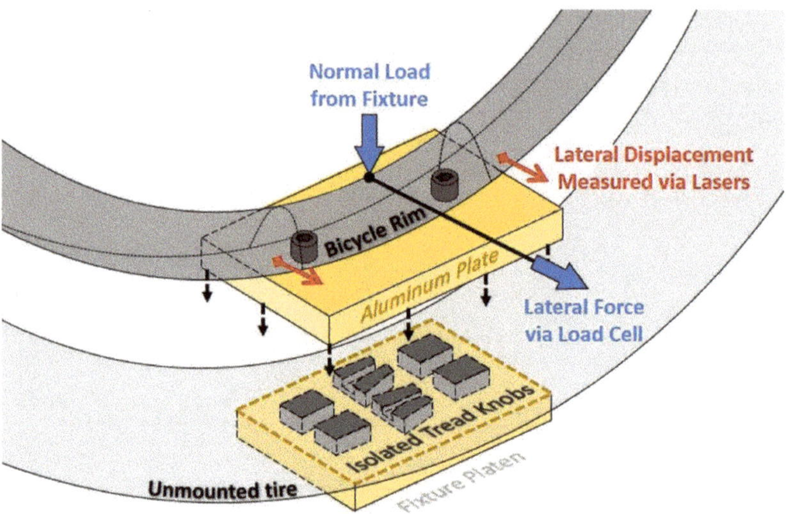

Figure 13. Schematic illustrating non-destructive measurement of lateral shear stiffness of tire tread knobs via loads applied to unmounted tire squeezed between appropriately sized aluminum plates.

Finally, per Equation (8), the tread shear stiffness was calculated as springs in parallel by multiplying the individual knob stiffness by the number of knobs evident in the tire's footprint.

$$K_{tread} = n_{knobs} * k_{knob}, \quad (8)$$

These respective stiffness values and the comparison of measured versus calculated total tire stiffness are shown for the 29 × 2.3″ knobby tire in Figure 14.

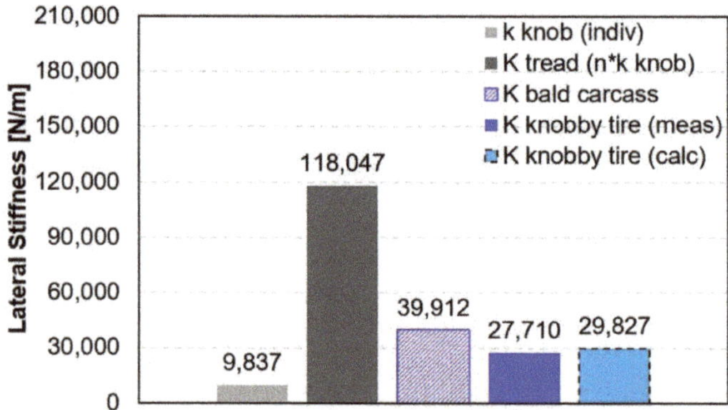

Figure 14. Bar chart of knob, tread, bald carcass, measured, and calculated total tire stiffness for 29 × 2.3″ knobby tire comparing relative stiffness magnitudes at 25 psi (1.7 bar).

As can be seen, the approximate total tire stiffness matched well (within 8%) with the measured lateral stiffness of the inflated knobby tire. Possible sources of variation included difficulty in determining exactly how many knobs, or fractions thereof, were included in the contact patch. This approach also assumed equal stiffness contribution from various knob typologies and ignored possible variation in load distribution across the patch. A similar investigation was not repeated for the file-tread pattern because of the impracticality of counting and measuring the stiffness of an integral number of the file-tread knobs.

3.3. Effect of Inflation Pressures and Rim Width

Tire inflation pressure is an important setup and tuning parameter for mountain bike applications. A happy compromise between traction, durability, protection for the rim, handling, and ride characteristics often depends on rider weight, bike fitment, terrain, even surface, or soil consistency, and of course, rider preference. Potentially for this reason, bicycle manufacturers, unlike automotive or motorcycle companies, rarely specify a specific recommended operating inflation pressure for each vehicle model and leave it to the end-user to experiment within the bounds typically designated on the tire sidewall. Thus, to better explore the potential operating range of various tires mentioned thus far in this study, additional measurements were taken at various inflation pressures. These results are reported in a format much like the previous sections.

3.3.1. Footprints

Figures 15 and 16 illustrate the effect of varying inflation pressure on the tire footprints for the 29 × 2.3″ knobby tire on 25 mm rim. The images show that as inflation pressure increased from left to right, the contact patch length and width became smaller and thus contact patch area was reduced. Interestingly, for the 29 × 2.3″ knobby tire at 10 psi, in the leftmost plot, the tire deformed enough for the shoulder knobs to contact the ground (even at zero camber), and thus the shape of the patch became significantly wider. As inflation pressure increased, fewer and fewer individual knobs remained in the contact patch.

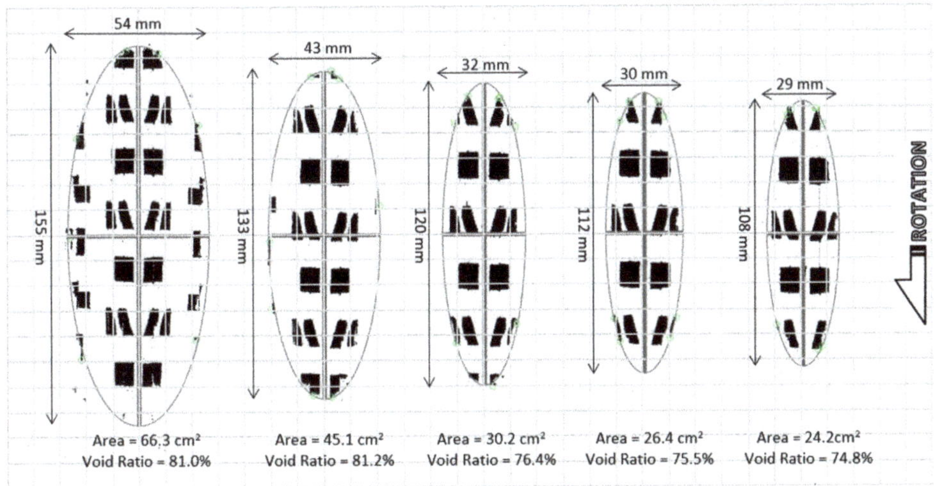

Figure 15. Footprints for 29 × 2.3″ knobby tire at inflation pressures of 10, 20, 30, 40, and 50 psi (0.7, 1.4, 2.1, 2.8, and 3.4 bar) from left to right on 25 mm inner width rim.

Figure 16 shows contact patch characteristics as functions of inflation pressure. The 29 × 2.3″ knobby, 29 × 2.3″ bald, 29 × 2.5″ file-tread, and 29 × 2.5″ bald tires were measured at eight pressure increments. This gave added resolution to various trends. The other three knobby tires in 27.5 × 2.8″, 29 × 3″, and 26 × 4″ are also shown, but with inflation pressure at just two levels, 10 psi (potential use case for lightweight riders or usage on soft or loose terrain), and the "nominal" pressure considered in previous sections. Finally, two additional datapoints of the 29 × 2.3″ knobby and 29 × 2.5″ file-tread tire on slightly narrower, 22 mm internal rim widths are shown to begin to explore those effects.

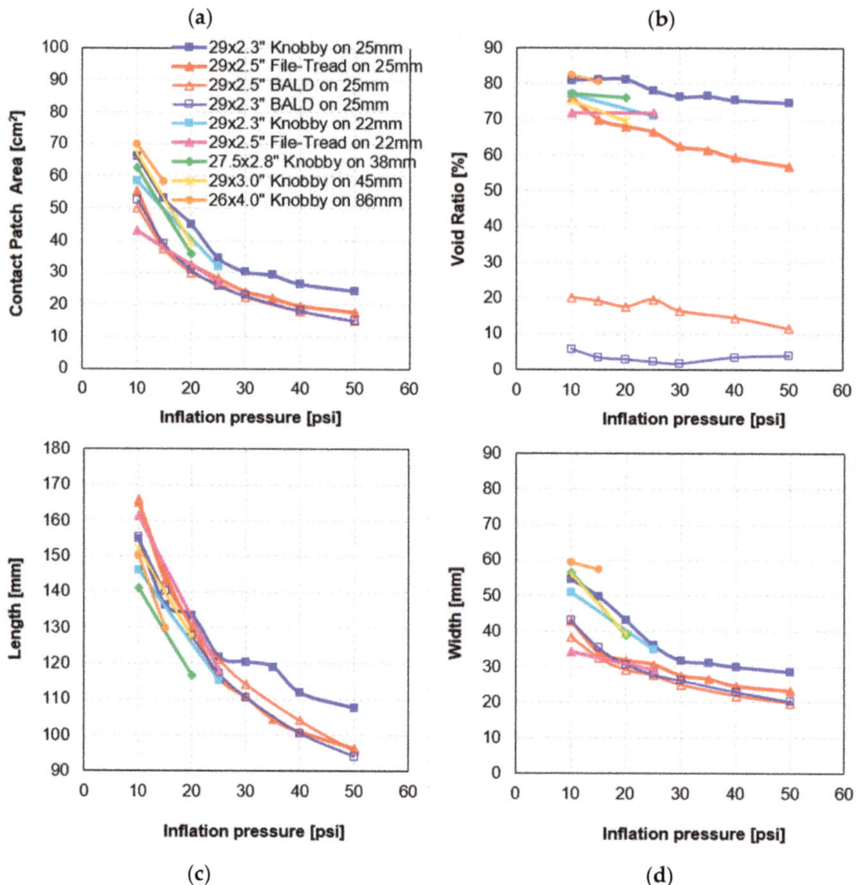

Figure 16. Tire footprint parameters versus inflation pressure for various bicycle tires with (**a**) contact patch area, (**b**) void ratio, (**c**) contact patch length and (**d**) contact patch width.

The upper left plot (Figure 16a) suggests the ranking of tires in terms of contact patch area varied somewhat with inflation pressure, but in general the contact patch area decreased in a seemingly nonlinear fashion with increasing inflation pressure. Here, the 26x4″ fat bike tire remained near the largest, whereas the 29 × 2.3″ knobby surpassed the plus tires between 10 and 20 psi. The contact patch area of the less-treaded tires in the form of the 29 × 2.5″ file-tread on the 25 mm rim, 29 × 2.3″ bald, and 29 × 2.5″ bald tire seemed slightly more sensitive in the lower pressure regime (from 10 to 15 psi), whereas as pressure was increased (beyond 20 psi), the curvature stabilized. The 29 × 2.3″ knobby on 22 mm rim and 29 × 2.5″ file-tread tire on 22 mm rims showed a small decrement in contact patch area with respect to their 25 mm rim variants.

The upper right plot (Figure 16b) depicts how the tread void ratio within the contact patch area varied between tires and with inflation pressure. It is quite clear that the knobby tires had higher void ratios than the 29 × 2.5″ file-tread on 25 mm rim at all but the lowest inflation pressure. The 26 × 4″ fat bike tire and 29 × 2.3″ knobby tire were both around 80% void ratio at low pressures, which for the 29 × 2.3″ knobby tire on 25 mm rim fell to 74% at 50 psi. Seemingly, scaling the knob size up for the 27.5 × 2.8″ and 29 × 3″ plus tires caused a small reduction in their void ratio with respect to the 29 × 2.3″ knobby tire on the 25 mm rim. The void ratio of the 29 × 2.5″ bald tire, which contained only the negative grooves, reduced from 20% to 10% with increasing inflation pressure, whereas the nearly slick 29 × 2.3″ bald tire hovered at a void ratio less than 5%.

The lower two plots show the contact patch length (Figure 16c) and width (Figure 16d) versus inflation pressure. Here, the length was generally at least twice the width or more. The change in lengths of most of the tires shown were similar, whereas the 26 × 4″ fat and 27.5 × 2.8″ plus tires were slightly shorter than others at the same inflation pressures. The 29 × 2.3″ knobby tire retained a slightly longer contact patch than the 29 × 2.5″ file-tread tire or their bald counterparts at inflation pressures above 25 psi (1.7 bar). Finally, in terms of width, there was a clear grouping between the wider contact patch of the knobby tires and the less-wide patch of the file-tread and bald variants.

3.3.2. Forces and Moments

Figure 17 shows the stiffness coefficients for slip and camber forces as well as self-aligning and twisting torques versus inflation pressure.

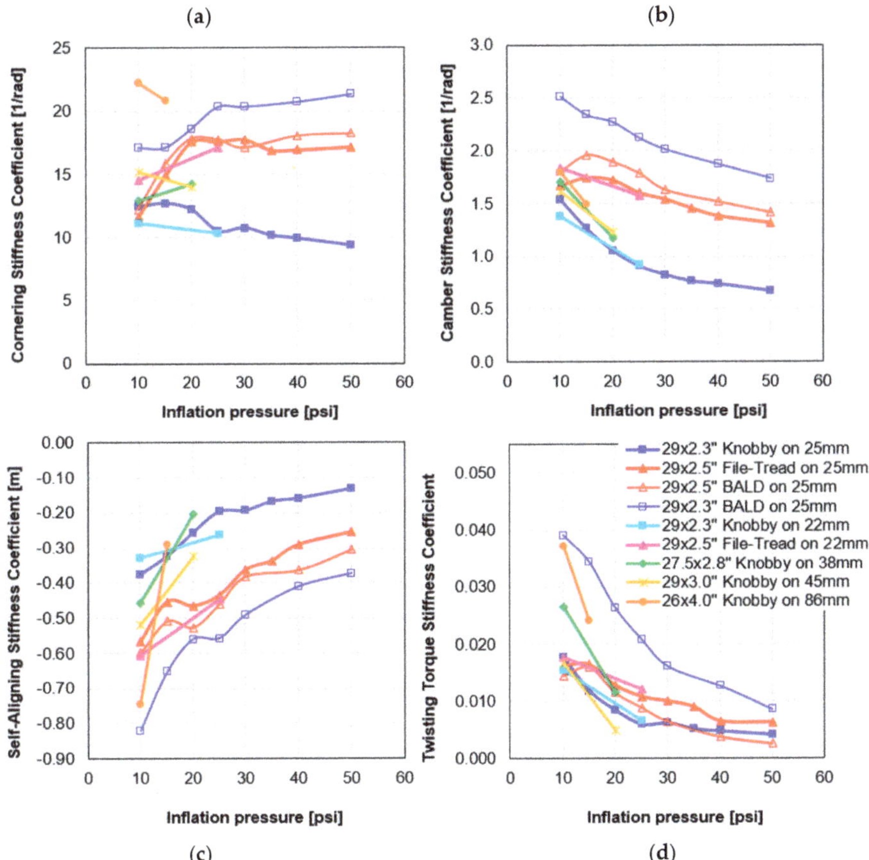

Figure 17. Force and moment fitted stiffness coefficients versus inflation pressure for various bicycle tires with (**a**) cornering stiffness, (**b**) camber stiffness, (**c**) self-aligning stiffness and (**d**) twisting torque coefficients.

The upper left plot (Figure 17a) displays various trends in cornering stiffness coefficient. First, large differences were apparent between the different tire sizes bounded on the upper end by the

26 × 4″ fat bike and 29 × 2.3″ bald tires, with upper values in excess of 20 1/rad, and on the lower end by the 29 × 2.3″ knobby tire, on either rim width, at around 10 1/rad. Aside from the 26 × 4″ fat tire, the less-treaded tires had higher cornering stiffness above 20 psi (1.4 bar). Additionally, above this pressure, the cornering stiffness of the tires shown seemed to vary little with further increase in inflation pressure. Below this pressure, there were some nonlinear and disparate behaviors where cornering stiffness of the 29 × 2.5″ file-tread increased rapidly with inflation pressure whereas that of the 29 × 2.3″ knobby tire decreased gradually. Across the limited pressures measured, the 27.5 × 2.8″ plus tire exhibited a different sensitivity (slope) than the other knobby tires.

The plot of normalized camber stiffness, shown in the upper right (Figure 17b) showed similar overall groupings. The less-treaded variants exhibited higher camber stiffness coefficients, with a decreasing trend as inflation pressure increased, and the 29 × 2.3″ bald tire topped the chart. Camber stiffness of the knobby tires fell off more quickly from 10 to 25 psi, whereas the 29 × 2.3″ knobby tire on a 25 mm rim eventually achieved a slope similar to its less-treaded counterparts, albeit at a lower value (below 1.0 1/rad). Rim width had minimal effect.

In the lower left plot (Figure 17c), the sign convention for the aligning torque should be kept in mind. Its value was negative because it acted to reduce the tire slip angle, and thus a larger stiffness magnitude resulted in a more negative value. Again, there was a grouping. Except for the 26 × 4″ fat bike tire at 10 psi, the less treaded tires exhibited more self-aligning stiffness, with increased inflation pressure, albeit with diminishing magnitude. The 29 × 2.3″ and 27.5 × 2.8″ knobby tires grouped fairly close together with the lowest magnitudes of self-aligning stiffness. It should be noted that although the units of normalized self-aligning stiffness were in meters, this value was different from the pneumatic trail. As described earlier, the pneumatic trail can be derived by dividing the tire self-aligning moment by the lateral force due to slip (at zero camber).

Normalized twisting torque stiffness, shown in the lower right plot (Figure 17d), showed a fairly rapid decrease with increasing inflation pressure up to about 30 psi (2.1 bar) for the measured tires. On the basis of the higher twisting torque stiffness magnitudes of the 29 × 2.3″ bald and 26 × 4″ tires, one would expect them to generate the most steering torque at low inflation pressures. For the 26 × 4″, this was likely related to the difference in peripheral velocity across such a wide patch, and for the 29 × 2.3″ bald tire it may have been related to the relatively small toroid radius and steep shoulder drop. Again, a minor change in rim width had minimal effect. Presumably, the steep slope of the 27.5 × 2.8″ and 29 × 3″ plus tires between 10 and 15 psi may have been related to the widely spaced shoulder knobs contacting at lower lean angles on the less inflated, softer, and more deformed tire.

3.3.3. Static Lateral Stiffness

Section 3.2.4 described modelling the tire static lateral stiffness as multiple individual knobs acting as springs in parallel that constituted the tread stiffness. The tread stiffness then combined with the bald carcass stiffness as springs in a series to yield the total lateral stiffness of the treaded tire. Now, in Figure 18, that simplified model is applied across a range of inflation pressures, yielding results comparable to the measured value of the 29 × 2.3″ knobby tire. The figure shows the individual knob stiffness as derived from the tread shear measurement multiplied by the number of knobs in contact. On the basis of the footprints shown in Figure 15, "n", the number of knobs in contact, plotted here on the secondary axis, decreased with inflation pressure. Hence, the product of those terms yielded the theoretical tread stiffness, which also decreased with inflation pressure. It is interesting to note that the individual knobs did not change in stiffness, but the fact that fewer of them were in contact reduced the effective tread stiffness. Meanwhile, the as-measured bald carcass stiffness was shown to increase with inflation pressure. Combining these two as springs in series resulted in a theoretical total tire lateral stiffness within 15% of the measured values across the range of inflation pressures.

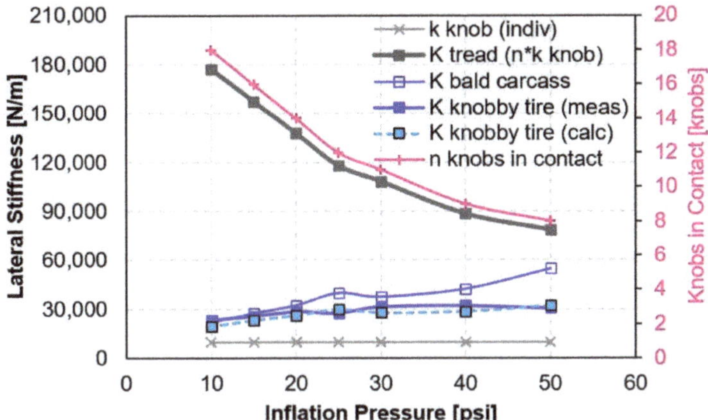

Figure 18. Knob, tread, bald carcass, measured, and calculated total tire stiffness, and knob count for 29 × 2.3″ knobby tire comparing stiffness magnitudes across inflation pressures.

As before, possible sources of the difference between calculated and measured total stiffness included difficulty in determining exactly how many knobs, or fractions thereof, were included in the contact patch and, for low inflation pressures, the participation of shoulder knobs, whose shear stiffness was not measured.

4. Conclusions

Characterization of four modern mountain bike tire sizes was carried out on appropriate rim widths at what were considered realistic nominal inflation pressures. Tire cross sections were compared, and tire radii and width values reported. Tire footprints at zero camber angle and realistic normal load were fit as ellipses and their area, geometry, void ratios, and tread patterns were discussed. Force and moment measurements were conducted using the measuring device at the University of Wisconsin–Milwaukee. A combination of Pacejka's Motorcycle Magic Tire Model and FastBike twisting torque polynomial were fitted to the data to yield parametric tire model coefficients. These fitted curves were then used to compare lateral force due to slip and camber, as well as self-aligning moment and twisting torque. Good resolution of the linear portion of the respective stiffness curves were obtained, whereas tire versus treadmill width limited identification of curvature and peak friction values to some extent. Static lateral and radial carcass stiffnesses were shown. The compilation of these results suggests appreciable differences in tire performance among 29 × 2.3″, 27.5 × 2.8″, 29 × 3″, and 26 × 4″ knobby tires.

Additional characterization was carried out to quantify the effects of tire tread knobs. A 29 × 2.3″ knobby tire was compared to a file-tread tire of similar size. The treads were then removed from both tires, and characterization of the resultant bald tires further informed this comparison. It was shown that file-tread and a tread pattern with negative grooves had appreciable void ratios and performance trade-offs that were different from a truly bald, slick tire. It was also shown that for the 29 × 2.3″ knobby mountain bike tire studied, the combination of tire tread shear stiffness, obtained by combining individual knobs as springs in parallel, with the inflated bald tire carcass as springs in series, yielded a lateral stiffness value similar to that measured for the total treaded tire.

Finally, the aforementioned footprint, force and moment, and static carcass lateral stiffness parameters for each tire were compared across a range of inflation pressures. Results suggest some differences between knobby and less-treaded tires, especially at higher inflation pressures. Inflation pressure influenced the number of knobs interacting with ground as the toroidal tires deformed. Incorporating this observation into the approximation of tire lateral stiffness as a combination of tread shear stiffness and carcass lateral stiffness as springs in series seemed promising. A brief comparison of a 25 mm and 22 mm rim suggested small relative differences for the two tires measured.

Ideally, this work supplements ongoing progress in bicycle modelling and simulation [12]. Parametrization of nonlinear tire properties, as shown here, may aide higher fidelity bicycle and even mountain bike handling and stability simulations. A further study looking at alternate tire sizes, specifications, and tread patterns; isolating individual knob typologies; and implementing a wider treadmill would be interesting. Applying similarity methods to link tire performance to width and radius may warrant additional study, whereas a design of experiments (DOE) approach focused on the effects of specific construction parameters would benefit from bicycle tire manufacturer support in building relevant test samples.

Author Contributions: Conceptualization, J.S. and A.D.; methodology, A.D.; software, A.D. and J.S.; validation, A.D. and J.S.; formal analysis, J.S. and A.D.; investigation, A.D. and J.S.; resources, A.D.; data curation, A.D. and J.S.; writing—original draft preparation, J.S. and A.D.; writing—review and editing, A.D. and J.S.; visualization, J.S.; supervision, J.S.; project administration, A.D.; funding acquisition, J.S. All authors have read and agreed to the published version of the manuscript.

Funding: This research received no external funding.

Conflicts of Interest: The authors declare no conflict of interest.

References

1. Statista: Number of Participants in Mountain/Non-Paved Surface Bicycling in the United States from 2011 to 2018 (in Millions). Available online: https://www.statista.com/statistics/763737/mountain-non-paved-surface-bicycling-participants-us/ (accessed on 3 April 2019).
2. Statista: Bicycle Sales in the United States by Category of Bike in 2017. Available online: https://www.statista.com/statistics/236150/us-retail-sales-of-bicycles-and-supplies/J.Y. (accessed on 3 April 2019).
3. Dressel, A. Measuring and Modeling the Mechanical Properties of Bicycle Tires. Ph.D. Thesis, The University of Wisconsin-Milwaukee, Milwaukee, WI, USA, 2013.
4. Pacejka, H.B. *Tyre and Vehicle Dynamics*, 2nd ed.; Butterworth and Heinemann: Oxford, UK, 2006.
5. *FastBike 6.3.1 User Manual*; Dynamotion: Padova, Italy, 2006; p. 36.
6. Doria, A.; Tognazzo, M.; Cusimano, G.; Bulsink, V.; Cooke, A.; Koopman, B. Identification of the mechanical properties of bicycle tyres for modelling of bicycle dynamics. *Veh. Syst. Dyn.* **2013**, *51*, 405–420. [CrossRef]
7. Cossalter, V. *Motorcycle Dynamics*, 2nd ed.; Lulu Press: Morrisville, NC, USA, 2006.
8. Gent, A.; Walter, J. *The Pneumatic Tire*; NHTSA: Washington, DC, USA, 2006.
9. Cossalter, V.; Doria, A. The relation between contact path geometry and the mechanical properties of motorcycle tires. *Veh. Syst. Dyn.* **2005**, *43*, 156–167. [CrossRef]
10. Wallaschek, J.; Wies, B. Tyre tread-block friction: Modelling, simulation and experimental validation. *Veh. Syst. Dyn.* **2013**, *51*, 1017–1026. [CrossRef]
11. Jansen, S.; Schmeitz, A.; Akkermans, L. *Study on Some Safety-Related Aspects of Tyre Use*; European Commission: Brussels, Belgium, 2014.
12. Klug, S.; Moia, A.; Verhagen, A.; Georges, D.; Savaresi, S. Control-Oriented modeling and validation of bicycle curve dynamics with focus on lateral tire parameters. In Proceedings of the 2017 IEEE Conference on Control Technology and Applications (CCTA), Mauna Lani, HI, USA, 27–30 August 2017; pp. 86–93.

Article

Study on the Generalized Formulations with the Aim to Reproduce the Viscoelastic Dynamic Behavior of Polymers

Andrea Genovese *, Francesco Carputo, Antonio Maiorano, Francesco Timpone, Flavio Farroni and Aleksandr Sakhnevych

Department of Industrial Engineering, University of Naples Federico II, Via Claudio, 21-80125 Naples, Italy; francesco.carputo@unina.it (F.C.); antonio.maiorano@unina.it (A.M.); francesco.timpone@unina.it (F.T.); flavio.farroni@unina.it (F.F.); ale.sak@unina.it (A.S.)

* Correspondence: andrea.genovese2@unina.it

Received: 29 February 2020; Accepted: 26 March 2020; Published: 28 March 2020

Abstract: Appropriate modelling of the real behavior of viscoelastic materials is of fundamental importance for correct studies and analyses of structures and components where such materials are employed. In this paper, the potential to employ a generalized Maxwell model and the relative fraction derivative model is studied with the aim to reproduce the experimental behavior of viscoelastic materials. For both models, the advantage of using the pole-zero formulation is demonstrated and a specifically constrained identification procedure to obtain the optimum parameters set is illustrated. Particular emphasis is given on the ability of the models to adequately fit the experimental data with a minimum number of parameters, addressing the possible computational issues. The question arises about the minimum number of experimental data necessary to estimate the material behavior in a wide frequency range, demonstrating that accurate results can be obtained by knowing only the data of the upper and low frequency plateaus plus the ones at the loss tangent peak.

Keywords: fractal derivative; viscoelastic models; polymers

---

1. Introduction

Viscoelasticity regards the study of materials exhibiting a time dependence behavior. The knowledge of the viscoelastic properties of these materials is at the basis of a correct analysis of structures, where the viscoelastic materials are employed, especially when dealing with dynamic phenomena and vibrations. Materials with viscoelastic behavior are widely employed in many different applications starting from biological tissues, such as the disks in the human spine [1] and skin tissue [2,3], up to civil and industrial systems such as elastomers. As regards the latter field of study, tall buildings are damped with viscoelastic materials to avoid instabilities and to face vibration problems caused by wind or earthquakes [4,5]. Viscoelastic dampers have long been used in the control of vibration and noise in aerospace structures and industrial machines. In many mechanical systems, such materials are widespread and used for damping and insulation scope [6], i.e., they are used in automobile bumpers, machine insulation, to protect computer drives from mechanical shocks and are also used in shoe insoles to reduce impact transmitted to the person and to increase comfort. Moreover, materials that behave elastically at room temperature often show significant viscoelastic properties when heated. In addition, tires are made of different polymers and elastomers with peculiar viscoelastic properties responsible for many tire/road interaction phenomena that affect vehicle dynamics in term of safety and performance. Friction performance [7,8], rolling resistance and wear are some examples of events in which the viscoelasticity of the tread plays a fundamental role.

The characterization of these materials involves two main phases: experimental tests and analytical modelling. A good understanding of the dynamical behavior of structures where viscoelastic materials are used is strictly dependent on their rheological properties and on some of their geometric parameters. These characteristics in turn depend on the temperature and on the induced excitation frequency quantities. Material models and simulation tools become fundamental to understand and to reduce uncertainties during the earliest phases of the material development, optimizing the design of composite and multi-layered material components and the manufacturing process. Furthermore, the possibility to employ realistic models of materials allows to make reliable predictions of the behavior of existing components as well as for those in the process of development.

To define a constitutive law for linear viscoelasticity, at least three different common approaches can be adopted: integral models, linear differential models and fractional derivative models. The basic foundation of linear viscoelasticity theory is the Boltzmann's superposition principle according to which every loading step makes an independent and additive contribution to the final state [9]. This idea can be used to formulate an integral representation of linear viscoelasticity. Nevertheless, a more convenient and widespread way to describe the linear viscoelasticity is the differential approach, where a combination of mechanical elements, ideal springs and dashpot, are used to describe the rheological properties. On one hand, spring is a perfect elastic element, following Hooke's law and behaving like an elastic solid; on the other hand, dashpot is a perfect viscous element, following Newton's Law. The methodology links stress and strain through linear differential equation. The simplest and most famous models currently employed are the Maxwell model, Kelvin–Voigt model and Zener model. However, Maxwell model fails in reproduction of the creep phenomenon and Kelvin–Voigt model fails in the description of the relaxation behavior [10]. Furthermore, in terms of frequency domain, both models are efficient only within a small frequency range, but they become insufficient within low and high frequency ranges, representing an important limit in case the specific application could require different boundary conditions in terms of excitation frequency to be applied on the same material model.

When the deformations of a viscoelastic material are neither small nor very slow, its behavior is no longer linear, and there is no universal rheological constitutive equation that can predict the response of the material to such a deformation. Hence, for modelling their behavior in these cases, finite strain-based viscoelastic models are required. Some notable studies can be found in literature regarding finite viscoelastic modelling of polymers. The finite strain thermo-viscoelastic constitutive model, proposed by Liao et al. [11] employs a non-linear evolution law based on the classical concept of the multiplicative decomposition of the deformation gradient using specific hyper-elastic parameters identified at reference temperature values. Following the Bergstrom–Boyce model for nonlinear finite rubber viscoelasticity [12], Hossain et al. proposed a modified version of the micro-mechanically motivated Bergstrom–Boyce viscoelastic model used along with a finite linear evolution law [13], while a model for rate-dependent phenomena in elastomers founded on finite strain theories in consistence with the natural laws of thermodynamics is proposed in [14].

In order to get around the specific characterization procedures, generalized models are often employed allowing to describe the viscoelastic behavior of the materials in a wide range of frequencies and time scales. The above models, consisting of a combination of a number of elementary springs and dashpots, generally known as the Generalized Maxwell model (GM) and generalized Kelvin–Voigt model (GKV), offer a good description of the real viscoelastic material but they consist of a considerable amount of appropriately connected spring and dashpot elements. This means that a set of differential equations must be solved to evaluate the dynamic behavior of the components under investigation [15], which could considerably complicate the mathematical formulation describing the dynamic state of the system and might significantly increase the computational load due to a larger set of motion equations to be solved. To overcome these issues, fractional models are becoming more and more popular because of their ability to describe the behavior of viscoelastic materials using a limited number of parameters with an acceptable accuracy over a vast range of excitation frequencies. Several authors have put the focus on the modelling approach based on fractional calculus allowing the characterization of rheological behavior of linear viscoelastic systems [16–20], pointing out several significant problems connected with the robust identification of model parameters.

The parameters' identification procedure can be classified according to the data domain employed (time-based or frequency-based) and according to the methods employed. Time-domain-based identification procedures can exploit creep and relaxation experimental tests [21], hysteresis loops of steady-state harmonic experimental tests [22] and impact tests [23]. Frequency-domain-based procedures can exploit magnitude and phase of the frequency response function [24] or the definition of storage modulus and loss factor [25,26]. Recently, Shabani et al. [27] have proposed a new frequency-domain-based identification procedure defining an ad-hoc testing setup. As concerns the classification by the identification method, an accurate literature review points out that the authors generally adopt a multitude of different procedures, i.e., Arikolu [26] and Zhou et al. [28] have proposed a genetic algorithm method, while procedure based on a curve fitting has been used by de Espindola et al. [25] and Sasso et al. [19]. A classical approach based on minimization of a proposed cost function has been explored by Yuan and Agrawal [29] while, more recently, ant colony optimization method [30] and spectral parameter estimation approach have been also investigated [31].

However, it must be highlighted that these references generally assess the goodness of the model behavior towards ideal and virtual materials acknowledging the potential of the mathematical formulation under study to describe a vast frequency range or the validation is obtained towards very limited experimental data of the material viscoelastic response. For this reason, to assess a generalized model and its robust parameter identification procedure to reproduce the experimental data of dynamic tests (in a wide range of frequency values), with a unique set of parameters, the aim of this work is to discover the optimum model formulation that can describe the real viscoelastic behavior with a minimum number of parameters. To this purpose, three completely different polymers have been considered to compare the response of two chosen models: the Generalized Maxwell model (GM) and the relative fractional derivative model (FDGM). For the both models, the parameters' optimum sets have been identified adopting a pole-zero formulation introduced by Renaud et al. in [32], limited to the case of the GM model. Furthermore, once the best model formulation has been discovered, a further study on the minimum amount of experimental data and their distribution within the frequency domain, necessary to correctly reproduce the materials' viscoelastic behavior in a wide frequency range with an acceptable accuracy, has been carried out.

The paper is organized as follows: in Section 2, the experimental data, employed as an experimental reference dataset for the study, are presented; in Section 3, the rheological models adopted are presented starting from their definition in the time domain and their re-definition in the frequency domain to obtain the relative pole-zero formulation; in Section 4, the parameters' identification procedure developed to achieve the fitting of the experimental data is presented; in Section 5, the model comparison in terms of the dynamic response and the ability to reproduce the viscoelastic behavior of the compounds under analysis are illustrated; in Section 6, a specific study to understand how the minimum number of experimental points should be acquired and distributed to identify the model parameter set in a robust way is reported.

2. Materials Experimental Data

The viscoelastic material is a deformable material with a behavior which lays between a viscous liquid and an elastic solid. Their behavior deviates from Hooke's elastic law, exhibiting both elastic and viscous characteristics at the same time. In this kind of materials, the stress–strain relationship is a function of time and the viscoelastic characteristics exhibit time-dependent behavior and load application speed at an established temperature value.

To better understand the mechanical behavior in viscoelastic materials, two main types of experiments are usually carried out: transient and dynamic. While static characterization regards the quasi-static application of load or deformation, transient and dynamic testing procedures concern the analysis of material response to a time applied deformation or load function (elongation or shear). Two important categories regarding the transient material testing are commonly performed: the creep experiment and stress-relaxation experiment.

In the creep experiment, the material is subjected to uniform load to analyze the strain time changes while during the stress-relaxation experiment, the material is subjected to a fixed deformation and the load required to maintain the deformation at a constant value is measured with time. The above transient tests allow to characterize the viscoelastic material reaction for each applied stress–strain load step.

Another important testing procedure class able to describe the viscoelastic behavior consists of the dynamic experiments. These tests are commonly employed to analyze the material reaction to cyclic stress (1) or strain applied:

$$\sigma(t) = \sigma_0 \sin \omega t \tag{1}$$

where ω represents the angular frequency of an applied sinusoidal stress and depends on the time. In quasi-elastic materials, the strain generated by the stress also exhibits a sinusoidal trend with the same phase of the applied load. On the opposite side, in viscoelastic materials, the strain reaction shows a time delay towards the applied stress, characterized by a phase angle δ. Therefore, the strain response is given by (2):

$$\varepsilon(t) = \varepsilon_0 \sin(\omega t - \delta) \tag{2}$$

Because of the phase displacement, the material dynamic stiffness can be considered as a complex variable E^* according to Euler's formulation (3):

$$\frac{\sigma(\omega)}{\varepsilon(\omega)} = E^* = E' + E'' \tag{3}$$

where E' is the storage modulus [Pa] and E'' is the loss modulus [Pa].

These quantities are deeply linked to the way the material dissipates a part of energy provided by means of a load/stress time function. They are related to the phase angle δ, according to the following Equation (4):

$$\frac{E''(\omega)}{E'(\omega)} = \tan \delta \tag{4}$$

The phase angle tangent δ is also called loss tangent. It is important to remark that all the quantities, referring to the viscoelastic behavior, are function of the frequency at which the sinusoidal load/deformation is applied and of the material temperature.

Hereafter, the output of dynamic mechanical analysis (DMA) [33] for the moduli characterization of three different polymers called A, B and C is reported. Such testing procedure prescribes an oscillatory deformation of the material sample with a constant strain or stress amplitude in order to measure the viscoelastic modulus. The oscillatory load is applied at different frequencies and then repeated at different temperatures. The master curves at a chosen reference temperature and over a wide range of frequencies is obtained exploiting the time–temperature superposition principle [34], which allows to shift the results measured at different temperatures. For each polymer, the real part of the Young's modulus and the tan δ are reported in Figure 1a,b, respectively, at the reference temperature of 120 °C.

The characteristics of the three polymers, which are summarized in the Table 1, show that these materials describe a wide range of application of polymers, having different characteristics from each other. In the table, T_g is glassy transition temperature, E_∞ is the glassy region dynamic modulus (the dynamic modulus at high frequency) and E_0 is rubbery region modulus (the dynamic modulus at low frequency).

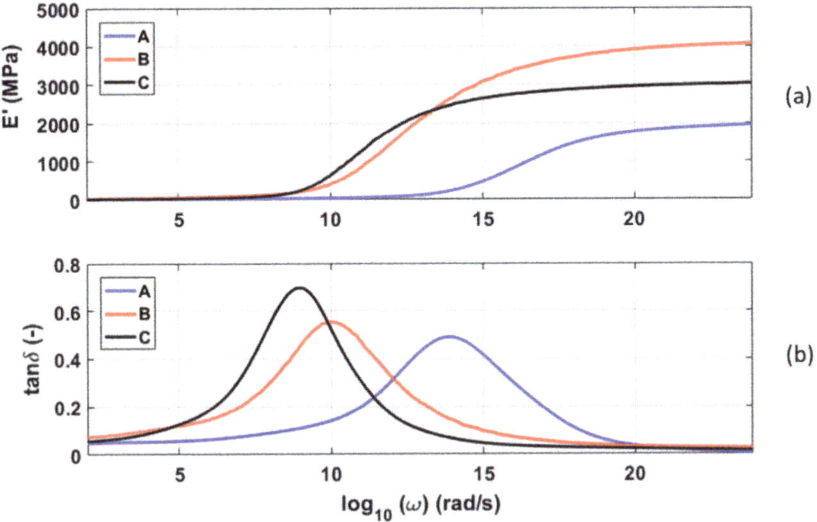

Figure 1. Storage modulus (a) and loss tangent (b) of rubber compounds A, B and C at a reference temperature T_{ref} =120°.

Table 1. Principal characteristics of the three studied polymers

POLYMER	T_g (°C)	E_∞ (MPa)	E_0 (MPa)
A	−27	1588	7.9
B	−16	4057	22.3
C	−10	3031	12.03

3. Rheological Models

For the purpose of understanding and describing of the material viscoelastic behavior, different mathematical models have been developed and are available in the literature. Since the experimental characterization of the viscoelasticity of the compounds is particularly important to properly feed the procedures of so many different application areas, the question arises about which could be the analytical model that is able to reproduce the viscoelastic behavior in an optimal and robust way employing the minimum number of parameters. In other words, the aim is to investigate the simplest constitutive model able to reproduce the experimental viscoelastic behavior in the widest range of operating frequencies.

The simple Maxwell or Kelvin models fail in representing the actual response of viscoelastic materials at low and high frequencies, respectively, while the generalized models are able to provide more reliable results. As this paper is focused on viscoelastic solids, the GM model and FDGM model are considered and refer to a spring in parallel with (respectively) Maxwell cells and fractional Maxwell cells as defined by Koeller [35]. Maxwell cells are composed of a spring and a dashpot arranged in series, while the fractional Maxwell element is obtained by replacing the dashpot elements with the so-called spring-pot elements. Figure 2 depicts the rheological model considered.

In the following chapters, the GM and FDGM models are mathematically described. Starting from the models' definition in the time domain, the frequency domain expressions are derived by means of the Fourier transform to obtain the pole-zero formulation of both the models under analysis.

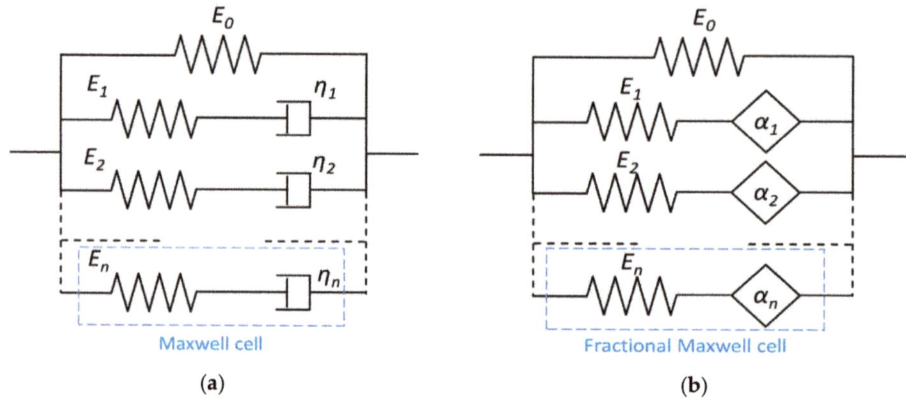

Figure 2. Models: (a) Generalized Maxwell, (b) Fractal derivative.

3.1. Generalized Maxwell Model

The differential equation for of the GM model formulation can be expressed in the general form as

$$\sum_{n=0}^{N} a_n \frac{d^n \sigma(t)}{dt^n} = \sum_{m=0}^{M} b_m \frac{d^m \varepsilon(t)}{dt^m} \tag{5}$$

where the parameters b_0=0 and N=M are assumed for the considered case of Maxwell model. The complex modulus can be derived transforming Equation (5) into the frequency domain. Calculating the Fourier transform of (5), the following expression for the complex moduli is derived:

$$E^*(i\omega) = E_0 + \sum_{k=1}^{N} \frac{i\omega E_k \eta_k}{E_k + i\omega \eta_k} \tag{6}$$

where the parameters E_k and η_k represent the springs stiffness and dashpots viscosity, respectively, as represented in Figure 2a, and ω is the angular frequency.

Renaud et al. [32] have demonstrated that the GM model, mathematically described in (6) in a frequency domain, can be equivalently expressed in the pole-zero formulations as in (7):

$$E^*(i\omega) = E_0 \prod_{k=1}^{N} \frac{1 + \left(\frac{i\omega}{\omega_{z,k}}\right)}{1 + \left(\frac{i\omega}{\omega_{p,k}}\right)} \tag{7}$$

where ω_z and ω_p are the zero and pole, respectively.

3.2. Fractal Derivative Generalized Maxwell Model

A generic constitutive equation for viscoelastic materials based on fractional derivative orders is expressed in the following form:

$$\sum_{n=0}^{N} a_n \frac{d^{\alpha_n} \sigma(t)}{dt^{\alpha_n}} = \sum_{m=0}^{M} b_m \frac{d^{\beta_m} \varepsilon(t)}{dt^{\beta_m}} \tag{8}$$

where α_n and β_m are the fractional derivative orders included between 0 and 1, N=M and b_0=0 for the considered case of Maxwell model. Turning to frequency domain, and assuming that $\alpha = \beta$, Equation (8) becomes:

$$E^*(i\omega) = E_0 + \sum_{k=1}^{N} \frac{(i\omega)^{\alpha_k} E_k \eta_k}{E_k + (i\omega)^{\alpha_k} \eta_k} \tag{9}$$

In analogy with the GM model, (9) can be equivalently expressed in the pole-zero formulations as in (10):

$$E^*(i\omega) = E_0 \prod_{k=1}^{N} \frac{1 + \left(\frac{i\omega}{\omega_{z,k}}\right)^{\alpha_k}}{1 + \left(\frac{i\omega}{\omega_{p,k}}\right)^{\alpha_k}} \tag{10}$$

4. Identification Procedure

The term *pole-zero identification* refers to the achievement of the poles and zeros of a linear, or linearized, system described by its frequency response. This is usually done using optimization techniques able to fit a given frequency response of the linear system towards a transfer function defined as the ratio of two polynomials [36,37]. This kind of linear system identification in the frequency domain has several applications in a wide variety of engineering fields and it can be constrained or unconstrained.

In this work, the pole-zero identification is performed adopting a constrained procedure taking advantage from the "fminmax" Matlab function that seeks the minimum of a nonlinear error function W of n real variables, the Zeros and the Poles. For the present identification problem, this function will be defined as the weighted sum of the normalized root mean square error (NRMSE) calculated for the storage modulus and the loss tangent as in (11):

$$W = (1-r)\cdot Err_E + r\cdot Err_{tan\delta} \tag{11}$$

where r is the weight factor and $Err_{tan\delta}$ and Err_E are defined by (12)

$$Err_{tan\delta} = \frac{\sqrt{\frac{\sum_{i=1}^{N}(tan\delta - tan\delta_{model})^2}{N}}}{mean(tan\delta)} \qquad Err_E = \frac{\sqrt{\frac{\sum_{i=1}^{N}(E - E_{model})^2}{N}}}{mean(E)} \tag{12}$$

Notice that this way of defining the error function allows to fit both modulus and phase at the same time. As concerns the weighting coefficient r, generally, it can be set equal to the more convenient value in order to have the more robust results. In the present paper, the weighting coefficient r has been assumed equal to 0.5 for all identifications in order to compare the results of different procedures.

The function W must be constrained imposing that the poles p and zeros z are alternate, according to the third property demonstrated by Bland [38] for linear dissipative systems (13):

$$z_1 < p_1 < z_2 < p_2 < z_3 < p_3 \ldots < z_i < p_i \tag{13}$$

As concerns the initial condition of the identification procedure, the initialization of the pole-zero parameters has been carried out by means the method proposed by Renaud et al. [32]. Dividing the frequency domain into N equal frequency subdomains each Pole Zero couple is given by (14):

$$\begin{cases} \omega_{z,i} = (1-r)\cdot\omega_{z,i}^{Modulus} + r\cdot\omega_{z,i}^{Phase} \\ \omega_{p,i} = (1-r)\cdot\omega_{p,i}^{Modulus} + r\cdot\omega_{p,i}^{Phase} \end{cases} \tag{14}$$

where, posing $\chi = \log_{10}(\omega)$, $\omega_{z,i}$ and $\omega_{p,i}$ for modulus and phase are calculated as in (15):

$$\begin{cases} \chi_{z,i}^{Modulus} = \chi_{c,i} - \frac{\alpha_i}{2} \\ \chi_{P,i}^{Modulus} = \chi_{c,i} + \frac{\alpha_i}{2} \end{cases} ; \quad \begin{cases} \chi_{z,i}^{Phase} = \chi_{c,i} - \frac{\alpha_i}{2} \\ \chi_{P,i}^{Phase} = \chi_{c,i} + \frac{\alpha_i}{2} \end{cases} \tag{15}$$

where $\chi_{c,i}$ is the medium frequency of a couple and α_i is evaluated considering the stiffing of the modulus $|E^*(\omega)|$ and the area under the phase curve the $\delta(\omega)$ defined respectively as in (16) and (17):

$$S = \alpha_i = \int_{\chi a,i}^{\chi b,i} \frac{d}{d\chi}\left(\log_{10}|E*(\chi)|\right)d\chi \tag{16}$$

$$A = \alpha_i \cdot \frac{\pi}{2} = \int_{\chi a,i}^{\chi b,i} \delta(\chi)d\chi \tag{17}$$

From the computational point of view, the convenience of the pole zero formulation in the identification of parameters of transfer functions lies in how the boundary conditions and the initial conditions of the optimization problem can be defined. As an example, using the formulation (6) of the GM model, the parameters to be identified are stiffness and damping coefficients. Hence, defining both the initial conditions and the boundary conditions is a hard task because they can vary between zero and very large numbers and as a function of the elements of the model itself. Using a pole-zero formulation, it is easier to define the initial conditions, i.e., by means of Equation (14). Moreover, the range of parameters to be identified (poles and zeros) is defined and limited by the frequency range of the experimental curves and this condition greatly facilitates the optimization algorithm efficiency.

5. Models Analysis

The identification procedure described in the previous paragraph has been employed to evaluate the capability of the GM and FDGM models to describe three completely different viscoelastic materials. Moreover, for each model, we need to understand the minimum number of parameters that can describe the dynamic characteristics of the considered viscoelastic materials. Therefore, to this purpose, the identification procedure has been performed for each model considering an increasing number of elements until the fitting of experimental data was not satisfactory.

To quantitatively compare the models presented in this paper, the NRMSE is evaluated for the storage modulus and the loss tangent quantities as defined in Equation (12). The purpose of these indicators is to quantify the goodness of the models' behavior towards the experimental data reproduction.

Figure 3 compares experimental data with the results obtained for the three compounds with the GM models composed by 15, 25 and 35 elements plus a spring in parallel (E_0 in Figure 2), while in Table 2, the relative calculated NRMSEs are reported.

Table 2. NRMSEs for Generalized Maxwell models composed by 15, 25 and 35 elements.

	Compound A		Compound B		Compound C	
GM model	NRMSE E′ (MPa)	NRMSE tanδ (-)	NRMSE E′ (MPa)	NRMSE tanδ (-)	NRMSE E′ (MPa)	NRMSE tanδ (-)
15 elements	0.0183	0.0876	0.0416	0.1146	0.0279	0.2358
25 elements	0.0038	0.0148	0.0306	0.0822	0.0087	0.2355
35 elements	0.0021	0.0053	0.0305	0.0789	0.0069	0.2352

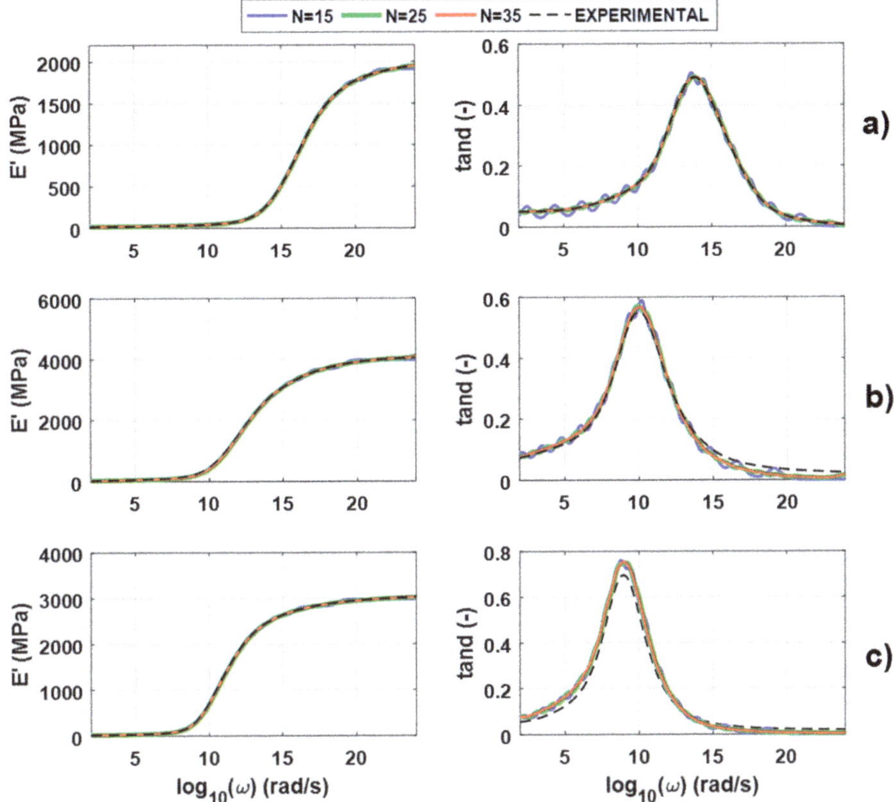

Figure 3. Storage modulus and loss tangent for compounds A (a), B (b) and C (c): experimental data (dashed line) vs. GM model with 15 (blue line), 25 (green line) and 35 (red line) elements.

The identified parameters of these models are detailed in Tables A1–A3 in Appendix A. Analyzing the models' results, it is evident that the GM model composed of 15 elements appears totally insufficient to describe the compounds' frequency response because it shows unphysical local oscillations although it seems to follow the real global trend. As highlighted by the NRMSE in Table 2, by increasing the number of Maxwell elements, the fitting of experimental data is improved as well. However, it should be noted that to have significant improvements in model results, several elements must be added with a consequent increase in the parameters to be identified.

The FDGM model results are reported in Figure 4. For each compound, the experimental data are compared with models composed by a spring (E_0 in Figure 2) in parallel with 3, 4 and 5 spring-pot elements. The corresponding NRMSEs are reported in Table 3, while the parameters of these models are detailed in Tables A4–A6 in Appendix A. It should be noted that the FDGM model is able to give an acceptable representation of the curves' shapes with a three-element model which is also consistent with the NRMSE values. Also in this case, increasing the number of model elements, the fitting of experimental data improve.

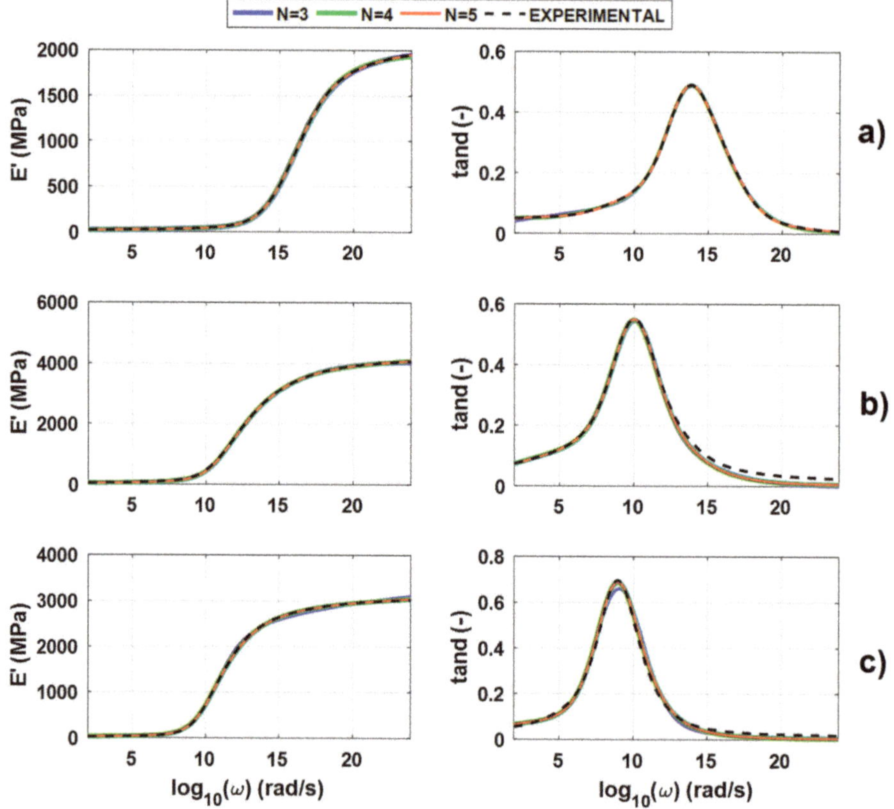

Figure 4. Storage modulus and loss tangent for compound A (a), B (b) and C (c): experimental data (dashed line) vs. FDGM model with three (blue line), four (green line) and five (red line) elements.

Table 3. NRMSEs for Fractal Derivative Generalized Maxwell models composed of 3, 4 and 5 elements.

	Compound A		Compound B		Compound C	
FDGM model	NRMSE E′ (MPa)	NRMSE tanδ (-)	NRMSE E′ (MPa)	NRMSE tanδ (-)	NRMSE E′ (MPa)	NRMSE tanδ (-)
3 elements	0.0028	0.0273	0.0187	0.059	0.0513	0.1054
4 elements	0.0059	0.012	0.0176	0.0574	0.0251	0.0822
5 elements	0.002	0.0038	0.0178	0.0568	0.0251	0.0821

A relative comparison between the three compounds shows that, regardless of the model used, compound A returns the best fitting results than the others for the same number of elements used. By mere comparison of normalized errors, we observed (in general) a good performance of both models adopted in describing the real compounds' behavior, but GM model needs a greater number of elements than the FDGM. While FDGM was able to provide good matching with the simple configuration with only ten unknowns, for the GM, it was necessary to build up a more complicated model by adding several elements. Indeed, in terms of parameters to identify, each element of GM requires two parameters (1 pole-zero couple), while for FDGM, each one needs three parameters (1 pole-zero couple and the fractal derivative order α). This means that a 15-elements GM implies 31 parameters to be identified; the 15 pole-zero couples and the E_0. Similarly, a three-elements FDGM needs ten parameters to be found; the three pole-zero couples, three fractal derivatives α_i and the E_0. Thus, comparing the normalized errors in Tables 2 and 3 and the relative curves, it is evident that the FDGM is able to fit the real viscoelastic behavior with the minimum number of parameters for the three compounds.

It is noteworthy that the number of parameters to be identified highly affects the computational load in terms of efficiency and effectiveness. In the reported study, the same function has been used for constrained minimization problems; thus, the nature of the adopted constitutive models in terms of complexity and mathematical formulation are responsible for the variation in efficacy (ability of providing good matching with experimental data) with reasonable computing time and robustness. In light of this, the comparison between the FDGM models with the GM models points out that the FDGM is preferable because of the limited number of parameters to be determined, which means a significant reduction in frequent mathematical difficulties encountered in the ill-posed problem of identification procedures from experimental data. It should be noted that, despite the great number of parameters to be identified especially in for GM model, taking advantages form the pole-zero formulation, no convergence problems have been encountered during the identification procedure.

6. Moduli Estimation with Partial Experimental Data

In the previous section, it has been demonstrated that the FDGM is an analytical model able to fit the real viscoelastic behavior with the minimum necessary number of parameters. Furthermore, the pole-zero formulation ensures a robust and fast parameter identification process starting from the experimental data of moduli in the frequency domain. Hence, the question arises about the minimum number of experimental data needed to reproduce the materials' behavior in a wide frequency range. To carry out this study, five frequency zones have been identified in the range in which the storage modulus and the loss tangent were defined for compound A, as depicted in Figure 5. Zones 1 and 5 are, respectively, representative of the lower and upper frequencies plateau of the storage modulus. Zone 3 identifies the peak of the loss tangent curve, while zones 2 and 4 describe the curvature changing of both curves.

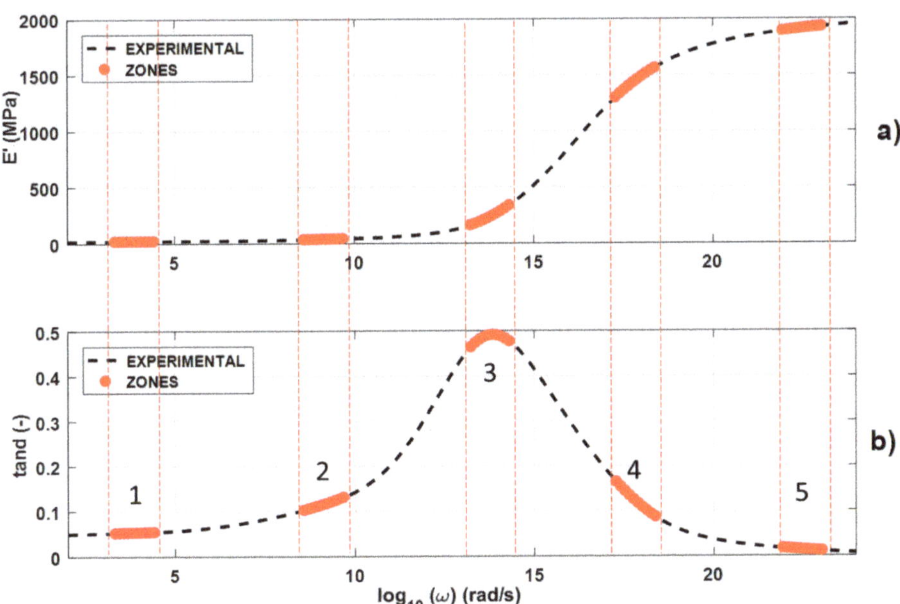

Figure 5. Definition of the five frequency zones.

The aim of the following analysis is to test the capability of the FDGM model in moduli estimation starting from a reduced set of experimental data on which to make the fitting. To pursue this aim, six different combinations of the zones defined in Figure 5 have been considered. For each combination, the parameters of the three-element FDGM have been identified and models results have been compared with the experimental ones in the entire frequency range. In Table 4, we summarize the zone combination and the relative fitting results in terms of NRMSE on the storage modulus and loss tangent while Figure 6 depicts the curves comparisons.

Table 4. NRMSE for Fractal derivative Generalized Maxwell models composed by 3, 4 and 5 elements.

Case	ZONES	NRMSE E′ (MPa)	NRMSE tanδ (-)
Combination A	1,2,3,4,5	0.004	0.035
Combination B	1,5	0.595	2.395
Combination C	1,3,5	0.004	0.066
Combination D	1,2,5	0.085	0.1695
Combination E	1,4,5	0.007	0.1640
Combination F	2,3,4	0.052	0.127

Adopting all the five zones to identify the parameters of the three-element FDGN model, it is possible to obtain an equally good fitting of the case in which all experimental data are take into account. This condition points out that the five zones chosen represent remarkable points of the curves that unequivocally determine the characteristics of a compound. Combination B results are totally insufficient to describe compound behavior, while combinations D and E fail principally in the fitting of the loss tangent. The latter result was predictable considering that, in these combinations, we did not take into account the information on the peak of the loss tangent. Combination F gives good results in the transition area of the curves, failing in both plateaus at lower and upper frequencies. Satisfying results can be achieved by combination C that, using only the data of the upper and low frequency plateaus plus the ones at the loss tangent peak, is able to fit the global compound behavior.

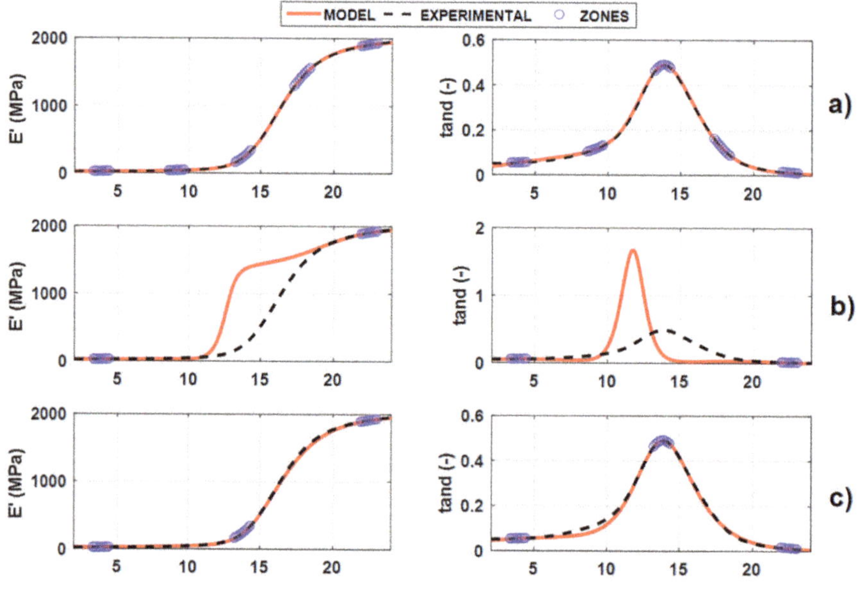

Figure 6. *Cont.*

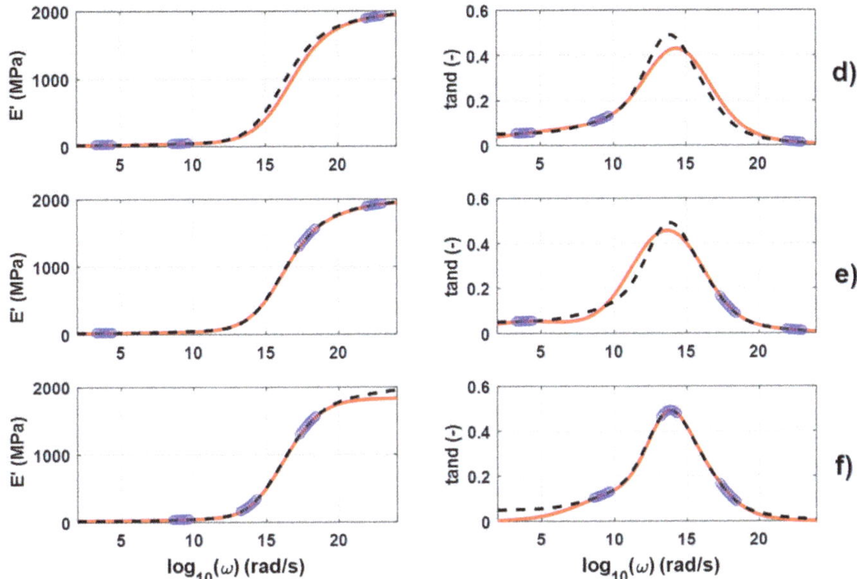

Figure 6. FDGM estimation of storage modulus and loss tangent for compound A adopting limited zones of experimental data.

7. Conclusions

The GM model and FDGM can be expressed adopting a pole–zero formulation. This formulation, that in logarithmic scale becomes a superposition of pole–zero couple behavior, allows to overcome the computational and convergence issues that occur in parameter identification of a rheological model in the time domain and frequency domain form. Thus, a constrained pole-zero identification procedure has been defined. The convenience of the pole-zero formulation in the identification of parameters of transfer functions lies in the determination of boundary conditions and initial condition of the optimization problem. Using this approach, an extensive study has been carried out to evaluate the capability of the GM and FDGM models in properly describing the viscoelastic behavior of three materials, the mechanical behavior of which was previously characterized by means of DMA tests. The analysis returns that both GM an FDGM models are able to fit the viscoelastic behavior of the compounds in the entire frequency range in which they are experimentally characterized, but the GM model needs a number of elements that are higher than the FDGM one. From a practical point of view, this means that FDGM is able to provide good matching with the simple configuration with only ten variables (three elements), while comparable results can be obtained with the GM model identifying more than 51 parameters (25 elements). Thus, comparing the normalized errors and the relative curves of the three compounds, it is observed that the FDGM is able to fit the real viscoelastic behavior with the minimum number of parameters.

The minimum number of experimental data needed to estimate the material behavior in a wide frequency range has also been investigated. The aim was to evaluate the capability of the FDGM model in moduli estimation starting from a reduced set of experimental data on which to make the fitting. An excellent prediction of the entire storage modulus and loss tangent can be obtained adopting data coming from the lower and upper frequencies plateau of the storage modulus, the peak of the loss tangent curve and the curvature changing of both curves. Furthermore, it is worth stressing that satisfying results can be achieved having also only the data available of the upper and low frequency plateaus plus the ones at the loss tangent peak.

Author Contributions: Conceptualization, A.G. and A.S.; methodology, A.G., A.S. and F.C.; software, F.C.; validation, A.M. and F.F.; formal analysis, F.T. and A.S.; investigation, F.C. and A.M.; resources, F.F. and F.T.; data curation, F.C. and A.G. All authors have read and agreed to the published version of the manuscript.

Funding: This research received no external funding.

Conflicts of Interest: Authors declare no conflict of interest.

Appendix A Parameters Values Obtained with Pole-Zero Identification Procedure

A.1. Generalized Maxwell Model Parameters

Table A1. Compound A parameters of GM models composed by 15, 25 and 35 elements.

Element	GM 15-elements		GM 25-elements		GM 35-elements	
E0 (MPa)	18.9		17.65		17.35	
	Pole	Zero	Pole	Zero	Pole	Zero
1	2.55	2.49	1.83	1.80	1.73	1.70
2	4.42	4.36	2.82	2.79	2.32	2.31
3	6.20	6.13	3.78	3.74	2.94	2.91
4	7.80	7.72	4.75	4.71	3.75	3.72
5	9.24	9.14	5.72	5.68	4.35	4.33
6	10.56	10.43	6.67	6.63	5.04	5.01
7	11.74	11.57	7.60	7.55	5.68	5.65
8	12.78	12.55	8.49	8.43	6.31	6.29
9	13.73	13.45	9.34	9.28	6.95	6.92
10	14.64	14.37	10.17	10.10	7.58	7.54
11	15.57	15.35	10.98	10.89	8.24	8.20
12	16.56	16.40	11.74	11.63	8.95	8.90
13	17.67	17.57	12.46	12.31	9.70	9.64
14	19.06	19.00	13.16	12.97	10.44	10.38
15	21.02	20.98	13.83	13.62	11.12	11.04
16			14.49	14.29	11.77	11.67
17			15.16	14.98	12.36	12.24
18			15.85	15.70	12.87	12.74
19			16.57	16.45	13.30	13.19
20			17.33	17.25	13.70	13.57
21			18.16	18.11	14.18	14.03
22			19.10	19.07	14.72	14.56
23			20.19	20.17	15.30	15.15
24			21.46	21.44	15.91	15.78
25			22.96	22.95	16.54	16.44
26					17.22	17.15
27					17.94	17.89
28					18.67	18.64
29					19.39	19.37
30					20.08	20.07
31					20.79	20.78
32					21.55	21.54
33					22.40	22.39
34					23.26	23.25
35					23.99	23.99

Table A2. Compound B parameters of GM models composed by 15, 25 and 35 elements.

Element	GM 15-elements		GM 25-elements		GM 35-elements	
E0 (MPa)	26.4		26.34		26.34	
	Pole	Zero	Pole	Zero	Pole	Zero
1	0.86	0.80	0.85	0.80	0.85	0.80
2	2.22	2.15	1.95	1.90	1.83	1.79
3	3.52	3.44	2.90	2.85	2.52	2.48
4	4.80	4.70	3.86	3.80	3.08	3.05
5	6.05	5.93	4.82	4.75	3.66	3.62
6	7.25	7.09	5.76	5.68	4.30	4.25
7	8.38	8.16	6.69	6.59	4.97	4.92
8	9.41	9.11	7.60	7.47	5.67	5.61
9	10.36	10.03	8.46	8.28	6.37	6.30
10	11.29	11.02	9.24	9.02	7.06	6.97
11	12.29	12.10	9.98	9.73	7.77	7.65
12	13.44	13.32	10.71	10.46	8.47	8.32
13	14.83	14.75	11.44	11.24	9.09	8.92
14	16.57	16.52	12.20	12.05	9.68	9.48
15	19.09	19.05	13.02	12.91	10.26	10.06
16			13.89	13.82	10.84	10.65
17			14.82	14.78	11.43	11.27
18			15.79	15.76	12.00	11.89
19			16.87	16.84	12.59	12.50
20			18.01	17.99	13.27	13.20
21			19.28	19.27	14.00	13.94
22			20.58	20.58	14.72	14.69
23			21.60	21.60	15.38	15.35
24			22.78	22.77	16.06	16.03
25			23.91	23.90	16.92	16.90
26					17.57	17.56
27					18.10	18.09
28					19.07	19.06
29					19.62	19.62
30					20.30	20.29
31					21.12	21.11
32					21.71	21.71
33					22.49	22.49
34					23.20	23.20
35					23.91	23.90

Table A3. Compound C parameters of GM models composed by 15, 25 and 35 elements.

Element	GM 15-elements		GM 25-elements		GM 35-elements	
E0 (MPa)	14.73		14.73		14.73	
	Pole	Zero	Pole	Zero	Pole	Zero
1	0.86	0.79	0.86	0.79	0.86	0.79
2	2.27	2.21	2.13	2.08	2.07	2.02
3	3.56	3.48	3.17	3.11	2.12	2.12
4	4.76	4.66	4.17	4.09	2.91	2.86
5	5.90	5.76	5.16	5.06	3.66	3.61
6	6.97	6.78	6.12	5.99	4.38	4.33
7	7.96	7.67	7.03	6.86	4.96	4.90
8	8.88	8.50	7.90	7.65	5.66	5.57
9	9.72	9.36	8.71	8.39	6.37	6.27
10	10.59	10.32	9.43	9.12	7.08	6.93
11	11.54	11.38	10.11	9.86	7.85	7.62
12	12.62	12.53	10.80	10.63	8.55	8.29
13	13.96	13.91	11.54	11.43	9.10	8.86
14	15.78	15.74	12.34	12.26	9.66	9.41
15	18.65	18.63	13.23	13.18	10.25	10.05
16			14.23	14.20	10.89	10.73
17			15.33	15.31	11.56	11.46
18			16.50	16.49	12.23	12.17
19			17.81	17.80	12.91	12.87
20			18.88	18.87	13.62	13.59
21			19.79	19.79	14.30	14.28
22			20.99	20.99	14.99	14.98
23			21.78	21.78	15.68	15.67
24			22.93	22.93	16.37	16.36
25			24.00	24.00	17.06	17.06
26					17.74	17.74
27					18.44	18.44
28					19.13	19.13
29					19.82	19.82
30					20.52	20.52
31					21.21	21.21
32					21.91	21.91
33					22.60	22.60
34					23.30	23.30
35					24.01	24.01

A.2. Fractal Derivative Generalized Maxwell Model Parameters

Table A4. Compound A parameters of FDGM models composed by 3, 4 and 5 elements.

Element	FDGM 3-elements			FDGM 4-elements			FDGM 5-elements		
E0 (MPa)	14.41			15.44			14.73		
	Pole	Zero	Gamma	Pole	Zero	Gamma	Pole	Zero	Gamma
1	12.45	5.62	0.12	2.08	1.96	0.45	3.63	1.66	0.16
2	14.53	12.45	0.40	12.60	7.37	0.14	10.55	9.06	0.24
3	16.06	14.53	0.34	14.51	12.60	0.41	15.09	12.44	0.42
4				16.16	14.51	0.32	16.67	16.05	0.40
5							19.00	18.34	0.18

Table A5. Compound B parameters of FDGM models composed by 3, 4 and 5 elements.

Element	FDGM 3-elements			FDGM 4-elements			FDGM 5-elements		
E0 (MPa)	24.82			22.41			25.19		
	Pole	Zero	Gamma	Pole	Zero	Gamma	Pole	Zero	Gamma
1	8.85	3.21	0.15	8.90	3.51	0.15	3.66	1.30	0.12
2	11.20	8.85	0.44	11.03	8.90	0.44	8.96	5.44	0.17
3	12.97	11.20	0.21	12.61	11.03	0.25	10.99	8.96	0.45
4				21.16	12.61	0.02	12.61	10.99	0.25
5							20.76	12.61	0.03

Table A6. Compound C parameters of FDGM models composed by 3, 4 and 5 elements.

Element	FDGM 3-elements			FDGM 4-elements			FDGM 5-elements		
E0 (MPa)	14.73			14.73			14.73		
	Pole	Zero	Gamma	Pole	Zero	Gamma	Pole	Zero	Gamma
1	7.46	2.26	0.12	7.62	2.23	0.12	3.66	1.30	0.12
2	10.70	7.46	0.47	10.09	7.62	0.49	8.96	5.44	0.17
3	24.45	10.70	0.02	11.13	10.09	0.32	10.99	8.96	0.45
4				24.45	11.13	0.02	12.61	10.99	0.25
5							20.76	12.61	0.03

References

1. Newell, N.; Little, J.; Christou, A.; Adams, M.; Adam, C.; Masouros, S. Biomechanics of the human intervertebral disc: A review of testing techniques and results. *J. Mech. Behav. Biomed. Mater.* **2017**, *69*, 420–434. [CrossRef] [PubMed]
2. Zhang, W.; Chen, H.Y.; Kassab, G.S. A rate-insensitive linear viscoelastic model for soft tissues. *Biomaterials* **2007**, *28*, 3579–3586. [CrossRef]
3. Peña, E.; Calvo, B.; Martínez, M.A.; Martins, P.; Mascarenhas, T.; Jorge, R.M.N.; Ferreira, A.; Doblaré, M. Experimental study and constitutive modeling of the viscoelastic mechanical properties of the human prolapsed vaginal tissue. *Biomech. Model. Mechanobiol.* **2010**, *9*, 35–44. [CrossRef] [PubMed]
4. Samali, B.; Kwok, K.C.S. Use of viscoelastic dampers in reducing wind- and earthquake-induced motion of building structures. *Eng. Struct.* **1995**, *17*, 639–654. [CrossRef]
5. Singh, M.P.; Moreschi, L.M. Optimal placement of dampers for passive response control. *Earthq. Eng. Struct. Dyn.* **2002**, *31*, 955–976. [CrossRef]
6. Park, S.W. Analytical modeling of viscoelastic dampers for structural and vibration control. *Int. J. Solids Struct.* **2001**, *38*, 8065–8092. [CrossRef]
7. Persson, B.N.J. Theory of rubber friction and contact mechanics. *J. Chem. Phys.* **2001**, *115*, 3840–3861. [CrossRef]
8. Genovese, A.; Farroni, F.; Papangelo, A.; Ciavarella, M. A Discussion on Present Theories of Rubber Friction, with Particular Reference to Different Possible Choices of Arbitrary Roughness Cutoff Parameters. *Lubricants* **2019**, *7*, 85. [CrossRef]
9. Mark, J.E. *Physical Properties of Polymers Handbook*; Mark, J.E., Ed.; Springer: New York, NY, USA, 2007.
10. Lakes, R.S. *Viscoelastic Materials*; Cambridge University Press: Cambridge, UK, 2009.
11. Liao, Z.; Hossain, M.; Yao, X.; Mehnert, M.; Steinmann, P. On thermo-viscoelastic experimental characterization and numerical modelling of VHB polymer. *Int. J. Non-Linear Mech.* **2020**, *118*, 103263. [CrossRef]
12. Dal, H.; Kaliske, M. Bergström-Boyce model for nonlinear finite rubber viscoelasticity: Theoretical aspects and algorithmic treatment for the FE method. *Comput. Mech.* **2009**, *44*, 809–823. [CrossRef]
13. Hossain, M.; Vu, D.K.; Steinmann, P. Experimental study and numerical modelling of VHB 4910 polymer. *Comput. Mater. Sci.* **2012**, *59*, 65–74. [CrossRef]

14. Amin, A.F.M.S.; Lion, A.; Sekita, S.; Okui, Y. Nonlinear dependence of viscosity in modeling the rate-dependent response of natural and high damping rubbers in compression and shear: Experimental identification and numerical verification. *Int. J. Plast.* **2006**, *22*, 1610–1657. [CrossRef]
15. Palmeri, A.; Ricciardelli, F.; De Luca, A.; Muscolino, G. State Space Formulation for Linear Viscoelastic Dynamic Systems with Memory. *J. Eng. Mech.* **2003**, *129*, 715–724. [CrossRef]
16. Bagley, R.L.; Torvik, J. Fractional calculus—A different approach to the analysis of viscoelastically damped structures. *AIAA J.* **1983**, *21*, 741–748. [CrossRef]
17. Pritz, T. Five-parameter fractional derivative model for polymeric damping materials. *J. Sound Vib.* **2003**, *265*, 935–952. [CrossRef]
18. Liu, J.G.; Xu, M.Y. Higher-order fractional constitutive equations of viscoelastic materials involving three different parameters and their relaxation and creep functions. *Mech. Time Dependent Mater.* **2007**, *10*, 263–279. [CrossRef]
19. Sasso, M.; Palmieri, G.; Amodio, D. Application of fractional derivative models in linear viscoelastic problems. *Mech. Time Dependent Mater.* **2011**, *15*, 367–387. [CrossRef]
20. Meral, F.C.; Royston, T.J.; Magin, R. Fractional calculus in viscoelasticity: An experimental study. *Commun. Nonlinear Sci. Numer. Simul.* **2010**, *15*, 939–945. [CrossRef]
21. Di Paola, M.; Pirrotta, A.; Valenza, A. Visco-elastic behavior through fractional calculus: An easier method for best fitting experimental results. *Mech. Mater.* **2011**, *43*, 799–806. [CrossRef]
22. Lewandowski, R.; Chorążyczewski, B. Identification of the parameters of the Kelvin–Voigt and the Maxwell fractional models, used to modeling of viscoelastic dampers. *Comput. Struct.* **2010**, *88*, 1–17. [CrossRef]
23. Fukunaga, M.; Shimizu, N. Comparison of fractional derivative models for finite deformation with experiments of impulse response. *J. Vib. Control* **2014**, *20*, 1033–1041. [CrossRef]
24. Khemane, F.; Malti, R.; Raïssi, T.; Moreau, X. Robust estimation of fractional models in the frequency domain using set membership methods. *Signal Process.* **2012**, *92*, 1591–1601. [CrossRef]
25. De Espíndola, J.J.; da Silva Neto, J.M.; Lopes, E.M.O. A generalised fractional derivative approach to viscoelastic material properties measurement. *Appl. Math. Comput.* **2005**, *164*, 493–506. [CrossRef]
26. Arikoglu, A. A new fractional derivative model for linearly viscoelastic materials and parameter identification via genetic algorithms. *Rheol. Acta* **2014**, *53*, 219–233. [CrossRef]
27. Shabani, M.; Jahani, K.; Di Paola, M.; Sadeghi, M.H. Frequency domain identification of the fractional Kelvin-Voigt's parameters for viscoelastic materials. *Mech. Mater.* **2019**, *137*, 103099. [CrossRef]
28. Zhou, S.; Cao, J.; Chen, Y. Genetic Algorithm-Based Identification of Fractional-Order Systems. *Entropy* **2013**, *15*, 1624–1642. [CrossRef]
29. Yuan, L.; Agrawal, O.P. A Numerical Scheme for Dynamic Systems Containing Fractional Derivatives. *J. Vib. Acoust.* **2002**, *124*, 321–324. [CrossRef]
30. Xiao, Z.; Haitian, Y.; Yiqian, H. Identification of constitutive parameters for fractional viscoelasticity. *Commun. Nonlinear Sci. Numer. Simul.* **2014**, *19*, 311–322. [CrossRef]
31. Dabiri, A.; Nazari, M.; Butcher, E.A. The Spectral Parameter Estimation Method for Parameter Identification of Linear Fractional Order Systems. In Proceedings of the 2016 American Control Conference (ACC), Boston, MA, USA, 6–8 July 2016; pp. 2772–2777.
32. Renaud, F.; Dion, J.-L.; Chevallier, G.; Tawfiq, I.; Lemaire, R. A new identification method of viscoelastic behavior: Application to the generalized Maxwell model. *Mech. Syst. Signal Process.* **2011**, *25*, 991–1010. [CrossRef]
33. Menard, K.P.; Menard, N. Dynamic Mechanical Analysis. In *Encyclopedia of Analytical Chemistry*; John Wiley & Sons, Ltd.: Chichester, UK, 2017; pp. 1–25.
34. Williams, M.L.; Landel, R.F.; Ferry, J.D. The Temperature Dependence of Relaxation Mechanisms in Amorphous Polymers and Other Glass-forming Liquids. *J. Am. Chem. Soc.* **1955**, *77*, 3701–3707. [CrossRef]
35. Koeller, R.C. Applications of fractional calculus to the theory of viscoelasticity. *J. Appl. Mech. Trans. ASME* **1984**, *51*, 299–307. [CrossRef]
36. Pintelon, R.; Schoukens, J. *System Identification: A Frequency Domain Approach, Second Edition*; Wiley: Hoboken, NJ, USA, 2012.

37. Collantes, J.M.; Mori, L.; Anakabe, A.; Otegi, N.; Lizarraga, I.; Ayllon, N.; Ramirez, F.; Armengaud, V.; Soubercaze-Pun, G. Pole-Zero Identification: Unveiling the Critical Dynamics of Microwave Circuits beyond Stability Analysis. *IEEE Microw. Mag.* **2019**, *20*, 36–54. [CrossRef]
38. Bland, D.R. *The Theory of Linear Viscoelasticity*; Pergamon Press: London, UK, 2019.

Article

Dynamics of Cylindrical Parts for Vibratory Conveying

Nicola Comand [1] and Alberto Doria [2,*]

[1] Department of Management and Engineering, University of Padova, 36100 Vicenza, Italy; nicola.comand@phd.unipd.it

[2] Department of Industrial Engineering, University of Padova, 35131 Padova, Italy

* Correspondence: alberto.doria@unipd.it; Tel.: +39-049-827-6803

Received: 20 January 2020; Accepted: 6 March 2020; Published: 11 March 2020

Featured Application: Motion of cylindrical parts in conveyors.

Abstract: Vibratory conveyors are widely used to feed raw materials and small parts to processing equipment. Up to now, most of the research has focused on materials and parts that can be modeled as point masses or small blocks. This paper focuses on the conveying of cylindrical parts. In this case, the rolling motion is an essential feature of conveyor dynamics. First, the dynamic equations governing the rolling motion are stated, and the effects of friction and rolling resistance coefficients on the behavior of the system are analyzed. Then, a non-linear numerical model is developed in MATLAB. It takes into account the transition between pure rolling and rolling with sliding and the impacts of the cylindrical part on the edges of the conveyor. Numerical results showing the effect of the operative parameters of the conveyor and of friction properties on the traveled distance are presented and discussed. Finally, a comparison between numerical and experimental results is presented.

Keywords: vibration; conveyor; friction; rolling; non-linear simulation

---

1. Introduction

The automation of industrial processes very often requires the handling of small [1–3] and micro items [4], which include mechanical and electrical components used in the assembly processes, finite products that have to be packaged, and row materials that have to be processed. In many industrial scenarios, these small items have to be transported, oriented, and sorted. Nowadays, vibratory systems are very common, because they allow handling small items in a versatile and efficient way. They have a simple and sturdy construction and are suited to handle dusty and hot materials as well. For these reasons, some researchers have studied the kinematics and dynamics of vibratory conveying, orienting, and sorting. In 1963, Taniguchi et al. [5] carried out a pioneering study that highlighted the role of friction and clearly stated the differences between the gliding and hopping motions of a machine part on a vibrating conveyor, in which the machine part was considered a point mass. The influence of directional friction characteristics was analyzed in [6], assuming a point mass model of the vibrating part. In [7], the mechanical part was modeled as a rigid body with six degrees-of-freedom (DOFs) and the contact mechanics were deeply analyzed; however, the rolling motion was neglected, since a block-shaped component was studied. The dynamics of a block moving along the spiral track of a vibratory bowl feeder were studied in [8], and the block was modeled as a point mass. Some recent studies dealt with the design of the conveyor [9–11]; also in these studies, the part was modeled as a point mass, and no rolling motion was considered.

Cylindrical parts, similar to pins and plugs, are very common in electrical and mechanical industry; also, a screw is better represented by a cylinder than by a block. Other manufacturing processes require the handling of cylindrical and quasi-cylindrical parts, e.g., the pharmaceutical industry has to convey cylindrical pills and phials. Cylindrical parts are processed by classical conveyors, if their axes of rotation form a small angle with the direction of conveying [12]. When this condition is not satisfied, they begin to roll down the track. Most of the scientific literature dealing with conveyors neglects this rolling motion [1], because sometimes, this phenomenon does not have a significant effect on the process. For example, in orientation processes, often a cylindrical part that can roll also has a wrong orientation and it has to be rejected both if it slides and if it rolls. Nevertheless, in some applications, the cylindrical part has to travel a larger distance, and the rolling motion strongly influences the process.

Thus, this paper focuses on the dynamics of cylindrical parts in a vibrating conveyor, with the aim of explaining the basic differences between the vibration conveying of a block and of a rolling cylinder. The scientific literature in this field is rather scarce. Some papers [13,14] analyzed the probabilities of natural resting aspects of cylinders and parts with curved surfaces in vibratory bowl feeders. In [15], a preliminary numerical analysis of the motion of a cylindrical part on a conveyor was carried out, and the possibility of rolling, sliding, and hopping motions was highlighted. This paper is an extension of the previous research [15] that includes analytical studies, a more detailed mathematical model that considers impacts, and a wide parametric analysis.

The paper is organized as follows. First, the dynamic equations governing the rolling motion of a cylindrical part are analytically derived, the differences between vibratory conveying of rolling and non-rolling parts are highlighted, and the parameters with the largest influence on the rolling motion are found. Then, a two-dimensional (2D) non-linear numerical model is developed that takes into account the transition between pure rolling and rolling with sliding and the impacts of the cylindrical parts with the edge of the conveyor. Series of numerical simulations are carried out, which show the effects of variations in the system's parameters on the motion of the cylindrical part. Finally, testing equipment composed of a vibrating conveyor and a vision system that monitors the motion of a cylinder is presented. Numerical results are compared with experimental results, and the benefits and limits of 2D simulations are discussed.

2. Theoretical Basis

There are very important differences between the vibration conveying of parts that can roll on the conveyor track and parts that cannot roll. A 2D model with the plane of motion perpendicular to the axis of the cylinder is enough to point out these differences.

Vibration conveying of non-rolling parts is a dynamic phenomenon dominated by dry friction. Figure 1a shows a small block on the track of the conveyor, which is tilted by angle θ with respect to the horizontal direction. A fixed reference frame with axis x parallel to the track and axis y perpendicular to track is introduced. R_x and R_y are the components of the reaction force exerted by the track surface on the block (the positive direction is represented); they are related by the static Coulomb friction coefficient μ_s and by the kinetic Coulomb friction coefficient μ_c. Acceleration ($\vec{a}$) of the conveyor is tilted by angle ψ with respect to the conveyor track. If the motion of the conveyor is harmonic with amplitude x_0 and angular frequency ω, the components of the acceleration are:

$$A_x = acos(\psi) = -x_0\omega^2 sin(\omega t)cos(\psi) \tag{1}$$

$$A_y = asin(\psi) = -x_0\omega^2 sin(\omega t)sin(\psi). \tag{2}$$

Angle ψ is essential for the vibration conveying of non-rolling parts. When acceleration has the direction shown in Figure 1a, the inertia force adds to the weight of the block, the reaction force R_y is large, the modulus of the demanded tangential reaction force $|R_x|$ is smaller than the maximum static friction force ($\mu_s|R_y|$), and the block cannot slide backwards. Conversely, when acceleration $\vec{a}$ changes its direction, the inertia force tends to cancel the weight force, R_y decreases, and the modulus of the demanded tangential reaction force $|R_x|$ may become larger than the maximum static friction force;

hence, the block moves, and forwards sliding begins. This condition repeats at every period of the harmonic motion and eventually the block moves forwards with respect to the track.

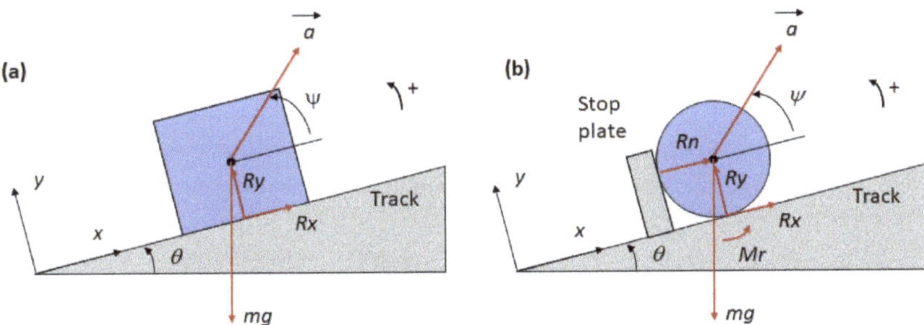

Figure 1. Block (**a**) and cylindrical part; (**b**) on the vibrating conveyor.

The scenario drastically changes when the part has a cylindrical shape, as depicted in Figure 1b. A stop-plate is introduced, and R_n is the reaction force exerted by the stop plate on the cylinder (the positive direction is represented), M_r is the rolling resistance torque (the positive direction is represented), m is the cylinder mass, and mg is the gravity force (the actual direction is represented).

First, the cylindrical part is assumed to be steady on the vibrating conveyor, and the dynamic conditions that make possible the beginning of the relative motion are analyzed. The equations of motion in the x and y directions are:

$$ma_x = -mgsin(\theta) + R_x + R_n \tag{3}$$

$$ma_y = -mgcos(\theta) + R_y. \tag{4}$$

In Equations (3) and (4), a_x and a_y are the two components of the acceleration of the center of mass of the cylindrical part.

When there is no relative motion, a_x and a_y coincide with the components of acceleration of the conveyor (A_x and A_y); therefore, the equations of motion become:

$$macos(\psi) = -mgsin(\theta) + R_x + R_n \tag{5}$$

$$ma\, sin(\psi) = -mgcos(\theta) + R_y. \tag{6}$$

The perpendicular reaction force con be calculated from Equation (6).

$$R_y = ma\, sin(\psi) + mgcos(\theta) \tag{7}$$

The contact between the cylinder and the track is unilateral; hence, $R_y \geq 0$. When R_y tends to zero, the cylinder separates from the track, and the hopping motion begins. The hopping motion can be avoided decreasing the vibration acceleration, and this solution leads to a noise reduction [9]. Thus, this kind of motion is not considered in the framework of this research.

The angular momentum equation about the center of mass of the cylindrical part is:

$$I_G\ddot{\varphi} = r\, R_x - u_{rs}R_y sgn\left(\dot{\varphi}\right). \tag{8}$$

Reaction force R_n does not appear in Equation (8), because it passes through the center of mass of the cylinder. I_G is the moment of inertia about the center of mass, r is the cylinder radius, u_{rs} is the coefficient of rolling resistance in static condition, $\ddot{\varphi}$ is the angular acceleration, $\dot{\varphi}$ is the angular velocity, and sgn is the signum function. Since the beginning of the rolling motion is considered, $\ddot{\varphi} = 0$ and

$sgn(\dot{\varphi}) = -1$, because after the beginning of relative motion, a negative rolling velocity will take place: $\dot{\varphi} = -v_r/R$. Tangential reaction force R_x can be calculated from the angular momentum equation:

$$R_x = -\frac{u_{rs}}{R} R_y. \tag{9}$$

The term $\frac{u_{rs}}{R}$ is the non-dimensional rolling friction coefficient in static condition, which is named f_{vs}.

The reaction force that the stop plate exerts on the cylinder (R_n) can be calculated from Equations (5) and (9). The contact between the cylinder and the stop plate is unilateral ($R_n > 0$). Hence, the condition $R_n \leq 0$ in conjunction with the adoption of the maximum rolling friction coefficient in Equation (9) defines the beginning of the relative motion between the cylinder and the conveyor track:

$$R_n = macos(\psi) + mgsin(\theta) + f_{vs} R_y \leq 0, \tag{10}$$

If Equation (6) is introduced into Equation (10), the following condition for the beginning of rolling motion is obtained:

$$\frac{a}{g} \leq -\frac{sin(\theta) + f_{vs}\cos(\theta)}{cos(\psi) + f_{vs}sin(\psi)}. \tag{11}$$

Figure 2 depicts the values of the static rolling friction coefficient (f_{vs}) that prevents the rolling motion against non-dimensional acceleration (a/g) for parametric values of ψ. Only negative values of acceleration are considered, since a negative acceleration is required to begin the rolling motion; the maximum value of non-dimensional acceleration is $a/g = -sin(\theta)/cos(\psi)$, since for larger values, the resultant of the gravity and inertia forces pushes the cylinder against the stop plate.

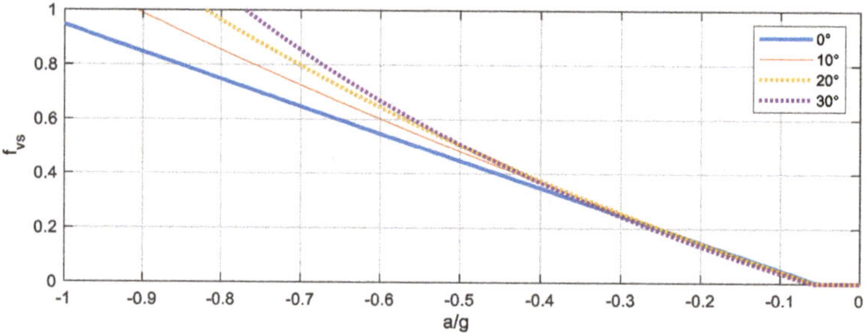

Figure 2. Static rolling friction coefficient that prevents the rolling motion, $\theta = 3°$ and parametric values of ψ.

Figure 2 shows that very large and unrealistic values [16] of static rolling friction coefficient (f_{vs}) are needed to prevent the beginning of the rolling motion. When a/g is lower than −0.35, the increase in angle ψ increases the value of f_{vs}.

Hypothetically, the cylindrical part excited by conveyor vibrations could begin a sliding motion as well. The following condition for the beginning of the sliding motion can be derived:

$$\frac{a}{g} \leq -\frac{sin(\theta) + \mu_s cos(\theta)}{cos(\psi) + \mu_s sin(\psi)}. \tag{12}$$

It is worth noticing that this condition is the same that holds true for non-cylindrical parts [1]. Equation (12) is similar to Equation (11), but f_{vs} is substituted by μ_s, which is the static Coulomb friction coefficient. For most of the materials used in industrial applications [16], the static Coulomb friction coefficient is much larger than the non-dimensional rolling friction coefficient in static condition.

Therefore, the negative acceleration needed to begin the sliding motion is much larger (in modulus) than the one needed to begin the rolling motion.

The evolution of the rolling motion is now analyzed. The equations of motion are:

$$ma_x = -mgsin(\theta) + R_x \tag{13}$$

$$mA_y = -mgcos(\theta) + R_y \tag{14}$$

$$I_G\ddot{\varphi} = r\,R_x - f_v R R_y sgn\left(\dot{\varphi}\right). \tag{15}$$

Moreover, during pure rolling, the following kinematic condition has to be fulfilled:

$$\ddot{\varphi} = -\frac{a_x - acos(\psi)}{R}. \tag{16}$$

R_y can be calculated from Equation (14); then, Equations (13), (15), and (16) can be solved to calculate R_x, a_x, and $\ddot{\varphi}$. Taking into account that for a cylinder $I_G = 0.5mR^2$, the following results are obtained:

$$R_x = \frac{1}{3}\,m(acos(\psi) + gsin(\theta)) + \frac{2}{3}f_v R_y sgn\left(\dot{\varphi}\right) \tag{17}$$

$$a_x = \frac{1}{3}\,acos(\psi) - \frac{2}{3}gsin(\theta) + \frac{2}{3}f_v\frac{R_y}{m}sgn\left(\dot{\varphi}\right) \tag{18}$$

$$\ddot{\varphi} = -\frac{1}{R}\frac{2}{3}\left(-acos(\psi) - gsin(\theta) + f_v\frac{R_y}{m}sgn\left(\dot{\varphi}\right)\right). \tag{19}$$

It is worth noticing that when the cylinder rolls forwards ($\dot{\varphi} < 0$), the rolling friction term in Equation (19) adds to the gravity term. The motion of the cylinder on the conveyor track can be studied integrating Equations (18) and (19); the input is $acos(\psi)$. Generally speaking, a numerical integration is needed (see Section 3) owing to the *sgn* function. Nevertheless, $f_v < 0.01$ in most industrial processes, and an approximation of the integrals of Equations (18) and (19) can be obtained assuming $f_v = 0$.

According to Equation (11), the forward motion begins when $a/g \le -sin(\theta)/cos(\psi)$. Since in normal conveying applications angle θ is rather small, another simplification can be made assuming that the forward rolling motion begins when $acos(\psi) = -x_0\omega^2 sin(\omega t)cos(\psi)$ becomes negative. This assumption simplifies the initial conditions, because in the harmonic motion vibration, the acceleration changes its sign when the displacement is zero and the velocity has the maximum positive value ($x_0\omega$).

The integration of Equation (18) with the above-mentioned simplifications and initial conditions leads to the following equations for the absolute velocity (v_x) and displacement (x) of the center of mass of the cylinder:

$$v_x = \frac{1}{3}\,\omega x_0 \cos(\omega t)cos(\psi) + \frac{2}{3}\,\omega x_0 cos(\psi) - \frac{2}{3}gsin(\theta)t \tag{20}$$

$$x = \frac{1}{3}x_0 \sin(\omega t)cos(\psi) + \frac{2}{3}\,\omega x_0 cos(\psi) - \frac{1}{3}gsin(\theta)t^2. \tag{21}$$

From the point of view of conveying, the most important parameter is the relative velocity v_r between the part and the track of the conveyor:

$$v_r = v_x - \,\omega x_0 \cos(\omega t)cos(\psi) \tag{22}$$

$$v_r = \frac{2}{3}\,\omega x_0 cos(\psi) - \frac{2}{3}\,\omega x_0 \cos(\omega t)cos(\psi) - \frac{2}{3}gsin(\theta)t. \tag{23}$$

The relative velocity includes a positive and constant term due to initial velocity ($x_0\omega$); a harmonic term due to the vibration of the conveyor plate, whose integral over a period ($T = 2\pi/\omega$) is zero; and a negative term due to gravity that increases (in modulus) with time.

If $\theta = 0$ (horizontal conveying), the average conveying velocity in the period ($\widetilde{v}_r$) is constant and equal to $\frac{2}{3}\ \omega x_0 cos(\psi)$. If $\theta > 0$ the average conveying velocity decreases, and at the nth period it assumes the value:

$$\widetilde{v}_r = \frac{2}{3}\left(\omega x_0 cos(\psi) - gsin(\theta)\frac{\pi}{\omega}(2n-1)\right)\quad n = 1, 2, \ldots N. \tag{24}$$

When $\widetilde{v}_r$ becomes negative, the cylindrical part rolls backwards and eventually impacts the stop plate. Figure 3 depicts the ratio between $\widetilde{v}_r$ and the maximum vibration speed ($x_0\omega$) versus period number (n) for $\theta = 3°$. In Figure 3a, the ratio between the vibration acceleration and gravity acceleration (a/g) is 1 and the parametric values of angle ψ are considered; the increase in ψ has a negative effect on the conveying speed. In Figure 3b, ψ is set to zero, and the parametric values of the ratio a/g are considered; the increase in vibration acceleration has a positive effect on vibration conveying.

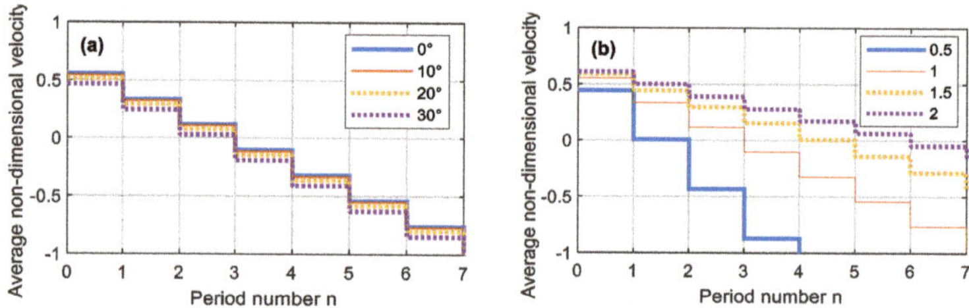

Figure 3. Average non-dimensional conveying velocity ($\widetilde{v}_r/x_0\omega$) versus period number, $\theta = 3°$, $a/g = 1$ and parametric values of ψ (**a**); $\psi = 0$ and parametric values of a/g (**b**).

Equation (24) holds true with the assumption $f_v = 0$. If rolling resistance is present, it contributes to decreasing the average conveying speed in the period.

During pure rolling motion, the reaction force R_x given by Equation (17) is needed to guarantee equilibrium. $|R_x|$ has to be smaller than the maximum static friction force ($\mu_s R_y$). Hence, the ratio $|R_x|/R_y$ has the meaning of static friction coefficient needed to guarantee equilibrium during pure rolling. Figure 4 shows the ratio $|R_x|/R_y$ against non-dimensional acceleration (a/g) for $\theta = 3°$, $f_v = 0$ and parametric values of ψ.

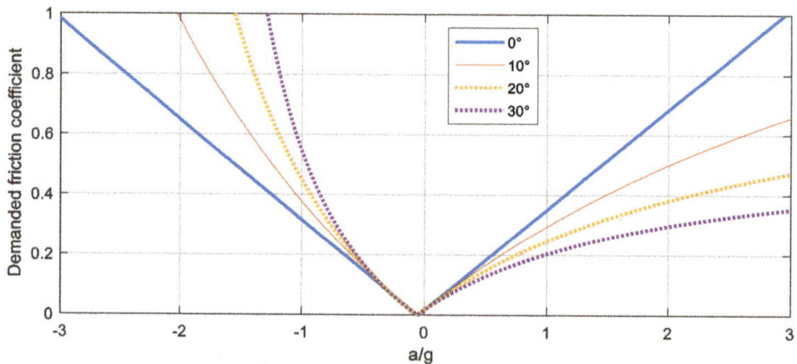

Figure 4. Friction coefficient needed to guarantee the pure rolling motion, $\theta = 3°$, $f_v = 0$, and parametric values of ψ.

With $\psi = 0$, the friction coefficient needed to guarantee the pure rolling motion is only slightly influenced by the versus of acceleration.

With $\psi > 0$, the friction coefficient needed to guarantee the pure rolling motion is significantly smaller when the conveyor acceleration is positive than when acceleration is negative. This phenomenon happens because when acceleration changes sign, the decrease in reaction force R_y (Equation (7)) is not compensated by the decrease in R_x (Equation (17)), and this effect becomes even more important when ψ is large. Since in most industrial material μ_s is in the range of 0.6 ÷ 1 [1], the transition from a pure rolling motion to a rolling with sliding can take place only in the presence of large negative accelerations of the vibration conveyor.

When the cylindrical part rolls backwards, it may impact the stop plate; in this case, the equation of motion in the x direction becomes:

$$ma_x = -mgsin(\theta) + R_x + F_{imp} \tag{25}$$

in which F_{imp} is the impact force, which has the same direction of R_n in Figure 1. F_{imp} strongly increases R_x; hence, the ratio $|R_x|/R_y$ may reach very large values, and the sliding motion can begin.

3. Numerical Model

The simulation of the motion of a cylindrical part on a vibrating plane is not a simple task due to the non-linear phenomena that take place, e.g., the frequent transition between states of motion and the impacts with the stop plate. In order to address the problem, a combination of a MATLAB (R2019b, MathWorks Inc., Natick, MA, USA, 2020) script and its simulation environment Simulink was adopted. The simulations are started from a main MATLAB script which initializes the variables at the initial conditions: the cylinder is touching the track and in contact with the stop plate, as depicted in Figure 1b. The cylinder is supposed to start with a pure rolling motion; therefore, the main script launches a Simulink simulation, which implements this case. When a proper dynamic condition is met, the simulation is stopped and the simulation of rolling with sliding motion is started by the main script. Such simulation is stopped when particular kinematic and dynamic conditions are met. The process is iterated until the desired simulation duration is reached.

The block diagram of the simulation is represented in Figure 5. The two transition conditions require special attention, because they rule the simulation. The transition from pure rolling to rolling with sliding is ruled by a dynamic condition. When the tangential reaction force R_x needed to guarantee the equilibrium of the pure rolling motion (Equation (17)) is larger than the maximum static friction force, the MATLAB main script stops the Simulink rolling model and starts the Simulink sliding model.

The MATLAB main script stops the Simulink sliding model when a dynamic and a kinematic condition are satisfied at the same time. The kinematic condition is the pure rolling condition, which is given by Equation (26).

$$v_x - V_x = -\dot{\varphi} \cdot R \tag{26}$$

This equation shows when there is a possible switching from rolling with sliding to pure rolling, but a dynamic condition is to be satisfied as well. This condition states that the tangential reaction force has to be smaller than the maximum static friction force.

One of the novelties of this numerical model with respect to previous papers [15] is the introduction of the impacts with the stop plate, which makes the integration of the equations more complex. Impact forces are calculated representing contact stiffness by means of a lumped spring, which is compressed only if the distance between the cylinder and the stop plate is smaller than a fixed threshold. For this reason, the simulation environment Simulink was chosen, which proved to be very intuitive for the integration of single sets of equations. Then, the problem was moved to the transition between the two cases. The development of a model entirely programmed in Simulink showed problems of flickering between different states and was not very effective. Therefore, it was decided to exploit the versatility of a simulation managed by a MATLAB script at the highest level with the power of the Simulink simulation environment at a lower level. The results lived up to expectations, although troubleshooting and a fine tuning of the parameters of the Simulink solver was needed.

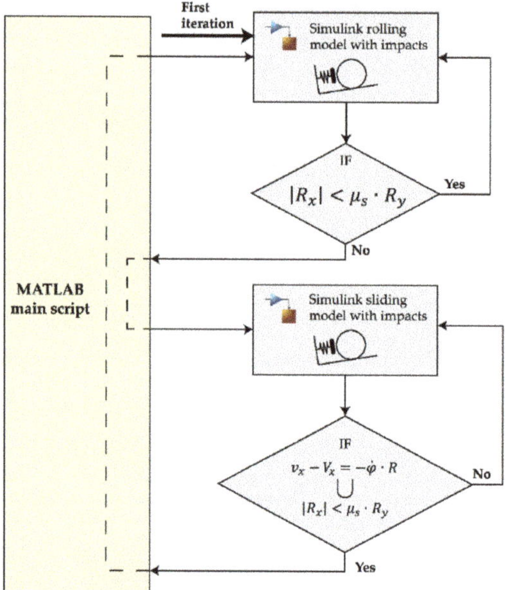

Figure 5. Logical flow of the simulation.

4. Numerical Results

Numerical simulations aimed at exploring the dynamics of vibration conveying of cylindrical parts beyond the limits of the analytical model of Section 2. In particular, numerical simulations made it possible to analyze the effect of the impacts of the cylindrical part on the stop plate.

A realistic reference case was defined, which is characterized by the following parameters of the vibratory conveyor: x_0= 0.0002 m, $\omega = 314$ rad/s (frequency 50 Hz), $\theta = 3°$, $\psi = 0°$. The maximum modulus of conveyor acceleration is 19.72 m/s^2 ($a/g = 2$).

The rolling part has a radius $r = 0.007$ m and mass $m = 0.096$ kg. The friction parameters are $f_v =$ 0.005, $\mu_c = 0.3$, and $\mu_s = 0.6$. The value of the rolling resistance is more than twice that of the typical rolling elements of bearings [16] and takes into account that the cylindrical part may be rough or dirty. The kinetic Coulomb friction coefficient value is inside the typical range of values found in vibratory conveyors [1,6,9]. The static Coulomb friction coefficient was set to be twice the kinetic coefficient [17]. Finally, the impact stiffness was set to $k = 1,000,000$ N/m; this value made it possible to obtain values of contact forces similar to the measured ones (about 60 ÷ 80 N).

The first parametric simulations aimed at analyzing the effect of angle ψ; Figure 6 shows numerical results in terms of the absolute displacement of the cylinder center of the mass. In the same figure, for comparison, the displacement of the conveyor is represented (along the direction tilted by ψ with respect to the x axis).

In the plots of Figure 6, two different phases can be identified. The first phase (phase 1), which ends at about 0.25 s and includes 10 ÷12 vibration periods, is the initial motion of the cylindrical part without impacts. As foreseen by theoretical analysis, the cylindrical parts begins its motion at $t = 0$ owing to the initial velocity given by the conveyor, reaches the maximum displacement, and eventually impacts the stop plate. The maximum displacement is rather small (about 0.003 m), and this value is in agreement with the theoretical predictions (Equation (21)).

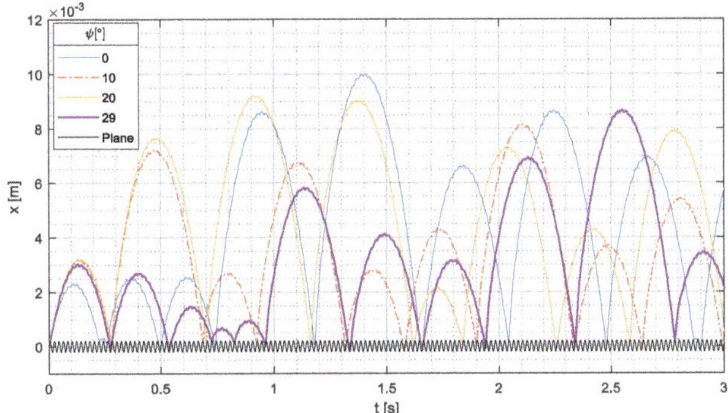

Figure 6. Numerical results, displacement of the cylindrical part along the track, $\theta = 3°$, $f_v = 0.005$, $\mu_c = 0.3$, $\mu_s = 0.6$, and parametric values of ψ.

In this phase, the increase in the angle ψ up to 20° leads to an increase in the traveled distance, a further increase leads to a reduction in the traveled distance. This particular behavior is due by the presence of two opposite effects.

The first effect can be understood looking at the sliding velocity (v_{sl}) of the cylindrical part:

$$v_{sl} = v_x - V_x + R\dot{\varphi} \tag{27}$$

which is plotted in Figure 7 with conveyor acceleration. If $\psi = 0$, the cylindrical part slides both when the conveyor acceleration is negative and positive (forwards and backwards sliding, respectively). Backwards sliding causes a reduction in the traveled distance. This phenomenon is in agreement with the analytical results of Figure 4, which show that for $\psi = 0$, the static friction coefficient needed to guarantee the pure rolling motion is larger than $\mu_s = 0.6$ both for negative (−2) and for positive (+2) values of non-dimensional acceleration.

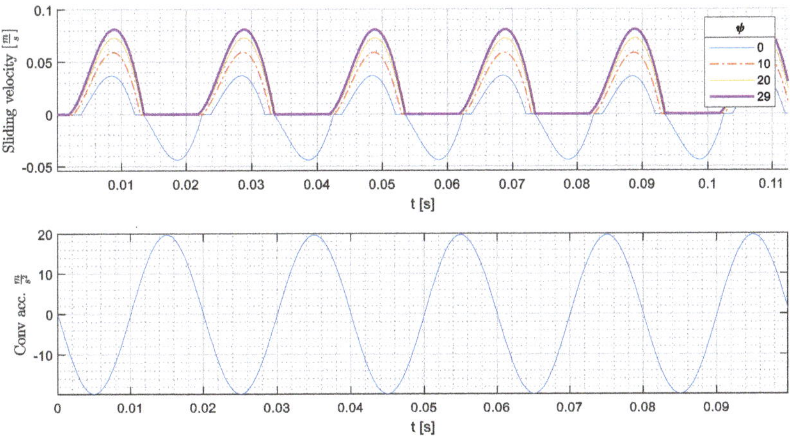

Figure 7. Numerical results, sliding velocity of the cylindrical part along the track, $t < 0.1$ s, $\theta = 3°$, $f_v = 0.005$, $\mu_c = 0.3$, $\mu_s = 0.6$, and parametric values of ψ.

Conversely, Figure 7 shows that if ψ increases, there is only forwards sliding, since the static friction coefficient is able to prevent backwards sliding when the conveyor acceleration becomes positive, and the cylindrical part behaves similar to a block. It is worth noticing that this behavior is in agreement with the analytical results of Figure 4.

The second effect is the decrease in the initial velocity of the cylindrical part when ψ increases; see Equation (20). In summary, for small values of ψ, the first effect is dominant, and the traveled distance increases, whereas for large values of ψ, the second effect reduces the traveled distance.

The second phase of the motion (phase 2, $t > 0.25$ s) is dominated by the impacts of the cylindrical part on the stop plate.

The cylindrical part reaches the maximum displacement in this phase, and this value chiefly depends on impact mechanics. If the relative velocity (Equation (22)) is large, because the cylindrical part and the conveyor move in opposition, the conveyor transfers a lot of energy to the cylindrical part, which has a large rebound after the impact. This phenomenon occurs at $t = 2.04$ for $\psi = 0°$. Figure 8, which is a zoom of Figure 6 of about $t = 2.04$ s, shows that the impact occurs when the conveyor has the maximum positive velocity, whereas the velocity of the cylindrical part is negative. Conversely, if the relative velocity is small, because the cylindrical part and the conveyor move in phase, the conveyor transfers less energy to the cylindrical part, and a smaller rebound occurs. This phenomenon takes place at $t = 0.49$ s for $\psi = 0°$, since Figure 9 shows that the cylindrical part impacts the conveyor when it has a negative velocity.

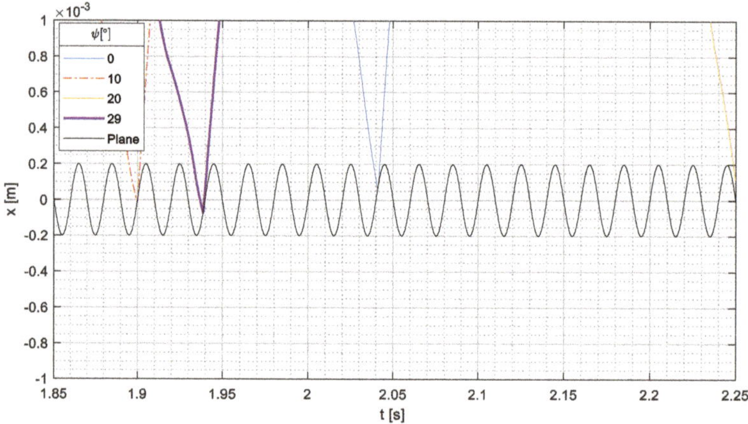

Figure 8. Numerical results, displacement of the cylindrical part along the track, impact with large relative velocity at $t = 2.04$, $\psi = 0°$, $\theta = 3°$, $f_v = 0.005$, $\mu_c = 0.3$, $\mu_s = 0.6$.

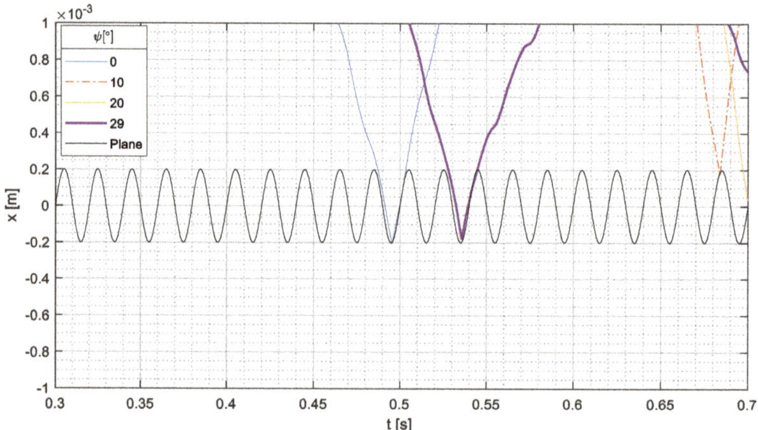

Figure 9. Numerical results, displacement of the cylindrical part along the track, impact with small relative velocity at $t = 0.49$, $\psi = 0°$, $\theta = 3°$, $f_v = 0.005$, $\mu_c = 0.3$, $\mu_s = 0.6$.

The angle of conveyor acceleration (ψ) has a negligible effect on the second phase (impact dominated) of the motion of the cylindrical part on the conveyor, since the relative velocity at the impact depends much more on the previous motion than on conveyor maximum velocity, which slightly decreases if ψ increases. Actually, after some impacts, the relative velocity at the impact changes randomly. Thus, the motion of the cylindrical part on the conveyor is similar to the motion of a bouncing ball that impacts a vibrating table, which may exhibit a chaotic behavior [18,19].

Figure 10 shows for the same cases of Figure 6 the sliding velocity of the cylindrical part and highlights that, owing to the large value of non-dimensional acceleration ($a/g = 2$), very often the cylinder slides on the conveyor track. The largest sliding velocities take place just after the impacts.

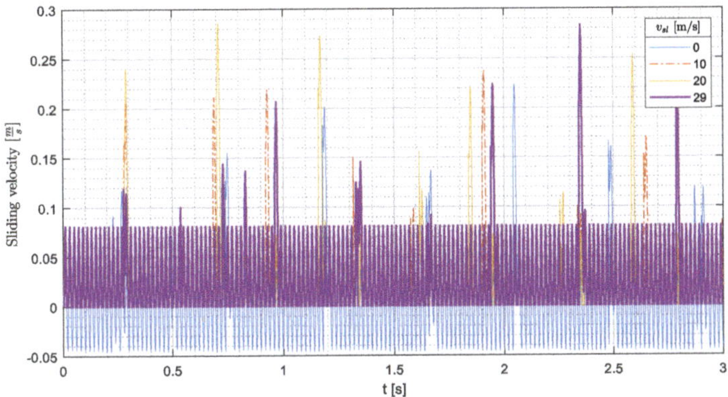

Figure 10. Numerical results, sliding velocity of the cylindrical part along the track, $\theta = 3°$, $f_v = 0.005$, $\mu_c = 0.3$, $\mu_s = 0.6$, and parametric values of ψ.

The second parametric simulation aimed at analyzing the effect of conveyor vibration frequency. Conveyor acceleration was kept equal to the previous value ($a/g = 2$); therefore, an increase in frequency led to a decrease in velocity according to this equation:

$$v = \frac{a}{\omega}. \tag{28}$$

Figure 11 clearly shows that the increase in the frequency has a negative effect on the motion of the cylindrical part both before and after the first impact. The first effect takes place because the initial velocity is smaller, while the second effect takes place because the relative velocity at the impact is smaller. If the vibration frequency decreases to 25 Hz, the traveled distance increases up to 0.043 m. It is worth noticing that even if the dynamics of conveying of block-shaped parts and cylindrical parts are very different, the decrease in vibration frequency has a positive effect in both cases [1].

Then, the effect of rolling friction coefficient was investigated by means of numerical simulations. The results, which are presented in Figure 12, show that realistic variations in f_v have a negligible effect on the conveying of cylindrical parts. Before the first impact (phase 1), the decrease in f_v causes a very small increase in the traveled distance. After the first impact (phase 2), the variation in the rolling friction coefficient changes the instants in which the most energetic impacts take place, but the traveled distance does not significantly change.

Figure 13 shows the effect of contact stiffness on the conveying of cylindrical parts. The contact stiffness does not affect phase 1, and it does not have a clear effect on the traveled distance in phase 2. This phenomenon agrees with physical intuition, because the contact stiffness is another factor that contributes to randomly varying the relative velocity at the impact, which is the main factor influencing the rebound of the cylindrical part.

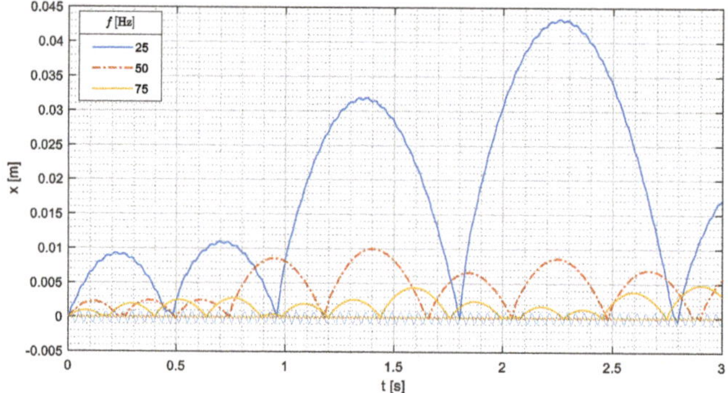

Figure 11. Numerical results, displacement of the cylindrical part along the track, $\theta = 3°$, $\psi = 0°$, $f_v = 0.005$, $\mu_c = 0.3$, $\mu_s = 0.6$, and parametric values of frequency.

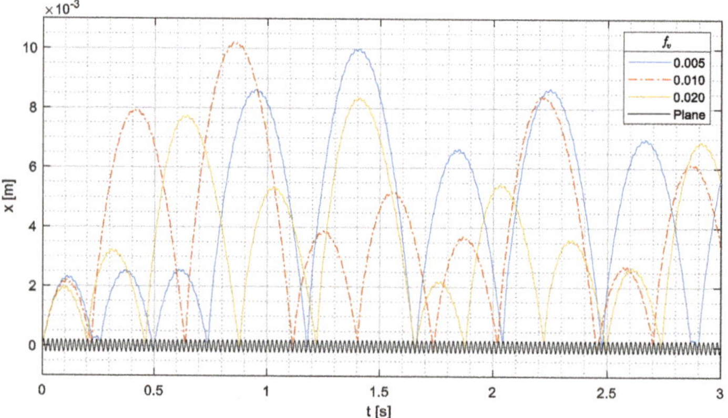

Figure 12. Numerical results, displacement of the cylindrical part along the track, $\theta = 3°$, $\psi = 0$, $\mu_c = 0.3$, $\mu_s = 0.6$, and parametric values of f_v.

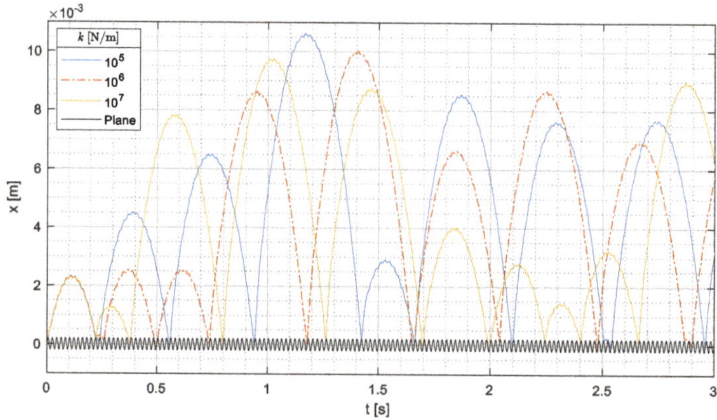

Figure 13. Numerical results, displacement of the cylindrical part along the track, $\theta = 3°$, $\psi = 0$, $f_v = 0.005$, $\mu_c = 0.3$, $\mu_s = 0.6$, and parametric values of contact stiffness.

Finally, the effect of friction coefficients was investigated.

Figure 14 shows that μ_s has a small effect on phase 1: if the static friction coefficient increases above 0.6, the traveled distance increases. This phenomenon can be explained looking at Figure 15, which shows the sliding speed for the same values of μ_s. With $\mu_s = 1$ during phase 1, the cylinder does not slide at all, and the traveled distance increases, since the pure rolling motion dissipates less energy than the rolling with sliding motion. It is worth noticing that this result agrees with the analytical results of Figure 4.

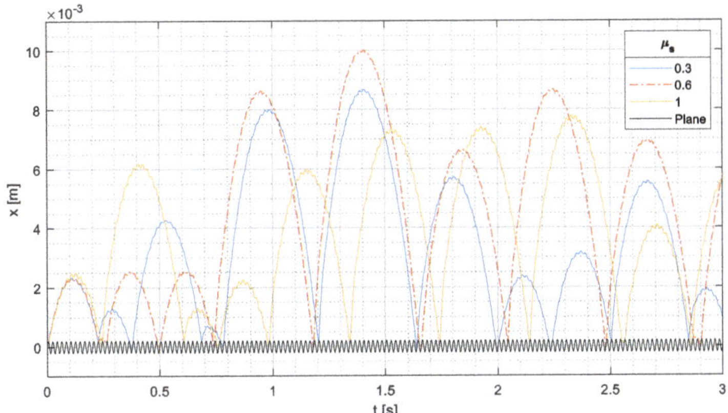

Figure 14. Numerical results, displacement of the cylindrical part along the track, $\theta = 3°$, $\psi = 0$, $f_v = 0.005$, $\mu_c = 0.3$, and parametric values of static friction coefficient μ_s.

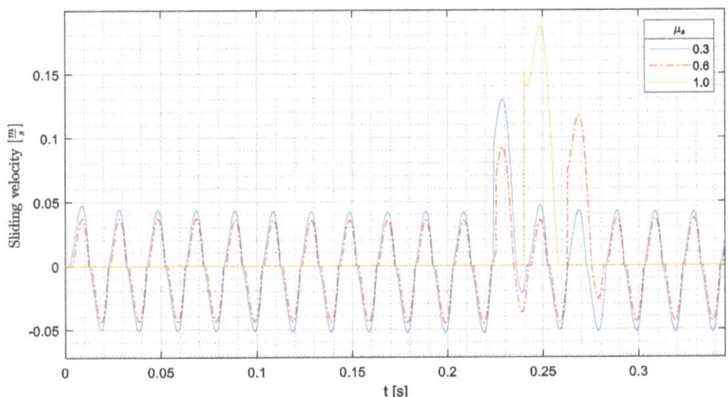

Figure 15. Numerical results, sliding velocity of the cylindrical part along the track, $\theta = 3°$, $\psi = 0$, $f_v = 0.005$, $\mu_c = 0.3$, and parametric values of static friction coefficient μ_s.

After the first rebound, the impacts take place at random instants and, again, the relative velocity at the impact has the largest effect.

The effect of a proportional variation of both friction coefficients (μ_c and μ_s) is presented in Figure 16. In phase 1, the cylinder with the highest μ_s reaches the largest distance because it rolls without sliding, as shown in Figure 17. After the first impact (phase 2), the motion is dominated by the relative velocity at the impact; nevertheless, a decrease in μ_c leads to an average increase in the traveled distance. With $\mu_c = 0.2$, there are 4 rebounds with a traveled distance larger than 0.008 m, whereas with $\mu_c = 0.4$, there is no rebound with a traveled distance larger than 0.008 m. This phenomenon occurs because for every value of μ_c and μ_s, there are large sliding velocities (Figure 17) after the impacts, and a reduction in μ_c leads to a reduction in energy dissipation.

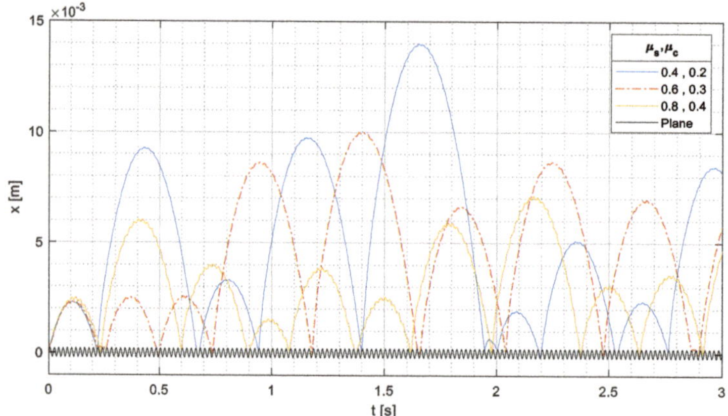

Figure 16. Numerical results, displacement of the cylindrical part along the track, $\theta = 3°$, $\psi = 0$, $f_v = 0.005$, and parametric values of static friction coefficients μ_c and μ_s.

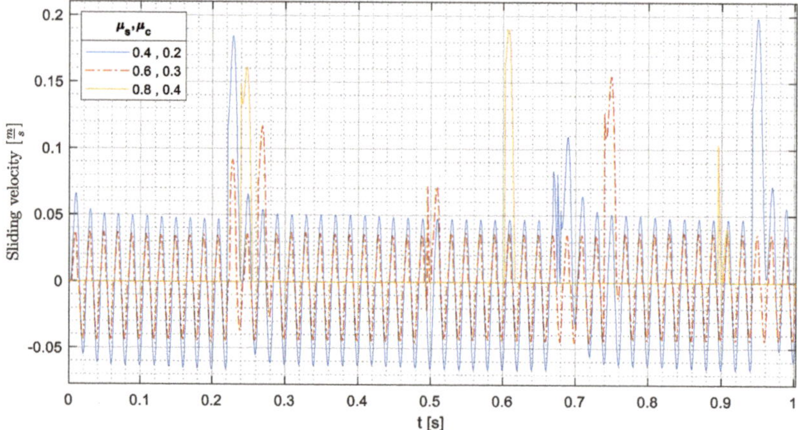

Figure 17. Numerical results, sliding velocity of the cylindrical part along the track, $\theta = 3°$, $\psi = 0$, $f_v = 0.005$, and parametric values of static friction coefficients μ_c and μ_s.

The numerical results presented in this section clearly highlight that only the decrease in the vibration frequency can significantly increase the distance traveled by the cylindrical part. Nevertheless, even the largest traveled distances foreseen by simulations (0.043 m), which corresponds to about six times the cylinder radius, is not enough to cover the distances requested in many automatic manufacturing processes. For this reason, the development of a saw-teeth grooved surface of the conveyor track is highly recommended (see Figure 18). In this case, a traveled distance equal to $5 \div 6$ times the cylinder radius is enough to move the cylindrical part from one grove to the following grove and to generate a sequential motion.

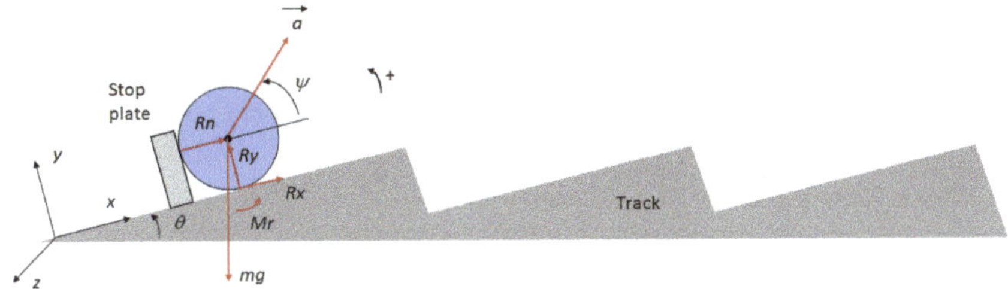

Figure 18. Conveyor with saw-teeth track.

5. Comparison between Numerical and Experimental Results

In order to validate the numerical model, some experimental tests were carried out at the Laboratory of Robotics and Mechatronics of Padova University, which is equipped with the linear vibratory conveyor shown in Figure 19.

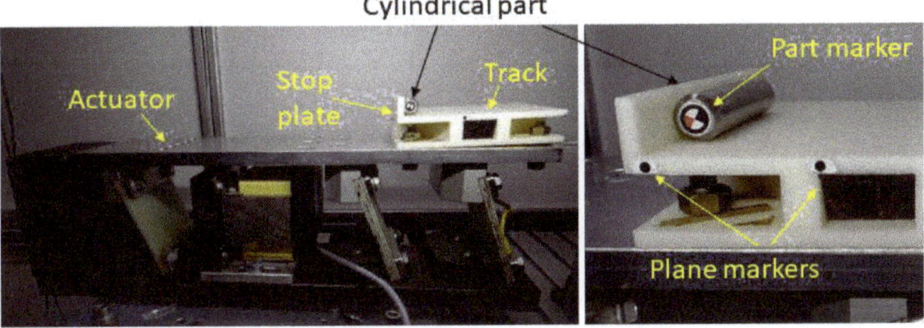

Figure 19. Vibratory conveyor for experimental tests.

A track (120 mm long) with a stop plate was printed with additive manufacturing and mounted on the conveyor. A single steel cylinder was placed on the track. The motion was monitored by an AVT Pike f-032b camera with a frame-rate of 470 frames/s. Both the track and the cylinder were equipped with markers that were used to reconstruct their trajectories, using the Hough transform [20]. The static friction coefficient of the cylinder on the track was measured as well, and resulted in $\mu_s = 0.7$.

Figure 20 makes a comparison between numerical and experimental results. The numerical model is able to capture the most important features of cylinder motion, since the experimental results show a sequence of chaotic rebounds of the cylinder on the stop plate. The largest rebounds predicted by the numerical model have about the same duration (0.5 s) of experimental rebounds, and the traveled distances are only a bit larger ($\approx$10%) than the measured values. This phenomenon may occur because, owing to the three-dimensional geometry of the system, the cylinder velocity may have a small component in the transverse direction (z in Figure 18).

The main difference between numerical and experimental results is that the numerical model predicts a smaller number of bad impacts, with a short rebound. This happens because the numerical model is two-dimensional and assumes an impact between the cylinder and the stop plate along a line. In reality, the system is three-dimensional, and often the contact begins when a corner of the cylinder impacts the stop plate. The energy transfer from the conveyor to the cylinder is less efficient and leads to a reduced traveled distance.

These phenomena could be analyzed by means of a 3D numerical model, which would be much more complex and cumbersome than the presented 2D model (a 2D simulation lasting 1 s requires a computation time of about 16 s). Nevertheless, it is worth noticing that a complete agreement between numerical and experimental results could be difficult to achieve even with a 3D simulation, since the system is non-linear and chaotic; thus, small differences between simulation and experimental parameters may lead to large differences in results [19,21]. In actual conveyors, there are many cylindrical parts, and their collisions further enhance the chaotic behavior of the system [22].

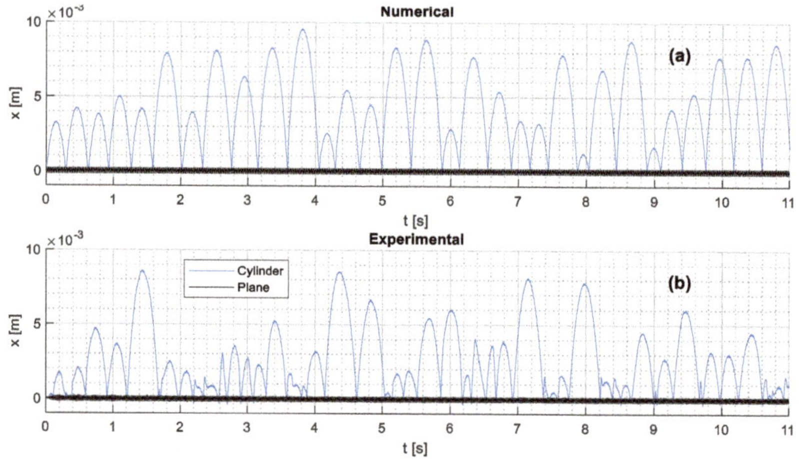

Figure 20. Comparison between numerical (**a**) and experimental (**b**) results. $\theta = 3°$, $\psi = 10°$, $f = 55$ Hz, $a/g = 2.254$, $k = 1{,}000{,}000$ N/m, $f_v = 0.005$, $\mu_c = 0.35$ and $\mu_s = 0.7$.

6. Conclusions

The dynamics of vibratory conveying of cylindrical parts are very different from the dynamics of conveying of block-shaped parts. Both analytical and numerical calculations showed that large distances cannot be traveled owing to the backwards rotation of the cylindrical part.

The impacts of the cylindrical part on the stop plate are an essential feature of this kind of motion and strongly increase the traveled distance. After some rebounds, the impacts of the cylindrical part on the stop plate of the vibratory conveyor take place at random instants, which correspond to random values of the relative velocity between the cylindrical part and the stop plate. The impacts with the largest relative velocity lead to the largest rebounds. Eventually, the motion becomes chaotic.

Numerical simulations made it possible to investigate the effect of the most important parameters on the vibratory conveying of cylindrical parts. When vibration acceleration is kept constant, a decrease in vibration frequency leads to a relevant increase in the traveled distance. The kinetic friction coefficient (μ_c) influences the traveled distance; when μ_c halves, the maximum traveled distance almost doubles. The other parameters have a small or negligible effect on the conveying process. This phenomenon occurs because the energy transferred by the impacts to the cylindrical part essentially depends on the relative velocity between the part and the stop plate, and the variations in ψ, f_v, k and μ_s have a small effect on the relative velocity but influence the instants in which the strongest impacts (with large relative velocity) take place.

Author Contributions: Conceptualization, N.C. and A.D.; methodology, N.C. and A.D.; software, N.C.; validation, N.C. and A.D.; formal analysis, N.C. and A.D.; investigation, N.C. and A.D.; writing—Original draft preparation, N.C. and A.D.; writing—Review and editing, N.C. and A.D.; supervision, A.D. All authors have read and agreed to the published version of the manuscript.

Funding: This research received no external funding.

Conflicts of Interest: The authors declare no conflict of interest.

References

1. Boothroyd, G. *Assembly Automation and Product Design*, 2nd ed.; Taylor & Francis: Boca Raton, FL, USA, 2005; pp. 29–45.
2. Rosati, G.; Faccio, M.; Barbazza, L.; Rossi, A. Hybrid fexible assembly systems (H-FAS): Bridging the gap between traditional and fully flexible assembly systems. *Int. J. Adv. Manuf. Technol.* **2015**, *81*, 1289–1301. [CrossRef]
3. Rosati, G.; Faccio, M.; Finetto, C.; Carli, A. Modelling and optimization of Fully Flexible Assembly Systems (F-FAS). *Assem. Autom.* **2013**, *33*, 165–174. [CrossRef]
4. Kritikou, G.; Aspargathos, N. Micro—Manipulation methods and assembly of hexagonal microparts on a programmable platform with electrostatic forces. *Int. J. Mech. Control* **2019**, *20*, 71–80.
5. Taniguchi, O.; Sakata, M.; Suzuki, Y.; Osanai, Y. Studies on vibratory feeder. *Bull. ISME* **1963**, *6*, 37–43. [CrossRef]
6. Okabe, S.; Yokoyama, Y.; Boothroyd, G. Analysis of vibratory feeding where the track has directional friction characteristics. *Int. J. Adv. Manuf. Technol.* **1988**, *3*, 73–85. [CrossRef]
7. Wolfsteiner, P.; Pfeiffer, F. Modeling, simulation, and verification of the transportation process in vibratory feeders. *ZAMM-J. Appl. Math. Mechanics/Zeitschrift Angewandte Mathematik Mechanik* **2000**, *80*, 35–48. [CrossRef]
8. Vilán, J.V.; Robleda, A.S.; Nieto, P.J.G.; Placer, C.C. Approximation to the dynamics of transported parts in a vibratory bowl feeder. *Mech. Mach. Theory* **2009**, *44*, 2217–2235. [CrossRef]
9. Ma, H.W.; Fang, G. Kinematics analysis and experimental investigation of an inclined feeder with horizontal vibration. *Proc. Inst. Mech. Eng. Part C J. Mech. Eng. Sci.* **2016**, *230*, 3147–3157. [CrossRef]
10. Buzzoni, M.; Battarra, M.; Mucchi, E.; Dalpiaz, G. Motion analysis of a linear vibratory feeder: Dynamic modeling and experimental verification. *Mech. Mach. Theory* **2017**, *114*, 98–110. [CrossRef]
11. Despotović, Z.; Šinik, V.; Janković, S.; Dobrilović, D.; Bjelica, M. Some specific of vibratory conveyor drives. In Proceedings of the the V International Conference Industrial Engineering and Environmental Protection 2015 (IIZS 2015), Zrenjanin, Serbia, 15–16 October 2015.
12. Edmondson, N.F.; Redford, A.H. Flexible parts feeding for flexible assembly. *Int. J. Prod. Res.* **2001**, *33*, 2279–2294. [CrossRef]
13. Boothroyd, G.; Ho, C. Natural resting aspects of parts for automatic handling. *Trans. ASME J. Eng. Ind.* **1977**, *1977*, 314–317. [CrossRef]
14. Lee, S.G.; Ngoi, B.K.A.; Lye, S.W.; Lim, L.E.N. An Analysis of the Resting Probabilities of an Object with Curved Surfaces. *Int. J. Adv. Manuf. Technol.* **1996**, *12*, 366–369. [CrossRef]
15. Comand, N.; Boschetti, G.; Rosati, G. Vibratory feeding of cylindrical parts: A dynamic model. In *Advances in Italian Mechanism Science*; Carbone, G., Gasparetto, A., Eds.; Springer: Berlin/Heidelberg, Germany, 2019; pp. 203–210.
16. Juvinall, R.C.; Marshek, K.M. *Fundamentals of Machine Component Design*; Wiley & Sons: Hoboken, NJ, USA, 2006; pp. 559–563.
17. Williams, G. *Engineering Tribology*; Cambridge University Press: Cambridge, UK, 2005; pp. 145–147.
18. May, R.M. Simple mathematical models with very complicated dynamics. *Nature* **1976**, *261*, 459–467. [CrossRef] [PubMed]
19. Tufillaro, N.B.; Albano, A.M. Chaotic dynamics of a bouncing ball. *Am. J. Phys.* **1986**, *54*, 939–944. [CrossRef]
20. Russ, J.C. *The Image Processing Handbook*; CRC Press: Boca Raton, FL, USA, 2016.
21. Naylor, M.A.; Sanchez, P.; Swift, M.R. Chaotic dynamics of an air-damped bouncing ball. *Phys. Rev. E* **2002**, *66*, 057201. [CrossRef] [PubMed]
22. Lee, C.-W.; Seireg, A.; Duffy, J. Chaotic behavior of a two mass bouncing system. In Proceedings of the 18th Annual ASME Design Automation Conference, Scottsdale, AZ, USA, 13–16 September 1992.

Article

A Real-Time Thermal Model for the Analysis of Tire/Road Interaction in Motorcycle Applications

Flavio Farroni [1,*], Nicolò Mancinelli [2] and Francesco Timpone [1]

1 Dipartimento di Ingegneria Industriale–Università degli Studi di Napoli Federico II, 80138 Napoli NA, Italy; francesco.timpone@unina.it

2 Vehicle Dynamics, Ducati Motor Holding spa-Ducati Corse Division, 40132 Bologna BO, Italy; nicolo.mancinelli@ducati.com

* Correspondence: flavio.farroni@unina.it; Tel.: +39-33-3374-2646; Fax: +39-08-1239-4165

Received: 4 February 2020; Accepted: 23 February 2020; Published: 28 February 2020

Abstract: While in the automotive field the relationship between road adherence and tire temperature is mainly investigated with the aim to enhance the vehicle performance in motorsport, the motorcycle sector is highly sensitive to such theme also from less extreme applications. The small extension of the footprint, along with the need to guarantee driver stability and safety in the widest possible range of riding conditions, requires that tires work as most as possible at a temperature able to let the viscoelastic compounds-constituting the tread and the composite materials of the whole carcass structure-provide the highest interaction force with road. Moreover, both for tire manufacturing companies and for single track vehicles designers and racing teams, a deep knowledge of the thermodynamic phenomena involved at the ground level is a key factor for the development of optimal solutions and setup. This paper proposes a physical model based on the application of the Fourier thermodynamic equations to a three-dimensional domain, accounting for all the sources of heating like friction power at the road interface and the cyclic generation of heat because of rolling and to asphalt indentation, and for the cooling effects because of the air forced convection, to road conduction and to turbulences in the inflation chamber. The complex heat exchanges in the system are fully described and modeled, with particular reference to the management of contact patch position, correlated to camber angle and requiring the adoption of an innovative multi-ribbed and multi-layered tire structure. The completely physical approach induces the need of a proper parameterization of the model, whose main stages are described, both from the experimental and identification points of view, with particular reference to non-destructive procedures for thermal parameters definition. One of the most peculiar and challenging features of the model is linked with its topological and analytical structure, allowing to run in real-time, usefully for the application in co-simulation vehicle dynamics platforms, for performance prediction and setup optimization applications.

Keywords: motorcycle tires; thermal modeling; performance optimization; real-time simulations

1. Introduction

A four-wheeled vehicle, even if often exerting a hyper static equilibrium, requires that its stability is guaranteed by an optimal adherence with road, allowing to satisfy safety, performance, and comfort requirements. If such evidence is fundamental for cars vehicle dynamics, the optimization of tire/road interaction becomes a key factor in motorcycles, and in particular considering racing ones, characterized by working with high roll angles and speed [1].

In a deeper analysis of tire/road interaction in motorsport sector, the focus moves necessarily to the optimization of the contact conditions in reference to the behavior of materials constituting tire tread and inner layers [2,3]. In particular, the adhesive bonding [4] and the power dissipated in the local indentation of road asperities [5], are highly influenced by the viscoelastic properties of tire polymers; such properties mainly vary with stress frequency [6], local displacement [7], and temperature [6,8].

The relationship between tire performance and temperature is a widely discussed topic [9–11], and racing tires, with the aim to exhibit an extra-ordinary frictional attitude, are designed adopting specific mixing of materials working at their best in a narrow thermal range. The challenge, for the driver and for the team engineers, is getting information on such thermal range to make sure that the tire spends the most of its time inside it, acting on proper vehicle setup, driving style and controls.

Moreover, several studies report that tire tread exhibits optimal grip depending on the temperature reached in its core layer [12,13], not reachable by measurement instruments for thermal monitoring, and that global tire stiffness is highly influenced by the thermodynamic conditions of the carcass [14], also not directly measurable. For this reason, and for the increasing request of tools able to reproduce with high reliability the contact with road in a vehicle dynamics environment, the development of a tire thermal model has become a necessity for the players requiring the predictivity in challenging simulation scenarios.

In literature, the first approaches to such issue are related to the modeling of the Fourier equations applied to a three-dimensional domain, in some cases coupled with a mechanical model of the tire [15,16], in other with stand-alone tools able to work together with other interaction formulations, like Pacejka's MF [17], or with FEM [18]. During the past years the focus has moved to highly discretized models able to work in real time [19], whose maximum level of complexity has been requested by motorcycle applications, for which tire contact patch moves along lateral direction of the tread, generating local stress [20] able to modify significantly the whole balance of energy with respect to car tires [19].

The paper describes the main structure and the approach followed in the thermodynamic modeling of a motorcycle racing tire, accounting for the parameterization of the diffusivity of the different layers, of the contact patch under variable working conditions, of the heat generation effects and of the conductive/convective interactions with the external environment, leading to a physical formulation for the real-time simulation of a virtual tire, conceived for a better knowledge of the mechanisms responsible for its forces exchange and for the development of performance optimization strategies linked to specific vehicle setup and boundary conditions management.

2. thermoRIDE–Tire Thermodynamic Model

thermoRIDE is a physical-analytical tire model, employable to study and understand all the phenomena concerning the tire during its interaction with both the external environment and the inner wheel chamber (inner air, rim, brakes, vehicle geometry, etc.,), as illustrated in Figure 1.

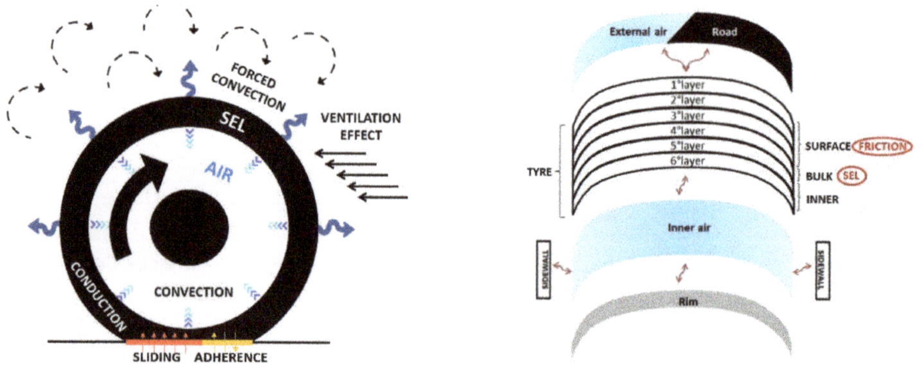

Figure 1. thermoRIDE model scheme.

thermoRIDE model takes into account the following physical phenomena:

- Heat generation within the tire structure due to:
    - Tire/road tangential interaction, known as FP (friction power);
    - Effect of tire cyclic deformation during the tire rolling, known as SEL (strain energy loss).
- Heat exchange with the external environment due to:
    - Thermal conduction between the tire tread and the road pavement;
    - Thermal convection of the tread surface with the external air;
    - Thermal convection of the inner liner surface with the inner air.
- Heat conduction between the tire nodes due to the temperature gradient.

2.1. Mathematical Model

thermoRIDE thermodynamic model is based on the use of the Fourier's diffusion equation applied to a three-dimensional domain.

It refers to energy contained within the system, excluding potential energy because of external forces fields and kinetic energy of motion of the system as a whole, and it keeps account of the gains and losses of energy of the system.

The law of heat conduction, also known as Fourier's law, states that the time rate of heat transfer through a material is proportional to the negative gradient in the temperature and to the area through which the heat flows. The differential form of Fourier's law shows that the local heat flux density $\vec{q}$ is equal to the product of thermal conductivity k and the negative local temperature gradient ∇T. The heat flux density is the amount of energy that flows through a unit area per unit time:

$$\vec{q} = -k * \nabla T \quad (1)$$

where:

- q is the local heat flux density, in $[\mathrm{W/m^2}]$;
- k is the material's conductivity, in $\left[\frac{W}{m*K}\right]$;
- ∇T is the temperature gradient, in $\left[\frac{K}{m}\right]$.

It is possible to obtain a parabolic partial differential equation from the Fourier's law, especially useful for the numerical integration problems in transient thermal conditions. An infinitesimal volume element $dV = dx * dy * dz$ is considered in order to derive the system of the diffusion equations for each part of the tire structure.

Since the change in the internal energy of a closed system is equal to the amount of heat supplied to the system, minus the amount of work done by the system on its surroundings, and the control volume is considered not deformable, the internal energy dU of the infinitesimal volume dV is given by:

$$dU = dm * c_v * T = \rho * dV * c_v * T \quad (2)$$

where the volume cannot do any work (dL=0), as assumed above. That is why the change in the internal energy dU is considered only to the amount of heat dQ added to the system.

The quantity dQ stands for the heat supplied to the system by its surroundings and it takes into account two different contributions:

- Heat exchanged $dQ_{EX} = -dt * \oint_{dS} \vec{q} \cdot \vec{n} * dS$ through the outer surface of the volume dV;
- Heat generated $dQ_G = \dot{q_G} * dV * dt$ inside it.

Therefore, the equation leads to:

$$dQ = dQ_{EX} + dQ_G \tag{3}$$

where:

- $\dot{q_G}$ is the heat amount generated per unit time and per unit volume, in $\frac{W}{m^3 * s}$;
- $\vec{n}$ is the normal unit vector respectively to the faces of the volume element.

For Gauss's divergence theorem, which postulates the equality between the flux of a vector field through a closed surface and the volume integral of the divergence over the region inside the surface, an integral taken over a volume $\oint_{dV} div\,\vec{q} * dV$ can replace the one taken over the surface bounding that volume $\oint_{dS} \vec{q} \cdot \vec{n} * dS$, as following:

$$dQ_{EX} = -dt * \oint_{dS} \vec{q} \cdot \vec{n} * dS = -dt * \oint_{dV} div\,\vec{q} * dV \tag{4}$$

In addition, the integral symbol can be avoided without affecting the physical meaning of the expression above, since the equation was carried out for an infinitesimal volume element dV:

$$dQ_{EX} = -dt * div\,\vec{q} * dV \tag{5}$$

Combining the Equations (4) and (5), it results:

$$dQ_{EX} = dt * div\,(k * \nabla T) * dV \tag{6}$$

In conclusion, summing up (2), (3), and (5), the energy balance equation is obtained for the infinitesimal volume dV:

$$\rho * dV * c_v * dT = \dot{q_G} * dV * dt + dt * div(k * \nabla T) * dV \tag{7}$$

which, divided both sides by the quantity $\rho * dV * c_v * dt$, defines the diffusion or heat equation of Fourier:

$$\frac{\partial T}{\partial t} = \frac{\dot{q_G}}{\rho * c_v} + \frac{div(k * \nabla T)}{\rho * c_v} \tag{8}$$

Equation (8) allows to obtain the three-dimensional distribution of temperature *T(x,y,z,t)*, once the boundary condition are specified.

The Fourier's heat equation governs the temperature variation in time in relation to a special thermal gradient, and shows how the temperature will vary over time because of the generative effects or because of the ones linked to the heat transport.

In general, the Fourier's law allows studying only stationary thermal phenomena, whereas the heat diffusion equation also admits transient states. The complexity of the phenomena under study and the degree of accuracy required have made it necessary to take into account the dependence of the thermodynamic quantities and in particular of the thermal conductivity on the temperature. Furthermore, the non-homogeneity of the tire has made it necessary to consider the variation of the above parameters also along the thickness.

2.2. Physical Model

2.2.1. Tire Structural Model

Depending on the tire peculiar characteristics (dimensions, diffusivity, and inertia), the tire discretization can vary considerably with the main purpose to satisfy both the representation of all the physical phenomena characterizing transient and steady-state tire thermal dynamic and the necessity to preserve the hard real-time requirement in all the tire operating conditions.

The default tire discretization along the radial and lateral directions are respectively illustrated in Figure 2 on the left and right sides. However, the motorcycle tire discretization along the ISO y direction can be freely modified up to 16 ribs (5 ribs default configuration is represented in the Figure 2) once the pre-initialized boundary conditions maps for the specific mesh configuration are available for the specific tire under analysis.

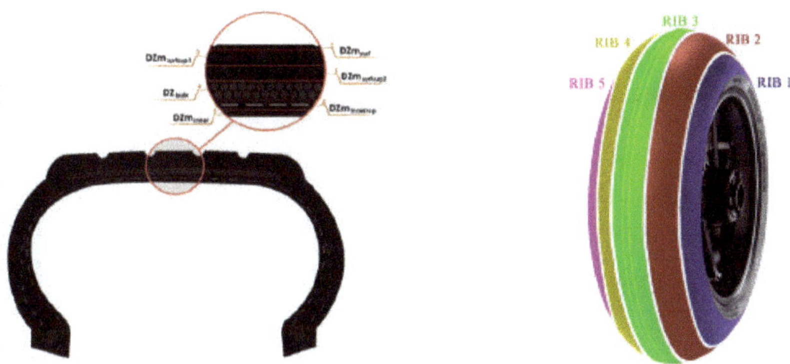

Figure 2. Example of thermoRIDE tire mesh configuration.

The default six layers of the tire structure along the radial direction are the followings:

- Tread surface, which is the most external part of the tread, the one which is in contact with the tarmac and the external air;
- Tread core is just below the surface and it is strictly connected with the grip level the tire is able to provide but it affects also the tire stiffness;
- Tread base is the deepest part of the tread, the last part before the belt; its temperature is more linked to the tire stiffness rather than the grip level;
- Belt is just below the Tread Base and it gives a big contribution to the SEL;
- Plies which is the last layer of the tire structure, it is another important contributor to the SEL, thanks to the energy dissipated by the friction among different plies and within the plies;
- Inner liner is the layer in contact with the inner air, which is not contributing to SEL dissipation neither linked to tire stiffness and grip.

2.2.2. Contact Patch Evaluation

The size and the shape of contact area are obtained by means of specifically developed test procedures, based on the pressure sensitive films or employing more complex models like multibody (MBD) or finished elements (FEA) ones.

Since the characterization procedures are linked to static application of vertical load F_z in different conditions of internal pressure P_i and wheel alignment configuration expressed in terms of camber γ, the contact patch configuration concerning its shape and extension refers to static load conditions in case of the employment of the sensitive film characterization methodology.

However, the instantaneous dynamic contact patch extension and shape can be rather different, because of particular transient conditions of wheel loading, centrifugal effect on the rolling tire, and viscoelastic tire intrinsic characteristics. The adoption of a MBD/FEA tire model able to fit both static and dynamic experimental data, respectively represented by vertical rim lowering and contact patch extension and shape in quasi-static conditions and strain energy loss cycle areas in different load and frequency conditions because of viscous properties of the tire structure, can constitute a valid instrument to implement the contact patch dynamic characteristics within the thermoRIDE model, by means of discretized areas associated to the control volume, as illustrated in Figure 3. Moreover, such models are capable to provide footprint areas depending on wheel travelling velocity or rolling frequency, affecting tire shape because of centrifugal effects.

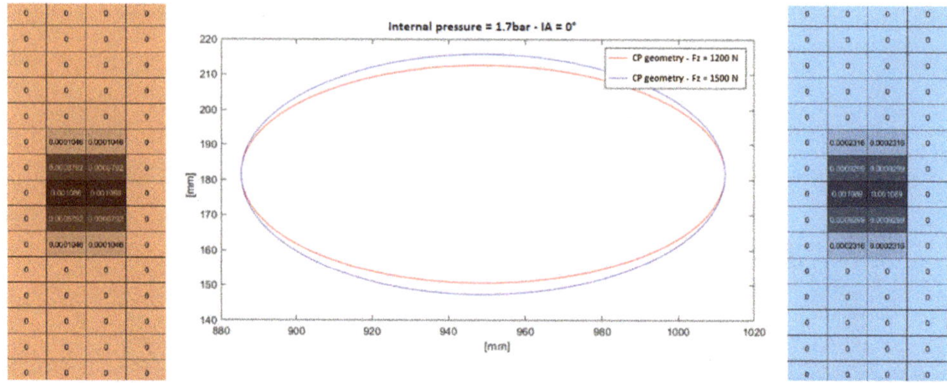

Figure 3. Example of thermoRIDE contact patch geometry representation.

At the current stage, thermoRIDE is able to adopt real contact patch areas from both the experimental activities (static, as said) or the MBD/FEM model outputs (static or dynamic, depending on the model), expressing thanks to parameters identification algorithms the footprint extension as a function of the main variables affecting its shape, as in the following:

$$A_{CP} = f(F_z, P_i, \gamma) \tag{9}$$

2.2.3. Heat Exchange with Road Surface

The thermal conductive exchange between the tread and the asphalt has been modeled through Newton's formula, schematizing the whole phenomenon by means of an appropriate coefficient of heat exchange. The term for such exchanges, for the generic i-th node will be equal to:

$$Q_C = H_c \cdot (T_r - T_i) \cdot \Delta X \cdot \Delta Y \tag{10}$$

where:

- H_c is the convective heat transfer coefficient, estimated for the track testing conditions, in $\frac{W}{m^2 \cdot K}$;
- T_r is the track temperature [K].

The heat generation at the tire-road interface is connected with the thermal power because of the tangential stresses that, in the sliding zone of the contact patch, dissipated in heat. Friction power can be associated directly to the nodes involved in the contact with the ground, and it is calculated as referred to global values of force and sliding velocity, assumed to be equal in the whole contact patch:

$$FP = \frac{F_x \cdot v_x + F_y \cdot v_y}{A} \tag{11}$$

2.2.4. Heat Exchange with External/Inside Air

The whole mechanism of the heat transfer between a generic surface and a moving fluid at different temperatures is described by natural and forced convection equations. The convection heat transfer is expressed by Newton's law of cooling, as before:

$$h_{conv}\cdot\left(T_{fluid} - T_i\right)\cdot\Delta X\cdot\Delta Y \tag{12}$$

Therefore, the heat exchange with the outside air is modeled by the mechanism of forced convection, occurring when there is relative motion between the motorcycle and the air, and by natural convection, when such motion is absent.

Natural convection is also employed to characterize the heat transfer of the inner liner with the inflating gas. The determination of the convection coefficient h, both forced h_{forc} and natural h_{nat}, is based on the classical approach of the dimensionless analysis.

Supposing the tire invested by the air similarly to a cylinder invested transversely from an air flux, the forced convection coefficient is provided by the following formulation [8,9]:

$$h_{forc} = \frac{K_{air}}{L}\cdot\left[0.0239\cdot\left(\frac{V_x\cdot L}{\nu_{air}}\right)^{0.805}\right] \tag{13}$$

in which:

- K_{air} is air conductivity, evaluated at an average temperature between the effective air one and outer tire surface one, in $\left[\frac{\mathrm{W}}{\mathrm{m}\cdot\mathrm{K}}\right]$;
- V_x is considered to be equal to the forward speed, in $\left[\frac{\mathrm{m}}{\mathrm{s}}\right]$;
- ν_{air} is the kinematic viscosity of air, in $\left[\frac{\mathrm{m}^2}{\mathrm{s}}\right]$;
- L is the characteristic length of the heat transfer surface, in [m];
- $T_{m,air}$ is the arithmetic mean between the temperatures of the tire outer surface and the external air in relative motion, in [K].

The values of h_{forc} evaluated with the above approach are close to those obtained by means of CFD simulations for a motorcycle tire, represented in Figure 4.

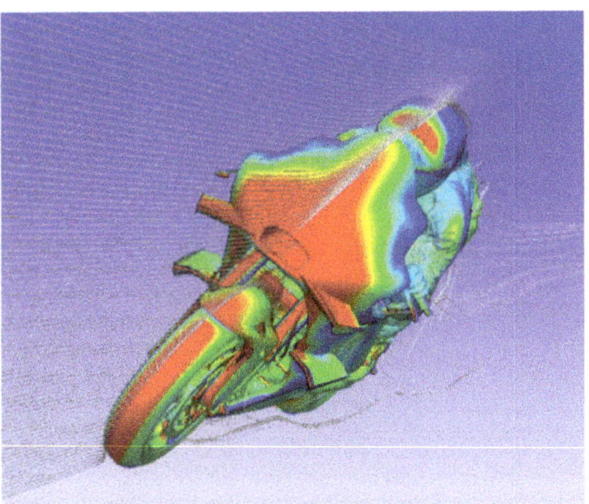

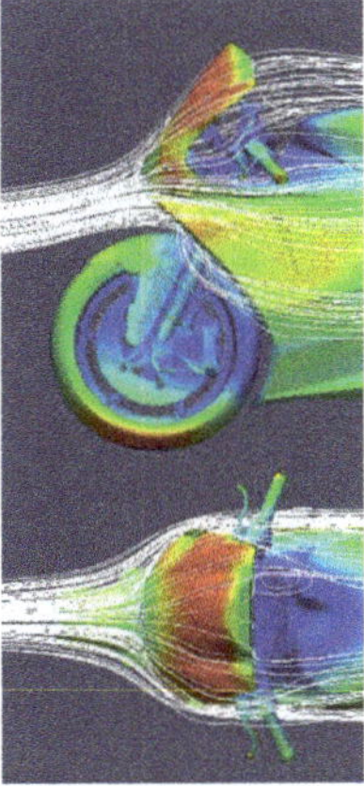

Figure 4. Example of motorcycle Computational Fluid Dynamics (CFD) simulations, useful for both front and rear tires convection parameterization.

The natural convection coefficient h_{nat}, however, can be expressed as:

$$h_{nat} = \frac{Nu \cdot k_{air}}{L} \quad (14)$$

in which, for this case:

$$Nu = 0.53 \cdot Gr^{0.25} \cdot Pr^{0.25} \quad (15)$$

2.2.5. Hysteretic Generative Term

The energy generated by the tire because of cyclic deformations is due to a superposition of several phenomena: intra-plies friction, friction inside singular plies, nonlinear viscoelastic behavior of all rubbery components, etc.

During the rolling, the entire tire is subjected to the cyclic deformations with a frequency corresponding to the tire rotational speed. During the motion, portions of tire, entering in sequence in the contact area, are subjected to deformations, which cause kinetic energy loss and heat dissipation.

The amount of heat generated by deformation (SEL) is estimated through experimental tests carried out deforming cyclically the tire in three directions (radial, longitudinal, and lateral).

Estimated energies do not exactly coincide with the ones dissipated in the actual operative conditions, as the deformation mechanism is different; it is however possible to identify a correlation between them on the basis of coefficients estimated from real data telemetry and from the specifically developed dynamic analysis involving MBD/FEA models.

At the current stage, the empirical SEL formulation is a function of the following parameters and it deeply depends on the tire characteristics:

$$SEL = f(F, \omega, \gamma, P_i) \quad (16)$$

whose further details and trends are available in [21] and where:

- F is the average interaction force at the contact patch, in [N];
- ω is the wheel rotation frequency, in [rad/s];
- γ is the wheel alignment camber angle, in [rad];
- P_i is the gauge pressure within the wheel internal chamber, in [bar].

2.2.6. Model Input/Output Interface

The input data required by the thermoRIDE model consists of the following telemetry channels, as summarized in Table 1.

Table 1. thermoRIDE model inputs.

Physical Quantity	Description
F_z	Vertical interaction force
F_x	Longitudinal interaction force
F_y	Lateral interaction force
v_x	Wheel hub longitudinal velocity
s_r	Slip ratio
s_a	Slip angle
ω	Wheel angular velocity
γ	Inclination angle
T_{air}	Ambient air temperature
T_{road}	Road pavement temperature

Some of these data result from the telemetry measurements available for different tracks and are preliminarily analyzed in order to check their reliability; others, such as in particular the ones related to structural and thermal characteristics of the tire, are estimated on the basis of measurements and tests conducted on the tires.

In addition to tread and inner liner temperature distributions, as reported in Table 2, the model also provides the thermal flows involving the tire, such as the flow due to the external air cooling, the one due to the cooling with the road, the one with the inflation air, as well as the flows due to friction, hysteresis, and exchanges between the different layers.

Table 2. thermoRIDE model outputs.

Physical Quantity	Description
$T_{treadSurf}$	Tread surface temperature
$T_{treadCore}$	Tread core temperature
$T_{treadBase}$	Tread base temperature
$T_{innerLiner}$	Inner liner temperature
$T_{innerAir}$	Internal air temperature
$P_{innerAir}$	Internal air pressure
$W_{extConvection}$	Ambient air convection
$W_{intConvection}$	Chamber air convection
$W_{extConduction}$	Road pavement conduction
$W_{longFriction}$	Longitudinal friction
$W_{latFriction}$	Lateral friction
W_{sel}	Strain Energy Loss

It has to be highlighted that, once per season, it is necessary to carry out an appropriate set of tuning factors to use it in a predictive manner.

The above procedure is linked with the vehicle specific configuration, connected with motorcycle setup and tires construction. Once the tuning phase is completed, known all vehicle data thermoRIDE inputs, the results obtained are in good agreement with the experimental data, with reference to the various operating conditions of the different tracks, allowing to adopt the model for analysis and supporting vehicle design determination.

3. Results

In the Figures 5 and 6, the temperature trends of all the tire layers of the thermodynamic model are illustrated (it must be highlighted that in the above figures the temperature values are dimensionless because of confidentiality agreements with industrial partners). It has to be clarified that the measured data in Figure 5 in red (acquired from external IR sensors) report the temperature of fixed points along the lateral direction of the tire tread, while the simulated ones, in blue, are related to each one of the 15 ribs available, in any moment of the run.

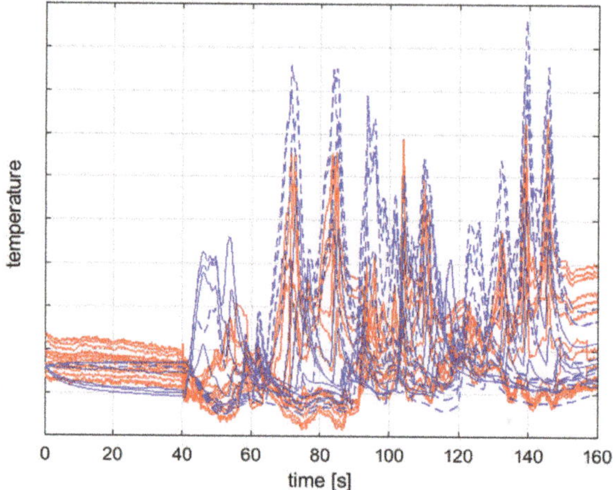

Figure 5. Comparison between the tread external temperatures obtained by means of the 16-ribs' configuration model (in blue) and the acquired ones (in red).

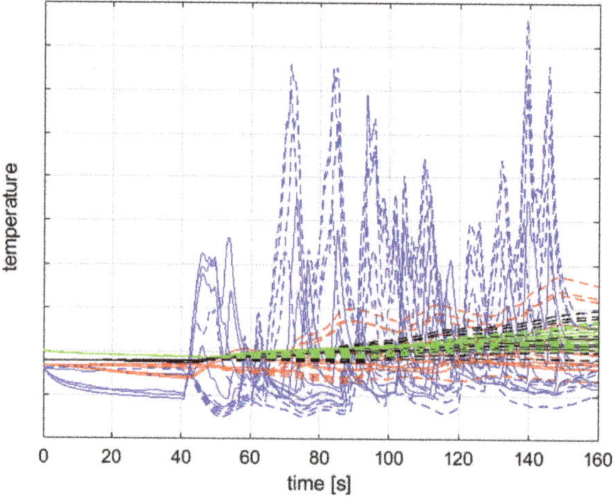

Figure 6. Tire temperatures of all the tire layers obtained by means of the thermoRIDE model: tread surface (in blue), tread core (in red), tread base (in black) and innerliner (in green).

The difference in the thermal shapes of the external layers is due to their position inside the tire structure: the tread layers, especially the surface one, are subjected to the instant thermal powers generated by the tire/road interaction and convective flows; meanwhile a slow temperature trend induced concurrently by the rolling fatigue effect and by the convective heat exchanges characterizes the internal layers' dynamics. That is why, the internal tire layers seem to have a slow temperature ascent during the rolling motion of the wheel, while the tread surface is characterized by an oscillating profile.

The ability to predict the interior temperature distribution, and thus the grip behavior of the tire, is fundamental in terms of the vehicle handling improvement and of the asset optimization according to highly variable outdoor testing conditions.

Vehicle optimal setup is deeply linked with the correct compound working conditions, as highlighted in Figure 7 for a reference automotive tire, not directly linked to the described motorcycle thermal model, but able to show expected physical trends. Such optimization is achievable with a proper suspension layout in terms of stiffness, compliance, and of suitable wheel alignment geometry. In particular, owing to the availability of information on the time by time tread core temperature, plotted on the x-axis, it becomes clear that the tire shows an optimal grip in a specific thermal range that results to be the variable for each different compound, useful to be known in order to set an optimal configuration of the vehicle in any condition.

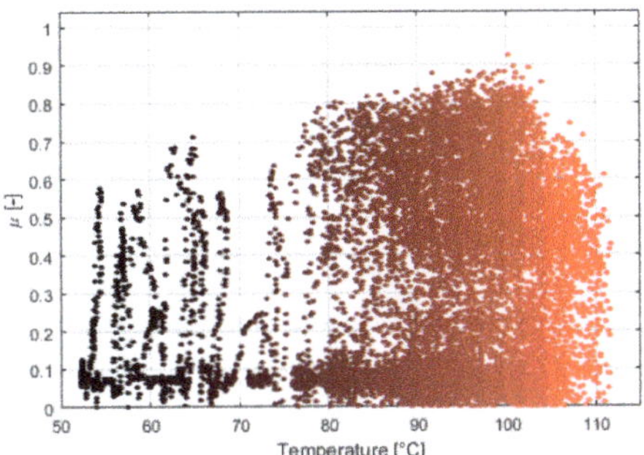

Figure 7. Tread core compound grip-temperature dependence.

Differences in terms of interaction characteristics (Figure 8), in which a considerable stiffness decrease because of a higher tire temperature is clearly appreciable, are shown both for longitudinal and lateral tire interaction curves.

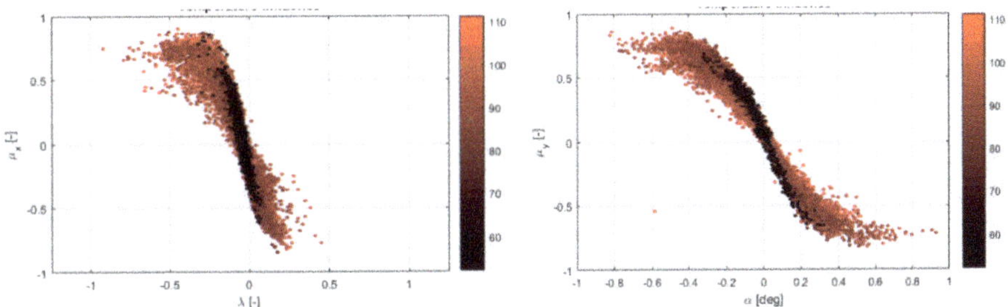

Figure 8. Temperature influence on tire interaction characteristics curves (longitudinal interaction on the left and lateral interaction on the right).

The influence of a large amount of parameters can be evaluated, such as, inclination angle, inflation pressure, different track and weather conditions, and the influence of the manufacturer's vehicle settings. To exploit the entire amount of grip available on the tire-road interface in order to preserve the highest level of handling performance preventing the tire from sliding, more and more physical phenomena regarding the vehicle and its subcomponents have to be taken into account within the integrated vehicle control systems.

4. Conclusions

The paper focuses on the development of a specific version of a tire thermal model for motorcycle applications. The main differences from the common thermodynamic models are related to the particular management of contact patch, responsible for the fundamental friction generation phenomena and for conduction with road, and that in motorcycle tires is characterized by particular elliptical and arched shape, in continuous motion along the lateral direction because of high roll/camber angle.

Such peculiar behavior required a dedicated modeling approach, coupled with the possibility to implement micro-hysteresis differences in each single rib, referred to the possibility to adopt different compounds in the same tire, very common in motorsport.

The results of simulations carried out by coupling the tire thermal model with a vehicle one have been reported, showing a significant good agreement with the experimental results coming from dedicated outdoor acquisitions on an instrumented testing motorcycle. Once validated the model calibration, the main advantages of a predictive and reliable thermal model are the evaluation of the thermal fluxes and temperature interesting the inner layers of the tire structure (deeply linked to grip and stiffness variations, highly influencing vehicle performances and ride/comfort dynamics) and the possibility to study the vehicle setup and aerodynamics, owing to the estimation of the exact amount of energy to be generated/subtracted to let the tire work in the optimal thermal range.

Further development stages are concerning the implementation of the tread wear mechanisms, correlated with the energy provided to the tire lap by lap and deeply influencing the temperature distribution during the race event. Moreover, the real-time availability of knowledge on local temperature, coupled with reliable interaction forces estimation, is a key factor to the realization of MiL (model in the loop), HiL (hardware in the loop), and DiL (driver in the loop) simulation scenarios.

Author Contributions: For research articles with several authors, a short paragraph specifying their individual contributions must be provided. The following statements should be used "conceptualization, F.F., N.M. and F.T.; software, F.F., N.M.; validation, N.M.; data curation, N.M..; writing—original draft preparation, F.F.; writing—review and editing, F.T. All authors have read and agreed to the published version of the manuscript.

Funding: This research received no external funding.

Conflicts of Interest: The authors declare that they have no conflict of interest.

References

1. Cossalter, V.; Lot, R.; Massaro, M. *Modelling, Simulation and Control of Two-Wheeled Vehicles*; John Wiley & Sons: Hoboken, NJ, USA, 2014; pp. 1–42.
2. Gargallo, L.; Radic, D. *Physicochemical Behavior and Supramolecular Organization of Polymers*; Springer Science & Business Media: Berlin/Heidelberg, Germany, 2009; pp. 43–162.
3. Smith, R.H. *Analyzing Friction in the Design of Rubber Products and Their Paired Surfaces*; CRC Press: Boca Raton, FL, USA, 2008.
4. Heinrich, G.; Klüppel, M. Rubber Friction, Tread Deformation and Tire Traction. *Wear* **2008**, *265*, 1052–1060. [CrossRef]
5. Farroni, F.; Russo, R.; Timpone, F. Theoretical and Experimental Estimation of the Hysteretic Component of Friction for a Visco-Elastic Material Sliding on a Rigid Rough Surface. *Int. Rev. Mech. Eng.* **2013**, *6*, 3.
6. Ferry, J.D.; Myers, H.S. Viscoelastic Properties of Polymers. *J. Electrochem. Soc.* **2007**, *41*, 53–62.
7. Radford, D.W.; Fitzhorn, P.A.; Senan, A.; Peterson, M.L. Application of Dynamic Mechanical Analysis to the Evaluation of Tire Compounds. *SAE Technical Pap. Ser.* **2002**, *111*, 2492–2496.
8. Kern, W.J.; Futamura, S. Effect of Tread Polymer Structure on Tyre Performance. *Polymer* **1988**, *29*, 1801–1806. [CrossRef]
9. Mavros, G. A Thermo-Frictional Tyre Model Including the Effect of Flash Temperature. *Veh. Syst. Dyn.* **2019**, *57*, 721–751. [CrossRef]
10. Segers, J. *Analysis Techniques for Racecar Data Acquisition*, 2nd ed.; SAE Technical Paper: Warrendale, PA, USA, 2014.
11. Sharp, R.S.; Gruber, P.; Fina, E. Circuit Racing, Track Texture, Temperature and Rubber Friction. *Veh. Syst. Dyn.* **2016**, *54*, 510–525. [CrossRef]
12. Angrick, C.; van Putten, S.; Prokop, G. Influence of Tire Core and Surface Temperature on Lateral Tire Characteristics. *SAE Int. J. Passeng. Cars Mech. Syst.* **2014**, *7*, 468–481. [CrossRef]
13. Farroni, F.; Sakhnevych, A.; Timpone, F. Physical Modelling of Tire Wear for the Analysis of the Influence of Thermal and Frictional Effects on Vehicle Performance. *J. Mater. Des. Appl.* **2017**, *231*, 151–161. [CrossRef]
14. Kelly, D.P.; Sharp, R.S. Time-Optimal Control of the Race Car: Influence of a Thermodynamic Tyre Model. *Veh. Syst. Dyn.* **2012**, *50*, 641–662. [CrossRef]

15. Ozerem, O.; Morrey, D. A Brush-Based Thermo-Physical Tyre Model and Its Effectiveness in Handling Simulation of a Formula SAE Vehicle. *J. Automob. Eng.* **2019**, *233*, 107–120. [CrossRef]
16. Février, P.; Blanco Hague, O.; Schick, B.; Miquet, C. Advantages of a Thermomechanical Tire Model for Vehicle Dynamics. *ATZ Worldw.* **2010**, *112*, 33–37. [CrossRef]
17. Farroni, F.; Giordano, D.; Russo, M.; Timpone, F. TRT: Thermo Racing Tyre a Physical Model to Predict the Tyre Temperature Distribution. *Meccanica* **2014**, *49*, 707–723. [CrossRef]
18. Wang, Z. Finite Element Analysis of Mechanical and Temperature Field for a Rolling Tire. In Proceedings of the IEEE 2010 International Conference on Measuring Technology and Mechatronics Automation, Changsha, China, 13–14 March 2010; IEEE: Piscataway, NJ, USA.
19. Farroni, F.; Russo, M.; Sakhnevych, A.; Timpone, F. TRT EVO: Advances in Real-Time Thermodynamic Tire Modeling for Vehicle Dynamics Simulations. *J. Automob. Eng.* **2019**, *233*, 121–135. [CrossRef]
20. Cossalter, V.; Doria, A. The Relation between Contact Patch Geometry and the Mechanical Properties of Motorcycle Tyres. *Veh. Syst. Dyn.* **2005**, *43*, 156–164. [CrossRef]
21. Farroni, F.; Rocca, E.; Timpone, F. A Full Scale Test Rig to Characterize Pneumatic Tyre Mechanical Behaviour. *Int. Rev. Mech. Eng.* **2013**, *7*, 841–846.

Article

Truck Handling Stability Simulation and Comparison of Taper-Leaf and Multi-Leaf Spring Suspensions with the Same Vertical Stiffness

Leilei Zhao [1], Yunshan Zhang [2], Yuewei Yu [1,*], Changcheng Zhou [1,*], Xiaohan Li [1] and Hongyan Li [3]

1 School of Transportation and Vehicle Engineering, Shandong University of Technology, Zibo 255000, China; zhaoleilei611571@163.com (L.Z.); xhan_lee@163.com (X.L.)
2 Shandong Automobile Spring Factory Zibo Co.,Ltd., Zibo 255000, China; dyn_keylab@163.com
3 State Key Laboratory of Automotive Simulation and Control, Jilin University, Changchun 130022, China; shanligongdaxue@126.com
* Correspondence: yuyuewei2010@163.com (Y.Y.); greatwall@sdut.edu.cn (C.Z.); Tel.: +86-135-7338-7800 (C.Z.)

Received: 16 December 2019; Accepted: 31 January 2020; Published: 14 February 2020

Abstract: The lightweight design of trucks is of great importance to enhance the load capacity and reduce the production cost. As a result, the taper-leaf spring will gradually replace the multi-leaf spring to become the main elastic element of the suspension for trucks. To reveal the changes of the handling stability after the replacement, the simulations and comparison of the taper-leaf and the multi-leaf spring suspensions with the same vertical stiffness for trucks were conducted. Firstly, to ensure the same comfort of the truck before and after the replacement, an analytical method of replacing the multi-leaf spring with the taper-leaf spring was proposed. Secondly, the effectiveness of the method was verified by the stiffness tests based on a case study. Thirdly, the dynamic models of the taper-leaf spring and the multi-leaf spring with the same vertical stiffness are established and validated, respectively. Based on this, the dynamic models of the truck before and after the replacement were established and verified by the steady static circular test, respectively. Lastly, the handling stability indexes for the truck were compared by the simulations of the drift test, the ramp steer test, and the step steer test. The results show that the yaw rate of the truck almost does not change, the steering wheel moment decreases, the vehicle roll angle obviously increases, and the vehicle side slip angle slightly increases after the replacement. Thus, the truck with the taper-leaf spring suspension has better steering portability, however, its handling stability performs worse.

Keywords: trucks; handling stability; lightweight; taper-leaf spring; suspension; step steer

1. Introduction

The handling stability is one of the extremely important performances that affect the driving safety of trucks. How to obtain the strong handling stability for improving the driving safety is an important issue in vehicle design [1–3]. The suspension is an important system connecting the truck frame and wheel [4]. The leaf spring is the most widely used elastic element in the suspension system for trucks. Moreover, it plays a crucial role in the handling stability for trucks.

At present, the lightweight design of trucks is of great importance to enhance the load capacity and reduce the production cost [5–7]. For leaf springs, the lightweight design of trucks is mainly reflected in replacing the multi-leaf spring with the taper-leaf spring, so as to reduce the trucks' weight, to improve the power performance, and to reduce the fuel consumption and the exhaust pollution [8–10]. In modern society, with the development of people's living standard, people have higher and higher requirements for the handling stability of trucks [11–13]. However, the changes of the handling stability after the replacement have not been revealed.

Due to the advantages of light weight and low noise, the taper-leaf spring is more and more used in trucks. At present, scholars mainly focus on its dynamic property, stress, and strain. Moreover, most of the studies of its mechanical properties mainly were conducted by using simulation software packages, such as Ansys, Nastran, Abaqus, and Adams. Duan et al. created the dynamic model of the tandem suspension equipped with the taper leaf spring for trucks based on Adams software [14]. In order to research the stress of a taper leaf spring, Moon et al. established a flexible multi-body dynamic model [15]. Wang et al. proposed a calculation method of the stiffness of the taper-leaf spring based on the combine superposition method and the finite difference method [16]. Zhou et al. analyzed the mechanical properties of the taper leaf spring considering the friction between the leaf springs on the basis of the FE (Finite Element) contact analysis [17]. To improve the kinematics characteristics of a midsize truck, Kim et al. selected the optimal combination parameters based on a vehicle model with a taper leaf spring [18].

Some scholars improved the performance of leaf springs from the perspective of materials. For example, Chandra et al. researched the high-temperature quality of the accelerated spheroidization on SUP9 leaf spring and the machining performance of Sup9 leaf spring can be significantly improved under high temperature [19]. Fragoudakis et al. optimized the development of 56SiCr7 leaf springs and micro-hardness measurements show surface degradation effects [20]. Kumar et al. optimized the key design parameters of EN45A flat leaf spring and the developed leaf spring program can be used to optimize various parameters of the leaf spring quickly and reliably [21]. Jenarthanan proposed carbon/glass epoxy composite as a leaf spring material and analyzed the leaf spring by using Ansys software [22]. Ozmen et al. proposed a new method on the basis of testing and simulation for the durability of leaf springs [23]. In their study, the finite element method and the multi-body simulation were used to calculate the fatigue life. Some scholars researched the fatigue life of leaf springs. For example, Duruş et al. proposed a method to predict the fatigue life of Z Type leaf spring and created an approach to validate the proposed method [24]. Bakir et al. researched the correlation of simulation, test bench, and rough road testing in terms of strength and fatigue life of a leaf spring [25]. Kong et al. conducted the failure evaluation of a leaf spring eye design under various load cases [26] and this study provides a valuable reference for preventing the failure of leaf spring in engineering design. Bi et al. carried out the fatigue analysis of leaf-spring pivots [27]. Malikoutsakis proposed the design and optimization procedure for parabolic leaf springs and made a multi-disciplinary optimization of the high performance front leaf springs [28]. In addition, some scholars researched the influences of the taper-leaf spring on the vehicle performances. For instance, Liu et al. analyzed the effects of the taper-leaf spring on the vehicle braking property, simulated the motion characteristics of the front less leaf spring suspension system and the motion law of the front axle jumping up with the wheel [29]. Liu et al. researched the main leaf center trajectory of the taper leaf spring and analyzed the suspension kinematics, and unreasonable toe-in angle was solved by optimizing the hard point of the plumbing arm [30].

In addition, some scholars focus on the influence of suspension designs on handling performance. Termous et al. a proposed coordinated control strategy to control the roll dynamics on the basis of active suspension systems [31]. Li et al. applied the hydraulically interconnected suspension to an articulated vehicle and proved that it can effectively improve the vehicle handling performance [32]. In order to improve the handling stability, Zhang et al. carried out the multi-objective optimization design of suspensions for electric vehicles [33]. Bagheri et al. carried out a multi-objective optimization on the double wishbone suspension to improve the vehicle handling stability [34]. A linear mathematical model and MATLAB model which included suspension K and C characteristics parameters were established by Li et al. [35]. Moreover, the accuracy of the mathematical model was validated. Ahmadian et al. discussed an application of magneto-rheological (MR) suspensions about vehicle handling stability and their results show that using MR suspensions can increase the speed at which the onset of hunting occurs as much as 50% to more than 300% [36].

The above-mentioned studies provide theoretical guidance for the application and promotion of the taper-leaf spring in vehicles. They also provide a useful reference for the theoretical research of the taper-leaf spring. They mainly focus on the mechanical properties of the taper-leaf itself. There is no research focusing on the changes of the handling stability after replacing the multi-leaf spring with the taper-leaf spring for trucks. The study of the changes will help to optimize the design of the suspension system and improve the handling stability for trucks. Thus, the changes need to be further researched.

The aim of this paper is to reveal the changes of the handling stability after replacing the multi-leaf spring with the taper-leaf spring for trucks. The main contributions of this paper are as follows: (1) An analytical method of replacing the multi-leaf spring with the taper-leaf spring was proposed and validated by test. (2) The dynamic models of the truck before and after the spring replacement were established and verified by tests, respectively. (3) The changes of the handling stability after replacing the multi-leaf spring with the taper-leaf spring were revealed by the simulations of the drift test, the ramp steer test, and the step steer test.

2. Modeling of the Taper-Leaf and the Multi-Leaf Spring

2.1. The Mechanical Model of the Multi-Leaf Spring

In our previous research [37], the mechanical model of the multi-leaf spring was created and validated by test, as shown in Figure 1. The half of the multi-leaf spring can be regarded as an elastic beam. It is fixed at one end. A concentrated load F is applied at the other end. It is assumed that under F, each piece does not separate from each other and two adjacent pieces have the same deflection at the contact point. w is the end displacement of the multi-leaf spring. n is the number of slices for the multi-leaf spring. b is the width of each piece. The length parameters of half of the spring pieces are L_1, $L_2, L_3, \ldots, L_{n-1}, L_n$ from large to small, respectively. Correspondingly, the thickness parameters of the half of the spring pieces are $h_1, h_2, h_3, \ldots, h_{n-1}, h_n$, respectively.

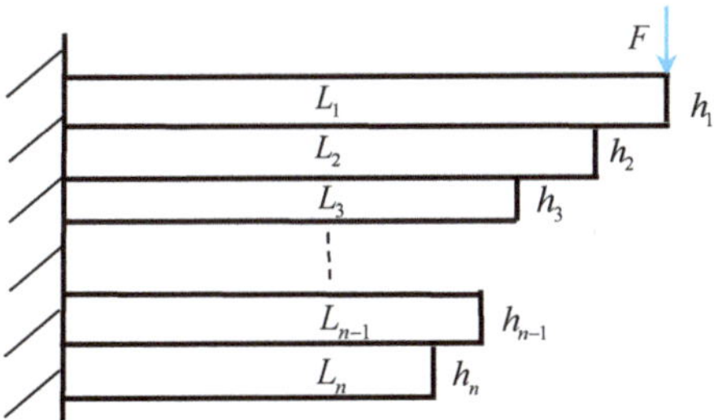

Figure 1. The mechanical model of the multi-leaf spring.

Based on the mechanical model in Figure 1, the stiffness of the half of the multi-leaf spring can be expressed as [37]:

$$K = \frac{bE}{4} / [\sum_{i=2}^{n} \frac{(L_1 - L_i)^3 - (L_1 - L_{i-1})^3}{h_1^3 + h_2^3 + \ldots + h_{i-1}^3} + \frac{L_1^3 - (L_1 - L_n)^3}{h_1^3 + h_2^3 + \ldots + h_{n-1}^3 + h_n^3}], \tag{1}$$

where E is the elastic modulus.

2.2. The Mechanical Model of the Single Taper-Leaf Spring

The half of the single taper-leaf spring is taken as the research object. Its mechanical model is shown in Figure 2. Similarly, the half of the single taper-leaf spring also can be regarded as an elastic beam. It is fixed at one end. A concentrated load F is applied at the other end. The geometric parameters and the coordinate system are marked in Figure 2. L is the half length of the single taper-leaf spring; h_2 is the thickness for $x \in [l_2, L]$; h_1 is the thickness for $x \in [0, l_1]$. $h(x)$ is the thickness for $x \in [l_1, l_2]$. γ is defined as the thickness ratio and $\gamma = h_1/h_2$.

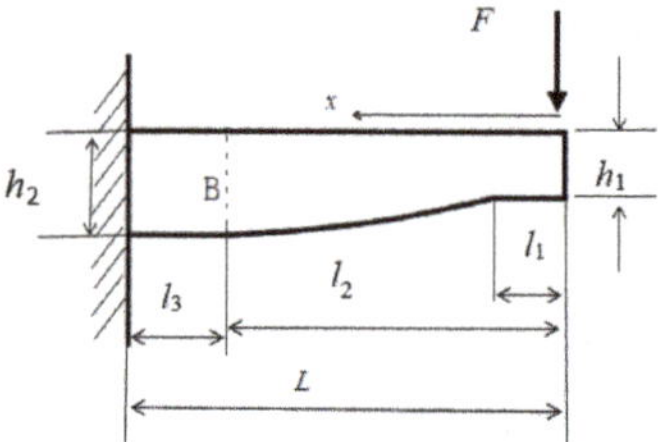

Figure 2. The mechanical model of the single taper-leaf spring.

Based on the model in Figure 2, when the load F is applied at the free end, the same normal stress at any position for $x \in [l_1, l_2]$ is [9]:

$$\sigma_x = \frac{6Fx}{bh^2(x)} \tag{2}$$

From Equation (2), the normal stress σ_B of Section B for $x = l_2$ can be expressed as:

$$\sigma_B = \frac{6Fl_2}{bh_2^2} \tag{3}$$

In order to meet the requirements of the equal stress at Section B for $x = l_2$, $\sigma_x = \sigma_B$ must be satisfied. According to Equations (2) and (3), $h(x)$ can be expressed as:

$$h(x) = h_2\sqrt{\frac{x}{l_2}} \tag{4}$$

According to Equation (4), h_1 can be expressed as:

$$h_1 = h_2\sqrt{\frac{l_1}{l_2}} \tag{5}$$

When the load F is applied at the free end, the deformation energy U can be expressed as:

$$U = \int_L \frac{(Fx)^2}{2EI}\mathrm{d}x = \int_0^{l_1} \frac{(Fx)^2}{2EI_1}\mathrm{d}x + \int_{l_1}^{l_2} \frac{(Fx)^2}{2EI_2}\mathrm{d}x + \int_{l_2}^{L} \frac{(Fx)^2}{2EI_3}\mathrm{d}x \tag{6}$$

where I_1, I_2, I_3 are the inertia moments at different thicknesses and $I_1 = \frac{bh_1{}^3}{12}$, $I_2 = \frac{bh^3(x)}{12}$, $I_3 = \frac{bh_2{}^3}{12}$, respectively.

According to the Castigliano Second Theorem [37], the end deformation of the spring can be expressed as:

$$y = \frac{\partial U}{\partial F} = \frac{4F\left[L^3 + l_2^3\left(1-\gamma^3\right)\right]}{Ebh_2^3} \tag{7}$$

Based on Equation (7), the stiffness of the half of the single taper-leaf spring can be expressed as:

$$K_s = \frac{Ebh_2^3}{4\left[L^3 + l_2^3\left(1-\gamma^3\right)\right]} \tag{8}$$

According to Equation (8), the design formula of the root thickness can be expressed as:

$$h_2 = \sqrt[3]{\frac{4K_s\left[L^3 + l_2^3(1-\gamma^3)\right]}{Eb}} \quad (9)$$

2.3. The Mechanical Model of the Taper-Leaf Spring Including n Pieces

The half of the taper-leaf spring including n pieces ($n = 2$ or 3) is taken as the research object. Its mechanical model is shown in Figure 3. Similarly, the half of the single taper-leaf spring also can be regarded as an elastic beam. It is fixed at one end. A concentrated load F is applied at the other end. The geometric parameters and the coordinate system are marked in Figure 3. The end displacement of the taper-leaf spring is y. γ_i is defined as the thickness ratio and $\gamma_i = h_{i1}/h_{i2}$. Moreover, all the design values of γ_i are the same in practical engineering application and $\gamma_i = \gamma$.

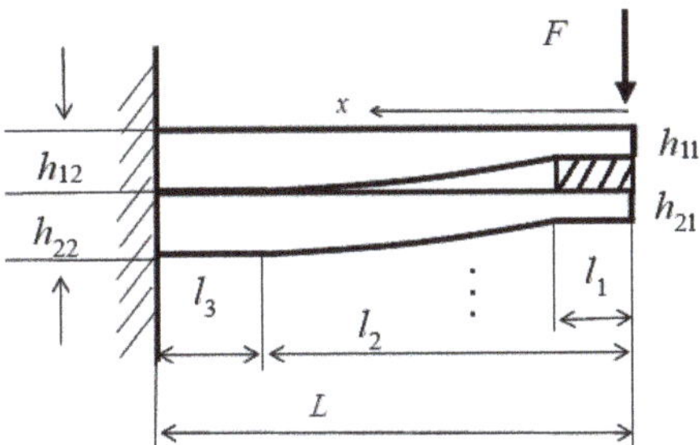

Figure 3. The mechanical model of the taper-leaf spring including n pieces.

Under the load F, each piece does not separate from each other at the free end. The free ends of the n pieces for the taper-leaf spring are equivalent to a whole body and the vertical displacement occurs under the load F. Thus, each piece has the same vertical displacement y at the free end, that is:

$$y_i = y \quad (10)$$

where y_i the displacement of the ith piece at the free end ($i = 1, 2$ or $i = 1, 2, 3$).

According to Equation (7), y_i can be expressed as:

$$y_i = \frac{4F_i\left[L^3 + l_2^3(1-\gamma^3)\right]}{Ebh_{i2}^3} \quad (11)$$

where F_i is the load shared by the ith piece.

Substituting Equation (11) into Equation (10), obtain the following:

$$\frac{F_1}{h_{12}^3} = \frac{F_2}{h_{22}^3} = \cdots = \frac{F_i}{h_{i2}^3} = \cdots = \frac{F_n}{h_{n2}^3} = \frac{F}{h_{e2}^3} \quad (12)$$

where h_{e2} is the equivalent thickness.

According to the model in Figure 3, the sum of the concentrated forces is equal to the concentrated load F, that is:

$$F_1 + F_2 + \cdots + F_n = F \quad (13)$$

According to Equations (12) and (13), h_{e2} can be calculated by:

$$h_{e2} = \sqrt[3]{h_{12}^3 + h_{22}^3 + \cdots + h_{n2}^3} \quad (14)$$

According to Equation (12), F_i can be calculated by:

$$F_i = \frac{h_{i2}^3}{h_{e2}^3} F \quad (15)$$

Under the load F, the maximum normal stress of the ith piece can be expressed as:

$$\sigma_{i\max} = \frac{6F_i L}{b h_{i2}^2} \quad (16)$$

Substituting Equation (15) into Equation (16), obtain the following:

$$\sigma_{i\max} = \frac{6FLh_{i2}}{b h_{e2}^3} \quad (17)$$

3. Analytical Method of Replacing the Multi-Leaf Spring with the Taper-Leaf Spring

This paper mainly research the changes of the handling stability after replacing the multi-leaf spring with the taper-leaf spring for trucks by comparison. Therefore, it is necessary to ensure that the two leaf springs have the same vertical stiffness. Only on this premise can the conclusions of the comparison be valuable. In addition, the design of the taper-leaf spring must meet the stress condition. In this section, an analytical method of replacing the multi-leaf spring with the taper-leaf spring under the precondition of the same vertical stiffness was proposed.

3.1. The Analytical Design Method of the Taper-Leaf Spring

For the taper-leaf spring, $\sigma_{i\max}$ must be not larger than the allowable stress $[\sigma]$. Thus, based on $\sigma_{i\max} \leq [\sigma]$ and Equation (17), the root allowable thickness of the ith piece can be expressed as:

$$[h_{i2}] = \frac{b h_{e2}^3 [\sigma]}{6FL} \quad (18)$$

According to Equation (14), when $h_{12} = h_{22} = \cdots = h_{n2}$, obtain the following:

$$h_{e2}^3 = n h_{i2}^3 \quad (19)$$

According to Equation (19), the number n of the spring pieces can be expressed as:

$$n = \frac{h_{e2}^3}{\left[h_{i2}^3\right]} \quad (20)$$

Based on Equation (20), rounding up n, obtain the number $[n]$ of the spring pieces meeting stress requirements.

Substituting $[n]$ into Equation (20), the design value of the root thickness of the ith piece can be calculated by:

$$h_{i2} = \frac{h_{e2}}{\sqrt[3]{[n]}} \quad (21)$$

According to Equation (5) and the definition of the thickness ratio γ, the design value of l_1 can be calculated by:

$$l_1 = l_2 \gamma^2 \quad (22)$$

According to the definition of the thickness ratio γ and h_{i2}, h_{i1} can be designed as:

$$h_{i1} = \gamma h_{i2} \quad (23)$$

3.2. The Design Flow

The design flow of the analytical method of replacing the multi-leaf spring with the taper-leaf spring under the precondition of the same vertical stiffness can be summarized as follows:

Step 1. By using Equation (1), calculate the stiffness K of the multi-leaf spring of the original vehicle.

Step 2. Let $K_s = K$ and by using Equation (9), calculate the equivalent thickness h_{e2}.

Step 3. Substituting h_{e2} into Equation (18), obtain the root allowable thickness $[h_{i2}]$.

Step 4. Substituting h_{e2} and $[h_{i2}]$ into Equation (20), rounding up n, obtain the number $[n]$ of the spring pieces meeting stress requirements.

Step 5. Based on $[n]$, by using Equation (21), obtain the root thickness h_{i2} of the ith piece.

Step 6. Based on γ and l_2, by using Equation (22), obtain the length l_1 of the end segment.

Step 7. Based on γ and h_{i2}, by using Equation (23), obtain the end thickness h_{i1} of the ith piece.

3.3. Case Study

A light truck was selected as the research object. Its rear suspension is equipped with the multi-leaf spring. Moreover, the number of slices for the multi-leaf spring is five. The specific parameters are as follows: b = 63.0 mm, h = 8.0 mm, L_1 = 525.0 mm, L_2 = 450.0 mm, L_3 = 350.0 mm, L_4 = 250.0 mm, L_5 = 150.0 mm, E = 206 GPa, $[\sigma]$ = 750.0 Mpa. In order to meet the requirements of the spring installation, l_3 was designed as 50 mm for the taper-leaf spring. According to $l_2 = L_1 - l_3$, l_2 was designed as 475 mm. Based on the analytical method of replacing the multi-leaf spring with the taper-leaf spring, the other design parameters values were determined. The key design parameters values of the taper-leaf spring are shown in Table 1. The prototype of the taper-leaf spring manufactured by Shandong Automobile Spring Factory is shown in Figure 4.

Table 1. The design parameters values.

Parameter	Value	Unit	Parameter	Value	Unit
n	3	pieces	h_{11}	5.6	mm
σ_{max}	460.2	MPa	h_{21}	5.6	mm
b	63.0	mm	h_{31}	5.6	mm
l_1	123.5	mm	h_{12}	11.0	mm
l_2	475.0	mm	h_{22}	11.0	mm
l_3	50.0	mm	h_{32}	11.0	mm

Figure 4. The prototype of the taper-leaf spring.

4. Dynamic Modeling and Test Verification

4.1. Dynamic Modeling and Verification of Leaf Springs

Adams is recognized as a professional software for the vehicle dynamics modeling. The dynamic models based on the software can truly reflect the dynamic performance of vehicles [38,39]. Adams/Leaf tool uses discrete beam elements to simulate leaf springs. In other words, each piece of leaf springs is divided into several sections, which are connected by flexible beams without mass. The flexible beam is a kind of flexible connection. The force and moment between two marked points of two elements are calculated by the theory of Timoshenko beam [40]. In this study, according to the structural parameters values of the multi-leaf spring and the taper-leaf spring, the simulation models were established based on Adams/Leaftool software, as shown in Figure 5. Because the friction force between the adjacent pieces is very small [37], it is ignored in the simulation models. Moreover, the previous tests in [37] show that the load-stiffness characteristics for multi-leaf springs and taper-leaf springs are almost linear. Thus, their vertical stiffness and the torsional stiffness are often used to verify the model accuracy [9].

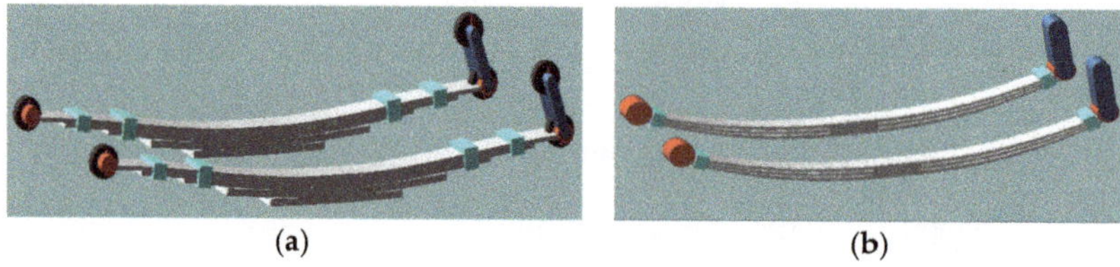

(a) (b)

Figure 5. The simulation models: (**a**) The multi-leaf spring; (**b**) the taper-leaf spring.

To verify the accuracy of the models, the vertical and torsional stiffness tests were conducted. The test equipment is the TYE-W400I hydraulic testing machine produced by Jinan Shidai Shijin Group, as shown in Figure 6. It uses the programmable logic controller. It can measure the load and the displacement through sensors. The load was measured by a NS-WL1 tension–compression sensor. The displacement was measured by a SDVG20 displacement sensor. The control system automatically collected and processed the test force and the deformation of leaf springs, and directly output the stiffness value through the microcomputer display screen. For the models in Figure 5, the vertical forces were applied to simulate the deformation, respectively. The simulated stiffness values from the simulation results were calculated. A comparison of the stiffness values is shown in Table 2.

Figure 6. The test equipment.

Table 2. A comparison of the stiffness values.

Type	**Vertical**			**Torsional**		
	Test (kN/m)	**Simulation (N/m)**	**Error (%)**	**Test (kNm/rad)**	**Simulation (kNm/rad)**	**Error (%)**
The multi-leaf spring	112.3	109.0	2.9	10.6	10.1	4.7
The taper-leaf spring	113.1	109.2	3.5	8.4	8.2	2.4

From Table 2, it can be seen that the vertical stiffness values of the multi-leaf spring and the taper-leaf spring are almost the same. The result shows that the taper-leaf spring can equivalently replace the original multi-leaf spring. From Table 2, it also can be seen that the torsional stiffness of the multi-leaf spring is obviously larger than that of taper-leaf spring. Moreover, the test values are close to the simulation values. The results show that the simulation models can effectively reflect the stiffness characteristics of the multi-leaf spring and the taper-leaf spring, respectively. In addition, the experimental results indicate that the weight of the taper-leaf spring is less 13.4% than that of the multi-leaf spring.

4.2. Dynamic Modeling of the Truck

The parameters values of the light truck are shown in Table 3. The geometric model of the light truck is shown in Figure 7a. The modeling process is as follows: firstly, the geometric model is imported into Adams through the "parasolid" data format; then, system constraints, loads, and drives are applied in Adams; finally, the simulation models of the multi-leaf spring and the taper-leaf spring are imported and assembled to establish the vehicle dynamic models, respectively. The dynamic model of the truck with the multi-leaf spring is shown in Figure 7b. The tire adopts the PAC2002 tire model. This tire model adopts the magic formula, which is suitable for the simulation analysis of stability with high accuracy [41,42]. The steering system is equipped with rack-and-pinion steering gear and the angular ratio of steering gear is 18.0.

Table 3. The parameters values of the light truck.

Parameter	Value	Parameter	Value
The unsprung mass for the front wheel /(kg)	46.4	The differential ratio	4.778
The front axle load /(kg)	884	The mass of the drive shaft /(kg)	8.3
The rear axle load /(kg)	1641	The arm length of the stabilizer /(mm)	204
The wheelbase/(m)	2.6	The bar length of the stabilizer /(mm)	910
The front tread /(m)	1.32	The front wheel radius /(m)	0.297
The rear tread /(m)	1.41	The rear wheel radius /(m)	0.297
The stiffness of the front suspension /(N/mm)	55	The front wheel stiffness /(kNm^{-1})	383.4
The unsprung mass for the rear wheel /(kg)	83	The rear wheel stiffness /(kNm^{-1})	383.4
The angular ratio of steering gear	18.0		

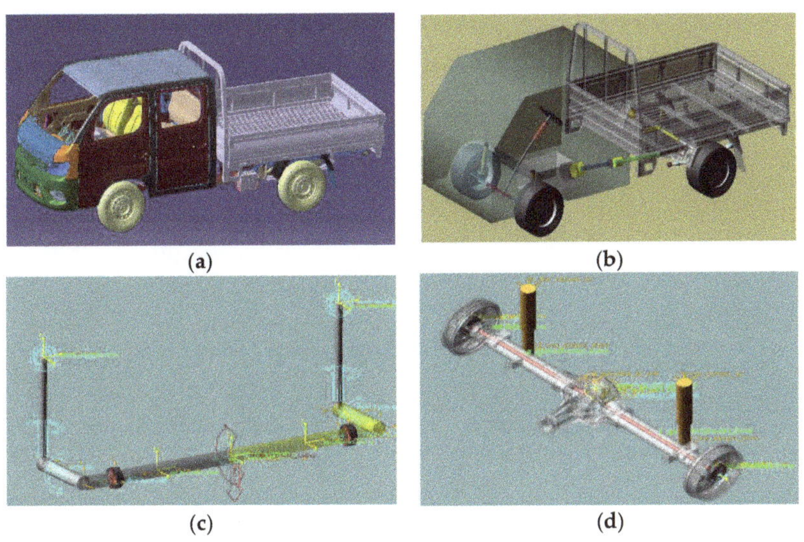

Figure 7. The light truck models: (**a**) The geometric model, (**b**) the dynamic model, (**c**) the sub-model of the stabilizer system, and (**d**) the sub-model of the rear axle with dampers.

4.3. Steady Static Circular Test Verification

Based on the standard GB/T6323.6-1994, the steady static circular tests were conducted. The specific test method is as follows: firstly, fix the steering wheel angle and the starting circle radius is 15.0 m; then, accelerate slowly at the longitudinal acceleration below 0.25 m/s^2 until the lateral acceleration reaches 6.5 m/s^2. Based on the two truck dynamic models, the steady static circular test were carried out, respectively. The simulation settings are the same as the real test conditions. A comparison of the curves of the body roll angle versus the lateral acceleration is shown in Figure 8. α_1

and α_2 are the absolute values of the sideslip angles of the front axle and the rear axle, respectively. In order to facilitate the test and analysis, researchers usually choose evaluation indicators to describe and evaluate the steady-state response of the vehicle according to their own habits. $\alpha_1 - \alpha_2$ is one of the most important evaluation indicators. If $\alpha_1 - \alpha_2 > 0$, the vehicle has under-steer characteristics; if $\alpha_1 - \alpha_2 = 0$, the vehicle has neutral steering characteristics; If $\alpha_1 - \alpha_2 < 0$, the vehicle has over-steer characteristics [43]. A comparison of the curves of $\alpha_1 - \alpha_2$ versus the lateral acceleration is shown in Figure 9. From Figure 9, it can be seen that the simulation results are close to the test results, which proves that the two truck models can reproduce the handling dynamic responses. Figure 9 also shows that the truck has slight over-steer characteristics.

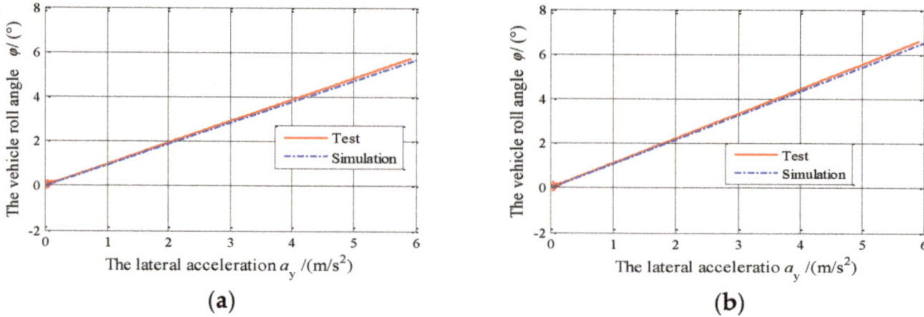

Figure 8. A comparison of the curves of the body roll angle versus the lateral acceleration for the truck with: (**a**) the multi-leaf spring, and (**b**) the taper-leaf spring.

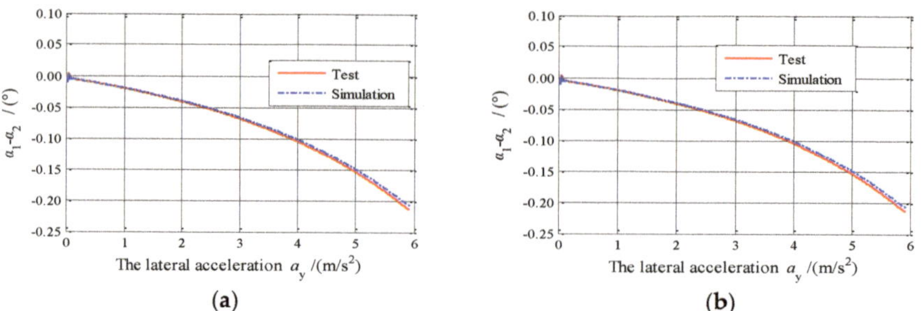

Figure 9. A comparison of the curves of $\alpha_1 - \alpha_2$ versus the lateral acceleration for the truck with: (**a**) The multi-leaf spring, and (**b**) the taper-leaf spring.

5. Handling Stability Simulation and Comparison

In general, handling stability tests of vehicles are very dangerous, such as the drift test [43,44]. In order to avoid the potential dangers of handling stability tests, this paper uses simulation methods to carry out the comparative analysis. The changes of the handling stability after replacing the multi-leaf spring with the taper-leaf spring were revealed by the simulations of the drift test, the ramp steer test, and the step steer test.

5.1. Simulation and Comparison of the Drift Tests

The drift test is an important method to study the transient responses of vehicles under the limit operating condition. In the drift simulation, the truck reaches a steady-state condition in the first 10.0 s. A steady-state condition is one in which the truck has the desired steer angle 360°, the initial throttle 20.0, and the initial velocity 40 km/h. In 1.0–4.0 s, Adams ramps the steering angle from an initial value to the desired value. It then ramps the throttle from 0.0 to the full throttle value 100.0 from 5.0 s to

10.0 s. Finally, it keeps the throttle fully open from 10.0 s to 15.0 s. The same simulation parameters were set for the two truck models with the multi-leaf spring and the taper-leaf spring, respectively. A comparison of the simulated dynamic responses is shown in Figure 10.

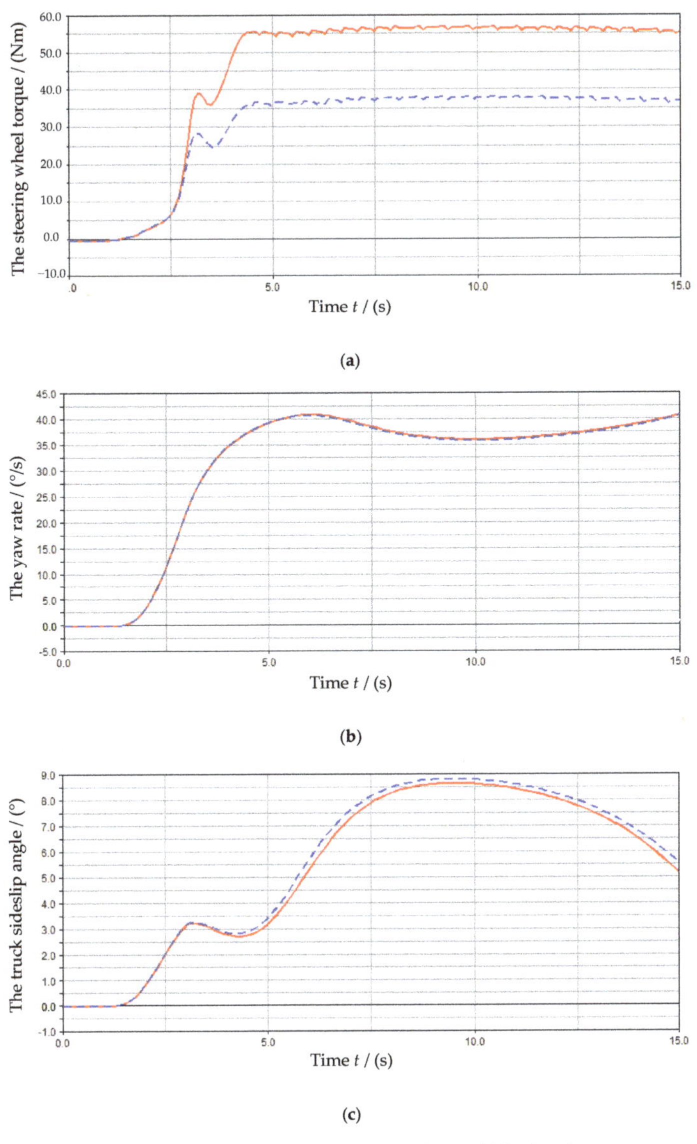

Figure 10. A comparison of the simulated dynamic responses obtained from the drift tests: (**a**) The steering wheel torque, (**b**) the yaw rate, and (**c**) the truck sideslip angle.

From Figure 10a, it can be seen that for the same input, when the steering angle reaches the set value, the steering wheel torque of the truck with the multi-leaf spring suspension is significantly larger than that of the truck with the taper-leaf spring suspension. The result shows that the truck with the taper-leaf spring suspension has better steering portability than the truck with the multi-leaf spring suspension. From Figure 10b, it can be seen that the two truck models have almost the same yaw rate. The result shows that replacing the multi-leaf spring with the taper-leaf spring has almost no effect on the yaw rate. From Figure 10c, it can be seen that when the steering wheel angle is increased to the set value, the sideslip angles on the basis of the two truck models are almost the same. With the increase of the throttle opening, the sideslip angles on the basis of the two truck models begin to deviate. The sideslip angle of the truck with the taper-leaf spring suspension is obviously larger than that of the truck with the multi-leaf spring suspension. The result shows that the handling of the truck with the taper-leaf spring suspension is poorer than that of the truck with the multi-leaf spring suspension. The main reason should be related to that the torsional stiffness of the taper-leaf spring is less 18.8% than that of the multi-leaf spring.

5.2. Simulation and Comparison of the Ramp Steer Tests

In a ramp-steer analysis, time-domain transient response metrics can be obtained [43]. The most important quantities to be measured are: the steering wheel torque, the yaw rate, the truck sideslip angle, and the truck roll angle. During a ramp-steer analysis, Adams ramps up the steering input from an initial value at a specified rate, as shown in Figure 11. The simulations of the ramp steer tests were conducted in Adams based on the two truck models, respectively. The simulation settings are as follows: the initial steer value = 0°, the initial velocity 40 km/h, the ramp = tan(θ) = 15.0, the start time = 3.0 s, and the end time = 15.0 s. A comparison of the simulated dynamic responses is shown in Figure 12.

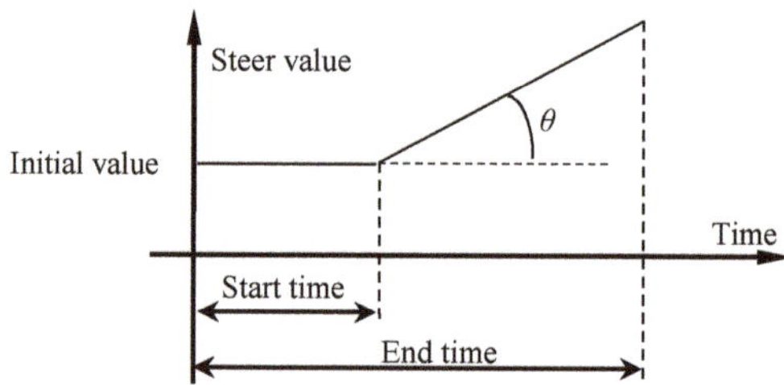

Figure 11. The curve of the input versus the time for the ramp steer test.

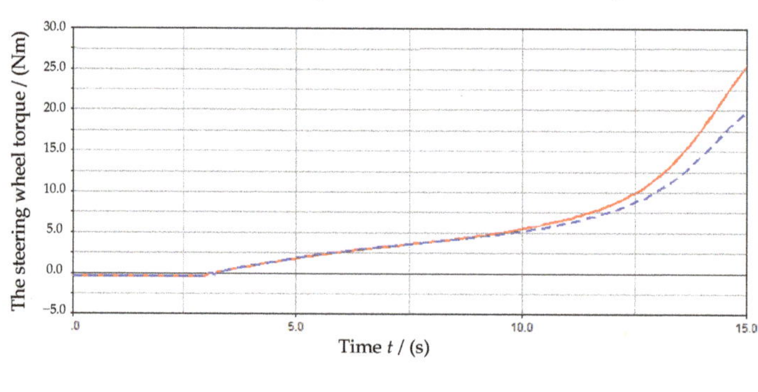

(a)

Figure 12. *Cont.*

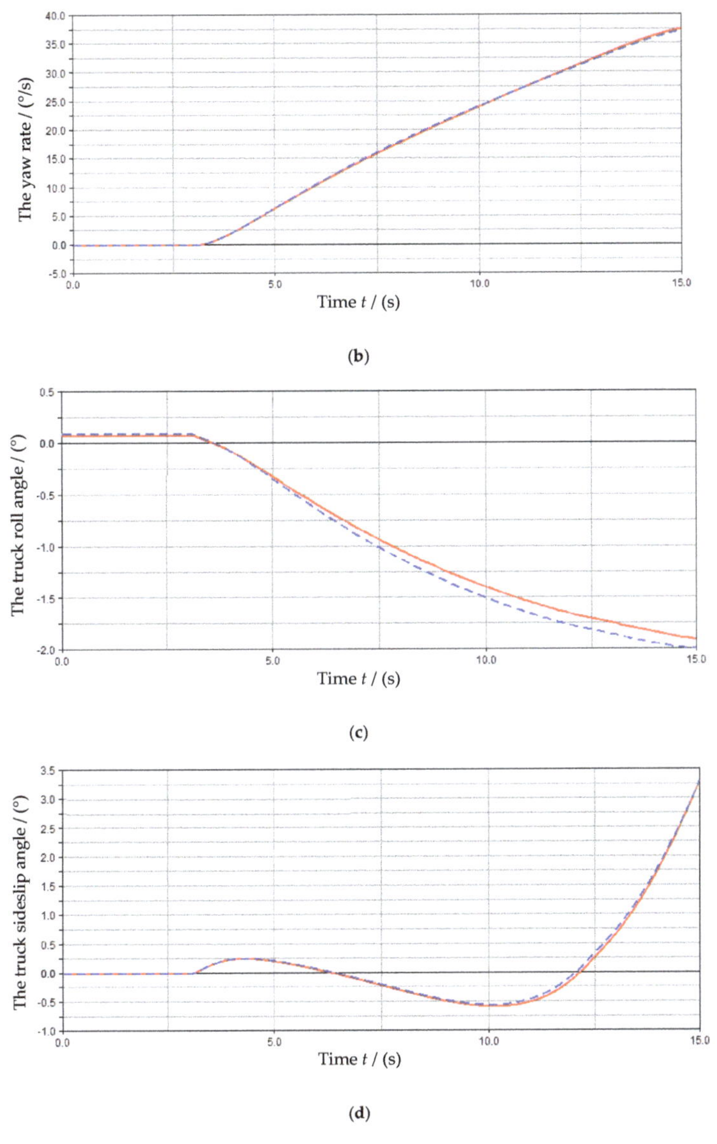

Figure 12. A comparison of the simulated dynamic responses obtained from the ramp steer tests: (**a**) The steering wheel torque, (**b**) the yaw rate, (**c**) the truck roll angle, and (**d**) the truck sideslip angle.

From Figure 12a, it can be seen that the steering wheel torques from the two truck models are almost the same from the beginning to about 10.0 s. Moreover, in the later stage, with the increasing of the steer angle, the steering wheel torque of the multi-leaf spring truck model increases more obviously. In other words, when the steer angle is very small, the two truck models show almost the same steering portability. However, when the steering angle is very large, the truck with the taper-leaf spring suspension has better steering portability. From Figure 12b, it can be seen that the two truck models show almost the same yaw rate. The result shows that replacing the multi-leaf spring with the taper-leaf spring has almost no effect on the yaw rate. From Figure 12c, it can be seen that the truck roll angles on the basis of the two truck models are almost the same from 0.0 s to 5.0 s. With the increase of the steer value, the truck roll angles on the basis of the two truck models begin to deviate. The truck roll angle of the truck with the taper-leaf spring suspension is obviously larger than that of the truck with the multi-leaf spring suspension. The results show that the truck with the multi-leaf spring suspension has the stronger stability. From Figure 12d, it can be seen that the two truck models show almost the same sideslip angle. The result shows that replacing the multi-leaf spring with the taper-leaf spring has almost no effect on the truck sideslip angle during the ramp steer test.

5.3. Simulation and Comparison of the Step Steer Tests

The purpose of the step steer test is to characterize the transient response behaviors of vehicles [44]. The steering input ramps up from an initial steer value to the maximum steer value, as shown in Figure 13. During the step steer test, the steady state of the truck changes from the straight driving to the circular motion and the transition between the two steady-state movements is the transient response. The simulations of the step steer tests were conducted in Adams based on the two truck models, respectively. According to Chinese GB/T6323.2-94, the simulation settings are as follows: the initial steer value = 0°, the step start time 2.0 s, the duration time 6.0 s, the end time 30.0 s, and the final steer value 100.0°. A comparison of the simulated dynamic responses is shown in Figure 14.

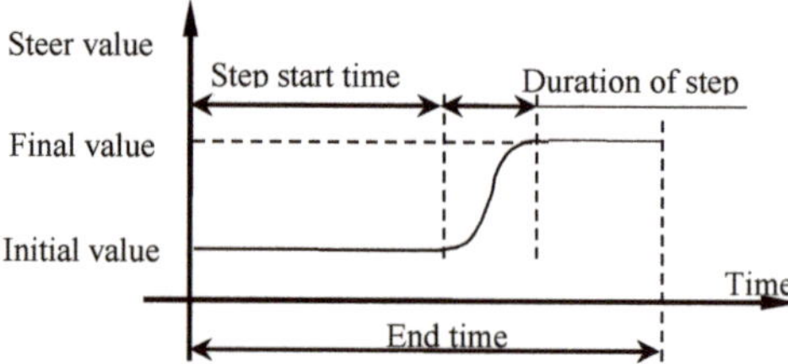

Figure 13. The curve of the input versus the time for the step steer test.

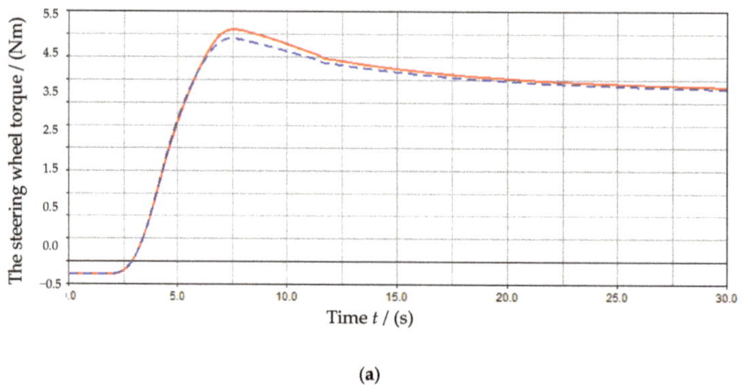

(a)

Figure 14. *Cont.*

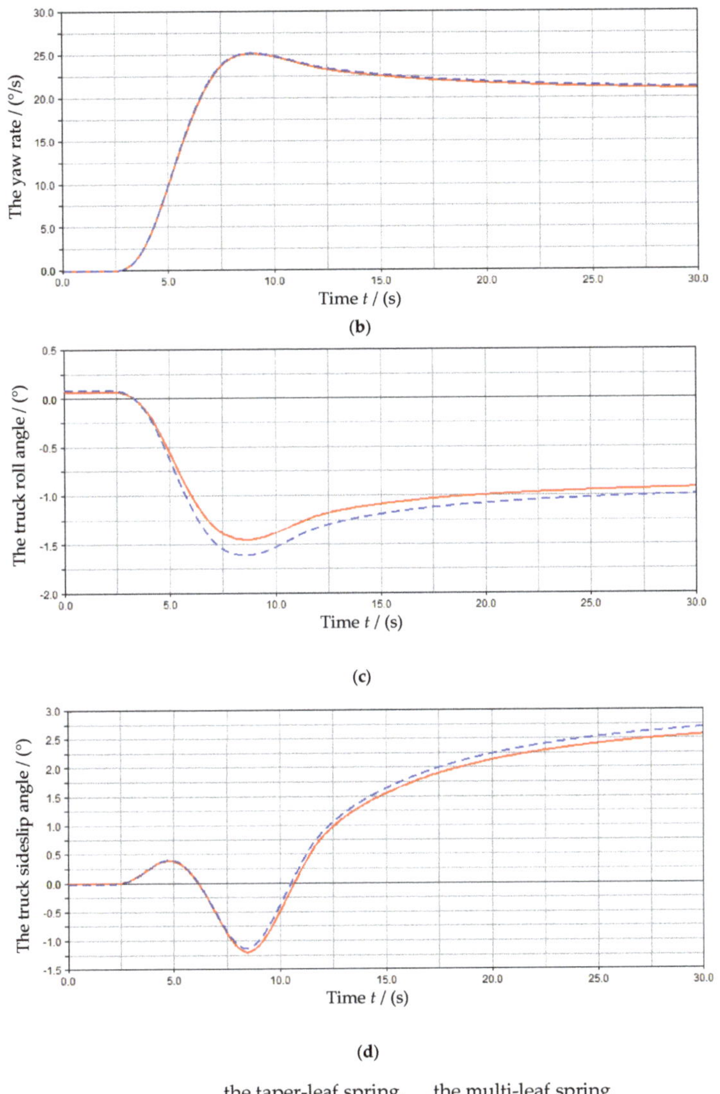

Figure 14. A comparison of the simulated dynamic responses obtained from the step steer tests: (**a**) The steering wheel torque, (**b**) the yaw rate, (**c**) the truck roll angle, and (**d**) the truck sideslip angle.

From Figure 14a, it can be seen that the steering wheel torques from the two truck models are basically the same from 0.0 s to about 6.5 s. The steering wheel torque of the truck with the taper-leaf spring suspension is slightly smaller than that of the truck with the multi-leaf spring suspension from 20.0 s to 30.0 s. Moreover, the steering wheel torque of the truck with the taper-leaf spring suspension is obviously smaller than that of the truck with the multi-leaf spring suspension from 6.5 s to 20.0 s. The results prove that the truck with the taper-leaf spring suspension has better steering portability during the step steer test. From Figure 14b, it can be seen that the two truck models show almost the same yaw rate during the step steer test. The result further shows that replacing the multi-leaf spring with the taper-leaf spring has almost no effect on the yaw rate. From Figure 14c, it can be seen that the truck roll angles on the basis of the two truck models are almost the same from 0.0 s to 3.0 s during the step steer test. With the increase of the steer value, the roll angle of the truck with the taper-leaf spring increases more obviously. The results show that the truck with the taper-leaf spring suspension has the poorer stability. Figure 14d illustrates that the sideslip angle of the truck with the taper-leaf spring suspension is obviously larger than that of the truck with the multi-leaf spring suspension. The result further shows that the truck handing performs worse after replacing the multi-leaf spring with the taper-leaf spring.

5.4. Simulation and Comparison of the Braking-in-Turn Tests

The braking-in-turn analysis is one of the most critical analyses encountered in everyday driving. This analysis examines path and directional deviations caused by sudden braking during cornering. Typical results collected from the braking-in-turn analysis include the steering wheel torque, the yaw rate, and the truck sideslip angle [40]. The simulations of the braking-in-turn tests were conducted in Adams based on the two truck models, respectively. According to the international standard ISO7975-85, the simulation settings are as follows: the brake deceleration 6.3 m/s^2, the lateral acceleration 5.1 m/s^2, the turn radius 30.0 m, the maximum brake duration 5.0 s, the initial straight-line distance 15.0 m. Figure 15 provides the simulation diagram of the braking-in-turn test based on Adams. Figure 16 provides a comparison of the simulated dynamic responses. From Figure 16a, it can be seen that the steering wheel torques from the two truck models are basically the same. Figure 16b depicts that there are almost no differences for the yaw rate. Figure 16c shows that the sideslip angle of the truck with the taper-leaf spring suspension is almost the same as that of the truck with the multi-leaf spring suspension. From Figure 16d, it can be seen that the two truck models show almost the same bushing longitudinal force at the spring eye during the braking-in-turn test. This proves that the replacement of the leaf spring has little impact on the in-train longitudinal forces during braking.

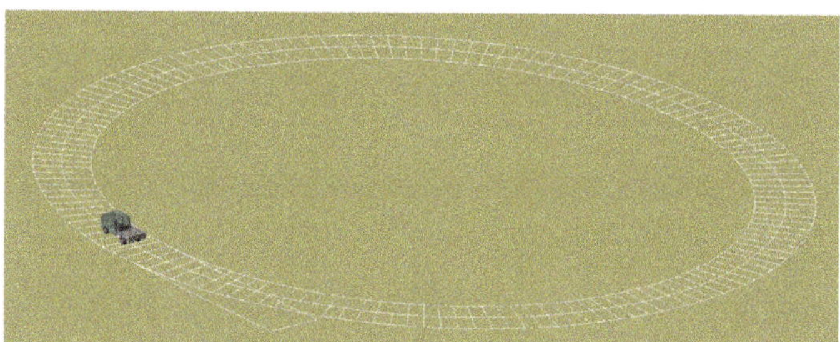

Figure 15. The simulation diagram of the braking-in-turn test based on Adams.

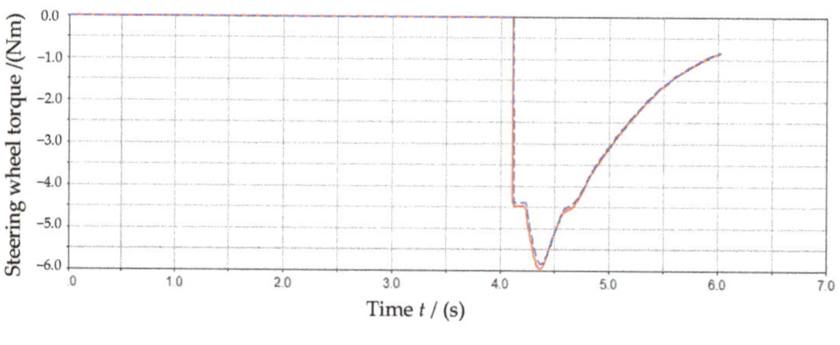

(a)

Figure 16. *Cont.*

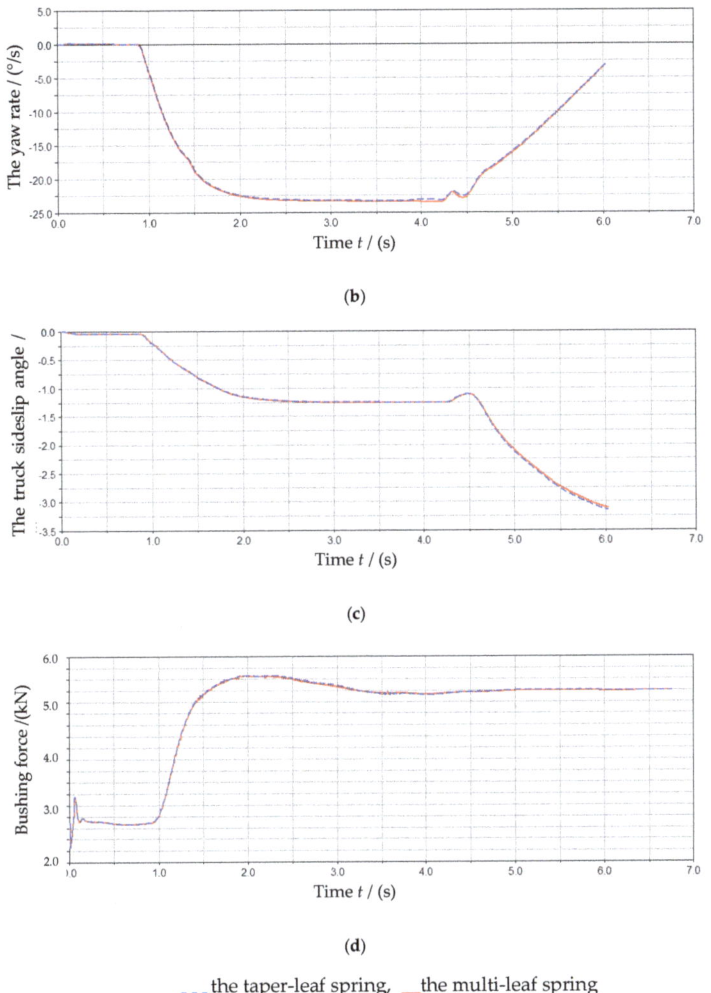

Figure 16. A comparison of the simulated dynamic responses obtained from the step steer tests: (**a**) The steering wheel torque, (**b**) the yaw rate, (**c**) the truck sideslip angle, and (**d**) the bushing longitudinal force at the spring eye.

6. Analysis on the Reason of Suspension Performance Differences for the Two Leaf Springs

In Section 5, the differences of the simulated dynamic responses for the handling stability should be related to that the torsional stiffness of the two leaf springs. To prove it, the influence of the stabilizer bar on suspension performance with the two leaf springs should be revealed. Thus, the drift simulation test was conducted. The stabilizer bar stiffness is increased by 1.2 times. The other simulation settings are the same as those in Section 5.1. A comparison of the simulated dynamic responses is shown in Figure 17. From Figure 17a, it can be seen that after increasing the stiffness of stabilizer bar, the steering wheel torque for the taper-leaf spring is closer to that for the multi-leaf spring with the original stabilizer bar. Figure 17b shows that the yaw rate for the taper-leaf spring is almost unchanged after increasing the stiffness of stabilizer bar. Figure 17c illustrates that the truck sideslip angle for the taper-leaf spring becomes smaller after increasing the stiffness of stabilizer bar. Moreover, after increasing the stiffness of stabilizer bar, the truck sideslip angle for the taper-leaf spring is more consistent with that for the multi-leaf spring with the original stabilizer bar. The comparison results show that the torsional stiffness of the two leaf springs have a certain impact on the handling stability.

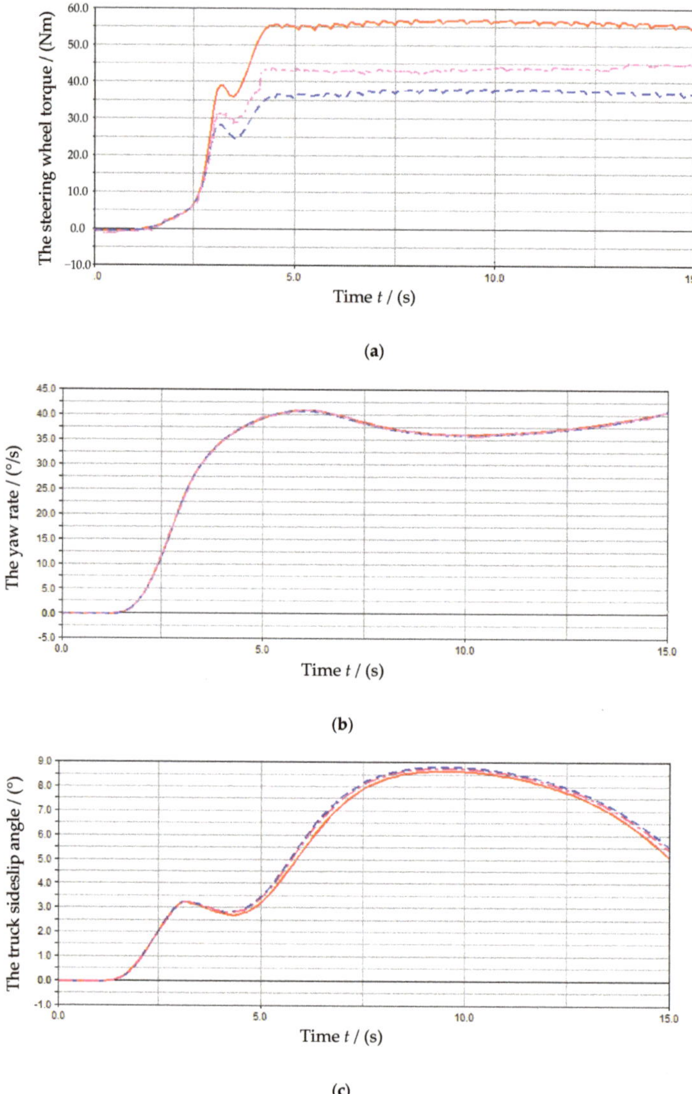

__the multi-leaf spring with the original stabilizer bar, - - - the taper-leaf spring with the original stabilizer bar, - - - the taper-leaf spring with increasing the stabilizer bar stiffness

Figure 17. A comparison of the simulated dynamic responses obtained from the drift tests: (**a**) The steering wheel torque, (**b**) the yaw rate, and (**c**) the truck sideslip angle.

7. Conclusions

To reveal the changes of the handling stability after the replacement, the simulations and comparison of the taper-leaf and the multi-leaf spring suspensions with the same vertical stiffness for trucks were conducted. The main innovations and achievements of this paper are as follows:

(1) An analytical method of replacing the multi-leaf spring with the taper-leaf spring for trucks was proposed.

(2) The dynamic models of the truck before and after the spring replacement were established.

(3) The changes of the handling stability after replacing the multi-leaf spring with the taper-leaf spring were revealed by the simulations of the drift test, the ramp steer test, and the step steer test.

Both the proposed method and the established model were validated by test. The simulation results show that the yaw rate of the truck almost does not change, the steering wheel moment decreases, the vehicle roll angle obviously increases, and the vehicle side slip angle slightly increases after the replacement. The main reason should be related to that the torsional stiffness of the taper-leaf spring is less than that of the multi-leaf spring.

This paper provides a useful reference for the design, modeling, and simulation of the taper-leaf spring suspension system for trucks.

Author Contributions: The corresponding author C.Z. proposed this research, Y.Z. designed the spring experiments; X.L. and Y.Y. performed the truck experiments; H.L. and L.Z analyzed the test data and conducted the simulations; C.Z. reviewed and edited the manuscript; and L.Z. wrote the paper. All authors have read and agreed to the published version of the manuscript.

Funding: This work was supported by the National Natural Science Foundation of China (51575325), Scientific Research Starting Foundation for Doctors (419067), and the School-Enterprise Cooperation Project (2018-KJ-4).

Conflicts of Interest: The authors declare no conflict of interest.

References

1. Sun, Y.; Guo, K.; Xu, Z. Research on electronic stability control system testing and evaluation methods for commercial vehicles. *Autom. Technol.* **2018**, *4*, 52–57.
2. Jaafari, S.M.M.; Shirazi, K.H. Integrated vehicle dynamics control via torque vectoring differential and electronic stability control to improve vehicle handling and stability performance. *J. Dyn. Syst. Meas. Control* **2018**, *140*, 13. [CrossRef]
3. Wan, Y.; Zhao, W.; Feng, R.; Ling, J.; Zong, C.; Zheng, H. Dynamic modeling and vehicle-liquid coupling characteristic analysis for tank trucks. *J. Jilin Univ.* **2017**, *47*, 353–364.
4. Zhao, L.; Yu, Y.; Zhou, C.; Wang, S.; Yang, F.; Wang, S. A hydraulic semi-active suspension based on road statistical properties and its road identification. *Appl. Sci.* **2018**, *8*, 740. [CrossRef]
5. Li, S.; Yang, S.; Chen, L. Investigation on cornering brake stability of a heavy-duty vehicle based on a nonlinear three-directional coupled model. *Appl. Math. Model.* **2016**, *40*, 6310–6323.
6. Santos, J.; Gouveia, R.M.; Silva, F.J.G. Designing a new sustainable approach to the change for lightweight materials in structural components used in truck industry. *J. Clean. Prod.* **2017**, *164*, 115–123. [CrossRef]
7. Liu, Y.; Sun, L. Lightweight design on auxiliary frame of mixer truck. *Int. J. U E Serv. Sci. Technol.* **2016**, *9*, 207–216. [CrossRef]
8. Ashwini, K.; Rao, C.V.M. Design and analysis of leaf spring using various composites-An overview. *Mater. Today Proc.* **2018**, *5*, 5716–5721. [CrossRef]
9. Wang, F.; Zhou, C.C.; Yu, Y.W. Analytical calculation method of stiffness and stress for few-chip root-intensive parabolic leaf spring. *J. Shandong Univ. Technol.* **2019**, *33*, 45–50.
10. Sert, E.; Boyraz, P. Optimization of suspension system and sensitivity analysis for improvement of stability in a midsize heavy vehicle. *Eng. Sci. Technol. Int. J.* **2017**, *20*, 997–1012. [CrossRef]
11. Gu, Y.; Dong, F. Handling stability simulation and test research on an 8×4 type construction truck. *Int. J. Control Autom.* **2016**, *9*, 403–410.
12. Huan, X.; Zhang, X. Numerical study of disturbance force effect on handing stability for electric-driven articulated truck in a straight-line. *Source UPB Sci. Bull. Ser. D Mech. Eng.* **2018**, *80*, 55–68.

13. Anchukov, V.; Alyukov, A.; Aliukov, S. Stability and control of movement of the truck with automatic differential locking system. *Eng. Lett.* **2019**, *27*, 131–139.
14. Duan, L.; Song, C.; Yang, S.; Fan, S.; Lu, B. High-precision modeling and simulation of the taper leaf spring of tandem suspension of commercial vehicles. *J. Mech. Sci. Technol.* **2016**, *30*, 3061–3067. [CrossRef]
15. Moon, I.D.; Yoon, H.S.; Oh, C.Y. A flexible multi-body dynamic model for analyzing the hysteretic characteristics and the dynamic stress of a taper leaf spring. *J. Mech. Sci. Technol.* **2006**, *20*, 1638–1645. [CrossRef]
16. Wang, C.; Shi, W.; Zu, Q. Calculation and analysis of stiffness of taper-leaf spring with variable stiffness. *SAE Tech. Pap.* **2014**. [CrossRef]
17. Zhou, S.; Huang, H.; Ouyang, L. Analysis and computation of taper leaf spring based on FE contact analysis. *Adv. Mater. Res.* **2013**, *705*, 516–522. [CrossRef]
18. Kim, B.M.; Kim, J.W.; Moon, I.D.; Oh, C.Y. Optimal combination of design parameters for improving the kinematics characteristics of a midsize truck through design of experiment. *J. Mech. Sci. Technol.* **2014**, *28*, 963–969. [CrossRef]
19. Chandra, H.; Pratiwi, D.K.; Zahir, M. High-temperature quality of accelerated spheroidization on SUP9 leaf spring to enhance machinability. *Heliyon* **2018**, *4*, e01076. [CrossRef]
20. Fragoudakis, R.; Savaidis, G.; Nikolaos, M. Optimizing the development and manufacturing of 56SiCr7 leaf springs. *Int. J. Fatigue* **2017**, *103*, 168–175. [CrossRef]
21. Kumar, K.; Aggarwal, M.L. Optimization of various design parameters for EN45A flat leaf spring. *Mater. Today Proc.* **2017**, *4*, 1829–1836. [CrossRef]
22. Jenarthanan, M.P.; Ramesh Kumar, S.; Venkatesh, G.; Nishanthan, S. Analysis of leaf spring using Carbon /Glass Epoxy and EN45 using ANSYS: A comparison. *Mater. Today Proc.* **2018**, *5*, 14512–14519. [CrossRef]
23. Ozmen, B.; Altiok, B.; Guzel, A.; Kocyigit, I.; Atamer, S. A novel methodology with testing and simulation for the durability of leaf springs based on measured load collectives. *Procedia Eng.* **2015**, *101*, 363–371. [CrossRef]
24. Duruş, M.; Kırkayak, L.; Ceyhan, A.; Kozan, K. Fatigue life prediction of Z type leaf spring and new approach to verification method. *Procedia Eng.* **2015**, *101*, 143–150. [CrossRef]
25. Bakir, M.; Ozmen, B.; Donertas, C. Correlation of simulation, test bench and rough road testing in terms of strength and fatigue life of a leaf spring. *Procedia Eng.* **2018**, *213*, 303–312. [CrossRef]
26. Kong, Y.S.; Abdullah, S.; Omar, M.Z.; Haris, S.M. Failure assessment of a leaf spring eye design under various load cases. *Eng. Fail. Anal.* **2016**, *63*, 146–159. [CrossRef]
27. Bi, S.; Li, Y.; Zhao, H. Fatigue analysis and experiment of leaf-spring pivots for high precision flexural static balancing instruments. *Precis. Eng.* **2019**, *55*, 408–416. [CrossRef]
28. Malikoutsakis, M.; Savaidis, G.; Savaidis, A.; Ertelt, C.; Schwaiger, F. Design, analysis and multi-disciplinary optimization of high-performance front leaf springs. *Theor. Appl. Fract. Mech.* **2016**, *83*, 42–50. [CrossRef]
29. Liu, Z.; Li, Y.; Hong, L. The effect of hinger on performance of front taper-leaf spring suspension of commercial truck. *J. Jilin Univ.* **2014**, *44*, 11–16.
30. Liu, N.; Zhao, D.; Gu, J.; Mu, P. Research on main leaf center trajectory of taper leaf spring and suspension kinematics analysis. *Automob. Technol.* **2019**, *11*, 42–46.
31. Termous, H.; Shraim, H.; Talj, R. Coordinated control strategies for active steering, differential braking and active suspension for vehicle stability, handling and safety improvement. *Veh. Syst. Dyn.* **2019**, *57*, 1494–1529. [CrossRef]
32. Li, H.; Li, S.; Sun, W. Vibration and handling stability analysis of articulated vehicle with hydraulically interconnected suspension. *J. Vib. Control* **2019**, *25*, 1899–1913. [CrossRef]
33. Zhang, L.; Zhang, S.; Zhang, W. Multi-objective optimization design of in-wheel motors drive electric vehicle suspensions for improving handling stability. *Proc. Inst. Mech. Eng. Part D J. Automob. Eng.* **2019**, *233*, 2232–2245. [CrossRef]
34. Bagheri, M.R.; Mosayebi, M.; Mahdian, A.; Keshavarzi, A. Multi-objective optimization of double wishbone suspension of a kinestatic vehicle model for handling and stability improvement. *Struct. Eng. Mech.* **2018**, *68*, 633–638.
35. Li, G.; Liu, F.; Tan, R.; Li, X.; Wang, T. Mathematical model of vehicle handling stability based on suspension K&C characteristics parameters. *China Mech. Eng.* **2018**, *29*, 1316–1323.
36. Ahmadian, M. Magneto-rheological suspensions for improving ground vehicle's ride comfort, stability, and handling. *Veh. Syst. Dyn.* **2017**, *55*, 1618–1642. [CrossRef]

37. Zhou, C.C.; Zhao, L.L. *Analytical Calculation Theory of Vehicle Suspension Elasticity*; Machinery Industry Press: Beijing, China, 2012.
38. Meng, Y.; Gan, X.; Wang, Y.; Gu, Q. LQR-GA controller for articulated dump truck path tracking system. *J. Shanghai Jiaotong Univ.* **2019**, *24*, 78–85. [CrossRef]
39. Gu, Z.; Mi, C.; Ding, Z.; Zhang, Y.; Liu, S.; Nie, D. An energy-based fatigue life prediction of a mining truck welded frame. *J. Mech. Sci. Technol.* **2016**, *30*, 3615–3624. [CrossRef]
40. Chen, J. *MSC. Adams-Examples of Technical and Engineering Analysis*; China Water Resources and Hydropower Press: Beijing, China, 2008.
41. Lu, Y.; Zheng, W.; Chen, E.; Zhang, J. Research on ride comfort and safety of vehicle under limited conditions based on dynamical tire model. *J. Vibroengineering* **2017**, *19*, 1241–1259. [CrossRef]
42. Bibin, S.; Pandey, A.K. A hybrid approach to model the temperature effect in tire forces and moments. *SAE Int. J. Passeng. Cars Mech. Syst.* **2017**, *10*, 25–37. [CrossRef]
43. Gupta, A.; Ghosh, B.; Balasubramanian, A.; Patil, S.; Raval, C. Reducing brake steer which causes vehicle drift during braking. *SAE Tech. Pap.* **2019**. No. 2019-26-0072. [CrossRef]
44. Yamauchi, S.; Kioka, W.; Kitano, T. Drifting motion of vehicles in tsunami inundation flow. *Procedia Eng.* **2015**, *116*, 592–599. [CrossRef]

Article

A Simplified Mathematical Model for the Analysis of Varying Compliance Vibrations of a Rolling Bearing

Radoslav Tomović

Mechanical Engineering Faculty, University of Montenegro, 81000 Podgorica, Montenegro; radoslav@ucg.ac.me

Received: 3 November 2019; Accepted: 10 January 2020; Published: 17 January 2020

Abstract: In this paper, a simplified approach in the analysis of the varying compliance vibrations of a rolling bearing is presented. This approach analyses the generation of vibrations in relation to two boundary positions of the inner ring support on an even and an odd number of the rolling element of a bearing. In this paper, a mathematical model for the calculation of amplitude and frequency of vibrations of a rigid rotor in a rolling bearing is presented. The model is characterized by a big simplicity which makes it very convenient for a practical application. Based on the presented mathematical model a parametric analysis of the influence of the internal radial clearance, external radial load and the total number of rolling elements on the varying compliance vibrations of rolling bearing was conducted. These parameters are the most influential factors for generating varying compliance vibrations. The results of the parametric analysis demonstrate that with the proper choice of the size of the internal radial clearance and external radial load, the level of the varying compliance vibrations in a rolling bearing can be theoretically reduced to zero. This result opposes the opinion that varying compliance vibrations of rolling bearing cannot be avoided, even for geometrically ideally produced bearing.

Keywords: rolling bearing; rigid rotor; internal radial clearance; number of rolling elements; vibration; varying compliance vibration; ball passage frequency; load

---

1. Introduction

The study of the motion of the rotational machine, i.e., rotor dynamics, is one of the most important areas in the engineering practice. The research of rotor dynamics finds its use in a broad spectrum of applications, from the big machines for energy production to those small ones found in the medical equipment. The mathematical models of a rotor and simulating the response to the activity of disturbing forces means a big aid in the developmental process of new structures and removal of problem related to already existing structures. The characteristics of a bearing where rotors are supported have a big influence on the rotor dynamic. The rolling bearings are the most used elements for the support of rotational machines. One of the most important problems related to rotational machines is the reduction of vibration and the increase of the accuracy of the spin of rotors supported on the rolling bearings, especially for the rotors operating with high rotational speeds. The rotation precision is often conditioned with the characteristics of vibration that are generated in the rolling bearings. Because of that, the problems related to the generation of vibrations in the rolling bearings should be treated with particular attention.

The main causes for vibration generation in rolling bearings are as follows [1]:

- The specific construction and the specific way of functioning of rolling bearing (varying compliance (VC) vibrations-structural vibrations),
- Micro and macro geometry errors of bearing elements (vibrations with technological origin),
- Damage of bearing elements (vibrations due to damage of bearings elements).

The construction of bearing affects the vibration generation in all the three mentioned cases. However, only in the (VC) vibrations, the construction of the bearing is the main cause of vibrations. Some hold the opinion that the VC vibrations in bearings cannot be avoided, even in ideally made rolling bearings and even in the absence of an external load [2,3]. In the other two cases, the vibrations occur because of various imperfections or damage which emerge during the fabrication or exploitation of bearings. With the correct production methods, maintenance, and exploitation, these vibrations can be reduced, and can even be avoided.

The first systematic researches varying compliance (VC) vibrations were done by Sunnersjö [4]. Sunnersjö studied the model of bearing considering the Hertzian contact and the clearance nonlinearities. The contact interaction between rolling elements and races is simplified as non-linear springs. Sunnersjö concluded that VC vibrations occur on the ball passage frequency (BPF) and its harmonics. Rahnejat and Gohar [5] later showed that even in the presence of elastohydrodynamic lubricating film between balls and races, a peak at the BPF appears in the spectrum.

The model proposed by Sunnersjö is a 2DOF model for bearing-rotor systems to study the varying compliance. This model and its modifications were later widely used in research by a large number of authors [5–15]. Mostly these are 2DOF models, however, 4DOF and 5DOF models were also used lately.

So, for example, Lynagh et al. [6] developed a two-dimensional model of the dynamic behavior of a rigid rotor in rolling bearings by observing the balance equation of bearing assembly as a system of elastically connected masses. Using this model, they have studied the problem of shaft rotation accuracy of highly precise machine tools at high rotation speeds. They also concluded that vibrations at BPF were dominant in the vibration spectrum. This observation was also confirmed by the experiment.

Tiwari M., Gupta K., and Prakash O. in [7] use the 2DOF model to study the effect of radial internal clearance on the dynamics behavior of a rigid rotor. They concluded that internal clearance is the biggest cause of unstable behavior of a rigid rotor and that subharmonic and chaotic behavior is strongly dependent on the size of the radial clearance. The study of the effect of clearance on a bearing's dynamic performance, quality, and operating life has gained a lot of attention lately because of the development of high-speed rotors [8]. Clearance non-linearity is different from most of the other non-linearities because it cannot be approximated by a mathematical series [7]. Upadhyay et al. [9] and Zhang et al. [10] also used a modified 2DOF model to study the nonlinear phenomena and chaotic behavior of a rigid rotor in a roller bearing.

The first model with 5DOF was developed by Aini et al. [11] and they applied it to the machine tool spindles. They have represented the bearings by nonlinear springs included elastohydrodynamic effects. This investigation was later refined by Li et al. [12], proposing a general dynamic model of the ball bearing-rotor system. They used the finite element method to model the contact stiffness. A little later, Wang et al. [13] propose their 5DOF dynamic model of the rotor system supported by angular contact ball bearings.

Five-degree-of-freedom models are used by many researchers to study the influence of various factors on the vibration generation of bearings. Thus, for example, Zhang et al. [14] study the effect of preload and varying contact angle on bearing vibration. Zhenhuan and Liqin [15] study the influence of rings misalignment on the dynamic characteristics of cylindrical roller bearings, and Yang et al. [16] influence of different ball number and rotor eccentricity on VC vibrations.

However, most of the above studies mainly merely verify the influence of the different factors on vibrations in the rotor-bearing system. Many studies only state that VC vibrations exist in the rotor-bearing system and that they occur on the BPF and that certain factors have an influence on the VC vibrations amplitude. Parametric analysis is very rarely performed.

The reason for this lies primarily in the great complexity of these models. All of these models are distinguished by their expressive non-linearity. Most of the mentioned models are encompass systems of differential equations, which describe the complex movement of the rigid rotor in rolling bearing, including the problem of nonlinear elastohydrodynamic contact between rolling elements and races. For solving systems of differential equations, most authors use the Newton–Raphson method, while for obtaining a vibration spectrum the fast Fourier's transformation (FFT) is commonly used. These models are extremely difficult for solving and they require a large number of complicated mathematical operations. Because of that, any parametric analysis would take a very long time.

In this connection, the model of Kryuchkov and Smirnov is particularly interesting because of its simplicity. Kryuchkov [17] investigated the VC vibrations of a rigid rotor in an ideal rolling bearing with an internal radial clearance and found that these vibrations can be represented by the equation of the absolute sinusoid. The frequency of these vibrations is equal to the ball passage frequency (BPF). Smirnov [18] further developed the research of Kryuchkov.

Using the research of Kryuchkov, in this paper, a new mathematical model for vibration response prediction of a rigid rotor in rolling bearing has been presented. The simple mathematical record makes the developed model very convenient for practical usage. The model is based on a new approach in the analysis of rolling bearing which is presented in the papers [19–22]. For the time being, the model considers only the VC vibrations of rolling bearing and it refers to an ideally made bearing with the radial contact and internal radial clearance.

Based on the presented model a parametric analysis of mutual influence between internal radial clearance and external radial load on the VC vibrations of rolling bearing is conducted. These two parameters are the most influential factors for generating VC vibrations for a given type of rolling bearing [23].

However, to understand and describe this model in a reasonable manner, as the first step, it is necessary to explain the basic mechanisms for the generation of the VC vibrations in rolling bearings.

2. Mechanisms for the Generation of Varying Compliance (VC) Vibrations in Rolling Bearings

Varying compliance (VC) vibrations in rolling bearings are a consequence of the bearing discrete structure and specific mode of operation. In literature, they are often called the primary bearing induced vibration [5–7] or structural vibrations [24,25]. In principle, VC vibrations appear based on two causes:

- Due to the periodic oscillations of the center of the bearing inner ring which resulting from the change in the angular position of the rolling elements. The internal radial clearance is the primary cause of these vibrations. These vibrations were analysed by Harris [1], Sunnersjö [4], Kryuchkov [17], Tomović et al. [20], and Datta and Farhang [24,25].
- Due to the periodic change of the rolling bearing stiffness. These vibrations are known as variable compliance vibrations. They are discussed by Harris [1], Sunnersjö [4], Kryuchkov [17], Rahnejat and Gohar [5], Lynagh et al. [6], Tiwari M., Gupta K., and Prakash O. [7], Upadhyay et al. [9], Zhang et al. [10], Li et al. [12], Wang et al. [13], Yang et al. [15], etc.

The rotor vibrations that are the direct consequence of an internal radial clearance influence and the discrete structure of a rolling bearing will emerge in all bearings with the internal radial clearance, both under the load and without it. The vibrations occurred due to a time-varying stiffness that will emerge only in the loaded rolling bearings [19].

2.1. Vibrations Caused by the Influence of Internal Radial Clearance

During the operation of the rolling bearings, in relation to the direction of the external load action, the set of rolling elements continuously oscillates between the two end positions of the support shown in Figure 1. This phenomenon is described in detail in [19–22], where the following terms were adopted for these boundary positions:

- Boundary case of inner ring support on an odd number of rolling elements (BSO) (Figure 1a),
- Boundary case of inner ring support on an even number of rolling elements (BSE) (Figure 1b).

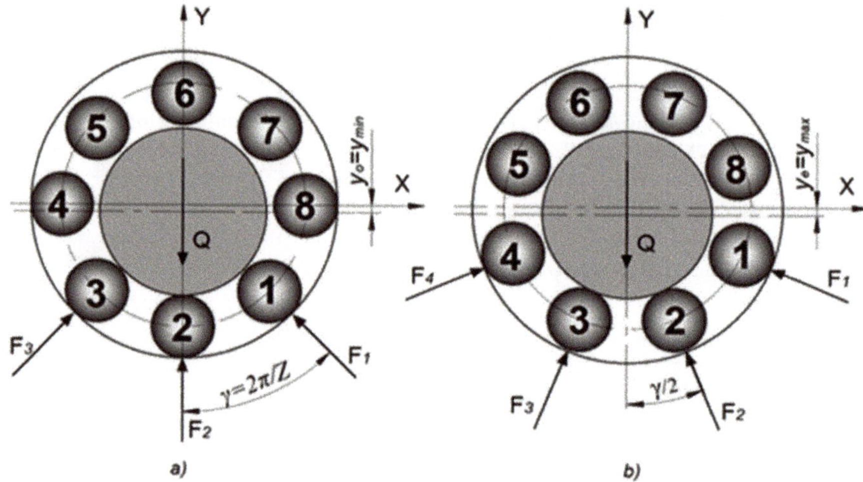

Figure 1. Boundary positions of inner rings support, (**a**) BSO position, (**b**) BSE position

BSO occurs when the center of one of the rolling elements coincides with the direction of the action of the external radial load. BSE occurs when the rolling elements are symmetrically positioned relative to the direction of the action of the external radial load. The minimum radial distance between the center of bearing rings appears in the BSO position, and in the BSE position the maximum radial distance appears.

The axis of the inner ring, in relation to the outer ring, will continuously oscillate between these two boundary positions of the support. These oscillations will occur in all bearings with an internal radial clearance, regardless of the number of rolling elements and the size of the external radial load.

2.2. Vibrations Caused by the Periodic Change of Rolling Bearing Stiffness

During the bearing operation, due to a contact load, in the contact between the rolling elements with the bearing rings, the elastic deformations occur. These deformations are non-linear and can be described according to *Hertz theory of the contact stresses and deformations* [1]:

$$\delta_i = \left(\frac{F_i}{K}\right)^{\frac{1}{n}}, \tag{1}$$

where δ_i is contact deformation at the place of i-th rolling element, F_i is load of i-th rolling element, K is effective stiffness coefficient due to Hertz's contact effect, n is the exponent that depends on bearing type (n = 3/2 for ball bearing and n = 10/9 for bearing with rollers).

With the change of position of the set of rolling elements, the load distribution in a bearing is changed, and with that intensity and the direction of contact forces and contact deformations between the rolling elements and rings also changes (Figure 2). All of this will cause a periodical change of the bearing stiffness [26,27]. The time-varying stiffness causes vibrations also in an ideally made bearing even in the absence of internal radial clearance. Moreover, the time-varying stiffness is considered to be a basic cause of the generation of the vibrations in the rolling bearings. The vibrations that occurred in this way are called variable compliance (VC) vibrations [28].

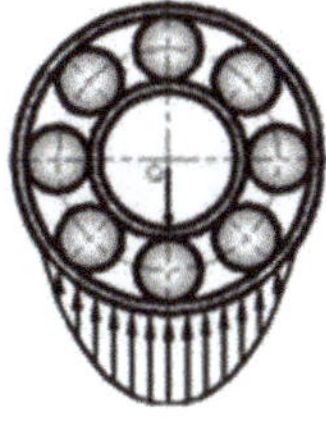 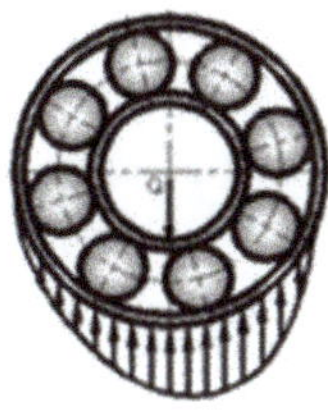 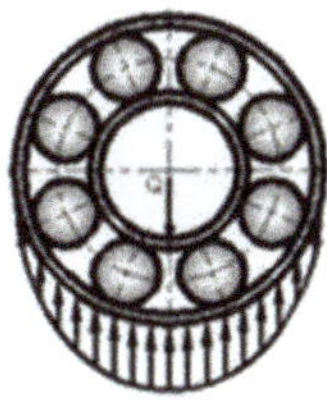 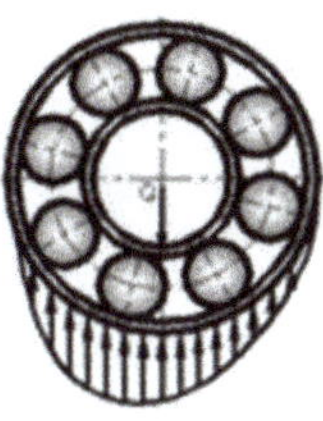 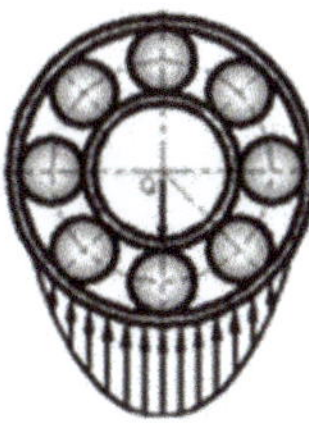

Figure 2. Periodic change of load distribution inside the rolling bearing.

3. The Equation of Rigid Rotor Vibration in Rolling Element Bearing

The equation for the rigid rotor vibrations in an ideal rolling bearing with the internal radial clearance was derived by Kryuchkov and Smirnov [17,18], in the form of an absolute sinusoid as:

$$y = \Delta\left|\sin\left(\frac{f_{bp}}{2}\cdot t\right)\right|, \quad (2)$$

where Δ is peak-to-peak (*pp*) amplitude, *t* is time, and f_{bp} is ball passage frequency (BPF). The derivation of Equation (2) is described in details in the reference under ordinal number [20].

According to the Equation (2), the vibration frequency of a rigid rotor in an ideal rolling bearing with the internal radial clearance is equal to a BPF. BPF represents the speed with which the rolling elements pass over some fixed point on an outer bearing race. According to [1] BPF is calculated as:

$$f_{bp} = \frac{z}{2}\cdot\omega\cdot\left(1 - \frac{d_b}{d_c}\cdot\cos\alpha\right), \quad (3)$$

where ω is the shaft frequency, d_b is the diameter of rolling elements, d_{c-} is the cage pitch diameter, α is the bearing contact angle.

The size of *pp*-amplitude (Δ) can be calculated as the difference of extreme values of the rotor displacement in two boundary positions of support, shown in Figure 1, according to the equation:

$$\Delta = y_e - y_o, \quad (4)$$

where y_e is displacement of the rotor in case *BSE* (Figure 1b), y_o is displacement of the rotor in case *BSO* (Figure 1a).

The rotor displacement is determined by the size of the relative motion between the rings during the bearing operation. According to [19], a relative movement of bearing rings directly depends on the size of internal radial clearance, load distribution in bearing, construction of bearing, the level of contact deformations and errors in the geometry of the contact surfaces.

4. Theoretical Analysis of Rigid Rotor Vibrations Amplitude in Rolling Bearings

4.1. Level of pp-Amplitude in an Unloaded Rolling Bearing

In the case of an unloaded bearing or a bearing loaded with relatively low loads, the contact deformations are insufficient to annul the influence of the internal clearance and the inner ring will be supported on one or two rolling elements, as shown in Figure 1. For this case of support in literature [29], a term *support system 1-2* was adopted.

The rotor displacement in boundary positions of the support is possible to be calculated by the help of the Equations [20]:

$$y_e = \frac{e}{2\cdot cos\gamma/2}, \quad (5)$$

$$y_o = \frac{e}{2}, \quad (6)$$

where e is internal radial clearance, γ is angular distance between the rolling elements.

According to the Equation (4) the level of the *pp*-amplitude in a case of the absence of external radial load, it will be equal to:

$$\Delta_1 = y_e - y_o = \frac{e}{2} \cdot \left(\frac{1}{\cos(\gamma/2)} - 1 \right). \quad (7)$$

The detailed derivation of the above equation is shown in [20].

4.2. *Impact of Load and Contact Deformations on Level of pp-Amplitude*

Due to the load and the occurrence of contact deformations, additional displacement and approach of the inner ring of the bearing relative to the external will occur and the new rolling elements will come into contact with the bearing rings. The support of bearing will continue to happen between the two boundaries positions: on the even (BSE) and the odd (BSO) number of the rolling elements. Only, depending on the size of contact deformations, the support system will move from *the support system 1-2* to *the support system 2-3*, *3-4*, *4-5*, etc., as shown in Figure 3 [19,21].

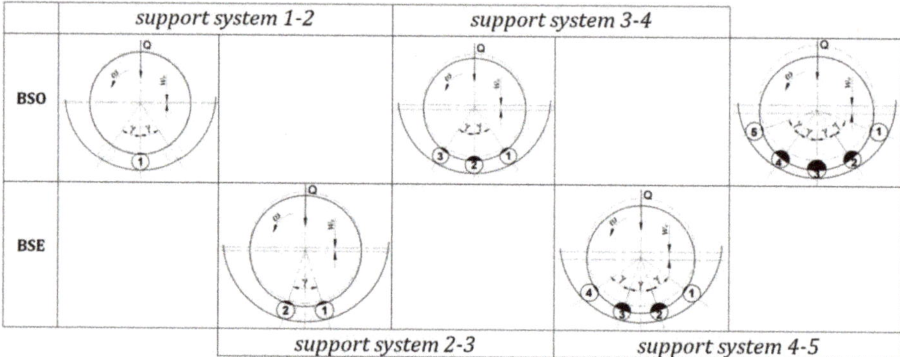

Figure 3. Support systems of rolling bearing.

The number of rolling elements that will come into contact with the bearing rings is directly dependent on two sizes, the inner radial clearance, and the external radial load, and can be easily determined by the procedure given in the literature [21,27].

The additional displacement of bearing rings will cause the further change of the *pp*-amplitude of rotor vibrations.

The total displacements of the rotor center will be equal to the sum of the displacements for unloaded bearing (Equations (5) and (6)) plus displacement due to the contact deformations of bearing elements (Figure 4). Hence, for the boundary positions of the supports on an even and an odd number of the rolling elements, the total displacement of the rotor center will be equal:

$$y_e = \frac{e}{2 \cdot \cos\gamma/2} + w_e, \quad (8)$$

$$y_o = \frac{e}{2} + w_o, \quad (9)$$

where w_e and w_o are rotor center displacements due to the contact deformations (bearing deflection) in the BSE and BSO cases of support system.

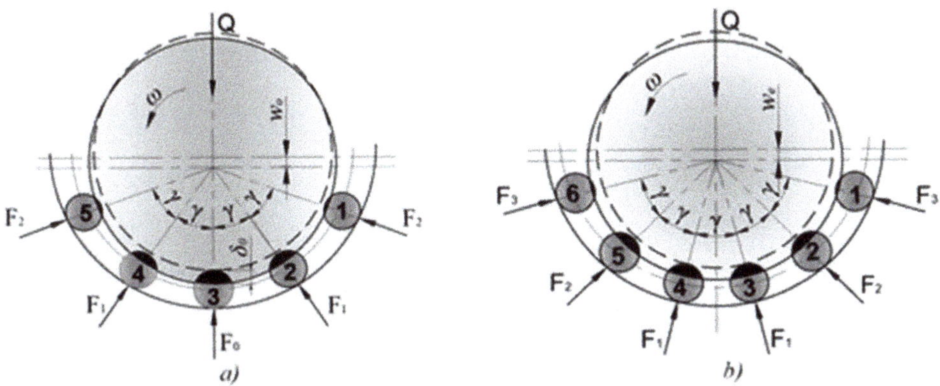

Figure 4. The influence of the external load and contact deformations on the relative displacement between the bearing rings, (**a**) BSO position of support, (**b**) BSE position of support.

According to the Equation (4), the amplitude of VC vibrations of the loaded rolling bearing can be calculated as a difference of the total center displacement for the boundary cases of the support, according to:

$$\Delta = y_e - y_o = \left(\frac{e}{2\cos(\gamma/2)} + w_e\right) - (e/2 + w_o), \tag{10}$$

$$\Delta = y_e - y_o = \frac{e}{2}\left(\frac{1}{\cos(\gamma/2)} - 1\right) - (w_o - w_e) = \Delta_1 - \Delta_2, \tag{11}$$

where $\Delta_2 = w_o$-w_e – pp- amplitude of the rotor vibrations due to the contact deformations of bearing elements. This amplitude is equal to a difference of the bearing deflection in two boundary support position: BSO (w_o) and BSE (w_e).

4.3. Radial Displacement of the Rotor Center Due to the Contact Deformations

In Figure 4 it is shown how due to contact deformations and relative displacement between the bearing rings, it comes to the annulment of the negative impact of the internal radial clearance and the entry of new rolling elements into the load transfer. The level of relative displacement and contact deformation is directly dependent on the size of the external radial load.

In [21], equations have been presented by which is possible to calculate the size of relative displacement of the bearing rings for BSO and BSE positions of support.

4.3.1. BSO Position of Support (Figure 4a)

The dependence between the bearing deflection (w_o) and external radial load (Q) for the BSO position according to [22] can be obtained from the following equation (Figure 4a):

$$Q = K \cdot \left[w_o^n + 2 \cdot \sum_{i=1}^{\frac{q-1}{2}} (w_o - a_{2i+1})^n \cdot \cos^{n+1}(i\gamma) \right], \tag{12}$$

where i is index of rolling element, K is effective coefficient of stiffness, w_o is bearing deflection, q is total number of rolling elements on which inner ring is supported, a_{2i+1} is boundary deflection of bearing when an i-th rolling element enters into the contact with the bearing rings, γ is angular spacing between rolling elements.

The bearing deflection that is necessary for entering of the *i-th* rolling element into the contact with the bearing rings in a BSO position can be calculated using the equation [21]:

$$a_{q,o} = \frac{e}{2} \cdot \left(\frac{1}{\cos \frac{q-1}{2}\gamma} - 1 \right) = t_{q,o} \cdot \frac{e}{2}, \tag{13}$$

where $t_{q,o}$ is the coefficient of the boundary deflection of bearing in the BSO position of inner ring support. The detailed derivation of the above equation and the value of coefficient $t_{q,o}$, can be found in the reference [19,21].

4.3.2. BSE position of support (Figure 4b)

The dependence between the bearing deflection (w_e) and external radial load (Q) for the BSE position according to [21] can be obtained from the following equation (Figure 4b):

$$Q = 2 \cdot K \cdot \sum_{i=1}^{\frac{q}{2}} (w_e - a_{2i,e})^n \cdot \cos^{n+1}(2i-1)\frac{\gamma}{2}, \tag{14}$$

where: i is index of rolling element, K is effective coefficient of stiffness, w_e is bearing deflection, q is total number of rolling elements on which inner bearing ring is supported, a_{2i} is boundary deflection of bearing when an *i-th* rolling element enters into the contact with the bearing rings, γ is angular spacing between rolling elements.

The necessary bearing deflection for entering of the *i-th* rolling element into the contact with bearing rings in the BSE position can be calculated as [21]:

$$a_{q,e} = \frac{e}{2} \cdot \left(\frac{1}{\cos \frac{q-1}{2}\gamma} - \frac{1}{\cos \frac{\gamma}{2}} \right) = t_{q,e} \cdot \frac{e}{2}, \tag{15}$$

where $t_{q,e}$ is the coefficient of the boundary deflection of bearing in the BSE position of inner ring support [19,21].

5. Parametric Analysis of the Amplitude of VC Vibration in Rolling Bearings

For the parametric analysis a radial single-row ball bearing 6206 with the internal radial clearance has been chosen. This bearing has nine balls altogether. According to [21], for this bearing, the maximum number of rolling elements that can emerge in the loaded zone is five. From the corresponding tables presented in [21], values of the coefficient of bearing boundary deflection t_q were read. These values are presented in Table 1.

Table 1. The coefficient of bearing boundary deflection t_q for ball bearing 6206.

	$q = 3$	$q = 4$	$q = 5$
t_q	0.3054	0.9358	4.7588

5.1. Numerical Procedure

Equation (11) gives a general pattern for the calculation of the amplitude of the rigid rotor vibration in a rolling bearing. According to this equation, the total amplitude of the VC vibrations (Δ) is equal to the difference of *pp*–amplitude of an unloaded bearing (Δ_1) and *pp*-amplitude due to the contact deformations (Δ_2), i.e.:

$$\Delta = \Delta_1 - \Delta_2. \tag{16}$$

The size of *pp*-amplitude in the case of an unloaded bearing can be obtained with the Equation (7). On the other side, the *pp*-amplitude due to the contact deformations is equal to the difference of the bearing deflection in the BSO and BSE positions, according to the equation:

$$\Delta_2 = w_o - w_e. \tag{17}$$

The values of the bearing deflection can be obtained by solving Equation (12) for the case of the support on an odd number of rolling elements, and/or Equation (14) for the case of the support on an even number rolling elements. Given equations are non-linear and can only be calculated numerically. The bisection method was used for equation solving. For that purpose, computer software that runs in the MATLAB programming environment has been developed [29].

5.2. Unloaded Rolling Bearing

In the case of an unloaded rolling bearing, the amplitude of the rotor vibration depends on the size of the clearance and the angular spacing of rolling elements (Equation (7)). As the angular spacing was directly determined by the number of rolling elements in a bearing, it can be said that this amplitude depends on the total number of rolling elements and the size of the internal radial clearance and that other characteristic of bearing do not influence on its size. This means that an unloaded radial bearing with the same number of rolling elements and the size of clearance will have the same level of vibrations regardless of the type and dimensions of its constituent parts.

The diagram in Figure 5 gives a three-dimensional dependence of the vibration amplitude on the number of rolling elements and the internal clearance in an unloaded rolling bearing.

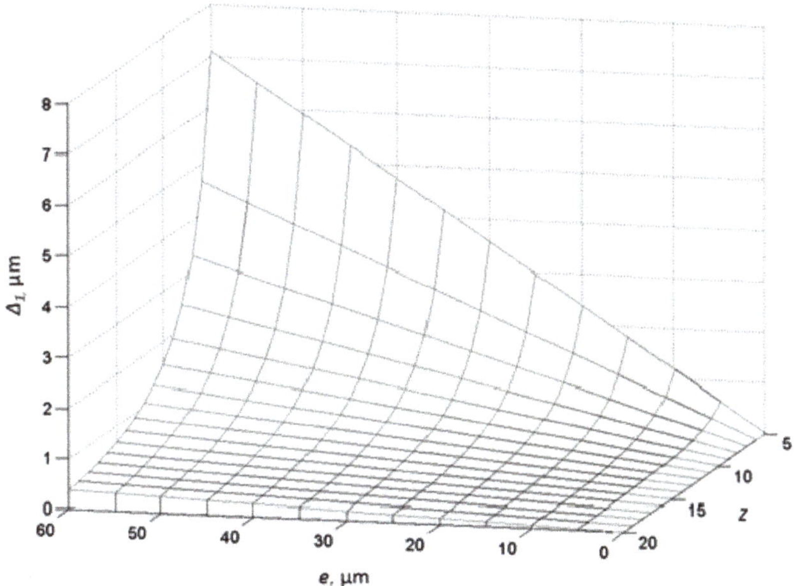

Figure 5. The vibrations amplitude of the unloaded rolling bearing depending on the size of the internal radial clearance and the total number of rolling elements.

From the analysis of the diagram by the increase of the internal clearance, the size of vibration amplitude linearly increases, and this increment is more pronounced when the number of rolling elements in a bearing is smaller. This means that more rotation precision and lower level of rotor vibrations are accomplished by the reduction of the angle between the rolling elements i.e., with the increase of the total number of rolling elements in a bearing. However, for construction reasons, the increase in the number of rolling elements is impossible over a certain limit. That means that the rotor vibrations in an unloaded rolling bearing are unavoidable and that this is a structural characteristic of a rolling bearing, which is impossible to avoid.

5.3. Loaded Rolling Bearing the Influence of Internal Radial Clearance and External Load

Figures 6–9 present the three-dimensional diagrams of the bearing deflection (w_o) and (w_e) in BSO and BSE positions, as well as the sizes of the vibration amplitude due to the contact deformations (Δ_2) and the total amplitudes of VC vibrations (Δ), depending on the size of the internal radial clearance and external radial load, for the 6206 ball bearing. The value of the internal radial clearance was varied in the range from 0 to 60 μm on the diagrams. That is nearly equal to the maximal recommended value of internal radial clearance for the 6206 ball bearing [30]. The value of the external radial load was varied up to a static load rating of the bearing 6206, which is $Q = 11200$ N.

The diagrams of the dependence of relative displacement of bearing rings (w_o) and (w_e) due to the contact deformations are given in Figures 6 and 7. The contact deformations of bearing elements increase together with the increase of the level of external radial load, but also with the increase of internal radial clearance size. However, the increase of the contact deformations is fairly more pronounced for the increase of the external radial load. The fact that the zones of various colors are almost parallel with the axis of internal radial clearance clearly indicates this. Also, it can be observed that there is a somewhat higher level of bearing deflection, at support on an odd (w_o) in relation to the support on an even (w_e) number of rolling elements. As the amplitude of the vibrations due to contact deformation (Δ_2) is equal to the difference of the bearing deflection in the BSO and BSE positions (Equation (17)), the level of this amplitude has a positive value for all the combination of the external radial load and internal radial clearance of bearing. This is also shown in the diagram shown in Figure 8.

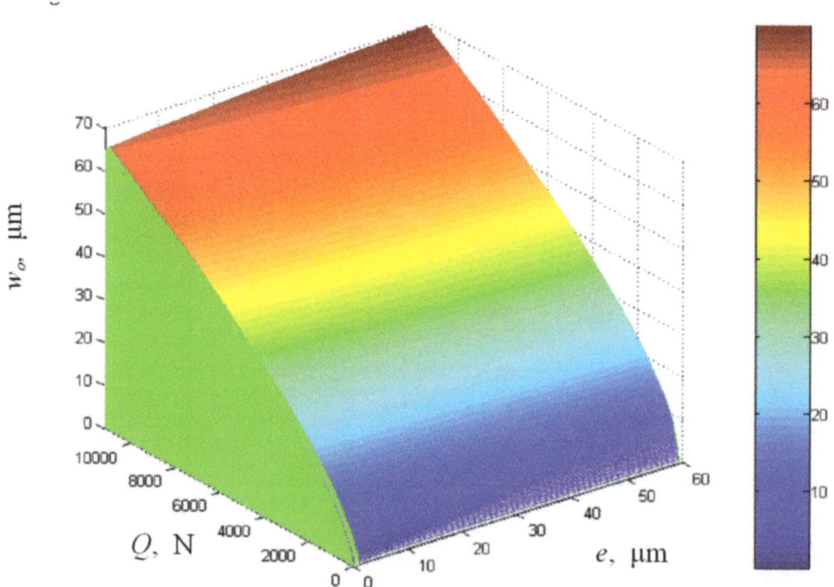

Figure 6. Three-dimensional dependence of bearing deflection (w_o) on the size of internal radial clearance (e) and level of external load (Q), for the BSO position of 6206-single-row ball bearing.

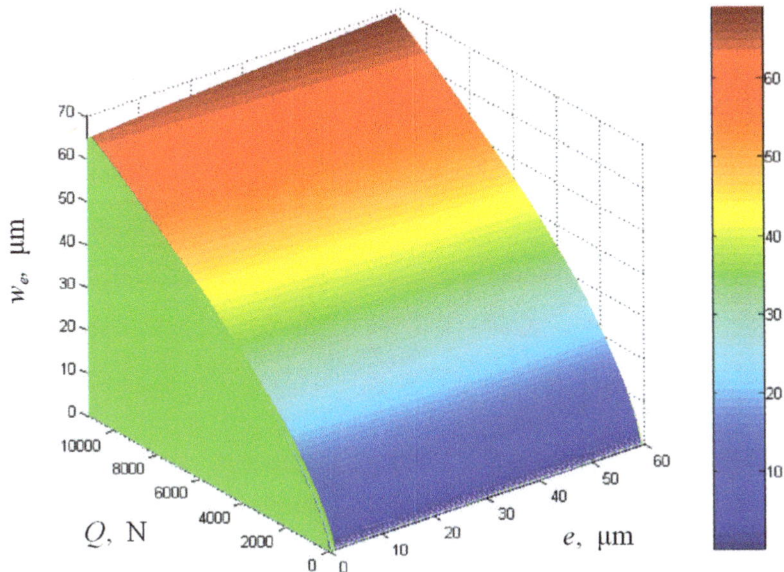

Figure 7. Three-dimensional dependence of bearing deflection (w_e) on the size of internal radial clearance (e) and level of external load (Q), for the BSE position of 6206-single-row ball bearing.

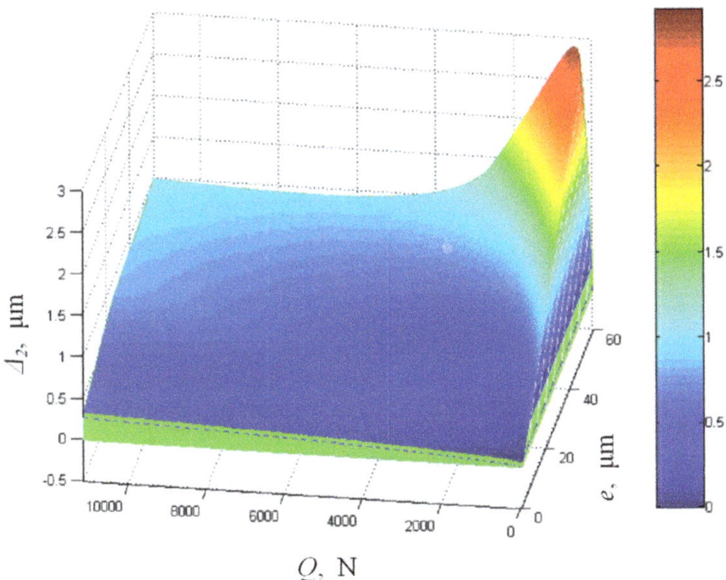

Figure 8. Three-dimensional dependence of vibration amplitude due to contact deformations (Δ_2) on the size of internal radial clearance (e) and level of external load (Q) for the 6206-single-row ball bearing.

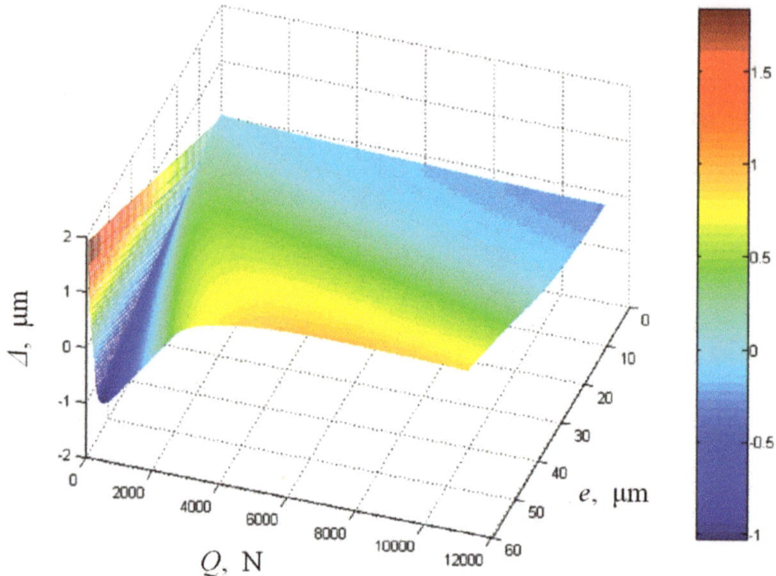

Figure 9. Three-dimensional dependence of the total amplitude of VC vibration (Δ) on the size of internal radial clearance (e) and level of external load (Q) for the 6206-single-row ball bearing.

In Figure 8 the change of the *pp*-amplitude due to the contact deformations (Δ_2) has been presented. In relation to the level of external radial load, we can discern two characteristic areas in this diagram. The first area is characterized by a high gradient of vibration amplitude due to the contact deformations (Δ_2). In this area, the support of the inner ring is carried out by *support systems 1-2*, or possibly by *support systems 2-3* (see Section 4.2). Because of the low intensity, the external radial load is not enough for the entering of a bigger number of rolling elements into contact with the bearing rings. The smaller number of rolling elements that transfer the external load, results in a less convenient load distribution into the bearing and in a sharper gradient of contact deformations. Hence, the steeper shape of the diagram of *pp*- amplitude Δ_2 in the direction of the external radial load in this area. On the other side, and increasing the clearance will also cause an increase in the level of the amplitude Δ_2 in this area, except for *support system 1-2*.

In *support systems 1-2*, the size of the clearance does not affect the vibration amplitude Δ_2. The vibration amplitude due to the contact deformation stays constant, regardless of the increase of internal radial clearance in *support systems 1-2*. This is also shown by the contour lines that are parallel to the axis that marks the value of the internal radial clearance (e). This area is characterized by a relatively low intensity of the external radial load.

The increase of external radial load significantly increases the level of the vibration amplitude due to the contact deformations. In the case of a boundary load that will cause the boundary deflection of bearing, the bearing support system will move from *support systems 1-2* to *support systems 2-3*. This phenomenon is explained in Section 4.2 and more comprehensively in the papers under the ordinal number [21,22]. Further increase in load will result in a negative gradient of the amplitude Δ_2.

In the second area, the high level of the external load results in a bigger number of rolling elements into contact with bearing rings. Hence, the load distribution between the rolling elements is fairly more uniform in this area. This results in the reduction of the vibration amplitude due to the contact deformations (Δ_2). The increase of the external radial load will reduce the amplitude level (Δ_2) until the point when an inner ring begins to support on a maximal possible number of active rolling elements. After this point, the amplitude will increase again.

The diagram of the total amplitude of VC vibrations in a rolling bearing has been presented in Figure 9. Unlike the vibration amplitude due to the contact deformations, the total amplitude of the VC vibrations can have both positive and negative values. The total amplitude has most values in the areas of the lower intensity of external radial load and the bigger values of internal radial clearance. These are the areas where, because of a low level of contact deformations, the load transfer is mostly based on *support systems 1-2*. The biggest influence on the rotor vibrations in this area has the size of the internal radial clearance. For the smaller values of clearance, the level of the total vibration amplitude is also smaller.

The external radial load has a favorable influence on the rotor vibrations only in the areas where the total vibration amplitude is positive. Namely, in the areas where emerges an unfavorable load distribution, i.e., in the areas where inner ring support on a smaller number of rolling elements, the level of the contact deformations can be very high so that the vibration amplitude due to contact deformations is bigger than amplitude Δ_1. In these areas, the value of the total amplitude is negative. In the areas where the total amplitude has a negative value, a bigger influence on rotor vibrations has contact deformations than the size of internal radial clearance. In the fields where the total amplitude is positive, a bigger influence on the rotor vibrations has the clearance in a bearing than the size of the external radial load.

5.4. Loaded Rolling Bearing the Influence of the Total Number of Rolling Elements and External Load

The diagrams in Figures 10–13. show the influence of the external radial load and the total number of rolling elements on the basic parameters that define the VC vibrations of the rolling bearing (bearing deflection (w_o) and (w_e), *pp*-amplitude due to contact deformations (Δ_2) and total *pp*-amplitude (Δ)). In order to exclude the influence of the internal bearing construction, the analysis was made with respect to the ratio of the external radial load and the effective bearing stiffness coefficient (Q/K). The analysis was made for bearings with an internal radial clearance of $e = 15$ μm.

It is clear that the increase of the external radial load causes an increase of the contact deformations in bearing and, consequently, an increase in the bearing deflection. However, for bearings with a larger number of rolling elements, this increase is much milder. The diagrams in Figures 10 and 11 confirm that an increase in the total number of rolling elements in the bearing significantly influences the reduction of bearing deflection.

In Figure 12, the change in *pp*-amplitude due to contact deformations (Δ_2) is shown. In relation to the total number of rolling elements in the bearing, the diagram in the figure can be divided into two characteristic areas. The first area refers to bearings with a small total number of rolling elements ($z \leq 6$). In this area, with increasing external radial load, the level of *pp*-amplitude (Δ_2) increases significantly. For these bearings, the maximum number of roller bearings that can be found in the load zone is $z_s = 3$, that is, the external load is transferred according to *the support system* 2-3. In Section 5.3, it is shown that in *the support system* 2-3, very high values of the *pp*-amplitude (Δ_2) occur and the total VC vibrations are more influenced by the contact deformations. In the second area, due to the larger number of rolling elements in the bearing, the contact deformations are smaller and the level of *pp*-amplitude Δ_2 decreases.

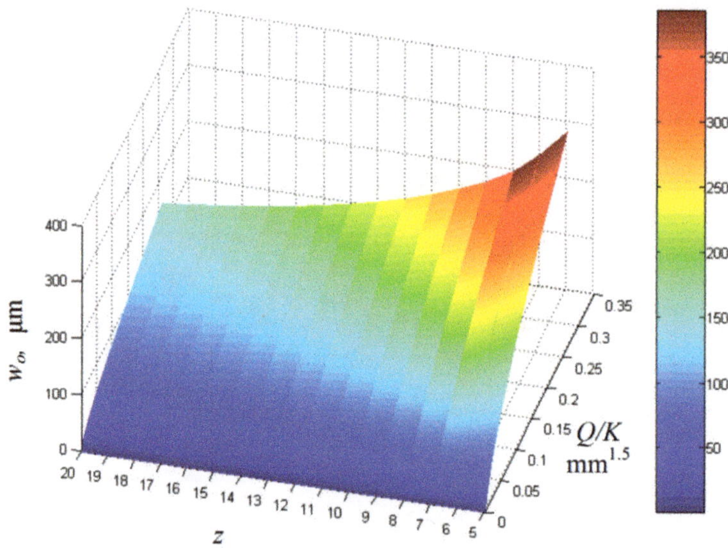

Figure 10. Three-dimensional dependence of bearing deflection (w_o) on the total number of rolling elements (z) and level of external load (Q), for the BSO position of rolling bearing.

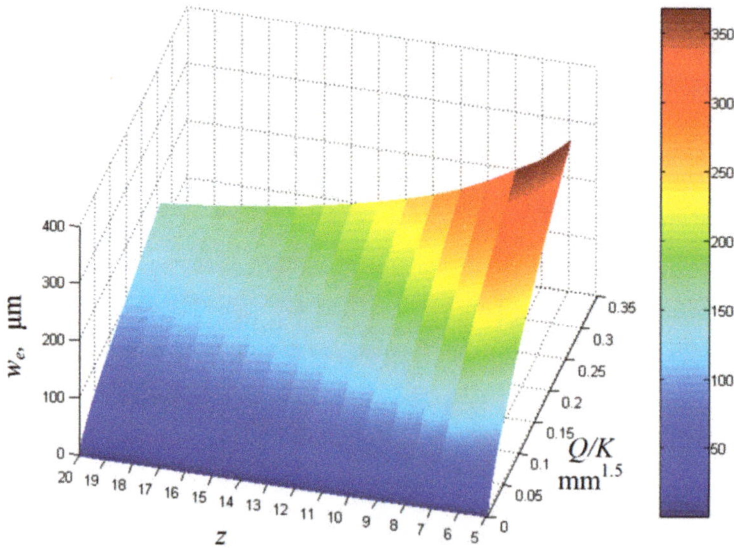

Figure 11. Three-dimensional dependence of bearing deflection (w_e) on the total number of rolling elements (z) and level of external load (Q), for the BSE position of rolling bearing.

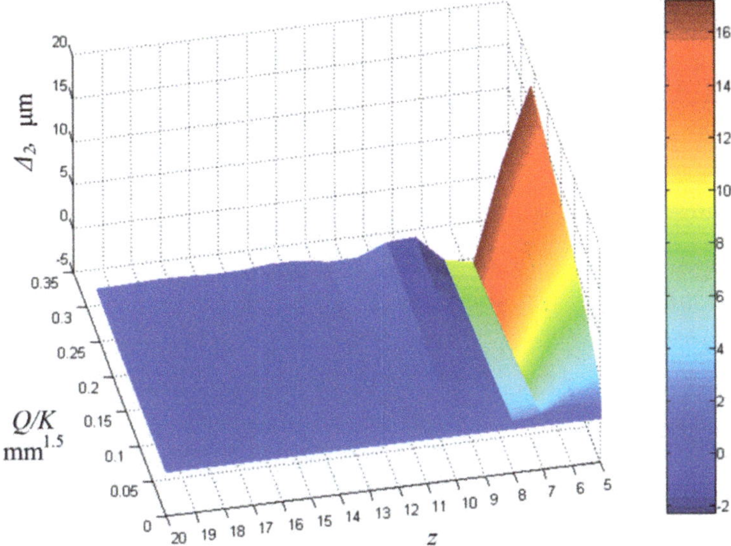

Figure 12. Three-dimensional dependence of vibration amplitude due to contact deformations (Δ_2) on the total number of rolling elements (z) and level of external load (Q).

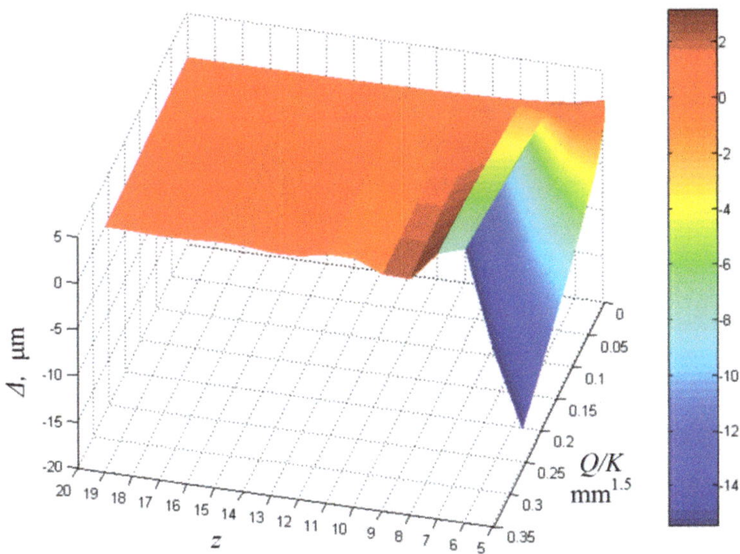

Figure 13. Three-dimensional dependence of the total amplitude of VC vibration (Δ) on the total number of rolling elements (z) and the level of external load (Q).

In Figure 13, the change of the total *pp*-amplitude of VC vibrations (Δ) is shown. Three characteristic areas can be distinguished in this diagram. The first two areas are characterized by quite high values of the total *pp*-amplitude. In the first area, the *pp*-amplitude takes values up to −15 μm and in the second area up to 8 μm. The first area refers to bearings with a total number of rolling elements $z \leq 6$. Due to the small number of rolling elements the contact deformations are large in this area. Therefore, in the first region, the influence of contact deformations on VC vibrations is predominant, and the total *pp*-amplitude (Δ) is negative. The second area refers to bearings with a total number of rolling elements from 7 to 9. Due to the larger number of rolling elements, a significant decrease in amplitude (Δ_2) occurs in this region, so high values of total *pp*-amplitude are a consequence of high values of amplitude Δ_1. The values of total *pp*-amplitude in this region are positive. The third area on the diagram is characterized by a relatively low value of total *pp*-amplitude. The value of *pp*-amplitude in this region decreases with the increase of the total number of rolling elements in the bearing. This confirms the conclusion that the greater total number of rolling elements in the bearing favorably affects the dynamic characteristics of the bearing.

6. Discussion

Shown diagrams demonstrate that there is a narrow link between the influence of internal radial clearance and the external radial load on the basic parameters that define the vibration response of a rolling bearing.

Depending on the combination of the level of external radial load and the size of internal radial clearance, the level of a bearing deflection is changed, as well as the number of rolling elements participating in the distribution of load in bearing. This also changes the vibration response of rolling bearing. This gives an opportunity that analysis of VC vibrations of rolling bearing can be also used as an important diagnostic parameter for the estimation of the internal radial clearance size and the uniformity of the internal load distribution, as well as the number of rolling elements which participate in the transfer of external load.

According to Equation (16), the vibrations that emerged due to the influence of internal radial clearance can be annulled by the vibrations that emerged due to the contact deformations. This is very important because by the appropriate combination of external radial load and internal radial clearance is possible to project the level of the VC vibrations of rolling bearing and reduce these vibrations to a minimum value.

7. Conclusions

Based on the results presented above, the following can be concluded:

- In this paper, a new simplified mathematical model for the calculation of amplitude of varying compliance vibrations is developed.
- The amplitude of the varying compliance vibrations in a rolling bearing is equal to the difference between vibration amplitude in an unloaded rolling bearing and the amplitude emerged due to the contact deformations.
- The vibration amplitude of an unloaded rolling bearing is determined by the size of the internal radial clearance and with the total number of rolling elements in a bearing.
- The vibration amplitude due to the contact deformations is equal to the difference of bearing deflection in two boundary positions of the support on an odd and an even number of rolling elements.
- The parametric analysis showed that the size of the internal radial clearance and the level of the external radial load have the biggest influence on the level of varying compliance vibrations.
- Higher values of the external radial load and the lower values of the internal radial clearance affect the VC vibrations level favorably.
- The greater total number of rolling elements in the bearing favorably affects the dynamic characteristics of the bearing
- The correct choice of the combination of the internal radial clearance and the external radial load can theoretically reduce the VC vibrations level to zero.

Funding: This research received no external funding.

Conflicts of Interest: The authors declare no conflict of interest.

Nomenclature

VC	varying compliance
BSO	boundary case of inner ring support on an odd number of rolling elements
BSE	boundary case of inner ring support on an even number of rolling elements
Δ	contact deformation
F	rolling element load
K	effective coefficient of stiffness due to Hertz's contact effect
n	exponent
y	rotor displacement
Δ	peak-to-peak (*pp*) amplitude of VC vibration
Δ_1	peak-to-peak (*pp*) vibration amplitude of unloaded bearing
Δ_2	peak-to-peak (*pp*) vibration amplitude due to the contact deformations
BPF	ball passage frequency
f_{bp}	ball passage frequency
t	time
z	total number of rolling elements
ω	shaft frequency
α	bearing contact angle
d_c	cage pitch diameter
d_b	diameter of rolling elements
w	bearing deflection
γ	angle between rolling elements
e	internal radial clearance
q	number of active rolling elements
a_q	boundary deflection of bearing
t_q	coefficient of boundary deflection of bearing
Q	total bearing load

Indexes

i	index of rolling element
e	boundary case of support onto an even number of rolling elements
o	boundary case of support onto an odd number of rolling elements

References

1. Harris, T.A.; Kotzalas, M.N. *Rolling Bearing Analysis*, 5th ed.; CRC Press: Boca Raton, FL, USA, 2007.
2. Cao, H.; Niu, L.; Xi, S.; Chen, X. Mechanical model development of rolling bearing-rotor systems: A review. *Mech. Syst. Signal Process.* **2018**, *102*, 37–58. [CrossRef]
3. Sharma, A.; Upadhyay, N.; Kankar, P.K.; Amarnath, M. Nonlinear dynamic investigations on rolling element bearings: A review. *Adv. Mech. Eng.* **2018**, *10*, 1687814018764148. [CrossRef]
4. Sunnersjö, C.S. Varying compliance vibrations of rolling bearings. *J. Sound Vib.* **1978**, *58*, 363–373. [CrossRef]
5. Rahnejat, H.; Gohar, R. The vibrations of radial ball bearings. *Proc. Inst. Mech. Eng. Part C Mech. Eng. Sci.* **1985**, *199*, 181–193. [CrossRef]
6. Lynagh, N.; Rahnejat, H.; Ebrahimia, M.; Aini, R. Bearing induced vibration in precision high speed routing spindles. *Int. J. Mach. Tools Manuf.* **2000**, *40*, 561–577. [CrossRef]
7. Tiwari, M.; Gupta, K.; Prakash, O. Effect of Radial Internal Clearance of a Ball Bearing on the Dynamics of a Balanced Horizontal Rotor. *J. Sound Vib.* **2000**, *238*, 723–756. [CrossRef]
8. Yakout, M.; Nassef, M.G.A.; Backar, S. Effect of clearances in rolling element bearings on their dynamic performance, quality and operating life. *J. Mech. Sci. Technol.* **2019**, *33*, 2037–2042. [CrossRef]
9. Upadhyay, S.H.; Harsha, S.P.; Jain, S.C. Analysis of Nonlinear Phenomena in High Speed Ball Bearings due to Radial Clearance and Unbalanced Rotor Effects. *J. Vib. Control* **2010**, *16*, 65–88. [CrossRef]
10. Zhang, Z.; Chen, Y.; Cao, Q. Bifurcations and hysteresis of varying compliance vibrations in the primary parametric resonance for a ball bearing. *J. Sound Vib.* **2015**, *350*, 171–184. [CrossRef]

11. Aini, R.; Rahnejat, H.; Gohar, R. A five degrees of freedom analysis of vibrations in precision spindles. *Int. J. Mach. Tools Manuf.* **1990**, *30*, 1–18. [CrossRef]
12. Le, Y.; Cao, H.; Niu, L.; Jin, X. A General Method for the Dynamic Modeling of Ball Bearing-Rotor Systems. *J. Manuf. Sci. Eng.* **2015**, *137*, 021016.
13. Wang, H.; Han, Q.; Zhan, D. Nonlinear dynamic modelling of rotor system supported by angular contact ball bearings. *Mech. Syst. Signal Process.* **2017**, *85*, 16–40. [CrossRef]
14. Zhang, X.; Han, Q.; Peng, Z.; Chu, F. A new nonlinear dynamic model of the rotor-bearing system considering preload and varying contact angle of the bearing. *Commun. Nonlinear Sci. Numer. Simul.* **2015**, *22*, 821–841. [CrossRef]
15. Zhenhuan, Y.; Liqin, W. Effects of axial misalignment of rings on the dynamic characteristics of cylindrical roller bearings. *Proc IMechE Part J: J. Eng. Tribol.* **2015**, *230*, 525–540. [CrossRef]
16. Yang, R.; Hou, L.; Jin, Y.; Chen, Y.; Zhang, Z. The varying compliance resonance in a ball bearing rotor system affected by different ball numbers and rotor eccentricities. *ASME J. Tribol.* **2018**, *140*, 051101. [CrossRef]
17. Крючков, Ю.С. Влияние зазора на вибрацию и шум подшипников качения. Вестник Машиностроения **1959**, *8*, 30–33. (In Russian)
18. Смирнов, В.А. Вибрационная Диагностика Подшипников Качения Двигателя НК-12СТГазоперекачивающего Агрегата ГПА-Ц-6,3. Available online: http://www.vibration.ru/12nks/12nks.shtml (accessed on 5 December 2010). (In Russian).
19. Tomović, R. Research of Rolling Bearings Construction Parameters Impact to the Condition of Their Working Correctness. Ph.D. Thesis, Faculty of Mechanical Engineering, University of Niš, Niš, Serbia, 2009.
20. Tomović, R.; Miltenović, V.; Banić, M.; Miltenović, A. Vibration response of rigid rotor in unloaded rolling element bearing. *Int. J. Mech. Sci.* **2010**, *52*, 1176–1185. [CrossRef]
21. Tomović, R. Calculation of the boundary values of rolling bearing deflection in relation to the number of active rolling elements. *Mech. Mach. Theory* **2012**, *47*, 74–88. [CrossRef]
22. Tomović, R. Calculation of the necessary level of external radial load for inner ring support on q rolling elements in a radial bearing with internal radial clearance. *Int. J. Mech. Sci.* **2012**, *60*, 23–33. [CrossRef]
23. Meyer, L.D.; Ahlgren, F.F.; Weichbrodt, B. An analytical model for ball bearing vibrations to predict vibration response to distributed defects. *ASME J. Mech. Des.* **1980**, *102*, 205–210.
24. Datta, J.; Farhang, K. A Nonlinear Model for Structural Vibrations in Rolling Element Bearings: Part I—Derivation of Governing Equations. *J. Tribol.* **1997**, *119*, 126–131. [CrossRef]
25. Datta, J.; Farhang, K. A Nonlinear Model for Structural Vibrations in Rolling Element Bearings: Part II—Simulation and Results. *J. Tribol.* **1997**, *119*, 323–331. [CrossRef]
26. Lazović, T.; Ristivojević, M.; Mitrović, R. Mathematical Model of Load Distribution in Rolling Bearing. *FME Trans.* **2008**, *36*, 189–196.
27. Tomović, R. Investigation of the Effect of Rolling Bearing Construction on Internal Load Distribution and the Number of Active Rolling Elements. *Adv. Mater. Res.* **2013**, *633*, 103–116. [CrossRef]
28. Mitrović, R.; Atanasovska, I.; Soldat, N. Dynamic behaviour of radial ball bearing due to the periodic variable stiffness. *Mach. Des.* **2015**, *7*, 1–4.
29. *MATLAB 2008b*; The MathWorks, Inc.: Natick, MA, USA, 2008.
30. *ISO 5753-1 Rolling Bearings—Internal Clearance—Part 1: Radial Internal Clearance for Radial Bearings*; International Organization for Standardization: Geneva, Switzerland, 2009.

Article

Tire Model with Temperature Effects for Formula SAE Vehicle

Diwakar Harsh [1] **and Barys Shyrokau** [2,*]

[1] Rimac Automobili d.o.o., 10431 Sveta Nedelja, Croatia; diwakar.harsh@rimac-automobili.com

[2] Department of Cognitive Robotics, Delft University of Technology, 2628CD Delft, The Netherlands

* Correspondence: b.shyrokau@tudelft.nl

Received: 22 October 2019; Accepted: 4 December 2019; Published: 6 December 2019

Abstract: Formula Society of Automotive Engineers (SAE) (FSAE) is a student design competition organized by SAE International (previously known as the Society of Automotive Engineers, SAE). Commonly, the student team performs a lap simulation as a point mass, bicycle or planar model of vehicle dynamics allow for the design of a top-level concept of the FSAE vehicle. However, to design different FSAE components, a full vehicle simulation is required including a comprehensive tire model. In the proposed study, the different tires of a FSAE vehicle were tested at a track to parametrize the tire based on the empirical approach commonly known as the magic formula. A thermal tire model was proposed to describe the tread, carcass, and inflation gas temperatures. The magic formula was modified to incorporate the temperature effect on the force capability of a FSAE tire to achieve higher accuracy in the simulation environment. Considering the model validation, the several maneuvers, typical for FSAE competitions, were performed. A skidpad and full lap maneuvers were chosen to simulate steady-state and transient behavior of the FSAE vehicle. The full vehicle simulation results demonstrated a high correlation to the measurement data for steady-state maneuvers and limited accuracy in highly dynamic driving. In addition, the results show that neglecting temperature in the tire model results in higher root mean square error (RMSE) of lateral acceleration and yaw rate.

Keywords: tire model; tire temperature; FSAE vehicle

---

1. Introduction

Formula SAE (FSAE) also known as formula student (FS) is a competition in which students design a single seat formula race car to compete against other FS teams from all over the world [1]. The competition motivates the students for innovative solutions to demonstrate their engineering talent and obtain new skills. It also allows the students to apply theoretical knowledge into practice in a dynamic and competitive environment.

Since 2001 students have joined Formula Student Team Delft [2] on an annual basis to participate in the FSAE competition, earlier in the combustion class and electric class from 2011. Aiming to improve the FSAE vehicle performance, the team collaborated with Apollo Vredestein B.V. (Enschede, The Netherlands) to develop original tires. The FSAE vehicle (2017 version) from FS Team Delft equipped with such tires is shown in Figure 1.

Figure 1. DUT17 Formula Society of Automotive Engineers (FSAE) vehicle at the Formula Student Germany competition [2].

To achieve the best FSAE vehicle performance, the team developed various tools to evaluate vehicle dynamics and to predict its behavior. However, predictions are based on estimated friction coefficient, neglecting the thermal effect on the tire performance. Frequently, lap simulation is based on a point mass or simplified vehicle models allowing for the development of the top-level concept of the new vehicle. It can quantify vehicle parameters such as the height of the center of gravity and lift coefficient. However, to design various vehicle components and, specifically, control algorithms, a full vehicle simulation including a comprehensive tire model is required.

For an accurate modeling of tire forces and moments, a physical, semi- or empirical tire model can be used [3]. The physical approach provides more insights regarding tire behavior and better represents the whole operation range; however, it lacks accuracy, specifically considering camber influence, conicity, plysteer, and other phenomena [4,5]. The empirical approach provides a higher modeling accuracy in the predefined tested range but requires extensive special tests. According to [6,7] the tire temperature, especially for racing application [8], has a significant effect on the force capability of the tire. Several advanced tire models using the brush element approach incorporated with thermal effects have been proposed [9–11]; however, their parametrization requires intensive test sessions, commonly unfeasible for FS teams. For a FSAE vehicle, less complex models incorporating temperature effect are proposed using the physical approach based on the brush model [12,13]. They demonstrate a good correlation with the experimental tire data collected using the indoor flat track tire test machine [14]. Also, the proposed physical-based models were compared to the original magic formula demonstrating close accuracy; however, original magic formula does not include the effect of temperature on the force capability of the tire.

Thus, the goal of the proposed study was to develop an empirical tire model incorporating thermal effects and to evaluate its performance compared to the experimental measurements. The main contribution of the study is the improvement of the accuracy of the empirical tire model using the proposed thermal model and modification of the magic formula to incorporate temperature effects.

The paper is structured as follows. Section 2 describes the experimental setup, test program, and experimental results. A thermal tire model is proposed in Section 3 discussing the basic thermal equations and comparing them to well-established thermal models. Section 4 presents the widely used magic formula with the extension related to the temperature effect. The complete vehicle model including thermal and modified magic formula is presented in Section 5; it includes the whole model simulation for steady-state and transient maneuvers compared to real tests and simulation without the temperature effect. The paper concludes with a discussion of the work and the recommendations for future research presented in Section 6.

2. Experimental Setup and Results

2.1. Setup and Test Program

The experimental tests were performed at Dynamic Test Center AG [15] located in Vaufellin, Switzerland. The test program is shown in Table 1 covering uncombined slip testing using sweep tests of slip angle α and wheel slip κ under various normal loads F_z for three tires: Apollo Single Compound, Apollo Double Compound, and Hoosier.

Table 1. Tire test program.

Measurement Type	α (rad)	κ (–)	F_z (N)
Pure cornering	sweep	0	600/1000
Pure acceleration/braking	0	sweep	

The measurements were performed outdoor using a test truck (Figure 2). To measure tire forces and moments, a Kistler wheel force transducer [16] was used. The measurement accuracy of the wheel force transducer is 1% of the full scale. For the investigated load range, it resulted in approx. 5% absolute measurement error including crosstalk. The normal load during the test was controlled using hydraulic suspension. The target normal load was obtained by the adjustment of hydraulic pressure. The wheel was driven through the axle unit from the motor for acceleration tests and using a reverse gear for braking. Another actuator was mounted on the tie-rod in order to generate the tire slip angle.

Figure 2. Test setup.

2.2. Pure Cornering Results

Tire lateral performance was assessed by performing sweep tests at a constant road wheel steering rate of 15°/s. The tests were conducted under various normal loads F_z of 600 and 1000 N correspondingly. During the tests, the vertical force was fluctuated due to road modulation and other irregularities, which as a result affected the lateral force F_y measurement. To obtain the lateral force, this influence was compensated using the normalized ratio between lateral and normal forces and then multiplying by mean vertical load. Figure 3a shows the effect of vertical load on the utilized friction coefficient μ. The results are aligned to the load sensitivity phenomena that utilized friction coefficient reduces as normal load increases. The test results of Hoosier and Apollo tires (double compound) are compared in Figure 3b. The peak utilized friction for the Hoosier tire was around 5% higher than the Apollo one; however, the average utilized friction was in the same range.

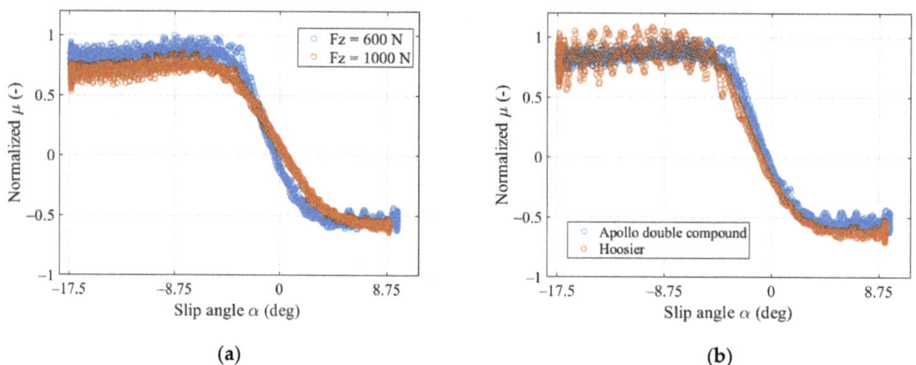

Figure 3. Normalized lateral friction μ vs. slip angle α: (**a**) different normal loads F_z (Apollo); (**b**) different tire types ($F_z = 600$ N).

2.3. Pure Acceleration and Braking Results

For the longitudinal force measurement, the wheel was lifted, accelerated, and then pressed onto the ground. The brake torque was applied to realize sweep tests of wheel slip while the forward velocity of the truck was kept constant. The effect of normal load on utilized friction is much less compared to lateral friction (Figure 4a). The performance of the Apollo tire was similar compared to the Hoosier one (Figure 4b). The difference was found to be 3% which can be related to the ambient and track conditions.

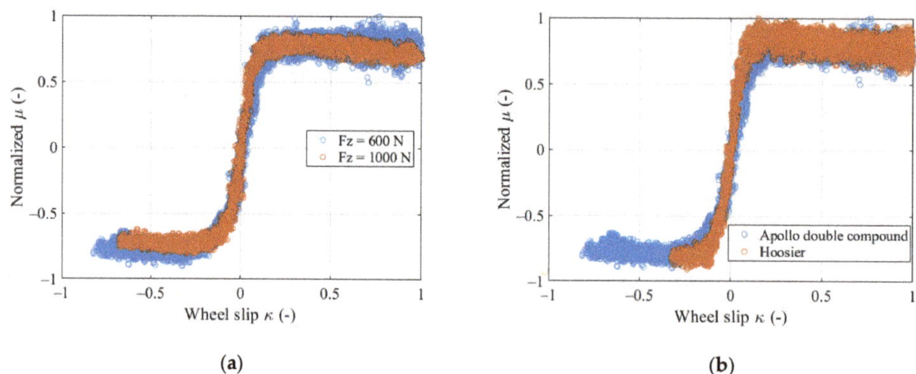

Figure 4. Normalized longitudinal friction μ vs. wheel slip κ: (**a**) different normal loads F_z (Apollo); (**b**) different tire types ($F_z = 600$ N).

2.4. Results Related to Temperature Effect

Both the force capability and the tire lifetime depend on tire temperature. Since the tire carcass is an elastic element, the temperature change will result in the modulus of elasticity of the rubber and therefore influence the cornering stiffness [17]. The effect of temperature on the normalized longitudinal and lateral friction is shown in Figure 5. It can be observed that the temperature affects the stiffness (curve slope) and the peak friction.

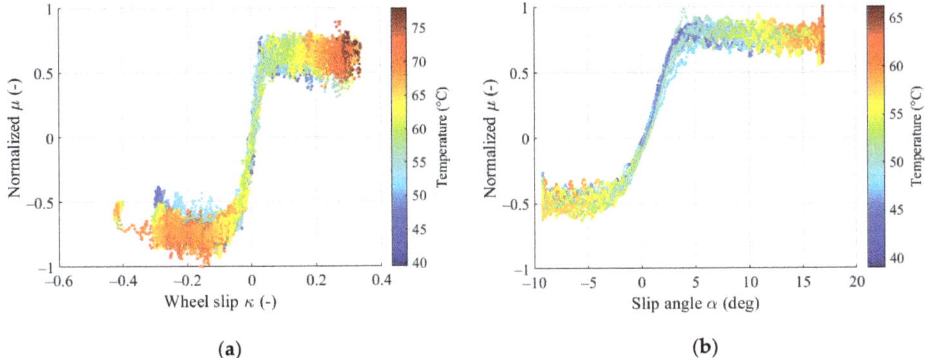

Figure 5. Temperature effects on tire forces for Apollo tire (F_z = 1000 N): (**a**) normalized longitudinal friction μ vs. wheel slip κ; (**b**) normalized lateral friction μ vs. slip angle α.

3. Thermal Model

3.1. Proposed Thermal Model

A lumped parameter approach is used according to [18] and the thermal model consists of three bodies such as the tread with temperature T_{tread}, the carcass with temperature $T_{carcass}$, and the inflation gas with temperature T_{gas}. The road surface with ambient air provides the boundary conditions with fixed temperatures T_{road} and T_{amb}. In contrast to the study [19], a single value of tread temperature was calculated without the need to define a complex temperature distribution across the contact patch area. The temperature change of tread, carcass, and gas is described as:

$$\begin{aligned} \dot{T}_{tread} &= \frac{Q_{sliding} - Q_{tread\to road} + Q_{carcass\to tread} - Q_{tread\to amb}}{S_{tread}M_{tread}} \\ \dot{T}_{carcass} &= \frac{Q_{damp} - Q_{carcass\to tread} - Q_{carcass\to amb} - Q_{carcass\to gas}}{S_{carcass}M_{carcass}} \\ \dot{T}_{gas} &= \frac{Q_{carcass\to gas}}{S_{gas}M_{gas}} \end{aligned} \quad (1)$$

where $Q_{sliding}$ is the heat flow due to sliding; $Q_{tread\to road}$ is the heat flow between tread and road; $Q_{carcass\to tread}$ is the heat flow between carcass and tread; $Q_{tread\to amb}$ is the heat flow between tread and ambient air; Q_{damp} is the heat flow due to carcass deflection; $Q_{carcass\to amb}$ is the heat flow between carcass and ambient air; $Q_{carcass\to gas}$ is the heat flow between carcass and inflation gas; S_{tread} is the specific heat capacity of carcass; $S_{carcass}$ is the specific heat capacity of carcass; S_{gas} is the specific heat capacity of inflation gas; M_{tread} is the tread mass; $M_{carcass}$ is the carcass mass; M_{gas} is the inflation gas mass.

According to [18], two heat generation processes should be considered:

- Due to carcass deflection

$$Q_{damp} = \left(E_x|F_x| + E_y|F_y| + E_z|F_z|\right)V_x \quad (2)$$

where E_x, E_y, and E_z are the carcass longitudinal, lateral, and vertical force efficiency factors; F_x, F_y, and F_z are the longitudinal, lateral, and normal forces; V_x is the longitudinal velocity.

- Due to sliding friction in the contact patch

$$Q_{sliding} = \mu_d F_z v_s \quad (3)$$

where μ_d is the dynamic friction coefficient; v_s is the sliding velocity.

To define the shift along the μ axis with compound temperature, the dynamic friction model [18] was modified. Based on the friction model [20], Equation (4) is proposed incorporating the shift along the μ axis due to temperature using the parameters μ and h as temperature dependent. The idea was taken from research [21] where the parameters of the friction models from [20] and [22] were made temperature-dependent. The final equation for the dynamic friction model is described:

$$\mu_d(T) = \mu_{base} + \left[\mu_{peak}(T) - \mu_{base}\right] e^{-(h(T)(\log_{10}(\frac{v_s}{V_{\max}}) - K_{shift}(T_{tread} - T_{ref})))^2} \tag{4}$$

The temperature-dependent peak friction coefficient μ_{peak} is defined:

$$\mu_{peak}(T) = a_1 T^2 + a_2 T + a_3 \tag{5}$$

where a_1, a_2, and a_3 are the tuning parameters for a second order polynomial function.

Dimensionless parameter h is related to the width of the speed range in which the friction coefficient varies significantly [23]. The following temperature-dependent adjustment is made similar to the adaptation of the compound shear modulus in [18]:

$$h(T) = b_1 \frac{e^{b_2 T_{tread}}}{e^{b_2 T_{ref}}} \tag{6}$$

where b_1 and b_2 are the tuning parameters.

The proposed modifications cover both sliding and non-sliding friction components. Temperature-dependent peak friction coefficient μ_{peak} corresponds to the peak friction properties of the tire compound. The parameter h affects the curve slope and results in the change of the shear modulus of the tire compound. The peak friction coefficient μ_{peak}, the lower limit of friction coefficient μ_{base}, and the parameter h are identified from measurement data. The simulation and experimental results for the obtained normalized lateral friction coefficient are shown in Figure 6. The model shows similar qualitative behavior observed in the experimental data at the slip angle above 4°.

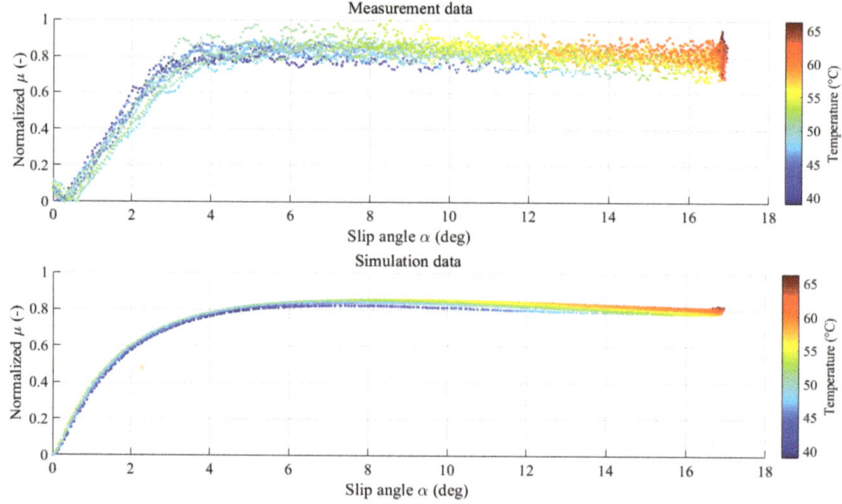

Figure 6. Comparison between measurement data and modified friction model.

Four types of heat flows are considered: (i) heat flow between tread and ambient air; (ii) heat flow between carcass and ambient air; (iii) heat flow between tread and carcass; (iv) heat flow between carcass and inflation gas.

Heat flow with the ambient air $Q_{tread \to amb}$ and $Q_{carcass \to amb}$ is defined as:

$$\begin{aligned} Q_{tread \to amb} &= H_{tread-amb}(T_{tread} - T_{amb}) \\ Q_{carcass \to amb} &= H_{carcass-amb}\left(T_{carcass} - T_{gas}\right) \end{aligned} \tag{7}$$

The heat flow coefficient $H_{carcass-amb}$ is constant and the heat flow coefficient $H_{tread\text{-}amb}$ is assumed to be a function of the longitudinal vehicle speed V_x and taken as a linear function [18]:

$$H_{tread-amb} = 2V_x + 10 \tag{8}$$

Heat flow between the tread and carcass $Q_{carcass \to tread}$ is calculated as:

$$Q_{carcass \to tread} = H_{carcass-tread}(T_{carcass} - T_{tread}) \tag{9}$$

Heat flow between the carcass and inflation gas $Q_{carcass \to gas}$ is defined as:

$$Q_{carcass \to gas} = H_{carcass-gas}\left(T_{carcass} - T_{gas}\right) \tag{10}$$

Using heat flow coefficient $H_{tread-road}$, heat flow between tread and road $Q_{tread \to road}$ can be found as:

$$Q_{tread \to road} = H_{tread-road} A_{cp}(T_{tread} - T_{road}) \tag{11}$$

The contact patch area, A_{cp}, is calculated assuming a constant width of the contact patch b and the variable contact patch length a which was adjusted compared to [18] to better match the data of the FSAE tires:

$$A_{cp} = 2ab = 0.12 p_{bar}^{-0.7}\left(\frac{F_z}{3000}\right)^{0.7} b \tag{12}$$

Since the volume of the gas is constant, the pressure is directly proportional to the Kelvin temperature and thus the inflation pressure p_{bar} can be calculated as:

$$p_{bar} = p_{cold}\frac{T_{gas} + 273}{T_{amb} + 273} \tag{13}$$

where p_{cold} is the pressure of the tire at ambient temperature.

3.2. Comparison with the Established Models

The Sorniotti model [12] describes an empirical model to estimate tire temperature as the function of the actual working conditions of the component. To evaluate the temperature effect on tire forces, a combination of the estimated temperature with a tire brush model [24] was used. The model was empirically tuned using experimental data to demonstrate the variation of tire performance as temperature function. The thermal model considers distinct thermal capacities related to the tread and carcass. The tread thermal capacity is related to the heat flux caused by the tire–road forces and carcass thermal capacity is affected by rolling resistance and exchanges heat with the external ambient. Other heat fluxes corresponded to the ambient and exchange between the two capacities. The original model [12] did not consider the heat flow between the road and tread. Therefore, the heat flow term $P_{tread,road}$ was introduced to the original model to improve the accuracy according to [25].

The model is described as:

$$\begin{aligned} C_{eq\_carcass}\frac{dT_{carcass}}{dt} &= \mathrm{P}_{rolling\_resis\tan ce} + P_{conduction} + P_{ambient,carcass} \\ C_{eq\_tread}\frac{dT_{tread}}{dt} &= \mathrm{P}_{Fx,tire} + \mathrm{P}_{Fy,tire} - P_{conduction} + P_{ambient,tread} + P_{tread,road} \end{aligned} \tag{14}$$

Power fluxes corresponding to the cooling flux due to the temperature difference between carcass and ambient $P_{ambient,carcass}$, and tire tread and ambient $P_{ambient,tread}$ are defined as:

$$\begin{aligned} P_{ambient,carcass} &= h_{carcass}(T_{ambient} - T_{carcass}) \\ P_{ambient,tread} &= h_{tread}(T_{ambient} - T_{tread}) \end{aligned} \tag{15}$$

Power fluxes related to conduction between tread and carcass $P_{conduction}$ and between tread and road $P_{tread,road}$ are calculated as:

$$\begin{aligned} P_{conduction} &= h_{conduction}(T_{tread} - T_{carcass}) \\ P_{tread,road} &= H_{tread-road}(T_{tread} - T_{road}) \end{aligned} \tag{16}$$

Compared to the Sorniotti model, the proposed thermal model due to a higher differential order should capture more dynamics related to heat flow and potentially produce better results.

The Kelly and Sharp model [18] states that in racing applications, the temperature of the tread significantly affects both the tire stiffness and the contact patch friction. It should be noted that the effect on the friction is higher. Since the rubber viscoelastic properties depend on temperature, the maximum performance on the racetrack is only available in a specific temperature range.

The tire model is also based on the brush model. Using the bristle stiffness c_p the adhesion part can be presented:

$$c_p = \frac{w_{cp} G_{tread}}{h_{tread}} \tag{17}$$

where w_{cp} is the contact patch width; h_{tread} is the tread height.

The shear modulus of the tread G_{tread} is defined as:

$$G_{tread} = \frac{G_{TA} - G_{\text{limit}}}{e^{-K_G T_{GA}}} e^{-K_G T_{tread}} + G_{\text{limit}}, \text{ where } K_G = \frac{\log(G_{TA} - G_{\text{limit}}) - \log(G_{TB} - G_{\text{limit}})}{T_{GB} - T_{GA}} \tag{18}$$

where G_{TA} is the reference shear modulus at temperature A; G_{TB} is the reference shear modulus at temperature B; G_{limit} is the limit shear modulus value at a high temperature; T_{GA} is the reference temperature A; T_{GB} is the reference temperature B.

The sliding part of the contact patch is described using the dynamic friction coefficient. The master curve of the friction coefficient μ_{mc} for the tread compound is assumed to have a Gaussian shape and it describes a friction curve with a Gaussian shape on a log frequency axis:

$$\mu_{mc} = \mu_{base} + \left[\mu_{peak} - \mu_{base}\right] e^{-(K_{shape}(\log_{10} v_s - K_{shift}(T_{tread} - T_{ref})))^2} \tag{19}$$

where μ_{base} and μ_{peak} are the lower and upper limits of the dynamic friction coefficient; K_{shape} is the master curve shape factor; v_s is the sliding velocity; K_{shift} is the master curve temperature shift factor; T_{tread} is the tread temperature; T_{ref} is the master curve reference temperature.

In addition, the model takes into account the influence of contact patch pressure on friction coefficient. It is assumed that friction coefficients are reduced linearly with contact patch pressure. The static and dynamic friction coefficients are defined as:

$$\begin{aligned} \mu_0 &= K_{cpp} \mu_{ref} \\ \mu_d &= K_{cpp} \mu_{mc} \end{aligned}, \text{ where } K_{cpp} = 1 - K_{pcpp} \frac{p_{cp}}{K_{refcpp}} \tag{20}$$

where p_{cp} is the contact patch pressure; K_{cpp} is the parameter of friction reduction rate; K_{pcpp} is the friction roll-off factor with contact patch pressure; K_{refcpp} is the reference contact patch pressure; μ_0 is the static friction coefficient; μ_d is the dynamic friction coefficient; μ_{ref} is the reference static friction coefficient.

Compared to the Kelly and Sharp model, the peak friction coefficient μ_{peak} is temperature-dependent in the proposed model, which could result in better accuracy.

3.3. Comparison Results

The simulation results obtained from the considered thermal models were compared with measurement data for various maneuvers. The initial values of coefficients were taken from [18,25] and further tuned using constrained parameter optimization.

To tune parameters, a non-linear least square method was applied to obtain a suitable approximation of temperature effect and was based on the error from the measurement defined as:

$$e_{meas} = \sqrt{\frac{\sum (T_{\text{model}} - T_{\text{measurement}})^2}{\sum T^2_{\text{model}}}} \tag{21}$$

It was checked that the fitting parameters provide a good estimation for different maneuvers. For example, the parameters were first optimized for lateral tests and later checked again for acceleration/braking tests. Then an average was taken for the tuning parameters to cover all the experimental tests and represent the whole driving envelope. The comparisons between the measured temperature, the temperature obtained from the Kelly and Sharp model, and the proposed thermal model are shown in Figure 7.

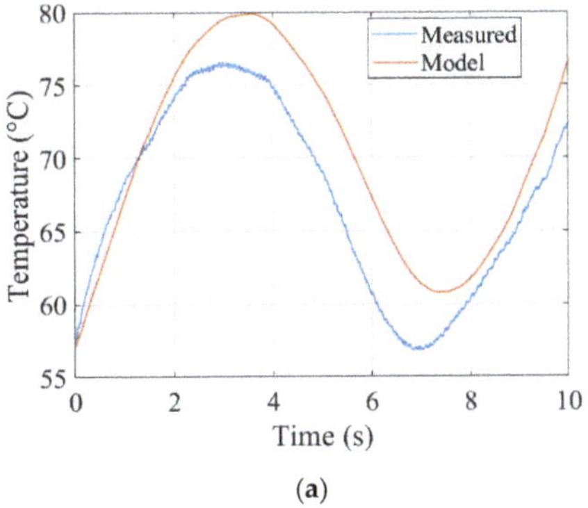

(**a**)

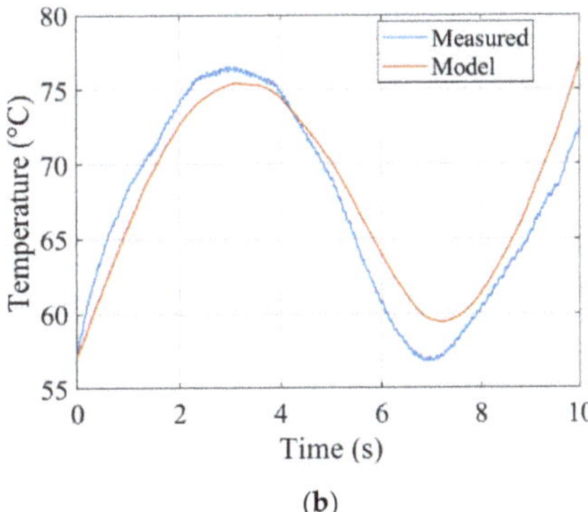

(**b**)

Figure 7. Temperature change in pure braking test: (**a**) results of the Kelly and Sharp model; (**b**) results of the proposed model.

The accumulated error for the compared thermal models is summarized in Table 2.

Table 2. Accumulated errors for compared thermal models.

	Sorniotti Model	Kelly and Sharp Model	Proposed Model
Pure acceleration	5.2%	4.7%	5.5%
Pure braking	2.5%	10%	2.9%
Pure cornering	3.2%	3.0%	2.6%

The results of the proposed thermal model and the compared models are close to the measurement data for all conducted tests. The proposed thermal model provides the smallest error for pure cornering. For pure braking the Sorniotti model shows the best performance and for pure acceleration the smallest error can be achieved using the Kelly and Sharp model. However, the conclusions regarding adaptability of the models to different normal loads cannot be extracted from Table 2. Since the Kelly and Sharp model considers the contact patch area depending on the normal load and air pressure, the model can adapt to different normal loads and the maximum error is limited to ~6%. Therefore, the models were also checked using measurement data for different load conditions. It was found that although the Sorniotti model has a good fit to the measured data, the error increases when the load condition is changed (Figure 8).

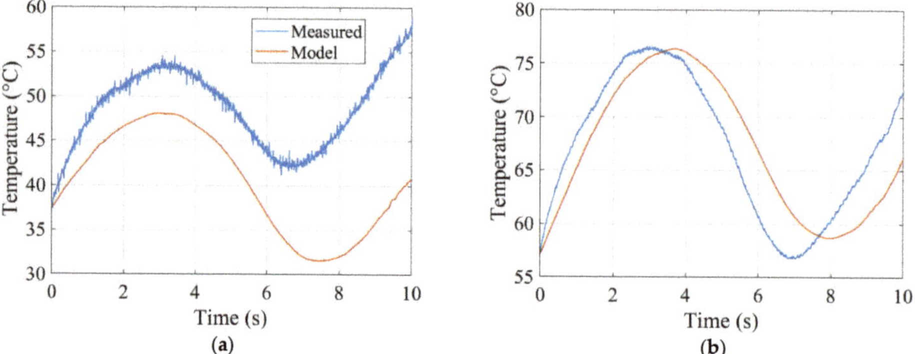

Figure 8. Temperature change in pure braking test using Sorniotti model: (**a**) normal load F_z of 600 N; (**b**) normal load F_z of 1000 N.

4. Modified Magic Formula Incorporating Temperature Effect

4.1. Baseline Parameter Estimation

Temperature plays a crucial role in tire performance. A large number of factors, such as the normal load, ambient conditions, etc., affect the temperature of the tread. Since the tread temperature profile was different on the two tires due to normal force fluctuation (two tires were tested simultaneously), it was necessary to align the measurements in such a way that the tires were in a similar temperature range.

Once the measurements were processed according to similar temperature ranges, the measurement data was converted according to TYDEX format [26]. This format enables the usage of MFTool software provided by TASS International (former TNO Automotive, Helmond, The Netherlands) [27]. The MFTool software is a commercial software to determine the coefficients of the empirical model, well known as the magic formula, from the pre-processed measurement data. It also allows to compensate for the variation in normal load and fits the curve for a nominal load. In the content of this study, MFTool was used to generate a tire property file for the extended magic formula (Appendix A).

4.2. Basic Magic Formula

The magic formula is an empirical model commonly used to simulate steady-state tire forces and moments. It is based on a special function instead of look-up tables or other various polynomial functions. By adjusting the function coefficients, the same special function can be used to describe longitudinal and lateral forces (sine function), and self-aligning moment (cosine function). The basic magic formula has the following form [28]:

$$\begin{aligned} y(x) &= D\sin(C\arctan Bx - E(Bx - \arctan Bx)) \\ Y(x) &= y(x) + S_v \\ x &= X + S_h \end{aligned} \tag{22}$$

where $Y(x)$ may represent longitudinal, lateral forces or self-aligning moment; X denotes slip angle α or longitudinal wheel slip κ; B is stiffness factor; C is shape factor; D is peak factor; E is curvature factor; S_h is horizontal shift; S_v is vertical shift.

The peak value is described by the peak factor D. The derivative at the origin provides BCD as a result and defines the longitudinal or cornering stiffness of the tire. The curve shape can be adjusted by both factors C and E, where C, called as shape factor, controls the "stretching" in the x direction and E, called the curvature factor, enables a local stretch or compression. The parameters S_h and S_v allow to model the influence of plysteer and conicity. The extended version of the magic formula [29] also covers effects of load dependency, combined slip, camber angle, inflation pressure, etc.

4.3. Proposed Modification for Temperature Effect

To change the peak friction and the stiffness of the force curve according to temperature variation, the magic formula was modified to affect the tire forces and moments. The proposed modifications are only demonstrated for the basic magic formula; meanwhile, they were integrated with the extended version of the magic formula. To represent temperature change, a delta temperature ΔT was defined:

$$\Delta T = \frac{T - T_{ref}}{T_{ref}} \tag{23}$$

where T_{ref} is reference temperature (the performance is known from measurements).

To evaluate the performance of the proposed modification, the normalized friction values from measurement data and simulation results were compared. The measured temperature was used as an input to the modified magic formula.

Longitudinal Force. Four temperature coefficients were introduced to capture the temperature effect on the longitudinal force. The coefficients related to temperature effect are:

T_{X1}	Linear temperature effect on longitudinal slip stiffness
T_{X2}	Quadratic temperature effect on longitudinal slip stiffness
T_{X3}	Linear temperature effect on longitudinal friction
T_{X4}	Quadratic temperature effect on longitudinal friction

To affect the longitudinal slip stiffness and longitudinal friction, the above-mentioned coefficients were used. The modified peak factor D_x and longitudinal slip stiffness K_x are:

$$\begin{aligned} D_x &= \left(1 + T_{X3}\Delta T + T_{X4}\Delta T^2\right) D_{xbase} \\ K_x &= \left(1 + T_{X1}\Delta T + T_{X2}\Delta T^2\right) K_{xbase} \end{aligned} \tag{24}$$

where D_{xbase} is the peak factor of pure longitudinal force D and K_{xbase} is the longitudinal slip stiffness K_x according to the basic magic formula.

As can be observed in Figure 9, the measurement data and simulation results of normalized longitudinal friction are close to each other. The quantitative comparison between simulation and experimental results for longitudinal force is summarized in Table 3. It also can be noted that temperature changes longitudinal stiffness and the longitudinal friction similar to experimental data. The main difference is related to the measurement noise and fluctuations due to road surface irregularities.

Table 3. Errors from measurements for different temperatures.

Temperature Range	Longitudinal Force F_x	Lateral Force F_y
$T < 50$ °C	-	18.35%
50 °C $\leq T <$ 60 °C	14.35%	11.53%
60 °C $\leq T <$ 70 °C	12.9%	4.9%
70 °C $\leq T <$ 80 °C	12.5%	-
80 °C $\leq T <$ 90 °C	10.4%	-
$T > 90$ °C	10%	-

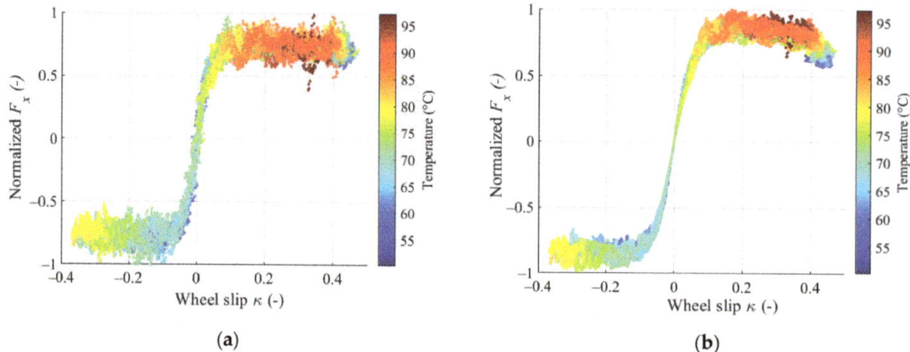

Figure 9. Normalized longitudinal force F_x: (**a**) experimental measurements; (**b**) modified magic formula.

Lateral Force and Self-Aligning Moment. Four temperature coefficients were introduced to incorporate the temperature effect in lateral force variation:

T_{Y1}	Temperature effect on cornering stiffness magnitude
T_{Y2}	Temperature effect on location of cornering stiffness peak
T_{Y3}	Linear temperature effect on lateral friction
T_{Y4}	Quadratic temperature effect on lateral friction

To affect the cornering stiffness and lateral friction respectively, the above-mentioned coefficients were used. The modified peak factor D_y and the cornering stiffness K_y are described:

$$\begin{aligned} D_y &= \left(1 + T_{Y3}\Delta T + T_{Y4}\Delta T^2\right)D_{ybase} \\ K_y &= (1 + T_{Y1}\Delta T)P_{KY1}F_{z0}\sin\left(\arctan\left(\frac{F_z}{P_{KY2}F_{z0}(1+T_{Y2}\Delta T)}\right)\right) \end{aligned} \quad (25)$$

where D_{ybase} is the peak factor of pure lateral force D in the basic magic formula; F_{z0} is the nominal normal load; P_{KY1} is the maximum value of stiffness K_{fy}/F_{z0}; P_{KY2} is the load at which K_{fy} reaches maximum value.

The results of normalized lateral force from measurement data and simulation results are shown in Figure 10 and demonstrate a similar behavior while temperature changes. The quantitative assessment between simulation and experimental results for lateral force is summarized in Table 3. The pneumatic trail would remain constant under the assumption of the same pressure distribution in a contact patch. Since the self-aligning moment is the product between the lateral force and the tire pneumatic trail, it will change only due to the lateral force variation. Therefore, there is no need to introduce the coefficients affecting the self-aligning moment as shown in Figure 11.

The error to compare the performance of the modified magic formula is defined as:

$$e_{meas} = 100\% \frac{\sqrt{(F_{simulation} - F_{measurement})^2}}{|F_{measurement}|} \quad (26)$$

Data for different temperature ranges were taken to evaluate the error values for the longitudinal and lateral forces, and the summary is presented in Table 3.

As it can be observed, the modified magic formula performs sufficiently close to the actual measurement data taking into account that the absolute measurement error was 5%. The largest errors occur at lower temperatures and a general trend of improved accuracy is seen as the temperature increases. For lower temperatures, the potential reason of the deviation is related to a high variance of measured data. The errors for the self-aligning moment are not shown due to the high variance in the measurement data related to the measurement setup.

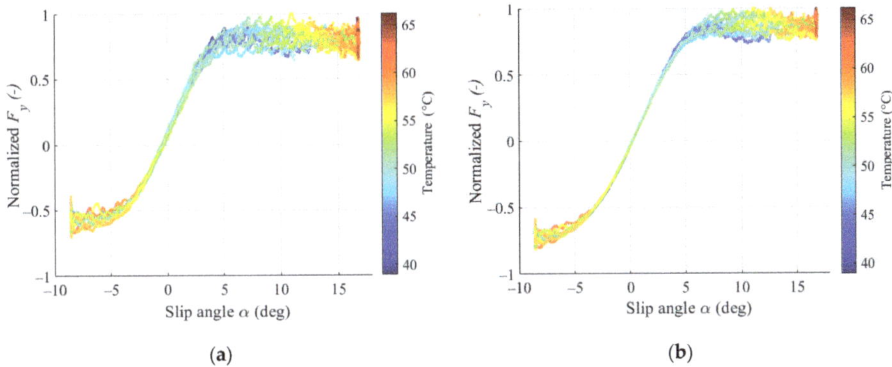

Figure 10. Normalized lateral force F_y: (**a**) experimental measurements; (**b**) modified magic formula.

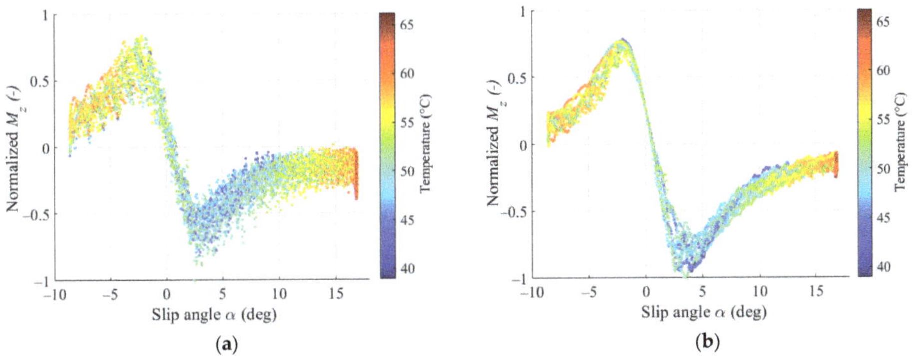

Figure 11. Normalized self-aligning moment M_z: (**a**) experimental measurements; (**b**) modified magic formula.

5. Integration into Full FSAE Vehicle Model

5.1. Full Vehicle Modeling

The following assumptions have been made for the multi-body model of the FSAE vehicle:

- All joints and bodies besides the tires and springs are considered as rigid. Since the suspension and chassis elements are designed to be stiff, compliance is neglected. Joint friction is omitted, except modeling of the transmission friction. For this purpose, damping in the revolute joints between the suspension and the wheels is added.
- Aerodynamic forces are applied on a constant center of pressure. Its position was calculated based on computational fluid dynamic simulations in steady-state conditions and changes according to the velocity, roll, pitch, etc.
- The characteristics of suspension springs and dampers are linear.

The model was developed in Simscape Multibody (Simulink toolbox). The coordinate system and axis orientation was according to the SAE coordinate system. Center of gravity position and mass-inertial characteristics of the chassis were obtained from the computer-aided design (CAD) model or/and actual measurements on the FSAE vehicle. The model includes four suspension subsystems connected to the chassis body. Using revolute joints from the upright center, the tire was connected to the suspension. The suspension geometry was imported from the CAD model corresponding to the actual FSAE vehicle. A steering system was connected to the front suspension and the chassis body through different hardpoints. More details regarding the vehicle modeling can be found in [30].

5.2. Model Structure

The completed structure of the simulation setup is shown in Figure 12. The tire states (position, velocity, acceleration) are calculated on each time step based on the multi-body vehicle model and transferred to the modified magic formula. According to kinematic states, the magic formula calculates the forces and moments. Then tire forces and moments are sent to the thermal model calculating the temperature and its change. Finally, tire forces and moments are transferred to the multibody vehicle model, while temperature and pressure are transferred to the tire model.

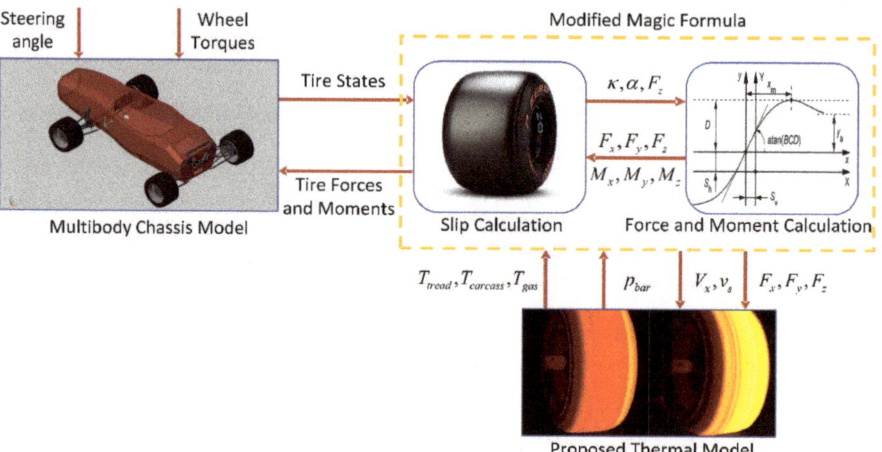

Figure 12. Model structure.

5.3. Validation

To validate the overall model, two test scenarios were considered. The first test scenario was a skidpad maneuver to evaluate a steady-state lateral behavior. The second test scenario was a full lap maneuver performed for a longer time.

Skidpad Maneuver. The skidpad course consists of two pairs of concentric circles in a figure of eight pattern. The centers of these circles are 18.25 m apart. The inner circles are 15.25 m in diameter and the outer circles are 21.25 m in diameter. The driving path is the 3 m wide path between the inner and outer circles [31].

To increase the signal-to-noise ratio without significant signal distortion, the raw sensor data was filtered using a Savitzky–Golay digital filter [32]. The filtered data was used for comparison with simulation results. The longitudinal and lateral accelerations, and yaw rate are shown in Figure 13.

The simulation shows more oversteered behavior compared to the measured data. The mismatch can be caused by the modeling of chassis roll stiffness. A real chassis has a limited roll stiffness; however, as mentioned, the chassis was assumed as a rigid body. Hence, the chassis was infinitely stiff in the simulation model. It could result in a higher yaw rate and increase the root mean square error (RMSE) of the yaw rate.

The longitudinal and lateral accelerations, and yaw rate were better aligned with the measurement data in the left turn circles. This observation can be explained by the fact that the elevation change in the skidpad circuit affected the measurements from the accelerometer sensor.

Figure 14 shows the evolution of tire temperature during skidpad maneuver for four wheels: front left (FL), front right (FR), rear left (RL), and rear right (RR). As can be observed, due to higher normal loads, the outer tires (initially left tires for the right turn) start heating up first. When the vehicle steered in another direction (at 15 s), the right tires start heating up.

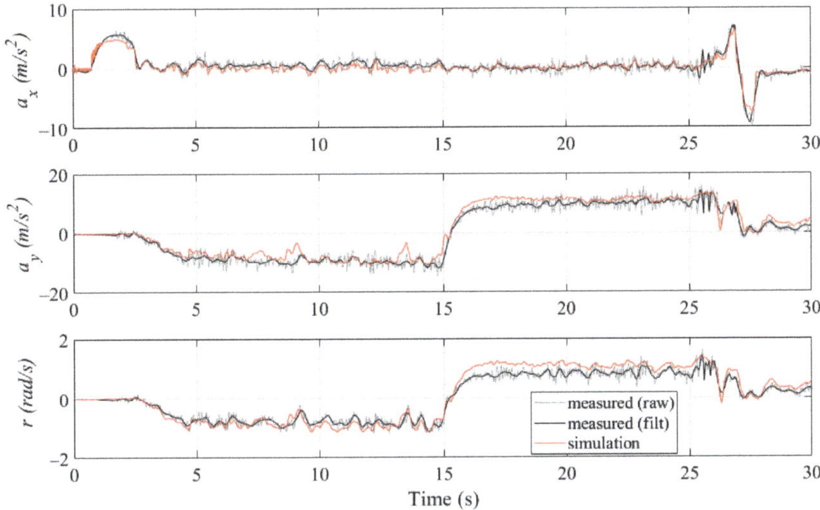

Figure 13. Longitudinal acceleration a_x, lateral acceleration a_y, and yaw rate r for skidpad.

Full Lap. To validate transient behavior of the simulation model, a full lap maneuver was investigated. The measurement data correspond to one of the endurance runs with a duration of ~22 km driving (18–20 laps). The simulation was conducted for one lap of the circuit performed approx. in 50 s.

In a similar way as was done for the skidpad scenario, the raw measurement data of longitudinal and lateral accelerations, and yaw rate, were filtered and the RMSE calculated. The RMSE for the longitudinal and lateral accelerations, as well as yaw rate are summarized in Table 4, including also the simulation results performed without the proposed model. The comparison between simulation results with a temperature effect and measurement data is shown in Figure 15. It can be noted that the yaw rate from the simulation results was aligned with the measured data; however, an offset can be observed which might relate to the above-mentioned difference in the chassis roll stiffness.

Table 4. The root mean square error (RMSE) for both simulation maneuvers.

	Skidpad Maneuver		**Full Lap Maneuver**	
Temperature effect	without	with	without	with
Longitudinal acceleration, m/s^2	0.62	0.65	1.50	1.42
Lateral acceleration, m/s^2	1.86	1.26	4.77	3.19
Yaw rate, rad/s	0.27	0.16	0.37	0.16

The RMSE for longitudinal acceleration is slightly higher compared to the skidpad scenario; however, the longitudinal acceleration in the skidpad maneuver was much lower compared to a full lap scenario. Thus, the simulation results of longitudinal dynamics using wheel torques as an input matched the longitudinal acceleration as sufficiently accurate.

The highest difference was found in lateral dynamics by the RMSE of lateral acceleration of 3.19 m/s^2. The potential reasons can be related to: (i) combined slip conditions; (ii) camber effect on lateral force, (iii) longitudinal velocity not tacked precisely. Due to a limited financial budget, the tire testing was performed only for pure longitudinal and lateral conditions. The coefficients corresponding to combined slip behavior were estimated based on similarly sized FS tires. It can be observed, that the maximum mismatch happens in a corner where the tires were cambered due to chassis roll and performed in combined slip conditions. Another aspect is the body slip of the actual vehicle which might affect the differences between the simulation and real FSAE vehicles.

Figure 16 shows the evolution of tire temperature during a full lap maneuver. The tread temperature steadily increased and its reduction corresponds to straight lines or long corners. It can also be observed that a higher temperature for right tires was achieved compared to the left tires. Due to the circuit orientation, the right tires have higher loads. In this case, the gas temperature increased and due to the deflections, the carcass heated up.

As can be concluded from Table 4, the RMSE of lateral acceleration was up to 30% and 33% in steady-state and transient maneuvers, correspondingly. Regarding the yaw rate, the RMSE was up to 41% in steady-state maneuver and up to 56% in the transient one. Therefore, the modeling of temperature effect has a significant effect on FSAE vehicle dynamics.

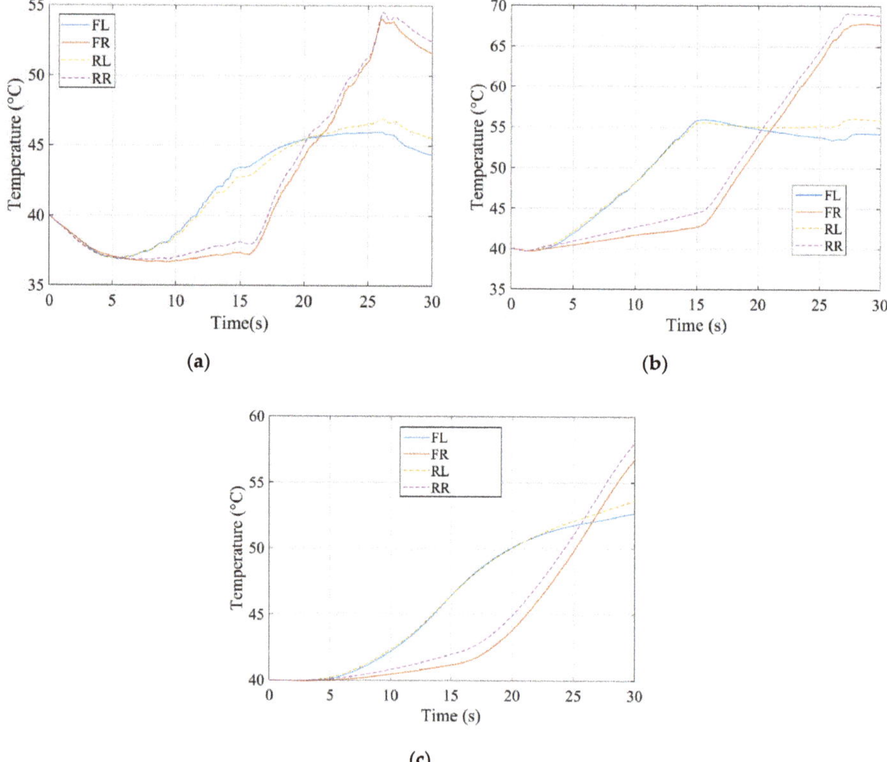

Figure 14. Evolution of tire temperatures for skidpad maneuver: (**a**) tire tread temperature; (**b**) tire carcass temperature; (**c**) tire inflation gas temperature.

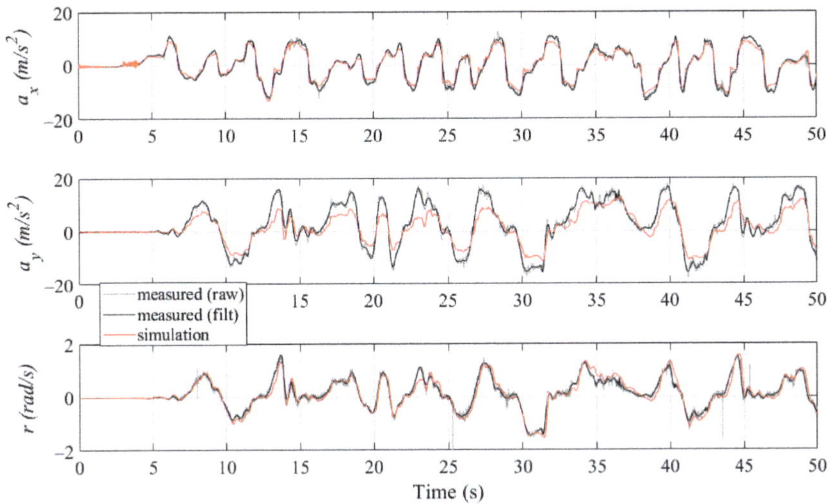

Figure 15. Longitudinal acceleration a_x, lateral acceleration a_y, and yaw rate r for a full lap.

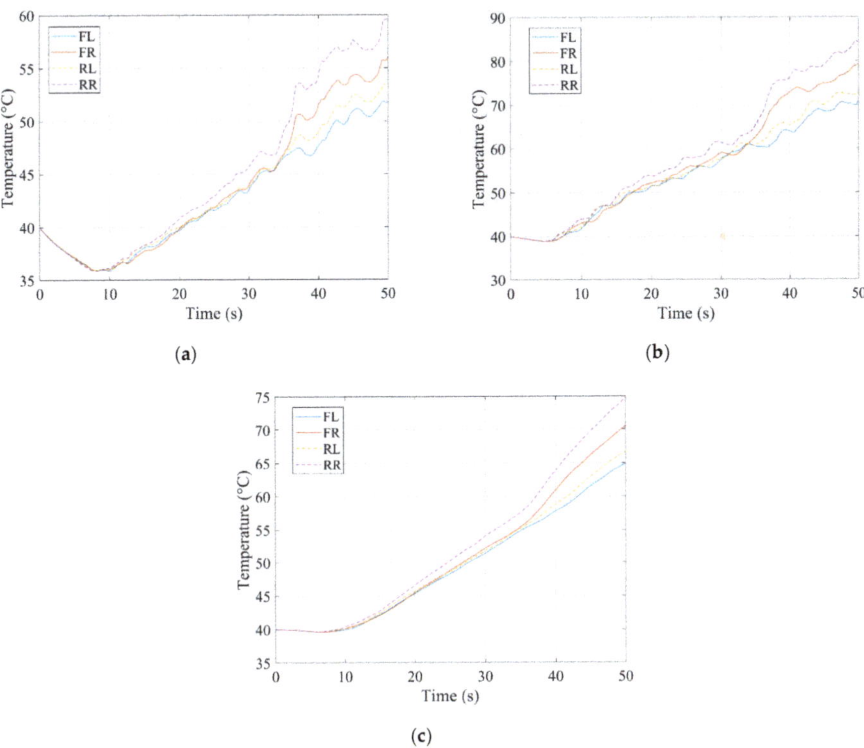

Figure 16. Evolution of tire temperatures for full lap: (**a**) tire tread temperature; (**b**) tire carcass temperature; (**c**) tire inflation gas temperature.

6. Conclusions

This paper proposes an empirical tire model (modified magic formula) coupled with a thermal model to incorporate temperature effects on the force capability of the tire. The thermal model was proposed to model the influence of temperature on peak friction and the shear modulus, allowing for a better representation of the temperature variation under different load conditions. The magic formula was modified introducing new coefficients allowing for the change in longitudinal and cornering stiffness, and peak factors. The simulation results of the modified magic formula with temperature sufficiently matched the variation in longitudinal and lateral force, as well as self-aligning moment, compared to the experimental tire tests. The error between simulation results and experimental data was in the range between 10% and 15%. However, limited accuracy with an error of 18.5% was achieved at lower temperatures due to a high variance in the measured data.

To evaluate the proposed tire model, the thermal and modified magic formula models were integrated to a multi-body vehicle model. The simulation results match the behavior of the real FSAE vehicle in steady-state maneuver. For transient maneuvers, limited accuracy was achieved due to estimated parameters or unmodeled dynamics such as tire behavior in combined slip conditions, neglect of camber effect on lateral force, etc. Comparing the proposed model with an empirical-based tire model without temperature effect, the RMSE of lateral acceleration was reduced by up to 30% and 33% in the steady-state and the transient maneuvers, correspondingly. For yaw motion, the RMSE of the yaw rate was reduced up to 41% in the steady-state maneuver and up to 56% in the transient one.

Further work related to tire testing which would result in a more precise identification of the magic formula parameters and thus improve the tire model. The following improvements should be considered: (i) combined slip conditions, (ii) camber influence on performance, and (iii) relaxation behavior.

Author Contributions: Conceptualization, D.H. and B.S.; testing, D.H.; modeling, D.H.; supervision, B.S.; validation, D.H.; writing—original draft, D.H.; writing—review and editing, B.S.

Funding: This research received no external funding.

Acknowledgments: The authors would like to thank Formula Student Team Delft, The Netherlands, for their support with tire and vehicle testing.

Conflicts of Interest: The authors declare no conflicts of interest.

Appendix A —Tire Property File

[MODEL]		
FITTYP	=62	$Magic Formula Version number
[DIMENSION]		
UNLOADED_RADIUS	=0.17	$Free tyre radius
WIDTH	=0.2165	$Nominal section width of the tyre
RIM_RADIUS	=0.127	$Nominal rim radius
RIM_WIDTH	=0.225	$Rim width
ASPECT_RATIO	=0.1986	$Nominal aspect ratio
[INERTIA]		
MASS	=2.825	$Tyre mass
IXX	=0.0538	$Tyre diametral moment of inertia
IYY	=0.1	$Tyre polar moment of inertia
BELT_MASS	=2.2	$Belt mass
BELT_IXX	=0.05	$Belt diametral moment of inertia
BELT_IYY	=0.08	$Belt polar moment of inertia
GRAVITY	=−9.81	$Gravity acting on belt in Z direction

```
[VERTICAL]
FNOMIN                  =600            $Nominal wheel load
VERTICAL_STIFFNESS      =95000          $Tyre vertical stiffness
VERTICAL_DAMPING        =50             $Tyre vertical damping
BREFF                   =8.4            $Low load stiffness e.r.r.
DREFF                   =0.27           $Peak value of e.r.r.
FREFF                   =0.07           $High load stiffness e.r.r.
Q_RE0                   =0.9974         $Ratio of free tyre radius with nominal tyre radius
[LONGITUDINAL_COEFFICIENTS]
PCX1                    =1.391          $Shape factor Cfx for longitudinal force
PDX1                    =1.5314         $Longitudinal friction Mux at Fznom
PDX2                    =−0.04906       $Variation of friction Mux with load
PEX1                    =0.4454         $Longitudinal curvature Efx at Fznom
PEX2                    =0.2192         $Variation of curvature Efx with load
PEX4                    =0.1665         $Factor in curvature Efx while driving
PKX1                    =43.63          $Longitudinal slip stiffness Kfx/Fz at Fznom
PKX2                    =4.4735         $Variation of slip stiffness Kfx/Fz with load
PKX3                    =0.023027       $Exponent in slip stiffness Kfx/Fz with load
PHX1                    =−0.003839      $Horizontal shift Shx at Fznom
PHX2                    =0.0044605      $Variation of shift Shx with load
PVX1                    =0.04359        $Vertical shift Svx/Fz at Fznom
PVX2                    =0.007515       $Variation of shift Svx/Fz with load
RBX1                    =10             $Slope factor for combined slip Fx reduction
RBX2                    =6              $Variation of slope Fx reduction with kappa
RCX1                    =1              $Shape factor for combined slip Fx reduction
[LATERAL_COEFFICIENTS]
PCY1                    =1.3318         $Shape factor Cfy for lateral forces
PDY1                    =1.6502         $Lateral friction Muy
PDY2                    =−0.14737       $Variation of friction Muy with load
PEY1                    =0.5            $Lateral curvature Efy at Fznom
PEY2                    =−9.1214E-7     $Variation of curvature Efy with load
PKY1                    =−85            $Maximum value of stiffness Kfy/Fznom
PKY2                    =5              $Load at which Kfy reaches maximum value
PKY4                    =1.7923         $Curvature of stiffness Kfy
PHY1                    =0.008          $Horizontal shift Shy at Fznom
PVY1                    =0.1            $Vertical shift in Svy/Fz at Fznom
PVY2                    =−0.001143      $Variation of shift Svy/Fz with load
RBY1                    =16             $Slope factor for combined Fy reduction
RCY1                    =1              $Shape factor for combined Fy reduction
[ALIGNING_COEFFICIENTS]
QBZ1                    =7              $Trail slope factor for trail Bpt at Fznom
QBZ2                    =2              $Variation of slope Bpt with load
QCZ1                    =1.2            $Shape factor Cpt for pneumatic trail
QDZ1                    =0.12           $Peak trail Dpt = Dpt*(Fz/Fznom*R0)
QDZ2                    =−0.02          $Variation of peak Dpt with load
QEZ1                    =−2.8           $Trail curvature Ept at Fznom
QEZ2                    =3              $Variation of curvature Ept with load
```

[TEMPERATURE_COEFFICIENTS]

TY1	=−0.25	$Temperature effect on cornering stiffness magnitude
TY2	=0.15	$Temperature effect on location of cornering stiffness peak
TY3	=0.25	$Linear temperature effect on lateral friction
TY4	=−0.1	$Quadratic temperature effect on lateral friction
TX1	=−0.25	$Linear temperature effect on slip stiffness
TX2	=0.15	$Quadratic temperature effect on slip stiffness
TX3	=0.25	$Linear temperature effect on longitudinal friction
TX4	=−0.1	$Quadratic temperature effect on longitudinal friction
TREF	=50	$Reference temperature

Note: the non-mentioned coefficients are equal to zero.

References

1. Formula SAE. Available online: https://www.fsaeonline.com (accessed on 10 October 2019).
2. Formula Student Team Delft. Available online: https://www.fsteamdelft.nl (accessed on 10 October 2019).
3. Ammon, D. Vehicle dynamics analysis tasks and related tyre simulation challenges. *Veh. Syst. Dyn.* **2005**, *43*, 30–47. [CrossRef]
4. Hirschberg, W.; Rill, G.; Weinfurter, H. User-appropriate tyre-modelling for vehicle dynamics in standard and limit situations. *Veh. Sys. Dyn.* **2010**, *38*, 103–125. [CrossRef]
5. Li, J.; Zhang, Y.; Yi, J. A hybrid physical-dynamic tire/road friction model. *J. Dyn. Syst. Meas. Control* **2013**, *135*, 011007. [CrossRef]
6. Yavari, B.; Tworzydlo, W.; Bass, J. A thermomechanical model to predict the temperature distribution of steady state rolling tires. *Tire Sci. Technol.* **1993**, *21*, 163–178. [CrossRef]
7. Clark, S.K. Temperature rise times in pneumatic tires. *Tire Sci. Technol.* **1976**, *4*, 181–189. [CrossRef]
8. Maniowski, M. Optimisation of driver actions in RWD race car including tyre thermodynamics. *Veh. Sys. Dyn.* **2016**, *54*, 526–544. [CrossRef]
9. Durand-Gasselin, B.; Dailliez, T.; Mössner-Beigel, M.; Knorr, S.; Rauh, J. Assessing the thermo-mechanical TaMeTirE model in offline vehicle simulation and driving simulator tests. *Veh. Sys. Dyn.* **2010**, *48*, 211–229. [CrossRef]
10. Pearson, M.; Blanco-Hague, O.; Pawlowski, R. TameTire: Introduction to the model. *Tire Sci. Technol.* **2016**, *44*, 102–119. [CrossRef]
11. Farroni, F.; Giordano, D.; Russo, M.; Timpone, F. TRT: Thermo racing tyre a physical model to predict the tyre temperature distribution. *Meccanica* **2014**, *49*, 707–723. [CrossRef]
12. Sorniotti, A. *Tire Thermal Model for Enhanced Vehicle Dynamics Simulation*; No. 2009-01-0441, SAE technical paper 2009; SAE: Warrendale, PA, USA, 2009.
13. Ozerem, O.; Morrey, D. A brush-based thermo-physical tyre model and its effectiveness in handling simulation of a Formula SAE vehicle. *IMechE Part D J. Automob. Eng.* **2019**, *233*, 107–120. [CrossRef]
14. Kasprzak, E.M.; Gentz, D. *The Formula Sae Tire Test Consortium-Tire Testing and Data Handling*; No. 2006-01-3606, SAE technical paper 2006; SAE: Warrendale, PA, USA, 2006.
15. Dynamic Test Center. Available online: https://www.dtc-ag.ch (accessed on 10 October 2019).
16. Kistler Wheel Force Transducer. Available online: https://www.kistler.com (accessed on 10 October 2019).
17. Milliken, W.F.; Milliken, D.L. *Race Car Vehicle Dynamics*; Society of Automotive Engineers Warrendale: Warrendale, PA, USA, 1995.
18. Kelly, D.; Sharp, R. Time-optimal control of the race car: Influence of a thermodynamic tyre model. *Veh. Sys. Dyn.* **2012**, *50*, 641–662. [CrossRef]
19. Fevrier, P.; Le Maitre, O. Tire temperature modeling: Application to race tires. In Proceedings of the VDI 13th International Congress Numerical Analysis and Simulation in Vehicle Engineering, Wurzburg, Germany, 27–28 September 2006; pp. 27–28.
20. Savkoor, A. Some aspects of friction and wear of tyres arising from deformations, slip and stresses at the ground contact. *Wear* **1966**, *9*, 66–78. [CrossRef]
21. Falk, K.; Nazarinezhad, A.; Kaliske, M. *Tire Simulations Using a Slip Velocity, Pressure and Temperature Dependent Friction Law*; Deutsche Simulia Anwenderkonferenz: Dresden, Germany, 2014.

22. Huemer, T.; Liu, W.N.; Eberhardsteiner, J.; Mang, H.A. A 3D finite element formulation describing the frictional behavior of rubber on ice and concrete surfaces. *Eng. Comput.* **2001**, *18*, 417–437. [CrossRef]
23. Lemaitre, J. Nonlinear Models and Properties. In *Handbook of Materials Behavior Models*, 2nd ed.; Elsevier: Amsterdam, The Netherlands, 2001.
24. Svendenius, J.; Gäfvert, M.; Bruzelius, F.; Hultén, J. Experimental validation of the brush tire model. *Tire Sci. Technol.* **2009**, *37*, 122–137. [CrossRef]
25. Van Rijk, S. Development and Modelling of a Formula Student Racing Tyre. Master's Thesis, Eindhoven University of Technology, Eindhoven, The Netherlands, 2013.
26. Oosten, J.; Unrau, H.; Riedel, A.; Bakker, E. TYDEX workshop: Standardisation of data exchange in tyre testing and tyre modelling. *Veh. Sys. Dyn.* **1997**, *27*, 272–288. [CrossRef]
27. *MF-Tyre. MF-Swift 6.2-Help Manual*; TNO Automotive: Helmond, The Netherlands, 2013.
28. Bakker, E.; Pacejka, H.B.; Lidner, L. A new tire model with an application in vehicle dynamics studies. *SAE Trans.* **1989**, 101–113. [CrossRef]
29. Pacejka, H.B.; Besselink, I. *Tire and Vehicle Dynamics*, 3rd ed.; Elsevier: Amsterdam, The Netherlands, 2012.
30. Harsh, D. Full Vehicle Model of a Formula Student Car: Full-Car Vehicle Dynamics Model Incorporating a Tyre Model Which Includes the Effects of Temperature on It's Performance. Master's Thesis, Delft University of Technology, Delft, The Netherlands, 2018.
31. Formula Student Germany e.V. Formula Student Rules. Available online: https://www.formulastudent.de/fsg/rules/ (accessed on 10 October 2019).
32. Schafer, R.W. What is a Savitzky-Golay filter? *IEEE Signal Process. Mag.* **2011**, *28*, 111–117. [CrossRef]

MDPI
St. Alban-Anlage 66
4052 Basel
Switzerland
Tel. +41 61 683 77 34
Fax +41 61 302 89 18
www.mdpi.com

Applied Sciences Editorial Office
E-mail: applsci@mdpi.com
www.mdpi.com/journal/applsci

www.ingramcontent.com/pod-product-compliance
Lightning Source LLC
LaVergne TN
LVHW071612170726
843515LV00009B/2378
9783036528700